Antitargets

Edited by
Roy J. Vaz and Thomas Klabunde

Methods and Principles in Medicinal Chemistry

Edited by R. Mannhold, H. Kubinyi, G. Folkers
Editorial Board
H. Timmerman, J. Vacca, H. van de Waterbeemd, T. Wieland

Previous Volumes of this Series:

M. Hamacher, K. Marcus, K. Stühler,
A. van Hall, B. Warscheid,
H. E. Meyer (eds.)

Proteomics in Drug Research
Vol. 28
2006, ISBN: 978-3-527-31226-9

D. J. Triggle, M. Gopalakrishnan,
D. Rampe, W. Zheng (eds.)

Voltage-Gated Ion Channels as Drug Targets
Vol. 29
2006, ISBN: 978-3-527-31258-0

D. Rognan (ed.)

Ligand Design for G Protein-coupled Receptors
Vol. 30
2006, ISBN: 978-3-527-31284-9

D. A. Smith, H. van de Waterbeemd,
D. K. Walker

Pharmacokinetics and Metabolism in Drug Design, 2nd Ed.
Vol. 31
2006, ISBN: 978-3-527-31368-6

T. Langer, R. D. Hofmann (eds.)

Pharmacophores and Pharmacophore Searches
Vol. 32
2006, ISBN: 978-3-527-31250-4

E. Francotte, W. Lindner (eds.)

Chirality in Drug Research
Vol. 33
2006, ISBN: 978-3-527-31076-0

W. Jahnke, D. A. Erlanson (eds.)

Fragment-based Approaches in Drug Discovery
Vol. 34
2006, ISBN: 978-3-527-31291-7

J. Hüser (ed.)

High-Throughput Screening in Drug Discovery
Vol. 35
2006, ISBN: 978-3-527-31283-2

K. Wanner, G. Höfner (eds.)

Mass Spectrometry in Medicinal Chemistry
Vol. 36
2007, ISBN: 978-3-527-31456-0

R. Mannhold (ed.)

Molecular Drug Properties
Vol. 37
2008, ISBN: 978-3-527-31755-4

Antitargets

Prediction and Prevention of Drug Side Effects

Edited by
Roy J. Vaz and Thomas Klabunde

WILEY-VCH

WILEY-VCH Verlag GmbH & Co. KGaA

Series Editors

Prof. Dr. Raimund Mannhold
Molecular Drug Research Group
Heinrich-Heine-Universität
Universitätsstrasse 1
40225 Düsseldorf
Germany
mannhold@uni-duesseldorf.de

Prof. Dr. Hugo Kubinyi
Donnersbergerstrasse 9
67256 Weisenheim am Sand
Germany
kubinyi@t-online.de

Prof. Dr. Gerd Folkers
Collegium Helveticum
STW/ETH Zurich
8092 Zurich
Switzerland
folkers@collegium.ethz.ch

Volume Editors

Dr. Roy J. Vaz
Sanofi-Aventis
Route 202-206
Bridgewater, NJ 08807
USA
Roy.Vaz@sanofi-aventis.com

Dr. Thomas Klabunde
Sanofi-Aventis Deutschland
Industriepark Höchst G878
65926 Frankfurt
Germany
Thomas.Klabunde@sanofi-aventis.com

Cover Illustration
The cover shows a model of the hERG ion channel in the lower left (chapter 4), a CYP structure, lower right (chapter 10), a pharmacophore model for the alpha1a adrenergic receptor, upper right (chapter 6) and a schematic of the intestinal epithelium with some transporters, upper left (chapter 15).

All books published by **Wiley-VCH** are carefully produced. Nevertheless, authors, editors, and publisher do not warrant the information contained in these books, including this book, to be free of errors. Readers are advised to keep in mind that statements, data, illustrations, procedural details or other items may inadvertently be inaccurate.

Library of Congress Card No.: applied for

British Library Cataloguing-in-Publication Data
A catalogue record for this book is available from the British Library.

Bibliographic information published by the Deutsche Nationalbibliothek
Die Deutsche Nationalbibliothek lists this publication in the Deutsche Nationalbibliografie; detailed bibliographic data are available in the Internet at <http://dnb.d-nb.de>.

© 2008 WILEY-VCH Verlag GmbH & Co. KGaA, Weinheim

All rights reserved (including those of translation into other languages). No part of this book may be reproduced in any form – by photoprinting, microfilm, or any other means – nor transmitted or translated into a machine language without written permission from the publishers. Registered names, trademarks, etc. used in this book, even when not specifically marked as such, are not to be considered unprotected by law.

Typesetting Thomson Digital, Noida, India
Printing betz-druck GmbH, Darmstadt
Binding Litges & Dopf GmbH, Heppenheim
Cover Design Grafik-Design Schulz, Fußgönheim

Printed in the Federal Republic of Germany
Printed on acid-free paper

ISBN: 978-3-527-31821-6

Contents

List of Contributors *XV*
Preface *XIX*
A Personal Foreword *XXI*

I	**General Aspects** *1*	
1	**Why Drugs Fail – A Study on Side Effects in New Chemical Entities** *3*	
	Daniela Schuster, Christian Laggner, Thierry Langer	
1.1	Introduction *3*	
1.2	Drugs Withdrawn from the Market between 1992 and 2006 Listed Alphabetically *4*	
1.2.1	Amineptine *4*	
1.2.2	Aminophenazone (Aminopyrine) *5*	
1.2.3	Astemizole *5*	
1.2.4	Bromfenac Sodium *6*	
1.2.5	Cerivastatin *6*	
1.2.6	Chlormezanone *6*	
1.2.7	Fenfluramine and Dexfenfluramine *7*	
1.2.8	Flosequinan *7*	
1.2.9	Glafenine *7*	
1.2.10	Grepafloxacin *8*	
1.2.11	Levacetylmethadol *8*	
1.2.12	Mibefradil *9*	
1.2.13	Rapacuronium Bromide *9*	
1.2.14	Rofecoxib *10*	
1.2.15	Temafloxacin *10*	
1.2.16	Troglitazone *10*	
1.2.17	Ximelagatran *11*	
1.3	Borderline Cases *12*	
1.4	Investigational Drugs That Failed in Clinical Phases from 1992 to 2002 *12*	

Antitargets. Edited by R. J. Vaz and T. Klabunde
Copyright © 2008 WILEY-VCH Verlag GmbH & Co. KGaA, Weinheim
ISBN: 978-3-527-31821-6

1.4.1	A Case Study: Fialuridine	12
1.4.2	A Recent Case Study: Torcetrapib	14
1.4.3	General Reasons for Project Failing in Clinical Phases I–III	15
1.5	Strategies for Avoiding Failure	16
1.6	An Unusual Case: The Revival of Thalidomide	17
	References	18

2   Use of Broad Biological Profiling as a Relevant Descriptor to Describe and Differentiate Compounds: Structure–In Vitro (Pharmacology-ADME)–In Vivo (Safety) Relationships   23
Jonathan S. Mason, Jacques Migeon, Philippe Dupuis, Annie Otto-Bruc

2.1	Introduction	23
2.1.1	Biological Profiling/Fingerprints and Drug Discovery Applications	23
2.1.2	Polypharmacology of Drugs	26
2.2	The BioPrint® Approach	28
2.2.1	BioPrint® – General	28
2.2.2	BioPrint® Assay Selection and Profile Description	28
2.2.3	Compounds in BioPrint®	30
2.2.4	BioPrint® In Vivo Data sets	31
2.2.4.1	Compound Details	31
2.2.4.2	ADR Data	31
2.2.4.3	Pharmacokinetics	31
2.2.4.4	Toxicity Data	32
2.3	Structure–In Vitro Relationships	32
2.3.1	Similarity, Chemotypes – What Is a Biologically Relevant Descriptor?	32
2.3.2	Using Biological Fingerprints as a Meaningful Descriptor for Drug Leads and Candidates	33
2.3.2.1	Differentiation of Leads	33
2.3.2.2	Analysis of Attrited Compounds	36
2.3.3	Structural versus Experimental Differentiation – Dependence on Structure-Derived Descriptor Used	37
2.3.4	Predictive Models from Pharmacological Data	40
2.3.5	Predictive Models from ADME Data – BioPrint® Learnings	41
2.4	Chemogenomic Analysis – Target–Target Relationships	41
2.5	In Vitro–In Vivo Relationships – Placing Drug Candidates in the Context of BioPrint®	42
2.5.1	Analyzing Potential ADR Liabilities Based on Individual Hits	43
2.5.2	Analyzing Potential ADR Liabilities Based on Profile Similarity	49
2.6	A Perspective for the Future	50
	References	50

II	**Antitargets: Ion Channels and GPCRs** 53
3	**Pharmacological and Regulatory Aspects of QT Prolongation** 55
	Fabrizio De Ponti
3.1	Introduction 55
3.2	hERG: Target Versus Antitarget 57
3.3	Pharmacology of QT Prolongation 58
3.3.1	Multiple Mechanisms Leading to QT Prolongation 59
3.3.2	hERG as the Key Mechanism for the Drug-Induced Long QT Syndrome 59
3.3.3	Pharmacogenetic Aspects 60
3.4	Significance of Drug-Induced QT Prolongation 61
3.4.1	Prolonged QT/QTc and Occurrence of TdP 61
3.4.2	Dose–Response Relationship for QT Prolongation 66
3.5	Regulatory Aspects of QT Prolongation 67
3.5.1	Regulatory Guidance Documents 67
3.5.2	Preclinical *In Vitro* and *In Vivo* Studies 68
3.5.3	Clinical Studies 72
3.6	Conclusions 76
	References 77
4	**hERG Channel Physiology and Drug-Binding Structure–Activity Relationships** 89
	Sarah Dalibalta, John S. Mitcheson
4.1	Introduction 89
4.2	hERG Channel Structure 90
4.3	hERG Potassium Channels and the Cardiac Action Potential 91
4.4	Mutations in hERG Are Associated with Cardiac Arrhythmias 93
4.5	Acquired Long QT Syndrome 94
4.6	Drug-Binding Site of hERG 95
4.7	Structural Basis for hERG Block 95
4.8	Alternative Mechanisms of Block 97
4.9	Role of Inactivation in hERG Block 98
4.10	Inhibition of hERG Trafficking by Pharmacological Agents 99
4.11	Computational Approaches to Predict hERG K^+ Channel Block 99
4.12	Conclusions 101
	References 102
5	**QSAR and Pharmacophores for Drugs Involved in hERG Blockage** 109
	Maurizio Recanatini, Andrea Cavalli
5.1	Introduction 109
5.2	Ligand-Based Models for hERG-Blocking Activity 110
5.2.1	3D QSAR 111
5.2.2	2D QSAR 113

5.2.3	Classification Models	116
5.3	Ligand-Derived Models in the Light of the hERG Channel Structure	119
5.4	Conclusions	121
	References	123

6 GPCR *Antitarget* Modeling: Pharmacophore Models to Avoid GPCR-Mediated Side Effects 127

Thomas Klabunde, Andreas Evers

6.1	Introduction: GPCRs as *Antitargets*	127
6.2	*In Silico* Tools for GPCR *Antitarget* Modeling	129
6.3	GPCR Antitarget Pharmacophore Modeling: The α_{1a} Adrenergic Receptor	130
6.3.1	Generation of Cross-Chemotype Pharmacophore Models	131
6.3.2	Description of Cross-Chemotype Pharmacophore Models	131
6.3.3	Validation of Antitarget Pharmacophore Models	132
6.3.3.1	Virtual Screening: Hit Rates and Yields	132
6.3.3.2	Virtual Screening: Fit Values and Enrichment Factors	134
6.3.4	Mapping of Pharmacophore Models into Receptor Site	135
6.3.5	Guidance of Chemical Optimization to Avoid GPCR-Mediated Side Effects	138
6.4	Summary	139
	References	140

7 The Emergence of Serotonin 5-HT$_{2B}$ Receptors as DRUG Antitargets 143

Vincent Setola, Bryan L. Roth

7.1	Receptorome Screening to Identify Drug Targets and Antitargets	144
7.2	Post-Receptorome Screening Data Implicate 5-HT$_{2B}$ Receptors in Drug-Induced VHD and PH	145
7.3	Drug Structural Classes and VHD/PH	149
7.4	Conclusions	150
	References	151

8 Computational Modeling of Selective Pharmacophores at the α_1-Adrenergic Receptors 155

Francesca Fanelli, Pier G. De Benedetti

8.1	Introduction	155
8.2	Ligand-Based and Receptor-Based Pharmacophore Modeling and QSAR Analysis	158
8.3	The General α_1-AR Pharmacophore	161
8.3.1	Ligand-Based Pharmacophore and Virtual Screening	161
8.3.1.1	Prazosin Analogues (2,4-Diamino-6,7-dimethoxyquinazoline Derivatives)	163
8.3.1.2	1,4-Benzodioxan (WB-4101) Related Compounds	166

8.3.1.3	Arylpiperazine Derivatives *167*	
8.3.1.4	Target and Antitarget Pharmacophore Modeling *168*	
8.4	Modeling the α_1-AR Subtype Selectivities of Different Classes of Antagonists *170*	
8.4.1	Supermolecule-Based Subtype Pharmacophore and QSAR Models *171*	
8.4.2	Ligand-Based Subtype Pharmacophores *179*	
8.4.3	Receptor-Based Subtype Pharmacophore and Ligand–Target/Antitarget Interaction-Based QSAR *181*	
8.5	Antitarget Modeling of Biogenic Amine-Binding GPCRs: Common Features and Subtle Differences *182*	
8.6	Conclusions *183*	
8.6.1	From Molecules to Pharmacophores to Descriptors to Models *183*	
8.7	Perspectives *184*	
8.7.1	Pharmacophore Combination Approach: From Lock and Key to Passe–Partout Model *184*	
	References *186*	
III	**Antitargets: Cytochrome P450s and Transporters** *195*	
9	**Cytochrome P450s: Drug–Drug Interactions** *197*	
	Dan Rock, Jan Wahlstrom, Larry Wienkers	
9.1	Introduction *197*	
9.1.1	CYP1A Subfamily *199*	
9.1.2	CYP2C Subfamily *200*	
9.1.3	CYP2D6 *201*	
9.1.4	CYP3A Subfamily *201*	
9.2	Reversible Inhibition *203*	
9.2.1	Bioanalytical Techniques *204*	
9.2.2	Reagent Selection and Reaction Conditions *206*	
9.2.3	Experimental Design *213*	
9.3	Irreversible Inhibition *217*	
9.3.1	Identification of Mechanism-Based Inhibitor *220*	
9.3.2	Characterization of Mechanism-Based Inhibition *222*	
9.3.3	Additional Mechanistic Tools for Investigating Mechanism-Based Inhibition *225*	
9.4	Conclusion *230*	
	References *231*	
10	**Site of Metabolism Predictions: Facts and Experiences** *247*	
	Ismael Zamora	
10.1	Introduction *247*	
10.2	Factors That Influence the Site of Metabolism Prediction by Cytochrome P450s *248*	

10.2.1	Chemical Reactivity *249*
10.2.2	Protein/Compound Interaction *249*
10.3	Methods to Predict the Site of Metabolism *250*
10.3.1	Knowledge-Based Methods *251*
10.3.2	From Protein Structure to Chemical Reactivity *251*
10.3.2.1	Docking *251*
10.3.2.2	MetaSite *252*
10.3.2.3	Quantum Chemistry *252*
10.4	The Influence of the Protein Structure on the Site of Metabolism *253*
10.4.1	Comparative Analysis Between the Different CYP Crystal Structures *253*
10.4.1.1	CYP1A2 *253*
10.4.1.2	CYP2C5 *255*
10.4.1.3	CYP2B4 *256*
10.4.1.4	CYP2C9 *256*
10.4.1.5	CYP2D6 *257*
10.4.1.6	CYP3A4 *257*
10.4.2	The Effect of the Structure in the Different Methods to Predict the Site of Metabolism *259*
10.5	Conclusions *260*
	References *262*

11 Irreversible Cytochrome P450 Inhibition: Common Substructures and Implications for Drug Development *267*
Sonia M. Poli

11.1	Introduction *267*
11.2	Overview *268*
11.2.1	Characteristics of Irreversible CYP Inhibition *268*
11.2.2	Quantitative Prediction of Drug–Drug Interaction Caused by Irreversible CYP Inhibitors *269*
11.2.3	Irreversible CYP Inhibition and Autoimmune Diseases *269*
11.3	Structural Features Often Responsible for Mechanism-Based CYP Inhibition *270*
11.3.1	Terminal Acetylenes (ω or $\omega-1$) *271*
11.3.2	Alkenes *271*
11.3.3	Furans and Thiophenes *273*
11.3.4	Secondary and Tertiary Amines *273*
11.3.5	Benzodioxoles (MDP) *273*
11.4	Conclusions *274*
	References *274*

12 MetaSite: Understanding CYP Antitarget Modeling for Early Toxicity Detection *277*
Yasmin Aristei, Gabriele Cruciani, Sergio Clementi, Emanuele Carosati, Riccardo Vianello, Paolo Benedetti

12.1	Introduction *277*

12.2	The CYPs as Antitarget Enzymes	278
12.3	The UGTs as Antitarget Enzymes	279
12.4	The MetaSite Technology	282
12.4.1	Mechanism-Based Inhibitors	285
12.4.2	Phase II Metabolism by UGTs	287
12.4.3	The Flowchart of the Overall Method	287
12.5	Conclusions	289
12.6	Software Package	290
	References	290

13 Orphan Nuclear Receptor PXR-Mediated Gene Regulation in Drug Metabolism and Endobiotic Homeostasis 293

Jie Zhou, Yonggong Zhai, Wen Xie

13.1	Cloning and Initial Characterization of PXR	293
13.2	PXR and Its Regulation of Drug-Metabolizing Enzymes and Transporters	295
13.2.1	PXR in Phase I CYP Enzyme Regulation	295
13.2.2	PXR in Phase II Enzyme Regulation	296
13.2.2.1	PXR in UGT Regulation	296
13.2.2.2	PXR in SULT Regulation	296
13.2.2.3	PXR in GST Regulation	297
13.2.3	PXR in Drug Transporter Regulation	298
13.3	Crosstalk Between PXR and Other Nuclear Receptors	298
13.4	Implications of PXR-Mediated Gene Regulation for Drug Metabolism and Pathophysiology	299
13.4.1	PXR in Drug Metabolism and Drug–Drug Interactions	300
13.4.1.1	PXR in Drug Metabolism	300
13.4.1.2	PXR in Drug–Drug Interactions	300
13.4.1.3	Enantiospecificity of PXR-Activating Drugs and its Implication in Drug Development	301
13.4.2	Endobiotic Function of PXR	302
13.4.2.1	PXR in Bile Acid Detoxification and Cholestasis	302
13.4.2.2	PXR in Bilirubin Detoxification and Clearance	303
13.4.2.3	PXR in Adrenal Steroid Homeostasis and Drug–Hormone Interactions	303
13.4.2.4	PXR in Lipid Metabolism	304
13.5	Species Specificity of PXR and the Creation of "Humanized" Mice	304
13.5.1	Challenges of using Rodents as Drug Metabolism Models	304
13.5.2	Species Specificity of the Rodent and Human PXR	305
13.5.3	Creation and Characterization of the hPXR "Humanized" Mice	305
13.5.4	Significance of Humanized Mice in Drug Metabolism Studies and Drug Development	307
13.6	Conclusions	307
	References	310

14	**Ligand Features Essential for Cytochrome P450 Induction** *317*
	Daniela Schuster, Theodora M. Steindl, Thierry Langer
14.1	Introduction *317*
14.2	Molecular Mechanisms Leading to P450 Induction *318*
14.2.1.1	Heteroactivation *318*
14.2.1.2	Activation of Nuclear Receptors *319*
14.2.2	Ligands Directly Inducing P450 Activity *319*
14.2.2.1	P450 2C9 Heteroactivators *319*
14.2.2.2	P450 3A4 Heteroactivators *322*
14.2.3	P450 Inducers Acting via Nuclear Receptors *322*
14.2.3.1	Pregnane X Receptor *322*
14.2.3.2	Constitutive Androstane Receptor *325*
14.2.3.3	Farnesoid X Receptor *326*
14.2.3.4	Liver X Receptors *328*
14.2.4	P450 Inducers Acting via Transcription Factors *330*
14.2.4.1	Aryl Hydrocarbon Receptor *330*
14.3	General Ligand Features Leading to NR Activation *332*
	References *333*

15	**Transporters and Drugs – An Overview** *341*
	Hartmut Glaeser, Martin F. Fromm, Jörg König
15.1	Introduction *341*
15.2	Organic Anion Transporting Polypeptides and Drug Transport *342*
15.3	Multidrug Resistance Proteins and Drug Transport *346*
15.4	Role of P-Glycoprotein for Drug Disposition *349*
15.5	Vectorial Drug Transport *352*
	References *355*

16	**Computational Models for P-Glycoprotein Substrates and Inhibitors** *367*
	Patrizia Crivori
16.1	P-Glycoprotein Structure, Expression, Mechanism of Transport and Role on Drug Pharmacokinetics *367*
16.2	*In Vitro* Models for Studying P-gp Interacting Compounds *369*
16.3	Computational Models for Predicting P-gp Interacting Compounds *371*
16.3.1	Ligand-Based Approach *372*
16.3.2	Protein and Ligand–Protein Interaction-Based Approaches *387*
16.4	Computational Models for Other Important Drug Transporters *389*
16.5	Conclusions *390*
	References *392*

IV	**Case Studies of Drug Optimization Against Antitargets** *399*	
17	**Selective Dipeptidyl Peptidase IV Inhibitors for the Treatment of Type 2 Diabetes: The Discovery of JANUVIA™ (Sitagliptin)** *401*	
	Scott D. Edmondson, Dooseop Kim	
17.1	Introduction *401*	
17.2	Selectivity of DPP-4 Inhibitors *402*	
17.3	α-Amino Acid Amide Series *406*	
17.3.1	Cyclohexylglycines and Related Derivatives *406*	
17.3.2	β-Substituted Phenylalanines as Potent and Selective DPP-4 Inhibitors *407*	
17.4	Early β-Amino Acid Amide DPP-4 Inhibitors *411*	
17.4.1	Discovery of JANUVIA™ (Sitagliptin) and Related Structure–Activity Relationships *413*	
17.5	Conclusions *416*	
	References *417*	
18	**Strategy and Tactics for hERG Optimizations** *423*	
	Craig Jamieson, Elizabeth M. Moir, Zoran Rankovic, Grant Wishart	
18.1	Introduction *423*	
18.1.1	Classification of Optimization Literature *423*	
18.1.1.1	Discrete Structural Modifications *424*	
18.1.1.2	Formation of Zwitterions *425*	
18.1.1.3	Control of log P *425*	
18.1.1.4	Attenuation of pK_a *425*	
18.2	Survey of Strategies Used to Diminish hERG *426*	
18.2.1	Discrete Structural Modifications *426*	
18.2.2	Formation of Zwitterions *434*	
18.2.3	Control of log P *440*	
18.2.4	Attenuation of pK_a *444*	
18.3	Summary and Analysis *448*	
18.3.1	Summary of Optimization Literature *448*	
18.3.2	Global Analysis of Reported Optimizations *448*	
18.3.3	Summary and Recommendations *449*	
18.3.4	Future Directions *450*	
	References *451*	
19	**Structure-Based *In Silico* Driven Optimization: Discovery of the Selective 5-HT$_{1A}$ Agonist PRX-00023** *457*	
	Oren M. Becker	
19.1	Introduction *457*	
19.2	Structure-Based *In Silico* Driven Multidimensional Optimization Paradigm *459*	
19.3	Clinical Candidate Selection Criteria *461*	

	19.4	Lead Identification *461*
	19.5	Optimization Round 1: Reducing Off-Target Activities *462*
	19.6	Optimization Round 2: Reducing Affinity to hERG *468*
	19.7	Conclusion *471*
		References *472*

Index *477*

List of Contributors

Yasmin Aristei
University of Perugia
Chemistry Department
Laboratory for Chemometrics
and Cheminformatics
Via Elce di Sotto 10
06123 Perugia
Italy

Oren M. Becker
Epix Pharmaceuticals Ltd.
3 Hayetzira St.
Ramat Gan 52521
Israel

Current address:
Dynamix Pharmaceuticals
12 Motza Ha-Qetana
Mevasserei Zion 90805
Israel

Paolo Benedetti
Molecular Discovery Ltd
215 Marsh Road
Pinner, Middlesex
HA5 5NE London
United Kingdom

Emanuele Carosati
University of Perugia
Chemistry Department
Laboratory for Chemometrics
and Cheminformatics
Via Elce di Sotto 10
06123 Perugia
Italy

Andrea Cavalli
Department of Pharmaceutical Sciences
University of Bologna
Via Belmeloro 6
40126 Bologna
Italy

Sergio Clementi
University of Perugia
Chemistry Department
Laboratory for Chemometrics
and Cheminformatics
Via Elce di Sotto 10
06123 Perugia
Italy

Patrizia Crivori
Nerviano Medical Sciences
Accelera, Pharmacokinetic & Modeling/
Modeling
Viale Pasteur 10
20014 Nerviano
Italy

Gabriele Cruciani
University of Perugia
Chemistry Department
Laboratory for Chemometrics
and Cheminformatics
Via Elce di Sotto 10
06123 Perugia
Italy

Sarah Dalibalta
University of Leicester
Department of Cell Physiology and
Pharmacology
Medical Sciences Building
University Road
Leicester LE1 9HN
United Kingdom

Pier G. De Benedetti
University of Modena and Reggio Emilia
Department of Chemistry and
Advanced Scientific
Computing Laboratory
Via Campi 183
41100 Modena
Italy

Fabrizio De Ponti
University of Bologna
Department of Pharmacology
Via Irnerio 48
40126 Bologna BO
Italy

Philippe Dupuis
Cerep SA
Le bois l'Eveque
86600 Celle L'Evescault
France

Scott D. Edmondson
Merck Research Laboratories
Department of Medicinal Chemistry
P.O. Box 2000, RY123-234
Rahway, NJ 07065
USA

Andreas Evers
Sanofi-Aventis Deutschland GmbH
Chemical Analytical Sciences
Drug Design
Building G838
65926 Frankfurt am Main
Germany

Francesca Fanelli
University of Modena and Reggio Emilia
Dulbecco Telethon Institute and
Department of Chemistry
Via Campi 183
41100 Modena
Italy

Martin F. Fromm
University of Erlangen-Nuremberg
Institute of Experimental and Clinical
Pharmacology and Toxicology
Fahrstr. 17
91054 Erlangen
Germany

Hartmut Glaeser
University of Erlangen-Nuremberg
Institute of Experimental and Clinical
Pharmacology and Toxicology
Fahrstr. 17
91054 Erlangen
Germany

Craig Jamieson
Organon Laboratories Ltd
Medicinal Chemistry
Newhouse
Lanarkshire
ML1 5SH
United Kingdom

Dooseop Kim
Merck Research Laboratories
Department of Medicinal Chemistry
P.O. Box 2000, RY800-C206
Rahway, NJ 07065
USA

List of Contributors

Thomas Klabunde
Sanofi-Aventis Deutschland GmbH
Science & Medical Affairs
Drug Design
Building G878
65926 Frankfurt am Main
Germany

Jörg König
University of Erlangen-Nuremberg
Institute of Experimental and Clinical
Pharmacology and Toxicology
Fahrstr. 17
91054 Erlangen
Germany

Christian Laggner
University of Innsbruck
Department of Pharmaceutical
Chemistry
CAMD Group
Innrain 52c
6020 Innsbruck
Austria

Thierry Langer
University of Innsbruck
Department of Pharmaceutical
Chemistry
CAMD Group
Innrain 52c
6020 Innsbruck
Austria

Jonathan S. Mason
Lundbeck Research
Ottiliavej 9
2500 Valby
Denmark

Jacques Migeon
Cerep SA
Le bois l'Eveque
86600 Celle L'Evescault
France

John S. Mitcheson
University of Leicester
Department of Cell Physiology and
Pharmacology
Medical Sciences Building
University Road
Leicester LE1 9HN
United Kingdom

Elizabeth M. Moir
Organon Laboratories Ltd
Medicinal Chemistry
Newhouse
Lanarkshire
ML1 5SH
United Kingdom

Annie Otto-Bruc
Cerep SA
Le bois l'Eveque
86600 Celle L'Evescault
France

Sonia Maria Poli
Addex Pharma
Non-Clinical Development
12, Chemin des Aulx
1228 Plan-Les-Ouates – GE
Switzerland

Zoran Rankovic
Organon Laboratories Ltd
Medicinal Chemistry
Newhouse
Lanarkshire
ML1 5SH
United Kingdom

Maurizio Recanatini
University of Bologna
Department of Pharmaceutical Sciences
Via Belmeloro 6
40126 Bologna
Italy

List of Contributors

Dan Rock
Amgen Inc.
1201 Amgen Court West
Seattle, WA 98119
USA

Bryan L. Roth
UNC Chapel Hill School of Medicine
Medicinal Chemistry
Burnett Womack Building; Room 8032
UNC Chapel Hill
Chapel Hill, NC
USA

Daniela Schuster
University of Innsbruck
Department of Pharmaceutical Chemistry
CAMD Group
Innrain 52c
6020 Innsbruck
Austria

Vincent Setola
INSERM
U616 Hôpital Pitié-Salpêtrière
Paris
France

Theodora M. Steindl
Inte:Ligand Softwareentwicklungs- und Consulting GmbH
Clemens-Maria-Hofbauer-G. 6
2344 Maria Enzersdorf
Austria

Riccardo Vianello
Molecular Discovery Ltd
215 Marsh Road, Pinner
Middlesex
HA5 5NE London
United Kingdom

Jan Wahlstrom
Amgen Inc.
1201 Amgen Court West
Seattle, WA 98119
USA

Larry Wienkers
Amgen Inc.
1201 Amgen Court West
Seattle, WA 98119
USA

Grant Wishart
Organon Laboratories Ltd
Medicinal Chemistry
Newhouse
Lanarkshire
ML1 5SH
United Kingdom

Wen Xie
University of Pittsburgh
Center for Pharmacogenetics
633 Salk Hall
Pittsburgh, PA 15216
USA

Ismael Zamora
Pompeu Fabra University
GRIB-IMIM
Dr Aiguader 80
E080003 Barcelona
Spain

Yonggong Zhai
Beijing Normal University
Biomedicine Research Institute and College of Life Science
Beijing 100875
China

Jie Zhou
University of Pittsburgh
Center for Pharmacogenetics
633 Salk Hall
Pittsburgh, PA 15216
USA

Preface

When the hammer hits your thumb in addition to, or instead of the nail - this clearly is an antitarget problem. In drug action and especially in drug-drug interactions, the situation is much more serious. There are numerous targets which should not be activated, induced, or inhibited by a drug. Adverse drug reactions (ADRs) are estimated to be one of the leading causes of morbidity and mortality in healthcare. Especially older patients receive up to 15 different medications, and even more, at the same time – no wonder that unfavorable drug-drug interactions happen, considering also the poor physical condition of such patients. In January 2000, the Institute of Medicine reported that in the US about 7000 deaths per year occur due to ADRs; the number may be even much larger, not to count the manyfold non-fatal side effects.

In defining the term "antitarget" one runs into some problems. Let us start with terfenadine, a non-sedating H1-antihistaminic. In its clinical studies it was obviously safe, with only minor side effects. However, after its broad therapeutic use it turned out to produce fatal arrhythmias. In addition to being an H1 antagonist, terfenadine is a potent hERG channel inhibitor. Under normal conditions, terfenadine is already metabolized in the intestinal wall; no measurable plasma levels and no cardiac side effects are observed. If CYP inhibitors prevent terfenadine metabolism, cardiac toxicity results. Thus, in addition to its H1 receptor antagonism, the compound inhibits the hERG channel, a most prominent antitarget. The terfenadine problem could be circumvented by replacing the compound by its active metabolite fexofenadine, which is not any longer a hERG channel inhibitor.

Already around 1990, David Bailey discovered that grapefruit (but not orange) juice significantly increases the bioavailability of some drugs, for example the calcium channel blockers felodipine and nifedipine. After some problems to publish this highly surprising observation, his manuscript was accepted by Lancet and many reports followed for other drugs. The whole story shall not be elaborated here in detail but it remains the question: is CYP3A4, which is inhibited by some flavonoid and furocoumarin constituents of grapefruit juice, an antitarget? It is: despite the fact that increase in bioavailability is a desirable effect, individual variation is too large to be of therapeutic value. Only in the case of clinically monitored drugs, for example cyclosporin, the co-medication of a CYP3A4 inhibitor, for example ketoconazole, is used to reduce the dose of this expensive drug.

Antitargets. Edited by R. J. Vaz and T. Klabunde
Copyright © 2008 WILEY-VCH Verlag GmbH & Co. KGaA, Weinheim
ISBN: 978-3-527-31821-6

Recently it was discovered that grapefruit juice also induces the expression of intestinal drug transporters. With fexofenadine, the safe replacement of terfenadine, the paradox situation results that bioavailability of this drug decreases (!) after intake of grapefruit juice, due to induction of an efflux pump, the organic anion-transporting polypeptide 1A2 (OATP1A2). Kirby and Unadkat commented this paradox by asking the question "Grapefruit juice, a glass full of drug interactions?"

Not only food constituents but also OTC drugs may cause significant drug-drug interactions. One of the well-known examples is St. John's Wort (*Hypericum perforatum* L.) extract, supposed to be beneficial against mild depression. However, in addition to the phototoxic agent hypericin, the plant contains hyperforin, the strongest inducer of CYPs (especially CYP3A4) and drug transporters. In this manner, self-medication with St. John's Wort reduces the bioavailability and thus the activity of several drugs, whereas doses of some drugs have to be reduced after discontinuation of this extract because CYP and transporter levels return to normal.

There are many more antitargets, prominent ones being several G protein-coupled receptors. Thus, it is high time that a book on antitargets becomes available and we, as Editors of the series Methods and Principles in Medicinal Chemistry, are very much indebted to Roy Vaz and Thomas Klabunde, leading scientists in antitarget research, for editing such a monograph. The book starts with an introduction on the reasons why drugs fail in the clinics or after market introduction and a chapter on ADME and side effects prediction. The main two sections contain several chapters on antitargets and side effects (hERG channel and GPCR antitargets) and on antitargets in ADME (CYPs and drug transporters). The book concludes with some case studies of drug optimization against antitargets.

We are very grateful to all chapter authors and we thank the publisher Wiley-VCH, especially Dr. Frank Weinreich, for the ongoing support of our series, and Dr. Nicola Oberbeckmann-Winter for her contributions in the preparation of this volume.

October 2007

Raimund Mannhold, Düsseldorf
Hugo Kubinyi, Weisenheim am Sand
Gerd Folkers, Zürich

A Personal Foreword

A single report of a drug reaction in a 39-year-old woman ultimately contributed to the removal of the allergy drug Seldane (terfenadine) from the market in 1998 [1]. Doctors at the National Naval Medical Center in Bethesda, Md., admitted the woman to the hospital because of fainting episodes. She had been prescribed Seldane (terfenadine) 10 days before. She also started using the prescription drug Nizoral (ketoconazole) for a vaginal yeast infection. That combination caused potentially fatal changes in her heart rhythm. The Food and Drug Administration (FDA) issued warnings indicating that ketoconazole interfered with terfenadine's metabolism, which resulted in increased levels of terfenadine in the blood and slowed its elimination from the body. The FDA also warned that a similar effect could occur if Seldane was taken with the antibiotic erythromycin.

Thus the first awareness of antitargets was brought to the forefront with the withdrawal of terfenadine. Ketoconazole is a strong inhibitor of CYP3A4, which is also the primary enzyme responsible for the clearance of terfenadine. The inhibition of CYP3A4 leads to the increase in concentration of terfenadine in the blood. Terfenadine itself is a blocker of the ion channel hERG (human ether-a-go-go related gene) and caused a prolonged QT, leading to Torsades de Pointes and possibly death. Also ketoconazole inhibits the efflux transporter P-glycoprotein (P-gp) or MDR1 (multidrug resistance protein), for which terfenadine is a substrate. Hence when co-administered with ketoconazole, the concentration of terfenadine in blood would be much higher than if taken without other drugs such as ketoconazole. Therefore inhibition of both P-gp and CYP3A4 could lead to drug-drug interactions and inhibition of hERG either by the compound itself or its metabolite.

The example of terfenadine shows that toxic effects can be either induced directly by the action of a drug or a drug metabolite on an antitarget like the hERG channel. In addition, certain transporters and metabolizing enzymes, like P-gp and CYP3A4, need also be considered as antitargets as blocking their activity can change the concentration of a co-administered drug or its metabolite in blood, thus causing drug-drug interactions and potential toxicity.

Adverse drug reactions (ADRs) cost approximately one hundred and thirty nine billion dollars annually [2–4] in the United States. This number is larger than the cost of cardiovascular or diabetic care. ADRs cause 1 out of 5 injuries or deaths per year to hospitalized patients and the mean length of stay, the cost and the mortality for patients admitted due to an ADR are double that for control patients. Many ADRs are due to off-target and antitarget interactions. Some of these have lead to withdrawal of the drug(s) from the marketplace.

Since terfenadine, there have been other market withdrawals of drugs. As shown in the first chapter of the book the main cause for the 16 drug withdrawals from 1992 to 2002 was toxicity, mainly cardiovascular toxicity or hepatotoxicity. Only recently Vioxx (rofecoxib) had to be withdrawn from the market. In contrast to terfenadine, where the molecular mechanism of its side effects has been fully understood, the underlying mechanism by which rofecoxib, a selective cyclooxygenase 2 inhibitor that exhibits cardiovascular effects is still unclear [5]. Pondimin (fenfluramine), a serotonergic anorectic, was withdrawn in 1997 due to the risk of development of primary pulmonary hypertension or valvular heart disease. The first case of fenfluramine associated valvular heart disease discovered 7 years after discontinuation of treatment and requiring double valve replacement 2 years later has just been reported [6]. For fenfluramine the mechanism by which it causes valvular heart disease has recently been uncovered showing a causal association between agonism on the G-protein coupled receptor (GPCR) 5-HT$_{2B}$ and valvular heart disease (Chapter 7 in the book).

According to FDA experts, discovering terfenadine's interactions with other drugs marked a significant advance. These and other discoveries improved the ability of the FDA and drug manufacturers to test for drug interactions and to investigate risks of heart rhythm abnormalities and other toxicities before drugs could be marketed. In addition, unraveling the mechanism of drug toxicities and identifying specific channels, receptors including nuclear receptors, transporters or enzymes as antitargets enabled establishment of *in vitro* test systems to monitor potential antitarget mediated side effects and toxicity in the drug discovery phase. The list of antitargets is still being compiled but the events that have lead to the discovery of the known antitargets has impacted the way research is conducted during drug discovery today.

In every family of biological targets there are antitargets. In this book, we have avoided discussion of kinases, which are still controversial as non-oncology targets and could probably command several chapters or volumes. Transporters and metabolizing enzymes like CYP450s that can mediate undesired drug-drug interactions have been mentioned before. In the area of potassium voltage-gated ion channels, there are therapeutic targets such as Kv1.3 and Kv1.5 but at the same time there are antitargets such as hERG. GPCRs form a large protein family that plays an important role in many physiological and patho-physiological processes. Especially the subfamily of biogenic amine binding GPCRs has provided excellent drug targets for the treatment of numerous diseases [7]. Although representing excellent therapeutic targets, the central role that many of the biogenic amine binding GPCRs play in cell signaling also poses a risk on new drug candidates which reveal side-affinities towards these receptor sites: These candidates bear the risk to interfere with the physiological signaling process and to cause undesired effects in preclinical or clinical studies. Besides the 5-HT$_{2B}$ receptor mentioned before, the α_{1A} adrenergic receptor, being a drug target for the treatment of benign prostatic hypertrophy (BPH), has been suggested as an antitarget at the same time that mediates cardiovascular side-effects of many drug candidates causing orthostatic hypotension, dizziness and fainting spells [8]. Other examples of GPCR antitargets are the muscarinic M1 receptor correlated with attention and memory deficits or the serotonin 5-HT$_{2C}$ receptor associated with weight gain. As shown in one of the introductory chapters of this book, correlation between *in vitro* affinity and *in vivo* adverse effects can currently be recognized by profiling, hundreds of drugs with known ADRs using large panels of pharmacological *in vitro* assays.

There are several references made to both off-targets as well as antitargets in the literature. In this book we will primarily be attempting to cover the topic of antitargets. Off-target activity the way we interpret it, is activity for a particular compound towards a target that was not anticipated, when it was synthesized or isolated. For example, compounds that are designed or synthesized for activity towards serine proteases (not for thrombosis) are not expected to have any activity towards serine protease targets such as thrombin or others in the coagulation pathway, which could be anti-thrombotic targets themselves. The term off-targets includes antitargets and the off-target activities could be beneficial or detrimental. Antitargets on the other hand are targets that are detrimental towards progression of the compound towards becoming a drug.

Within recent years the understanding of the molecular interactions between antitargets and drugs or drug candidates has tremendously increased allowing *in silico* antitarget models to be established. 3-dimensional structures of several antitargets (often in complex with inhibitors) are now available either derived by homology modeling (e.g. the hERG channel or GPCRs) or by protein crystallography (e.g. cytochrome P450s). Structural chemical motifs often associated with antitarget interactions (e.g. for cytochrome P450 binding or inhibition) have been captured in knowledge databases. Computational models like 3D-pharmacophore or 3D-QSAR models (e.g. for GPCRs, hERG, CYPs, P-gp) have been established to not only recognize antitarget affinities in chemical lead series but also to guide the chemical optimization of these leads towards development candidates lacking undesired antitarget side affinities and thus potential side effects or toxicities. These models are captured – together with introductory chapters on the biological aspects – in the second (focusing on ion channels and GPCRs) and the third section of this book (describing antitargets mediating drug-drug interactions).

Examples of optimization of selectivity towards the antitargets have been well described in the recent literature such as illustrated in Gao et al. [9] and Kuduk *et al.* [10]. In the last section we have tried to include very specific case studies of successful drugs for which optimization of selectivity towards specific or general antitargets were successfully negotiated, e.g. for Januvia (chapter 17), a recently released DPP4 inhibitor or for PRX-00023, a selective 5-HT$_{1A}$ agonist currently in phase IIb clinical trials (chapter 19).

We would first thank the editors of the series for enabling this volume in the Methods and Principles in Medicinal Chemistry series. We would like to sincerely thank all chapter authors for making this book a reality. We would like to acknowledge their great enthusiasm in preparing their manuscripts and the high quality of their contributions. It has been a pleasure working with each and every one of them. The editors are also grateful to Frank Weinreich, Nicola Oberbeckmann-Winter and the staff of Wiley-VCH for their excellent support in the production of this book. We also thank the Sanofi-Aventis Discovery Management for enabling this book. We thank our families for putting up with us during the last few months.

October 2007

Roy Vaz, Bridgewater
Thomas Klabundl, Frankfurt

References

1 Monahan, B.P., Ferguson, C.L., Kileavy, E.S., Lloyd, B. K., Troy, J., Cantilena, L.R. (1990) Torsades de Pointes occurring in association with Terfenadine use. *The Journal of the American Medical Association*, **264**, 2788–2790.

2 Johnson J.A., Bootman, J.L. (1995) Drug-related morbidity and mortality. A cost-of-illness model. *Archives of Internal Medicine*, **155** (18), 1949–1956.

3 Leape, L.L., Brennan, T.A., Laird, N., Lawthers, A.G., Localio, A.R., Barnes, B.A., Hebert, L., Newhouse, J.P., Weiler, P.C., Hiatt, H.H. (1991) The nature of Adverse events in hopitalized patients. Results of the Harvard Medical Practice Study II. *The New Journal of Medicine*, **324** (6), 377–394.

4 Classen D.C., Pestotnik, S.L., Evans, R.S., Lloyd, J.F., Burke, J.P. (1997) Adverse drug events in hospitalized patients. Excess length of stay, extra costs, and attributable mortality. *The Journal of the American Medical Association*, **277** (4), 301–306.

5 Zarraga, I.G.E., Schwarz, E.R. (2007) Coxibs and Heart Disease. *Journal of the American College of Cardiology*, **49** (1), 1–149.

6 Greffe, G., Chalabreysse, L., Mouly-Bertin, C., Lantelme, P., Thivolet, F., Aulagner, G., Obadia, J-F. (2007) Valvular Heart Disease associated with Fenfluramin Detected 7 Years after Discontinuation of Treatment. *The Annals of Thoracic Surgery*, **83**, 1541–1543.

7 Klabunde, T., Hessler, G. (2002) Drug design strategies for targeting G-protein-coupled receptors. *A European Journal of Chemical Biology*, **3**, 928–944.

8 Kehne, J.H., Baron, B.M., Carr, A. A., Chaney, S. F., Elands, J., Feldman, D. J., Frank, R. A., van Giersbergen, P. L., McCloskey, T. C., Johnson, M.P., McCarty, D. R., Poirot, M., Senyah, Y., Siegel, B. W., Widmaier, C. (1996) Preclinical characterization of the potential of the putative atypical antipsychotic MDL 100,907 as a potent 5-HT2A antagonist with a favorable CNS safety profile. *The Journal of Pharmacology and Experimental Therapeutics*, **277**, 968–981.

9 Gao, Y.D., Olson, S.H., Balkovec, J.M., Zhu, Y., Royo, I., Yabut, J., Evers, R., Tan, E.Y., *et al.* (2007) Attenuating pregnane X receptor (PXR) activation: A molecular modeling approach. *Xenobiotica*, **37** (2), 124–138.

10 Kuduk, S.D., DiMarco, C.N., Chang, R.K., Wood, M.R., Schirripa, K.M., Kim, J.J., Wai, J.M.C., DiPardo, R.M. *et al.* (2007) Development of Orally Bio-available and CNS Penetrant Biphenyla-minocyclopropane Carboxamide Bradykinin B1 Receptor Antagonists. *Journal of Medicinal Chemistry*, **50**, 272–282.

I
General Aspects

1
Why Drugs Fail – A Study on Side Effects in New Chemical Entities

Daniela Schuster, Christian Laggner, Thierry Langer

1.1
Introduction

Drug development is a long and cost-intensive business. Only after years of lead identification, chemical optimization, *in vitro* and animal testing can the first clinical trials be conducted. Unfortunately, many projects still fail in this late stage of development after a considerable amount of money has been spent. According to estimates, preapproval costs for a new drug exceed US$ 800 million [1].

Approximately 10% of new chemical entities (NCEs) show serious adverse drug reactions (ADRs) after market launch. Such events usually result in 'new black box warnings' by the US Food and Drug Administration (FDA), label change or market withdrawal. The most common causes for these actions are hepatic toxicity, hematologic toxicity and cardiovascular toxicity [2]. Reasons for such ADRs, which are identified only after NCEs are launched on the market, include the narrow spectrum of clinical disorders and participating patient profiles in clinical studies as well as the fact that serious ADRs are often rare and that the number of patient exposures required to identify such occurrences sometimes may range over a few millions [3].

To avoid the occurrence of ADRs in the future, specific trials to detect them should therefore be conducted before an NCE is launched on the market. Before this can be done, however, the major reasons leading to the withdrawal of drugs and termination of NCE-to-drug development should be identified and analyzed.

In this chapter, reasons why 17 drugs were withdrawn from the Western market between 1992 and 2006 are discussed and facts on 63 terminated clinical development projects presented, so as to identify the most common reasons for the failure of drugs in this late stage of drug development. This analysis is then compared with two previous related studies published more than 18 years ago by Prentis *et al.* [4] and Kennedy [5].

The study by Prentis *et al.* [4] included an analysis of 198 NCEs, developed between 1964 and 1985 by British pharmaceutical companies but had not been marketed for reasons presented in Figure 1.1. Kennedy [5] further analyzed these data and noticed that a high number of anti-infective drug development projects were all terminated

---

Antitargets. Edited by R. J. Vaz and T. Klabunde
Copyright © 2008 WILEY-VCH Verlag GmbH & Co. KGaA, Weinheim
ISBN: 978-3-527-31821-6

1 Why Drugs Fail – A Study on Side Effects in New Chemical Entities

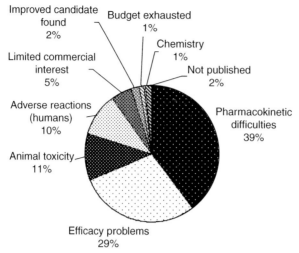

Figure 1.1 Reasons for drug development termination from 1964 to 1985 (n = 198).

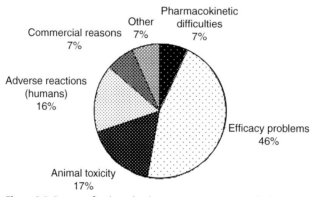

Figure 1.2 Reasons for drug development termination, excluding anti-infectives (n = 121).

because of pharmacokinetic difficulties. He therefore excluded the anti-infective NCEs from the statistics and presented the facts as illustrated in Figure 1.2.

1.2
Drugs Withdrawn from the Market between 1992 and 2006 Listed Alphabetically

1.2.1
Amineptine

The atypical tricyclic antidepressant amineptine (Survector) is an indirect dopamine agonist, which selectively inhibits dopamine uptake and induces its release, with additional stimulation of the adrenergic system. Its antidepressant effects are similar to those of other tricyclic antidepressant drugs. However, it acts more rapidly, is better tolerated and has little cardiovascular, analgesic or anorectic effects [6]. Amineptine was launched on the market in 1978 for the treatment of dysthymia and was marketed by Merck and Servier [7]. Microcystic and macrocystic acne and toxicomania were observed as its common side effects. Also, a major risk of addiction was reported [8]. More females than males were found to suffer from acneiform eruptions resulting from amineptine therapy. The very florid, only slightly inflammative, retentional acne lesions with multiple comedones and microcystic and macrocystic lesions appeared mainly on the face, ears and neck of the patients. The incidence of acne was dose-related [9]. In comparison with other tricyclic antidepressants, amineptine – as tranylcypromine – was considered to have a clinically significant liability to cause addiction, which was attributed to its dopaminergic and stimulant properties [10]. In addition, cases of hepatotoxicity were reported, thought to be related to oxidation of amineptine forming a reactive metabolite. Especially, patients of the extensive metabolizer phenotype and those with an increased susceptibility to amineptine reactive metabolites – probably related to a genetic deficiency in a cell defense mechanism – suffered from liver injury [11]. The overconsumption of amineptine was expressed by the appearance of neuropsychotic disorders such as agitation, confusion, anxiety and insomnia, as well as weight loss and severe cutaneous reactions (acne) [12]. In 1999, France, Italy and other countries decided to suspend the marketing authorization for amineptine. In the United States of America, amineptine has never been approved. All regulatory authorities withdrew this substance because of the risk of its potential abuse and dependence on it [13] – in other words, addiction. In some developing countries, however, amineptine is still in use; although hepatotoxicity associated with amineptine, along with acne eruption and anxiety, and the availability of safer antidepressants have all made amineptine therapeutically less useful [6].

1.2.2
Aminophenazone (Aminopyrine)

In December 1999, the FDA suspended aminophenazone, an analgetic, antipyretic and antirheumatic substance. The drug had been introduced way back in 1887 and used for over 100 years in the clinic. Aminophenazone caused agranulocytosis, a condition characterized by a decrease in the number of granulocytes – a type of white blood cells – and lesions on the mucous membrane and skin. Some of the cases of agranulocytosis were fatal [14]. Another reason for suspending this drug from the market was its ability to react with nitrite-containing food, thus forming carcinogenic nitrosamines [15]. Aminophenazone is a metabolite of metamizole, an analgesic, which is still marketed in Germany and Austria.

1.2.3
Astemizole

This second-generation histamine H1 receptor blocker was put on the world market in the 1980s for the relief of symptoms associated with seasonal allergic rhinitis and chronic idiopathic urticaria [14]. Janssen Pharmaceuticals marketed this drug under the brand name Hismanal. The major benefit of these antihistamines was their nonsedating character. Although this new class of medication was launched as highly selective and specific H1 antagonists, several of them were later found to cause prolongation of the QT interval in the electrocardiogram (ECG) and therefore induce severe cardiac arrhythmias [16]. The manufacturer withdrew the drug from the market in June 1999 [14]. Although its proarrhythmic effects are known, a second compound of this generation – terfenadine (Seldane) – is still available on the market [17].

1.2.4
Bromfenac Sodium

In 1998, Wyeth-Ayerst Laboratories announced to withdraw their nonsteroidal antiinflammatory drug (NSAID) Duract (bromfenac sodium) [18]. This drug had been submitted in 1994 and approved in July 1997 for the short-term management of acute pain. The use was indicated to 10 days or less, as there was a higher incidence of liver enzyme elevations in patients treated in long-term clinical trials [19]. The company withdrew the drug after postmarketing reports implicated it in severe hepatic failure that lead to four deaths and eight liver transplants. All but one of these 12 cases involved patients using Duract for more than 10 days. The exception involved a patient with preexisting significant liver disease. The company decided that steps to limit the use of a potent NSAID pain reliever such as Duract to just 10 days would not be feasible or effective and therefore withdrew the product [19].

1.2.5
Cerivastatin

Ever since lovastatin was launched in 1987 as the first potent HMG-CoA reductase inhibitor to reduce altered low-density lipoprotein (LDL) levels, statins are widely used to prevent thrombotic events. In 1997, Bayer launched cerivastatin (Lipobay, Baycol), a 50–200-fold more potent HMG-CoA reductase inhibitor than other statins [20]. At that time, it was known that the use of statins can lead to rhabdomyolysis, a severe and potentially life-threatening condition. This risk is increased when statins are taken along with fibrates, for example, gemfibrozil, another group of substances used to treat blood lipid disorders [21]. Although the package insert of cerivastatin carried several warnings concerning the combination with fibrates, there were 52 deaths reported under treatment with cerivastatin. In the United States alone, 31 of these deaths were counted, of which 12 occurred as a result of using cerivastatin in combination with gemfibrozil [20]. In August 2001, Bayer withdrew cerivastatin because of the serious side effects or even deaths associated with the improper use of the drug [21].

1.2.6
Chlormezanone

Since 1958, chlormezanone was marketed by Sanofi-Winthrop, under the brand name Trancopal, as a centrally acting muscle relaxant in lower back pain [22,23]. Evidence of the efficacy of chlormezanone was, however, limited and of poor quality. In comparison to other centrally acting muscle relaxants in lower back pain, there were no studies performed according to current standards. Considering the overall safety profile of chlormezanone, the most relevant risk involving the use of the drug was concluded to be life-threatening cases of toxic epidermal necrolysis and other bullous reactions [24]. Chlormezanone was shown to produce an increased relative risk of toxic epidermal necrolysis, also known as Stevens–Johnson syndrome [24]. Serious skin reactions may also occur with other muscle relaxants, but the available data suggested that the relative frequency of serious skin reactions was greater with chlormezanone [25]. In November 1996, Sanofi-Winthrop announced to discontinue this drug in accordance with the European Agency for the Evaluation of Medicinal Products (EMEA) report [26].

1.2.7
Fenfluramine and Dexfenfluramine

A dextrorotatory enantiomer of fenfluramine, dexfenfluramine (*Isomeride*) was introduced on the market in 1992, as its racemic fenfluramine (Ponderax), a serotonergic substance, had already been used for 25 years to treat obesity. However, like most of the anorexigenic drugs, it showed benefit for short-term treatment but failed to show significant benefit over placebo in long-term studies. Dry mouth, headache, fatigue, drowsiness and gastrointestinal disorders were reported as side effects [27]. When in 1997, such serious side effects as pulmonary hypertension and changes in the heart valves – especially in patients taking *fen-phen* (a combination of fenfluramine and phentermine) – were found, the FDA suspended both substances. This action came after physicians who had evaluated patients taking these two drugs with echocardiograms found that approximately 30% of the patients evaluated had abnormal echocardiograms, even though they had no symptoms [28].

1.2.8
Flosequinan

Flosequinan has a positive inotropic effect and shows a tendency to increase the heart rate, atrioventricular conduction in patients with atrial fibrillation and neurohormonal activation. Although the precise mechanisms involved have remained unclear up to now [29], this drug has been used to treat congestive heart failure (CHF). The FDA approved flosequinan (Manoplax) in 1993. However, the drug was withdrawn a year later because the PROFILE (prospective randomized flosequinan longevity evaluation) study indicated that flosequinan had adverse effects on survival, and that beneficial effects on the symptoms of heart failure did not last beyond the first 3 months of therapy, after which patients on the drug had a higher rate of hospitalization than patients taking a placebo [14].

1.2.9
Glafenine

A nonulcerogenic analgetic, antipyretric and nonsteroidal anti-inflammatory drug, glafenine was introduced in 1965 and was marketed under the brand names

Glifanan (Roussel Diamant) or Adalgur. Its main adverse effects included severe allergic reactions leading to anaphylactic shock, intrarenal crystallization of glafenine metabolites after long-term use and – in rare cases – liver toxicity [30,31]. Glafenine-induced anaphylactic shock was reported several times in the literature and was associated with cutaneous and respiratory manifestations [32]. In a study covering 20 years of drug surveillance in the Netherlands, glafenine was associated most often with probable or possible anaphylactic reactions (326 of the 992 reported cases) [33]. However, it was not until 1990 that Belgium ordered its withdrawal, the first country to do so. Two years later, the manufacturers withdrew glafenine worldwide. Some generic versions of this drug, though, may still be available in developing countries [34].

1.2.10
Grepafloxacin

Ever since its marketing began in August 1997, Raxar (grepafloxacin), an oral fluoroquinolone antibiotic, had been prescribed to an estimated 2.65 million patients in over 30 countries for a variety of infections including pneumonia, bronchitis and some sexually transmitted infections [35]. When severe cardiovascular events were reported implicating the drug, Glaxo Wellcome, its manufacturer, stated that it was no longer convinced the benefits of grepafloxacin outweighed the potential risk and considering that alternative antibiotics were available in the market withdrew the drug in 1999 [35]. Grepafloxacin binds to the human ether-a-go-go-related gene (hERG) potassium ion channel, which is responsible for repolarization in cardiac cells. Blocked by substrates such as fluoroquinolone antibiotics, repolarization is delayed, and a prolongation of the QT interval in the ECG can be observed. This symptom is also known as *torsade de pointes* (twist of points). As a result, cardiac arrhythmias can occur [36].

1.2.11
Levacetylmethadol

Levacetylmethadol, a synthetic opioid receptor agonist, was approved for the management of opiate dependence by the FDA in 1994 and was marketed as Orlaam by Roxane Laboratories since 1995. The European Commission granted a marketing authorization for the European Union to Sipaco Internacional Lda. in July 1997. However, in December 2000, following 10 cases of life-threatening cardiac disorders including ventricular rhythm disorders such as *torsade de pointes* reported in patients treated with levacetylmethadol, the EMEA recommended suspension of the marketing authorisation for Orlaam. The market suspension in the European Union took place in March 2001 [37]. However, levacetylmethadol remained on the US market as an orphan drug for the management of opioid dependence in patients who failed to show acceptable response to other adequate treatments. Meanwhile, Roxane Laboratories received more reports of severe cardiac-related adverse effects, including arrhythmias because of QT-interval prolongation

(15 cases), *torsade de pointes* (8 cases) and cardiac arrest (6 cases), as well as other cardiac-related adverse effects, such as arrhythmias, syncope and angina. In August 2003, Roxane Laboratories discontinued the sale and distribution of Orlaam, and levacetylmethadol is no longer available in the Western market [38].

1.2.12
Mibefradil

This calcium channel blocker, used to treat essential hypertension and stable angina pectoris, was marketed by Roche (as Posicor) and Astra Medica (as Cerate) [39]. As mibefradil turned out to be a potent inhibitor of certain liver enzymes of the cytochrome P450 (CYP) family, multiple drug interactions were observed. Mainly CYP 3A4 and CYP 2D6, two subtypes of the cytochrome P450 enzyme family, were found to be inhibited by this drug. This inhibition does not allow other substrates of CYP 3A4 or CYP 2D6 to be metabolized at the usual rate, which results in higher than usual concentrations of these substrates in the plasma. In such circumstances, the probability of side effects rises sometimes to a life-threatening level. Other kinds of interactions include interactions with terfenadine, cyclosporin A and metoprolol. Concomitant use of these drugs with mibefradil was either contraindicated or bound to a dose reduction of these drugs [39]. Cardiac effects have been reported as another side effect of mibefradil, while it has been associated with slowing down or complete suppression of sinoatrial node activity. Ventricular rates have been found to be as low as 30–40 beats per minute (bpm). Many patients were symptomatic. As this adverse effect occurred mainly in elderly patients who were on concomitant β-blocker therapy, it was warned against combining these substances. Use of mibefradil in patients suffering from sick sinus syndrome and possessing no pacemaker was contraindicated.

The FDA approved mibefradil for angina in August 1997. Until its removal from the market, over 25 interacting drugs had been identified, several labeling changes with additional warnings and contraindications were performed and continued reports of adverse effects from interacting drugs were received. As mibefradil showed no special benefits compared to other agents [40], Roche decided in June 1998 to withdraw Posicor from the market. In August 1998, Asta Medica followed suit with the removal of Cerate [41].

1.2.13
Rapacuronium Bromide

For the past 30 years, there have been efforts to find a nondepolarizing muscle relaxant to replace succinylcholine for endotracheal intubation. The goal has been to develop a fast-acting, short-duration drug without the side effects of succinylcholine such as bradycardia, rhabdomyolysis and malignant hyperthermia [42]. The injectable aminosteroid rapacuronium bromide (Raplon), launched by Organon Inc. in August 1999, seemed to be a promising substitute for succinylcholine. In comparison with other neuromuscular blocking agents, it had a rapid onset and a short duration of action. However, a pharmacologically active metabolite seemed to be responsible for a delay in spontaneous recovery after repeated bolus doses or infusions [43]. Several case reports and observations indicated that rapacuronium bromide could induce severe and potentially life-threatening bronchospasm in certain patients. The precise cause of incidence was unknown, but it seemed to be more frequent in patients with respiratory afflictions [42]. Histamine release, muscarinic receptor (M2) antagonism and cholinergic facilitation via airway stimulation were suggested as possible mechanisms for bronchospasm [44]. On 27th March 2001, Organon Inc. voluntarily withdrew Raplon [45].

1.2.14
Rofecoxib

The FDA approved this selective cyclooxigenase (COX)-2 inhibitor (Vioxx) for the treatment of pain and inflammation in 1999. This NSAID demonstrated to have a lower risk of side effects such as gastrointestinal ulcers and bleeding than nonselective COX inhibitors, for example, ibuprofen. In 2004, a long-term study of Vioxx in patients at increased risk of colon polyps was halted because of an increased cardiovascular risk (heart attack, stroke) in the rofecoxib group. Subsequently, Merck withdrew the drug from the world market at the end of September 2004 [46].

1.2.15
Temafloxacin

In late January 1992, Abbott Laboratories received the FDA approval for temafloxacin (Omniflox), a fluoroquinolone broad-spectrum antibiotic for treatment of infections with gram-negative pathogens, for example pulmonary infections. The drug had been marketed before in several European countries such as Germany and the United Kingdom. It was marketed in the United States from mid-February till 5th June 1992. However, Abbott halted all marketing and further distribution of this drug worldwide [47], as within 3 months of its use, the FDA received some 50 reports of serious adverse events including three deaths. These side effects included hypoglycemia in elderly patients as well as a constellation of multisystem organ involvement characterized by hemolytic anemia, frequently associated with renal failure, markedly abnormal liver function tests and coagulopathy [48]. There were also a substantial number of reports about allergic reactions, some causing life-threatening respiratory distress [49].

1.2.16
Troglitazone

Pfizer introduced troglitazone in 1997 and marketed it as Rezulin, an oral treatment for type 2 diabetes. Warner Lambert Co., which was acquired by Pfizer in June 2000, discontinued marketing this drug in March 2000. In January 2003, the company withdrew its new drug application [49]. In Europe, troglitazone was never marketed. Its use was associated with a markedly increased risk of acute idiopathic liver injury and acute liver failure [50].

1.2.17
Ximelagatran

The first orally bioavailable thrombin inhibitor prodrug ximelagatran was introduced on the German market in June 2004 and subsequently marketed as Exanta. Approved for the prevention of thromboembolic events such as venous thromboembolism (VTE) [51], its use was restricted to 11 days as elevated liver enzymes had been reported from clinical trials. On 14th February 2006, Astra Zeneca withdrew Exanta from the market because of a report of serious liver injury from the EXTEND clinical trial. This trial investigated whether ximelagatran could also be used in extended VTE prophylaxis after orthopedic surgery up to 35 days and so involved a longer duration of therapy than was currently approved for marketing [52].

Table 1.1 summarizes the reasons for the withdrawal of drugs in the past 14 years.

Figure 1.3 shows the percentage of substances withdrawn for toxicity reasons and illustrates the toxicity profile of these drugs.

Especially cardiovascular toxicity and hepatotoxicity played a crucial role in the decisions for withdrawing drugs from the market. These data suggest that there are

Table 1.1 Drugs withdrawn from the market from 1992 to 2006.

Drug name	Reason	Detailed information
Amineptine	Toxicity	Hepatotoxicity, addiction and neuropsychotic disorders
Aminophenazon	Toxicity	Cardiovascular (agranulocytosis)
Astemizole	Toxicity	Cardiovascular (hERG block)
Bromfenac sodium	Toxicity	Hepatotoxicity
Cerivastatin	Toxicity	Locomotor system (rhabdomyolysis)
Chlormezanone	Toxicity	Serious skin reactions
Fenfluramine	Toxicity	Cardiovascular
Flosequinan	Lack of efficacy	
Glafenine	Toxicity	Anaphylactic reactions
Grepafloxacin	Toxicity	Cardiovascular (hERG block)
Levacetylmethadol	Toxicity	Cardiovascular (QT-interval prolongation, arrhythmias, angina)
Mibefradil	Toxicity	Hepatotoxicity, cardiovascular and interactions via the CYP450 enzymatic system
Rofecoxib	Toxicity	Cardiovascular (heart attacks, stroke)
Rapacuronium bromide	Toxicity	Bronchospasm
Temafloxacin	Toxicity	Multiple side effects
Troglitazone	Toxicity	Hepatotoxicity
Ximelagatran	Toxicity	Hepatotoxicity

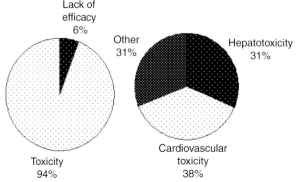

Figure 1.3 Reasons for the market withdrawal of the investigated drugs ($n = 17$, left). Toxicity profile of the withdrawn drugs (right).

currently no reliable test models for such toxicity problems available, for otherwise these drugs would never have entered the market at all.

1.3
Borderline Cases

To show a more complete picture on failed drugs, drugs that have been removed from the market and have been reintroduced later on with some restrictions as well as drugs that have been suspended from most drug markets but still remain on the market in only a few countries are listed in Table 1.2. As these drugs remain available on the Western market, they are not included in our statistics.

1.4
Investigational Drugs That Failed in Clinical Phases from 1992 to 2002

1.4.1
A Case Study: Fialuridine

Nucleoside analogues such as acyclovir, didanosine or zidovudine have been most widely evaluated as potential therapeutics for chronic hepatitis B. After gaining a better insight into hepatitis B virus (HBV) infection, a second generation of nucleoside analogues was developed, containing lamivudine, famciclovir and fialuridine. Fialuridine (1-(2-deoxy-2-fluoro-β-D-arabinofuranosyl)-5-iodouracil, FIAU) led to prompt and marked suppression of serum HBV DNA levels in two first phase II trials [60]. These two previous dose-finding studies – performed over a 2- and a 4-week course – led to the conclusion that longer therapy courses would be more efficacious in achieving a sustained loss of viral DNA in patients. In March 1993, a 6-month course of

Table 1.2 Drugs suspended from most but not all markets in the European Union and the United States.

Drug name	Indication	Important ADRs	Year	References
Alosetron (Lotronex)	Irritable bowel syndrome	Serious gastrointestinal events, especially ischemic colitis and constipation	2000	[53]
Cisapride (Propulsid)	Gastroesophagal reflux disease	Cardiovascular (QT-interval prolongation, arrhythmias)	2000	[54]
Phenylpropylamine (e.g. Proin)	Nasal decongestant, weight control	Hemorrhagic stroke	2000	[55,56]
Sertindole (Serdolect)	Schizophrenia	Cardiovascular (QT-interval prolongation, potentially fatal arrhythmias)	2000	[57]
Terfenadine (Seldane)	Allergy	Cardiovascular (potentially fatal heart condition)	1998	[58]
Tolcapone (Tasmar)	Parkinson's disease	Hepatic toxicity	1998	[59]
Trovafloxacin (Trovan)	Antibiotic	Hepatic toxicity	1999	[59]

fialuridine started. After 13 weeks, however, the study was terminated immediately when hepatic failure and lactic acidosis occurred in one of the 15 patients. The following side effects had been reported earlier in the study: intermittent crampy abdominal pain, paresthesias in the feet, fatigue, nausea, numbness and tingling in the feet or hands, crampy lower abdominal pain, constipation and mild thrombocytopenia. Three patients had already discontinued therapy with fialuridine between weeks 10 and 12 [60].

Severe fialuridine-induced toxicity included the following:

(i) Hepatic failure and lactic acidosis: Seven out of 15 patients showed varying degrees of hepatic failure and lactic acidosis. Common symptoms in these patients were fatigue, nausea, constipation, abdominal pain, steadily worsening jaundice, decreasing hepatic synthetic function, gradually worsening prothrombin times and increasing serum levels of ammonia and lactate. In two patients, the hepatic failure and lactic acidosis were rapidly progressive, so they had to be transferred to liver-transplantation centers. However, they both died 22 and 36 hours, respectively, after the transplantation. The other five patients showed a more gradual progression of hepatic failure and lactic acidosis. Their condition worsened over a period of several weeks. Two patients died of hemodynamic collapse caused by pancreatitis and lactic acidosis before liver transplantation could be performed on them. Another patient died of complications of pancreatitis after transplantation. The remaining two patients underwent a successful liver transplantation and survived. During 24 months of follow-up, they have had only little evidence of residual fialuridine-induced toxicity. However, a mild peripheral neuropathy remained [60].

(ii) Pancreatitis: The seven patients who showed severe hepatotoxicity had biochemical evidence of pancreatitis. Three of these patients had severe abdominal pain and clinically apparent pancreatitis, which ultimately led to their death. All five patients who died showed evidence of pancreatitis at autopsy [60].

(iii) Neuropathy and myopathy: Five of the seven patients with severe hepatotoxicity also had symptoms or signs of peripheral nerve injury. They reported paresthesias and dysesthesias, mild to moderate in severity, in the feet or toes. Two patients reported muscle pain or weakness [60].

Several mechanisms of fialuridine-induced hepatotoxicity have been suggested: fialuridine and its metabolites inhibit mitochondrial DNA replication, leading to decreased mitochondrial DNA and mitochondrial ultrastructural defects [61]. Another mechanism suggested lies in pyruvate oxidation inhibition [62].

1.4.2
A Recent Case Study: Torcetrapib

The recent years have seen the success of statins like Lipitor (atorvastatin) as hypolipidemic agents that help treating cardiovascular disease primarily by lowering low-density lipoproteins ('bad cholesterol') levels. Another novel strategy is to tackle the same problem by elevating high-density lipoproteins (HDL or 'good cholesterol') levels via inhibition of cholesteryl ester transfer protein (CETP).

On 2nd December 2006, Pfizer, whose CETP inhibitor torcetrapib (CP-529414) was the first compound aimed at this target, announced that its development had to be stopped in phase III. Of 7500 patients who were given a combination of torcetrapib and atorvastatin, 82 had died, while in the other arm of the study, where patients were given atorvastatin alone, only 51 of 7500 patients had died. There was also a significant rise in blood pressure observed in the group that received torcetrapib.

Many people had high hopes in this new class of therapeutics, which should act against the number one cause of death and disability in the United States and most European countries. While this chapter is being written, there is still a lot of speculation going on as to why this promising heart drug failed, but there is some support that CETP remains a rewarding drug target [63,64]. Both Roche and Merck, which have their own, structurally unrelated CETP inhibitors under development (Roche: JTT-705/R1658, Phase IIb; Merck: MK-859, Phase II), claim that their compounds do not elevate blood pressure.

1.4.3
General Reasons for Project Failing in Clinical Phases I–III

The termination of fialuridine and torcetrapib can be considered as 'worst case scenarios' in drug development. In over 50% of all cases, project termination occurs due to less spectacular reasons such as a lack of efficacy or liberation–absorption–distribution–metabolism–excretion (LADME) problems (Figure 1.4) [65].

Taking together all three clinical phases, it has been observed that most drug candidates fail because of a lack of efficacy. The second major problem is toxicity that leads to the termination of approximately one third of all projects. More details on other reasons are given in Figure 1.5.

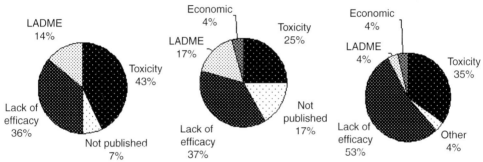

Figure 1.4 Reasons for project termination in clinical phases I (left), II (middle) and III (right) from 1992 to 2002.

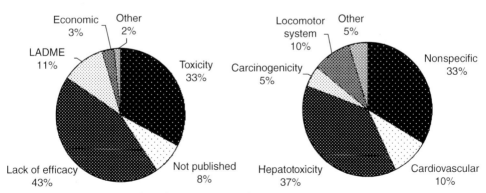

Figure 1.5 General reasons (left) and toxicity issues (right) leading to project termination in clinical phases I–III from 1992 to 2002.

1.5
Strategies for Avoiding Failure

The recent example of torcetrapib, where the world's largest drug company lost its potentially biggest drug, underlines the continued need for strategies to avoid failure during drug development. Despite the increased efforts in synthesis and testing via combinatorial synthesis and high-throughput screening (HTS) methods, the number of new drugs being approved is slowly declining and companies' drug pipelines are drying up. A recent study showed that the number of drugs that are the first to act at a new target has been relatively constant over the past 20 years [66]. The trend toward mergers and acquisitions (M&A) continues, though M&A have failed to provide long-term solutions in the past years, and only less than half of the new medical entities approved by the FDA in 2006 originated from the top 10 pharma companies [67,68].

Although HTS can process up to a million compounds per day, it has a high possibility of producing both false-negative and false-positive results. Replicate measurements in combination with statistical methods and careful data analysis may help to identify and reduce such errors [69].

In accordance with the credo 'fail early, fail cheap', there is a growing trend toward applying parallel screening and so-called *in vitro* safety pharmacology profiling methods at an earlier stage now [70]. Since the early 1990s, several contract research organizations such as Cerep, Euroscreen, MDS and Novascreen have emerged that offer profiling services against hundreds of different targets, to rapidly assess specificity and off-target effects, including absorption, distribution, metabolism, excretion and toxicity (ADMET) properties, for the selection of new lead candidates and for compound optimization. High-content screening, using whole cells, provides the possibility to monitor multiple targets or pathways and to directly identify toxic effects [71,72].

At the same time, *in silico* methods are expected to gain largely in importance with improved models that are able to address different aspects of ADMET [73], including metabolic effects at cytochromes [74,75] and nuclear receptors [76] as well as effects at antitargets such as certain GPCRs [77] and the hERG potassium channel [78]. Efforts to simulate *in vivo* effects by predicting drug disposition from demographic, physiological, genomic and *in vitro* data are also under way [79]. Our group recently presented the promising concept of parallel virtual screening with pharmacophoric methods, which allow for the fast profiling of molecules against a large number of drug targets and antitargets *in silico* [80,81]. A similar concept has been presented by Cleves and Jain [82].

One of the problems in creating reliable computer models for certain targets and antitargets is the lack of available data on inactive compounds, because this is widely regarded as an unimportant information. Furthermore, results from large-scale HTS are usually not available to the public scientific community. A notable project that may help to improve this situation is currently underway as part of the Molecular Libraries Initiative at NIH, screening more than 100 000 compounds against multiple targets and making the results available to the public via PubChem [83]. Other publicly accessible sites providing resources for *in silico* screening are listed in a recent paper by Strachan et al. [84].

1.6
An Unusual Case: The Revival of Thalidomide

The story of thalidomide (Contergan, Thalidomid) is certainly the most prominent example of a drug that had to be withdrawn from the market because of its severe side effects. First introduced in 1957 as a sedative that lacked both the addictive properties of barbiturates and the risk of death by accidental or intentional overdosing, it was marketed in 46 countries worldwide, including many European countries, Canada and Australia. It was also prescribed to pregnant women as an antiemetic for morning sickness. By the end of 1961, it was found that thalidomide caused severe teratogenic effects, with strong fetal malformations like phocomelia (shortened or absent arms and legs). The drug was quickly withdrawn from the market, but in between 5000 and 12 000 deformed babies were born worldwide, with an unknown number of aborted fetuses. It was found that as little as a single dose of 50 mg, taken between day 35 and 49 after the last menstrual period, could produce the characteristic birth defects.

In 1964, it was discovered by chance that thalidomide is also one of the most highly active agents in the treatment of erythema nodosum leprosum (ENL), a complication of leprosy. Later on, it was found that it is also an effective treatment for multiple myeloma and related plasma cell disorders. The use of thalidomide (Thalidomid, Celegene Corporation) was approved by the FDA for the treatment of ENL in 1998, and, in combination with dexamethasone, for the treatment of multiple myeloma in 2006. Because of the above-mentioned high risk of teratogenic effects after just one single dose of thalidomide, severe restrictions were imposed on its distribution. For this purpose, the System for Thalidomide Education and Prescribing Safety (STEPS) program was established to ensure that every patient is well informed about the associated risks and that no uncontrolled distribution of the drug is possible. The history and current use of thalidomide is summarized in some recent reviews [85–87].

New derivatives of thalidomide have been developed, such as lenalidomide (CC-5013, marketed as Revlimid) – approved by the FDA in 2005 for patients with myelodisplastic syndrome – and actimid (CC-4047), which is planned to enter phase II trials as an orally available treatment for myelofibrosis and sickle cell anemia, both by Celegene. These immunomodulatory drugs have a potency between 2000 and 20 000 times higher than thalidomide in inhibiting the drug's primary target tumor necrosis factor α, while the teratogenic effect is most likely to be retained [88].

Abbreviations

ADR	adverse drug reaction
CETP	cholesteryl ester transfer protein
EMEA	European Agency for the Evaluation of Medicinal Products
ENL	erythema nodosum leprosum
FDA	Food and Drug Administration

HBV	hepatitis B virus
HTS	high-throughput screening
LADME	liberation–absorption–distribution–metabolism–excretion
NCE	new chemical entity
NSAID	nonsteroidal anti-inflammatory drug

References

1 DiMasi, J.A., Hansen, R.W. and Grabowski, H.G. (2003) The price of innovation: new estimates of drug development costs. *Journal of Health Economics*, **22**, 151–185.

2 Lasser, K.E., Allen, P.D., Woolhandler, S.J., Himmelstein, D.U., Wolfe, S.M. and Bor, D.H. (2002) Timing of new black box warnings and withdrawals for prescription medications. *The Journal of the American Medical Association*, **287**, 2215–2220.

3 Graham, G.K. (2002) Postmarketing surveillance and black box warnings (To the Editor). *The Journal of the American Medical Association*, **288**, 955–956.

4 Prentis, R.A., Lis, Y. and Walker, S.R. (1988) Pharmaceutical innovation by the seven UK-owned pharmaceutical companies (1964–1985). *British Journal of Clinical Pharmacology*, **25**, 387–396.

5 Kennedy, T. (1997) Managing the drug discovery/development interface. *Drug Discovery Today*, **2**, 436–444.

6 WHO Amineptine (INN). www.who.int/medicines/library/qsm/who-edm-qsm-2000-5/amineptine.doc, last accessed on 24th November 2003.

7 Amineptine (Survector): the Smart Tricyclic Antidepressant. BLTC. www.amineptine.com, last accessed on 9th January 2007.

8 BIAM Amineptine Chlorhydrate. BIAM. www.biam2.org/www/Sub1826.html#Present, last accessed on 9th January 2007.

9 De Gálvez Aranda, M.V., Sánchez Sánchez, P., Alonso Corral, M.J., Bosch García, R.J., Gallardo, M.A. and Herrera Ceballos, E. (2001) Acneiform eruption caused by amineptine. A case report and review of the literature. *Journal of the European Academy of Dermatology and Venereology*, **15**, 337–339.

10 Haddad, P. (2000) Do antidepressants have any potential to cause addiction? *Journal of Psychopharmacology*, **14**, 300–307.

11 Larrey, D., Berson, A., Habersetzer, F., Tinel, M., Castot, A., Babany, G., Letteron, P., Freneaux, E. Loeper, J., Dansette, P. and Pessyre, D. (1989) Genetic predisposition to drug hepatotoxicity: role in hepatitis caused by amineptine, a tricyclic antidepressant. *Hepatology*, **10**, 168–173.

12 Documentary List of Main Dangerous Drugs Banned and Withdrawn in the Foreign Markets. www.drugavoid.com/dangerousdrugs.html, last accessed on 9th January 2007.

13 WHO Pharmaceuticals: Restrictions in Use and Availability. World Health Organization. http://whqlibdoc.who.int/hq/2001/EDM_QSM_2001.3.pdf, last accessed on 9th January 2007.

14 Dotzel, M.M. (2000) Additions to the list of drug products that have been withdrawn or removed from the market for reasons of safety or effectiveness. *Federal Register*, **65**, 256–258.

15 Mutschler, E. (2001) *Arzneimittelwirkungen*, 8 Auflage ed., Wissenschaftliche Verlagsgesellschaft mbH, Stuttgart.

16 Paakkari, I. (2002) Cardiotoxicity of new antihistamines and cisapride. *Toxicology Letters*, **127**, 279–284.

17 Pharmainformation Editorial. www.uibk.ac.at/c/c5/c515/info/info17-1.html;mark=109,18,28#WN_mark, last accessed on 15th December 2002.

18 FDA Wyeth-Ayerst Laboratories Announces the Withdrawal of Duract from the Market. www.fda.gov/bbs/topics/ANSWERS/ANS00879.html, last accessed on 12th December 2003.

19 FDA Duract Voluntarily Withdrawn. www.fda.gov/medwatch/safety/1998/duract2.htm, last accessed on 7th March 2003.

20 Sparing, R., Sellhaus, B., Noth, J. and Block, F. (2003) Rhabdomolyse unter Cerivastatin-Monotherapie. *Der Nervenarzt*, **74**, 167–171.

21 Bayer Geschäftsbericht 2001. www.bayer.de/geschaeftsbericht2001/gb/de/report/printview.php?page1=lipo.html&page2_lipo2.html, last accessed on 28th February 2003.

22 FDA from the Medwatch Office – Summaries of "Dear Health Professional" Letters and Other Safety Notifications. www.fda.gov/medbull/mar97/medwatch.html, last accessed on 16th December 2002.

23 Hunnius, (1998) *Pharmazeutisches Wörterbuch*, Walter de Gruyter, Berlin, New York.

24 EMEA Final Opinion of the Committee for Proprietary Medicinal Products Pursuant to Article 12 of Council Directive 75/319/EEC as Amended for Medicinal products Chlormezanone. www.emea.eu.int/pdfs/human/phv/EN/037597en.pdf, last accessed on 16th December 2003.

25 Roujeau, J.C., Kelly, J.P., Naldi, L., Rzany, B., Stern, R.S., Anderson, T., Auquier, A., Bastuji-Garin, S., Carreria, O., Locati, F., Mockenhaupt, M., Paoletti, C., Shapiro, S., Shear, N., Schöpf, E. and Kaufman, D.W. (1995) Medication use and the risk of Stevens–Johnson syndrome or toxic epidermal necrolysis. *The New England Journal of Medicine*, **333**, 1600–1607.

26 Sanofi, W.P. Discontinuation of Trancopal (chlormezanone). www.fda.gov/Medwatch/safety/trancopa.htm, last accessed on 16th December 2003.

27 Pharmainformation Appetitzügler. www.uibk.ac.at/c/c5/c515/info/info7-4.html, last accessed on 15th December 2002.

28 FDA FDA Announces Withdrawal of Fenfluramine and Dexphenfluramine (Fen-Phen). www.fda.gov/cder/news/phen/fenphenpr81597.htm, last accessed on 5th March 2003.

29 van Veldhuisen, D.J. and Poole-Wilson, P.A. (2001) The underreporting of results and possible mechanisms of "negative" drug trials in patients with chronic heart failure. *International Journal of Cardiology*, **80**, 19–27.

30 Tomorrow's Advice. Documentary List of Main Dangerous Drugs Banned and Withdrawn in the Foreign Markets. www.drugavoid.com/dangerousdrugs.html, last accessed on 16th March 2003.

31 BIAM GLIFANAN 200 mg comprimés (arrêt de commercialisation). www.biam2.org/www/Spe5764.html, last accessed on 16th March 2003.

32 Davido, A., Hallali, P. and Boutchnei, T. (1989) Shock caused by glafenine. Apropos of 7 complications. *Revue de Medecine Interne*, **10**, 113–117.

33 van der Klauw, M.M., Wilson, J.H. and Stricker, B.H. (1996) Drug-associated anaphylaxis: 20 years of reporting in The Netherlands (1974–1994) and review of the literature. *Clinical and Experimental Allergy*, **26**, 1355–1363.

34 Herxheimer, A. Side Effects: Freedom of Information and the Communication of Doubt. www.cssentialdrugs.org/edrug/archive/199601/msg0002.php, last accessed on 16th March 2003.

35 Glaxo, W. Glaxo Wellcome Voluntarily Withdraws Raxar (Grepafloxacin). www.gsk.com/press_archive/gw/1999/

press_991027pr.htm, last accessed on 16th March 2003.
36 Kang, J., Wang, L., Chen, X.L., Triggle, D.J. and Rampe, D. (2001) Interactions of a series of fluoroquinolone antibacterial drugs with the human cardiac K^+ channel HERG. *Molecular Pharmacology*, **29**, 122–126.
37 Wathion, N. (2001) *EMEA Public Statement on the Recommendation to Suspend the Marketing Authorisation for Orlaam (levacetylmethadol) in the European Union*, EMEA, London.
38 Schobelock, M.J. (2003) *Product Discontinuation Notice – Orlaam (Levomethadyl Hydrochloride Acetate) Oral Solution 10 mg/mL, CII*, FDA, Columbus, OH.
39 Roche Fachinformation (Zusammenfassung der Produkteigenschaften) POSICOR, "Roche" 50 mg-Filmtabletten, 1998.
40 Ross, D. Telithromycin (Ketek): General Safety Profile. www.fda.gov/ohrms/dockets/ac/01/slides/3746s_05_Ross/sld041.htm, last accessed on 17th March 2003.
41 Berthold, H. (1998) Rückruf von Cerate (Mibefradil) angeordnet; Arzneimittelkommission der Deutschen Ärzteschaftf.
42 Goudsouzian, N.G. (2001) Rapacuronium and bronchospasm. *Anesthesiology*, **94**, 727–728.
43 Sparr, H.J., Beaufort, T.M. and Fuchs-Buder, T. (2001) Newer neuromuscular blocking agents: how do they compare with establishes agents? *Drugs*, **61**, 919–942.
44 Stuth, E.A.E., Stucke, A.G. and Setlock, M.A. (2002) Another possible mechanism for bronchospasm after rapacuronium. *Anesthesiology*, **96**, 1528–1529.
45 Shapse, D. (2001) Voluntary Market Withdrawal, FDA.
46 FDA (2004) FDA issues public health advisory on Vioxx as its manufacturer voluntarily withdraws the product. in *FDA News*, FDA.
47 (1994) FDA TEMAFLOXACIN (Omniflox): Withdrawn from Market, FDA.
48 Kemper, E. (1992) Recalling the Omniflox (Temafloxacin) Tablets, FDA.
49 Woodcock, J. (2003) Withdrawal of approval of a new drug application. *Federal Register*, **68**, 1469.
50 Graham, D.J., Drinkard, C.R. and Shatin, D. (2003) Incidence of idiopathic acute liver failure and hospitalized liver injury in patients treated with troglitazone. *The American Journal of Gastroenterology*, **98**, 175–179.
51 (2004) AstraZeneca first launch for AstraZeneca's Exanta (TM) (ximelagatran): the first oral anticoagulant in new class of direct thrombin inhibitors (DTIs), AstraZeneca.
52 (2006) AstraZeneca. AstraZeneca decides to withdraw Exanta (TM), AstraZeneca.
53 DuBose, R. (2002) Lotronex tablets to be re-introduced for women with severe diarrhea-predominant IBS, GSK.
54 FDA. (2000) Janssen Pharmaceutica stops marketing cisapride in the US. in *FDA Talk Paper*, FDA, Rockville, MD.
55 FDA (2000) FDA issues public health warning on phenylpropanolamine. in *FDA Talk Paper*, FDA.
56 FDA (2005) Phenylpropanolamine-containing drug products for over-the-counter human use; tentative final monographs. *Federal Register*, **70**, 75988–75998.
57 Dubitsky, G.M. (2000) Clinically important QT interval prolongation with three antipsychotics: thioridazine, pimozide, sertindole, FDA.
58 Fleischer Kupec, I. (1998) Seldane and generic terfenadine withdrawn from market. in *FDA Talk Paper*, FDA, Rockville, MD.

59 Winkler, H. (1999) Kürzliche Marktrücknahme von Medikamenten: eine Analyse. *Pharmainformation*, **14**, 1.

60 McKenzie, R., Fried, M.W., Sallie, R., Conjeevaram, H., DiBisceglie, A.M., Park, Y., Savarese, B., Kleiner, D., Tsokos, M., Luciano, C., Pruett, T., Stotka, J.L., Straus, S.E. and Hoofnagle, J.H. (1995) Hepatic failure and lactic acidosis due to fialuridine (FIAU), an investigational nucleoside analogue for chronic hepatitis B. *The New England Journal of Medicine*, **333**, 1099–1105.

61 Lewis, W., Levine, E.S., Griniuviene, B., Tankersley, K.O., Colacino, J.M., Sommadossi, J.P., Watanabc, K.A. and Perrino, F.W. (1996) Fialuridine and its metabolites inhibit DNA polymerase gamma at sites of multiple adjacent analog incorporation, decrease mtDNA abundance, and cause mitochondrial structural defects in cultured hepatoblasts. *Proceedings of the National Academy of Sciences of the United States of America*, **93**, 3592–3597.

62 Honkoop, P., Scholte, H.R., deMan, R.A. and Schalm, S.W. (1997) Mitochondral injury. Lessons from the fialuridine trial. *Drug Safety*, **17**, 1–7.

63 Tall, A.R., Yvan-Charvet, L. and Wang, N. (2007) The failure of torcetrapib: was it the molecule or the mechanism? *Arteriosclerosis, Thrombosis, and Vascular Biology*, **27**, 257–260.

64 Nicholls, S.J., Tuzcu, E.M., Sipahi, I., Grasso, A.W., Schoenhagen, P., Hu, T., Wolski, K., Crowe, T., Desai, M.Y., Hazen, S.L., Kapadia, S.R. and Nissen, S.E. (2007) Statins, high-density lipoprotein cholesterol, and regression of coronary atherosclerosis. *The Journal of the American Medical Association*, **297**, 499–508.

65 Schuster, D., Laggner, C. and Langer, T. (2005) Why drugs fail – a study on side effects in new chemical entities. *Current Pharmaceutical Design*, **11**, 3545–3559.

66 Overington, J.P., Al-Lazikani, B. and Hopkins, A.L. (2006) How many drug targets are there? *Nature Reviews. Drug Discovery*, **5**, 993–996.

67 Frantz, S. (2006) Pipeline problems are increasing the urge to merge. *Nature Reviews. Drug Discovery*, **5**, 977–979.

68 Owens, J. (2007) 2006 drug approvals: finding the niche. *Nature Reviews. Drug Discovery*, **6**, 99–101.

69 Malo, N., Hanley, J.A., Cerquozzi, S., Pelletier, J. and Nadon, R. (2006) Statistical practice in high-throughput screening data analysis. *Nature Biotechnology*, **24**, 167–175.

70 Whitebread, S., Hamon, J., Bojanic, D. and Urban, L. (2005) In vitro safety pharmacology profiling: an essential tool for successful drug development (Keynote review). *Drug Discovery Today*, **10**, 1421–1433.

71 Branca, M.A. Screen Dreams: The Promise of Parallel Screening. Pharma DD www.pharmadd.com/archives/Sept2006/screen_dreams.asp, last accessed on 14th February 2007.

72 Haney, S.A., LaPan, P., Pan, J. and Zhang, J. (2006) High-content screening moves to the front of the line. *Drug Discovery Today*, **11**, 889–894.

73 van de Waterbeemd, H. and Gifford, E. (2003) ADMET *in silico* modelling: towards prediction paradise? *Nature Reviews. Drug Discovery*, **2**, 192–204.

74 de Groot Marcel, J. (2006) Designing better drugs: predicting cytochrome P450 metabolism. *Drug Discovery Today*, **11**, 601–606.

75 Schuster, D., Laggner, C., Steindl, T.M. and Langer, T. (2006) Development and validation of an *in silico* P450 profiler based on pharmacophore models. *Current Drug Discovery Technologies*, **3**, 1–48.

76 Schuster, D., Steindl, T.M. and Langer, T. (2006) Predicting drug metabolism induction *in silico*. *Current Topics in Medicinal Chemistry*, **6**, 1627–1640.

77 Klabunde, T. and Evers, A. (2005) GPCR antitarget modeling: pharmacophore models for biogenic amine binding

GPCRs to avoid GPCR-mediated side effects. *Chembiochem: A European Journal of Chemical Biology*, **6**, 876–889.

78 Aronov, A.M. (2005) Predictive *in silico* modeling for hERG channel blockers. *Drug Discovery Today*, **10**, 149–155.

79 Rostami-Hodjegan, A. and Tucker, G.T. (2007) Simulation and prediction of *in vivo* drug metabolism in human populations from *in vitro* data. *Nature Reviews. Drug Discovery*, **6**, 140–148.

80 Steindl, T.M., Schuster, D., Wolber, G., Laggner, C. and Langer, T. (2006) High-throughput structure-based pharmacophore modelling as a basis for successful parallel virtual screening. *Journal of Computer-Aided Molecular Design*, **20**, 703–715.

81 Steindl, T.M., Schuster, D., Laggner, C., Chuang, K., Hoffmann, R.D. and Langer, T. (2007) Parallel Screening and Activity Profiling with HIV Protease Inhibitor Pharmacophore Models. *Journal of Chemical Information and Modeling*, **47** (2), 563–571.

82 Cleves, A.E. and Jain, A.N. (2006) Robust ligand-based modeling of the biological targets of known drugs. *Journal of Medicinal Chemistry*, **49**, 2921–2938.

83 Austin, C.P., Brady, L.S., Insel, T.R. and Collins, F.S. (2004) Molecular biology: NIH molecular libraries initiative (Policy forum). *Science*, **306**, 1138–1139.

84 Strachan, R.T., Ferrara, G. and Roth, B.L. (2006) Screening the receptorome: an efficient approach for drug discovery and target validation. *Drug Discovery Today*, **11**, 708–716.

85 Rajkumar, S.V. (2004), Thalidomide: tragic past and promising future. *Mayo Clinic Proceedings*, **79**, 899–903.

86 Franks, M.E., Macpherson, G.R. and Figg, W.D. (2004) Thalidomide. *The Lancet*, **363**, 1802–1811.

87 Teo, S.K., Stirling, D.I. and Zeldis, J.B. (2005) Thalidomide as a novel therapeutic agent: new uses for an old product. *Drug Discovery Today*, **10**, 107–114.

88 Kumar, S. and Rajkumar, S.V. (2006) Thalidomide and lenalidomide in the treatment of multiple myeloma. *European Journal of Cancer*, **42**, 1612–1622.

2
Use of Broad Biological Profiling as a Relevant Descriptor to Describe and Differentiate Compounds: Structure–*In Vitro* (Pharmacology–ADME)–*In Vivo* (Safety) Relationships

Jonathan S. Mason, Jacques Migeon, Philippe Dupuis, Annie Otto-Bruc

2.1
Introduction

2.1.1
Biological Profiling/Fingerprints and Drug Discovery Applications

This chapter deals with the concept of 'biological fingerprints' that have been described as a better way to describe compounds of biological interest. It includes examples of how these descriptors are far more powerful than structure-based descriptors in both differentiating compounds and enabling the selection of the best lead compounds, and can provide a way to investigate *in vitro–in vivo* relationships such as for ADRs (adverse drug reactions).

Drug profiling is a strategy aimed at substantially improving the drug discovery process [1]. Results presented in this chapter are from the BioPrint® project that began in 1997 as a focused effort to build a high-quality data set linking the knowledge domains of computational chemistry, *in vitro* biology and clinical effects, leveraging the experience and competences of *in vitro* profiling data production, analysis and interpretation that have been the core business of Cerep for more than 15 years. BioPrint® [2–6] has been successful in bringing fundamentally new insights into each of these knowledge domains. The principle of the approach is to systematically profile (full matrix *in vitro* percent inhibition and IC_{50}s) drugs and related compounds against a broad range of pharmacological and ADME(T) [absorption, distribution, metabolism, excretion (toxicity)] targets. Coupled to this wherever available is curated *in vivo* data on compounds. Figure 2.1 illustrates the process. The pharmacological space covered includes receptors (e.g. G-protein-coupled receptors – GPCRs and nuclear hormone receptors – NHRs), ion channels, transporters and enzymes (e.g. proteases, kinases). The selection of assays in the Cerep BioPrint® [2–6] approach has evolved to give not only a reasonable sampling of the drugable genome (see Figure 2.2, BioPrint® assays in cyan), but also a focus on assays that may be important for side effects. It has benefited from the

Antitargets. Edited by R. J. Vaz and T. Klabunde
Copyright © 2008 WILEY-VCH Verlag GmbH & Co. KGaA, Weinheim
ISBN: 978-3-527-31821-6

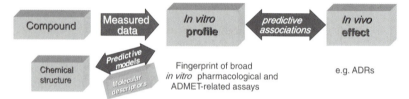

Figure 2.1 BioPrint® approach: the *in vitro* data are all measured in a consistent manner with full dose response for any activity >30% at 10 μM; the *in vivo* data are curated from available data, supplemented by custom measured data from collaborators.

input of therapeutic area and safety pharmacology scientists from several major pharmaceutical company partners. Another consideration was hit rate, as assays with very low hit rates provide too little a signal to be able to differentiate compounds, although assays for which an activity would be considered significant are included.

An important concept in this approach is that the actual assays, or combinations of the assays, may be surrogates for a far larger set of targets of interest. This principle

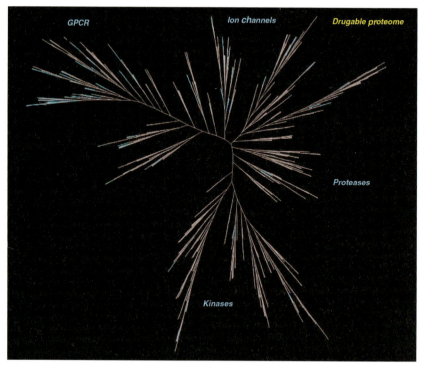

Figure 2.2 BioPrint® assays (cyan) mapped onto the drugable proteome.

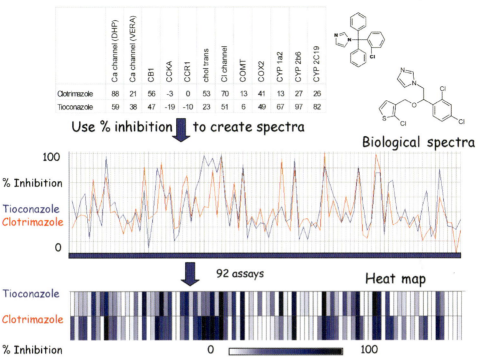

Figure 2.3 Biological spectra approach of Fliri et al.

was used in the affinity fingerprint method, discussed by Stanton and Cao [6] in which a small 'diverse' (orthogonal) set of protein targets and compounds was used to model the activity of a new protein. The broad profile can be used as a type of spectral analysis of the molecule (e.g. using percent inhibition, so that all assays have a continuous numerical value, see Figure 2.3), as described by Fliri et al. [7–9] as a new approach to understand the 'proteome interaction potential' for small molecules, termed 'biospectra', which allows for the grouping of compounds and their potential properties.

In this chapter, the full BioPrint® approach is described, as available from Cerep in terms of both the data set and the ability to have new compounds profiled and the results provided in the context of the BioPrint® data set, including the known *in vivo* side effects of near neighbors in this biological space (see Section 2.5). The results for the differentiation of hit/lead compounds (see Section 2.3.2.1) sometimes use a subset of the 70–100 pharmacological assays that provide the maximum signal. Usually a decision on future work prioritization could be clearly made from the data from these subsets, saving time and money. For key reference/tool compounds, a full profile was used and is recommended to be used, as unexpected off-target activities may be found that cannot usually be predicted.

2.1.2
Polypharmacology of Drugs

An early key finding of the BioPrint® analysis of drugs was that many drugs have polypharmacology. Some of this was expected (e.g. drugs for psychosis), but the diversity of off-target activities of many drugs (at levels sometimes close to the 'primary' activity) was an interesting insight into their potential pharmacological activities and ADME (T) properties. Even for drugs with polypharmacology as part of their action/efficacy, activity on all the targets may not be desirable, and may be responsible for undesired side effects etc. Figure 2.4 shows BioPrint® data for drugs (y-axis, activity fingerprint is the row) against assays (x-axis). The data have been hierarchically clustered using both sets of data and are color coded, with red indicating high binding affinity, yellow medium and blue-green. It can be noted that drugs of a particular therapeutic class tend to cluster together, although later experience showed that compounds with a desired profile could be identified that clustered into a different

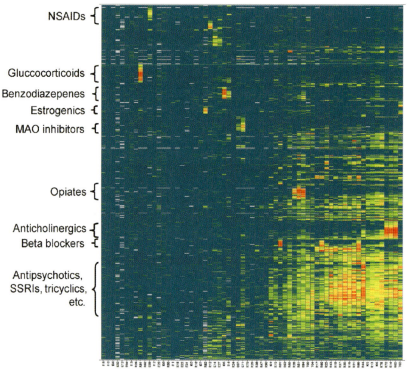

Figure 2.4 Compounds [2000 drugs and related] versus Assays [70 pharmacological from BioPrint®] database, pIC$_{50}$]. Biological assays are on the x-axis, and compounds on the y-axis (i.e. a row contains the biological fingerprint of a compound as heat map). Hierarchical clustering has been performed on both axes: compounds by their fingerprint of biological activities and targets in a chemogenomic way by the fingerprint of the activities of the same set of compounds for each target. The activities are shown in the form of a heat map, with red most active and blue-green inactive. Compounds tend to cluster into therapeutic area and this is marked on the left.

Figure 2.5 Biological fingerprint for clozapine showing the results for assays with a percent inhibition >90% at 10 µM (upper line, red). The IC_{50}s are shown as color coded on the lower line (red < 100 nM, orange < 1 µM, yellow < 5 µM).

part of space (see Section 2.3.2.1). Figure 2.5 shows the polypharmacology of clozapine, with the partial BioPrint® biological fingerprint showing only assays with a percentage inhibition >90% at 10 µM on the upper heat map (all red) and the related IC_{50}s below is color coded with red being the most active: red (<100 nM).

Figure 2.6 shows a histogram of a number of different targets (assays) against which drugs (1388) in the BioPrint® database were found to be active (using a cutoff of 50% inhibition at 10 µM).

The question, 'Can we rationally design promiscuous drugs?' was discussed in a recent paper by Hopkins et al. [5]. Different views of the data linking structural

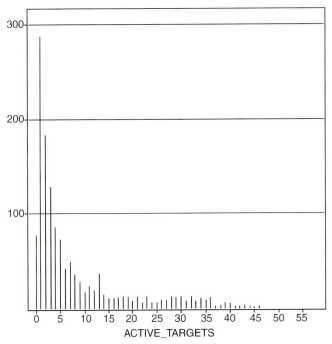

Figure 2.6 Compound target activity distribution ('promiscuity') for 1388 drugs profiled in BioPrint® assay panel (with 50% inhibition at 10 µM taken as active), shown as a histogram. The number of active compounds is shown along the y-axis and the number of targets along the x-axis (adapted from ref. [6]).

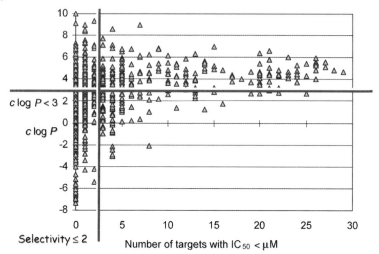

Figure 2.7 Selectivity ('promiscuity', x-axis) in terms of compound target activity numbers for 1098 drugs from BioPrint® profiled in the BioPrint® assay panel (with <1 μM IC$_{50}$ taken as active), versus $c\log P$ (hydrophobicity, y-axis) (adapted from ref. [5]).

properties and activity on multiple targets ('promiscuity') were discussed. One type of analysis shows that smaller (and thus likely simpler in terms of 3D pharmacophoric patterns) compounds hit more assays. An analysis of the BioPrint® data set showed that compounds that are active at <1 μM on more than 10 targets generally are more lipophilic, with a $c\log P > \sim 3$ (see Figure 2.7).

2.2
The BioPrint® Approach

2.2.1
BioPrint® – General

BioPrint® consists of a large database and a set of tools with which both the data and the models generated from the data can be accessed. The database contains structural information, *in vivo* and *in vitro* data on most of the marketed pharmaceuticals and a variety of other reference compounds. The *in vitro* data generated consist of panels of pharmacology and early ADME assays. The *in vivo* data consist of ADR data extracted from drug labels, mechanisms of action, associated therapeutic areas, pharmacokinetic (PK) data and route of administration data.

2.2.2
BioPrint® Assay Selection and Profile Description

The BioPrint® profile has been set up to deliver early assessment of the potential off-target and ADME(T) properties of lead compounds. To quickly provide coherent data to support the decision-making process, the profile has to generate reliable and consistent data in a rapid and regular manner. The BioPrint® profiles are stored in a database, which provides a context to interpret and cluster new data, and can be used to develop predictive models and design chemical libraries etc. Assays are selected primarily for their scientific interest, with due consideration given to their robustness and the quality of the data. The pharmacological part of the BioPrint® profile is designed for an optimal coverage of relevant biological diversity. Assay selection is based on several criteria including phylogenetic analysis, the concept of the 'drugable' proteome, coverage of major and important therapeutic areas, experience and skills on the targets proposed, the level of *in vivo* information available and constraints due to technical issues and high data quality standards. The ADME(T) assays in the BioPrint® profile are a large subset of the available assays. They cover physicochemical and metabolic properties and thus address the potential *in vivo* disposition of the test compounds. Other members of the ADME(T) panel address toxicity issues and the potential for drug/drug interactions at the levels of both metabolism and P-glycoprotein transport.

For the selection of the pharmacology assay panel, a rational molecular approach based on phylogenetic analysis and clustering of targets was followed to select targets to build a diverse profile. In this approach, a low degree of sequence homology between targets should correspond to the greatly differing affinity for a given ligand. The BioPrint® profile started in 1999 with the selection of about 50 different pharmacological targets. Since then, between 5 and 20 new assays have been added each year (see [2] for the profile in 2003). These additions have leveraged the knowledge and experience of Cerep and the major pharmaceutical company partners/collaborators for the BioPrint® project, as discussed next. For the newest additions, more recent and detailed phylogenetic analysis for human G-protein-coupled receptors and kinases has been used [10,11]. Currently there are over 200 *in vitro* assays in BioPrint®. There are 175 pharmacology assays whose targets include 91 receptors, 60 enzymes, 22 ion channels and 6 transporters. The BioPrint® panel also includes 29 ADME assays. Whenever possible, assays based on human sources are preferred (around 70%).

The BioPrint® profile represents a subset of the 'drugable' genome based on the clustering of sequence homology or on the molecular function of different human genes [6–8,12,13] (see Figure 2.2). Receptors and enzymes are highly represented in the 'drugable' genome. The approach where targets are chosen to cover all the most important therapeutic areas yields similar results. In different studies on the therapeutic target classes [13–15], receptors, mainly GPCRs, and enzymes are the major classes. Ion channels, transporters and nuclear receptors are also well represented. The BioPrint® profile matches this distribution: 67% are receptors, 24% enzymes, 13% ion channels, 3% transporters and 2% nuclear receptors. Among receptors, G-protein-coupled receptors are the most widely represented targets (99). Forty of them are listed in the work of Gozalbes *et al.* [16]. Monoamine and, more generally, neurotransmitter receptors are highly represented in the BioPrint® profile due to sustained interest in them. Many enzymes included in the profile are proteases and kinases, 13 and 8 targets, respectively. These enzymes are involved in diseases such as cancer and inflammation and are targets of interest to drug discovery. Recently, kinases have emerged as new targets of interest. However, not all have yet been characterized because of the large number of family members. The recent proof of concept of kinase inhibitors shows them as a new class of anticancer drugs [17] or immune system regulators [18].

The BioPrint® profile is also enriched with an empirical approach based on knowledge and experience. This experience includes an in-depth understanding of the use of biological data to describe relationships between compound structure, *in vitro* and *in vivo* data. Including assays for which links between *in vivo* and *in vitro* effects are well documented in the literature was of specific interest. Targets showing clear association between specific receptor or enzyme interactions and adverse effects were preferred. For example, muscarinic acetylcholine receptors and adrenergic receptors have been clearly implicated in *in vivo* cardiovascular effects. However, the addition of less understood targets has also been of interest because compounds with well-characterized *in vitro* effects can help us to understand the *in vivo* roles of those targets. Therefore, also included are targets such as the sigma receptor, the melatonin receptor, various kinases and metalloproteinases and several others.

Finally, to increase diversity and leveraging Cerep experience as a pharmacology service company, targets with high (e.g. AT2, B2, ETA), middle (e.g. AT1, CCKB, CB2) and low hit rate (e.g. Alpha2C, D3, 5-HT$_{2C}$) were selected, increasing the fingerprint information content and to accumulate enough positive and negative data to develop and validate predictive models with enough accuracy.

Among the representative and diverse biological assays selected, initially preferred were those measuring direct interaction of the compound tested with only one target, making the use and interpretation of data easier. Therefore, radioligand binding and isolated enzyme assays were preferred to functional or cell-based assays. Currently data from many new functional assays are being added to better understand the interactions detected by the binding assays. A high level of data quality is ensured by selection of assays of highest homogeneity and robustness, all of which have been developed and validated using the same criteria.

To verify that there was no obvious redundancy in the assay data, a subset of 80 assays and 1600 compounds was analyzed using the KEM approach [19–21] developed by Ariana [22], a systematic rule-based method that identifies *all* relations of the type A → B, A and B → C and not D. No such noncontradicted relations were found, even using very large activity bins, showing that there is signal in all the assay data, including those from related assays.

2.2.3
Compounds in BioPrint®

The very first goal was to create a learning database by profiling compounds with well-characterized *in vivo* effects, both therapeutic and off-target. The typical profile of those compounds displays activity on their respective therapeutic target(s), if included in the profile, and the presence or absence of off-target activities. Most human adverse effects are caused by off-target activities. Therefore, the collection of profiles is a tool of choice to correlate *in vitro* activities to the presence and frequency of adverse effects.

The BioPrint® collection includes over 2500 compounds. More than half of these compounds (60%) are marketed drugs, 5% tested in clinic but not marketed, 1% in clinical trials, 2% withdrawn from market and 20% are standard pharmacology references. A good number of the standard pharmacology reference compounds were one time drug candidates and therefore 'drug-like'. The remaining 12% compounds are veterinary drugs, drug metabolites, development compounds, synthesis intermediates, drug impurities, fungicides, herbals, nutraceuticals, herbicides, insecticides, food additives and preservatives, pharmaceutical aids and diagnostic and research reagents.

Collaborators have profiled their own compounds in the identical assays, to produce enhanced proprietary data sets. These are generally project and attrited compounds. Examples of results using data from these compounds in the context of BioPrint® compounds are given in Section 2.3.2.1 for the differentiation of hits/leads and in Section 2.3.2.2 for the analysis of attrited compounds.

2.2.4
BioPrint® *In Vivo* Data sets

The *in vivo* data sets can be classified as compound details, adverse drug reaction data (ADR), pharmacokinetic data and animal toxicity data.

2.2.4.1 Compound Details
Among the compound details collected are the following: drug indication, therapeutic area, market status including country and approval status, trade names, routes of administration and chemical class. For most of these data types, there is information on more than 1200 drugs.

2.2.4.2 ADR Data
The adverse drug reaction data set contains data on over 1160 drugs. The natural language of drug labels, binned by drug label frequency categories, is mapped to standard ADR terms such as Costart terms and Meddra terms. A more controlled ADR clinical incidence data set (529 drugs) is also available. In this case, trial data are presented using the actual frequency percentages and include placebo controls.

2.2.4.3 Pharmacokinetics
Also included in *in vivo* data is a set of human (90% of drugs) and animal pharmacokinetic (30% of drugs) data. While the *in vitro* data are generated in-house (Cerep), pharmacokinetic data are gathered from the literature. A variety of different parameters are covered including absolute bioavailability, oral absorption, clearance, volume of distribution, half-life, protein binding and excretion information.

2.2.4.4 Toxicity Data

BioPrint® also includes a small animal toxicity data set. Blood chemistry and organ toxicity data have been gathered for more than 200 compounds.

2.3
Structure–In Vitro Relationships

2.3.1
Similarity, Chemotypes – What Is a Biologically Relevant Descriptor?

Although compounds can be, and generally are, described by molecular structure-based descriptors/fingerprints, the relevance of this to the broad biological activity of a molecule is not clear. As many compounds are synthesized as analogues of an active compound, where the core 'pharmacophore' for activity is often retained, similarity in structure can be seen to correlate to similarity in biological activity. More recent publications [23] showed that there is only a 30% chance of a compound that is >0.85 (Tanimoto on Daylight fingerprints [24]) similar to an active is itself active.

To demonstrate this, Figure 2.8 shows the comparison of similarity for Daylight structural and biological fingerprints created from a panel of 154 assays from the BioPrint® database (measured by pairwise Tanimoto distance for 347 drugs with MW 200–600; 60 031 points) [6]. Figure 2.8a shows the overall scatter plot of the

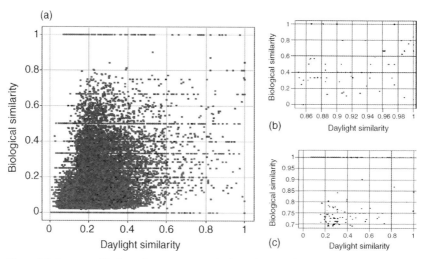

Figure 2.8 Compound biological activity similarity versus chemical structure similarity for 347 drugs from the BioPrint® data set. Pairwise Tanimoto distance (0–1, 1 = identical) using Daylight structural fingerprints was used for structural similarity (x-axis) versus from biological activity fingerprints from 154 BioPrint® assays (active = $IC_{50} < 100\,\mu M$, y-axis): (a) All data (b) Enlarged view of the region where Daylight fingerprint similarity is >0.85. (c) Enlarged view of the region where biological activity similarity is >0.7 (adapted from ref. [6]).

whole data set, demonstrating, at least for these measures, the poor correlation ($R^2 = 0.13$) of general biological and chemical (structural) similarity. Looking in detail at the region where the structural similarity is >0.85, where neighborhood behavior would be expected to be stronger, this lack of correlation is still observed (see Figure 2.8b) with a correlation coefficient R^2 of 0.05, supporting the finding of Martin et al. noted above. Similarly, looking at the region of high biological fingerprint similarity (Tanimoto > 0.7), the correlation is still poor with a correlation coefficient R^2 of only 0.04.

The question of whether chemical similarity infers biological similarity does not have a simple answer, as there are many possible definitions of chemical similarity that can use 2D or 3D structures. Horvath and Jeandenans [25,26] investigated which chemical/structural fingerprints best correlate with biological similarity. Many descriptors were studied (2D topology, shape, three- and four-point pharmacophores and fuzzy bipolar pharmacophores) as well as similarity metrics. They found 3D descriptors to have improved performance in replicating the biological similarities calculated from a smaller panel of 42 targets. 'Fuzzy pharmacophores' [27] [counts of the number of feature pairs (hydrophobes, aromatic groups, hydrogen bond donors/acceptors and positive/negative charges) separated by a binned distance] were able to correlate best with the biological similarity and four-point pharmacophoric fingerprints worked less well when structures had different conformational spaces constraints.

As illustrated in the next section, the use of biological fingerprints, such as from a BioPrint® profile, provides a way to characterize, differentiate and cluster compounds that is more relevant in terms of the biological activity of the compounds. The data also show that different *in silico* descriptors based on the chemical structure can produce quite different results. Thus, the selection of the *in silico* descriptor to be used, which can range from structural fragments (e.g. MACCS keys), through structural motifs (Daylight keys) to pharmacophore/shape keys (based on both the 2D structure via connectivity and from actual 3D conformations), is very important and some form of validation for the problem at hand should be performed.

2.3.2
Using Biological Fingerprints as a Meaningful Descriptor for Drug Leads and Candidates

2.3.2.1 Differentiation of Leads
A major challenge in the drug discovery process is the selection of a development candidate that has the best chance of survival and is differentiated from other compounds. Ideally, the attrition risk should be minimized, but at least the risk should be orthogonalized as much as possible for multiple candidates, so as to avoid multiple compound attrition for the same unexpected cause. The key to achieving this goal is to identify leads that have the best chance of becoming a suitable development candidate, with the most desired profile of biological and ADME(T) related properties. It is important to prioritize which leads are of most interest, and to progress the hits/leads that have the best chance of survival and are differentiated, ideally with an improved profile, from earlier candidates (in-house or competitor).

An early identification of the best leads is critical, and systematic biological profiling, as with Cerep BioPrint®, enables this progress to a great extent. Experience with using this approach in a major pharmaceutical company (Pfizer) has confirmed this. One component of the issue can be stated that it is better to identify the best (not just potency) lead series than to try and find the best candidate from a possibly suboptimal series, which can happen at late stages of lead optimization, where it is very difficult to make major changes to the lead series chemistry.

An example of this is illustrated in Figure 2.10 from a project at Pfizer [28], where four potent hit/lead series were identified from high-throughput screening (HTS), together with a reference compound in clinical development at the time, Duloxetine (see Figure 2.9). They all had the desired activities (polypharmacology) but from BioPrint® analyses were shown to have quite different overall biological profiles (fingerprints), that is, off-target activities with many activities significant. This was unpredictable from a simple structural analysis: core structures are shown in Figure 2.10 to illustrate this.

A clustering using a standard 2D structural fingerprint, the Daylight fingerprint, is also shown in Figure 2.10. Such an analysis, commonly used to reduce compound lists, could lead to wrong selections and missing the optimal 'clean' series ('orange' compound in Figure 2.10). Clustered with the reference compound is the compound represented in dark blue, but this compound actually has a differentiated biological profile. The most different cluster from the structural analysis contains compounds represented in yellow and orange. If the 'yellow' compound were selected to represent this part of structural 'diversity', a series with substantial polypharmacology and similar off-target issues to the reference compound would be worked on and the interesting 'orange' compound missed. Thus, the 'yellow' and 'green' compounds may be chosen for further work to represent a structurally diverse selection, yet both have similar profiles to the reference (blue) compound, so this approach would fail. The 'yellow' piperidine and 'orange' piperazine compounds may be considered to be relatively similar, yet the small nitrogen atom shift produces the most differentiated compound in biological space, moving into a space more occupied by opiates. This selection hypothesis was actually fully validated, as the project was active during the validation stage of the BioPrint® approach in the company, so all the lead series were followed up with further chemistry, armed with the knowledge of key selectivity assays. The 'SAR out of the selectivity problem' approach failed for all the series close to the reference compound. The 'orange' piperazine series stayed relatively clean and moved forwards toward clinical development.

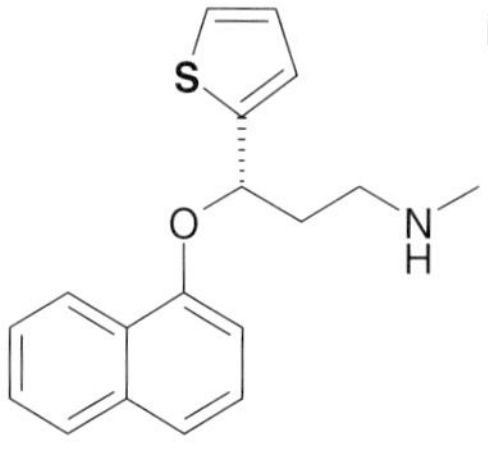

Figure 2.9 Structure of Duloxetine.

2.3 Structure–In Vitro Relationships | 35

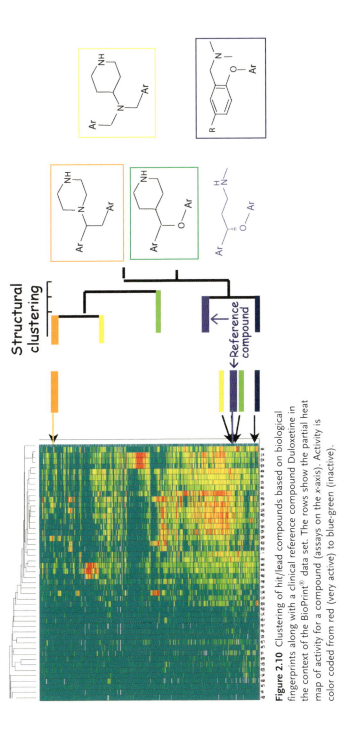

Figure 2.10 Clustering of hit/lead compounds based on biological fingerprints along with a clinical reference compound Duloxetine in the context of the BioPrint® data set. The rows show the partial heat map of activity for a compound (assays on the x-axis). Activity is color coded from red (very active) to blue-green (inactive).

BioPrint® biological profiling was then applied systematically to the analysis of hits/leads, to enable better data-driven decisions/prioritization. Many insights into unexpected activities (outside of the target class) were obtained. There was usually clear differentiation of different hits/leads, clustering by the profile together with other compounds in the BioPrint® database (including company compounds) aiding this and providing further insights. The off-target activity difference can be quite dramatic, changing from promiscuous to quite clean, for example for a pair of compounds with 2–4 nM activity, both with a distinctive cyclic base–linker–bicyclic polyheteroaromatic–(substituent1)–substituent2 system, by removal of an oxygen atom (alkoxy to alkyl) in substituent1, removal of nitrogen atoms in substituent2 (pyrimidine to phenyl) and addition of a further nitrogen to the polynitrogen bicycle [29].

2.3.2.2 Analysis of Attrited Compounds

An analysis of more than 130 preclinical candidates that had attrited during further development showed the failure of the chemotype approach (i.e. that a compound of the same/similar chemotype will have similar risks of attrition and that a structurally diverse chemotype will offer the best approach to minimize attrition risk) and 2D structure-based methods to be able to effectively differentiate compounds [29]. Thus, the risk of failing or succeeding in development is not related to being of the same 'chemotype', and differentiation by this method may not be the most effective way; dangers are both that a valuable series/chemotype could be discarded because of one bad result and that a structurally different compound may actually have similar off-target effects (e.g. due to the 'decoration' versus the scaffold).

Analysis of principal components space from structure-based descriptors and properties of a set of compounds developed for four different targets was able to cluster/differentiate compounds of a particular therapeutic activity. However, those failing in development (gray compounds in Figure 2.11) could not be differentiated from those that had not attrited (at the time of the analysis). These were the compounds that had met typical quality criteria, thus obvious structural alerts would not be expected to be present. These alerts are clearly of value, in terms of avoiding substructures etc., known to cause problems. They may be, but are not generally, associated with a 'chemotype' (an ambiguous term, but generally defined by a common core structural motif/scaffold).

An analysis of the attrited compound set using the BioPrint® biological fingerprints showed a richness of off-target pharmacology and CYP inhibition for many of the compounds. Figure 2.12 shows the activity for these compounds (Figure 2.12a) against 15 assays that were hit multiple times by these compounds compared to the same assays for the general drug set (Figure 2.12b). There is clearly an increased hit rate, with multiples of these assays being hit (but not always the same ones).

A closer analysis of some compound sets showed that very similar compounds can have quite different patterns of off-target activities, even if these are sometimes quite weak. Figure 2.13 shows the biological fingerprints for the compounds developed against a serotonin receptor, with three attrited compounds (toxicity issue) shown at the bottom. The polypharmacology, and diversity of this, both within related receptors and transporters, and generally in the BioPrint® profile, is evident. Figure 2.14 shows the profiles for a set of statins, of two structural classes, with a

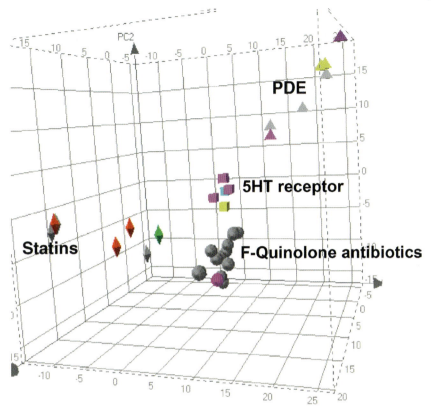

Figure 2.11 Plot of compounds developed for different target classes based on a principal components analysis (PCA) of 2D structure-based property fingerprints. Compounds are coded according to their target class (triangle, PDE; square, 5HT receptor; diamond, statin; circle, F-quinoline antibiotics) and clinical status at the time (gray, ok; yellow, clearance issue; red, CYP3A4 inhibition issue; purple, toxicity issue).

discontinued or withdrawn compound shown for each class. Although the off-target effects are relatively weak, they are quite different for very similar structures. Figure 2.15 shows the partial BioPrint® profile for a set of compounds developed against the same enzyme target. The richness and diversity of the polypharmacology of many of the compounds are clearly seen (red squares). The targeted primary activity was part of the profile, and is shown in a box as the last column on the right. Included are compounds that had attrited during development.

2.3.3
Structural versus Experimental Differentiation – Dependence on Structure-Derived Descriptor Used

The goal of differentiation, such as via clustering using descriptors derived from chemical structures, is to produce an end result as close as possible to the result that

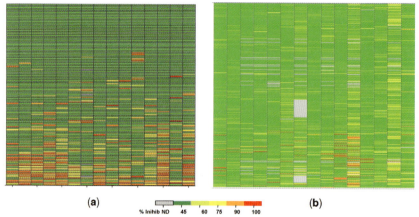

Figure 2.12 (a) Analysis of >130 attrited compounds using the BioPrint® assays. The 15 targets hit by more than two attrited compounds (>50% inhibition at 10 μM) are shown. (b) Activity of the drug compound set against the 15 assays shown in (a).

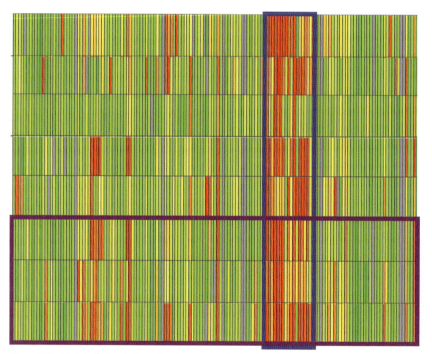

Figure 2.13 BioPrint® (partial) profile of compounds developed for activity against a serotonin receptor. Three compounds that attrited for some type of toxicity issue are shown at the bottom in a purple box. The serotonin receptors and transporters are highlighted with a blue box.

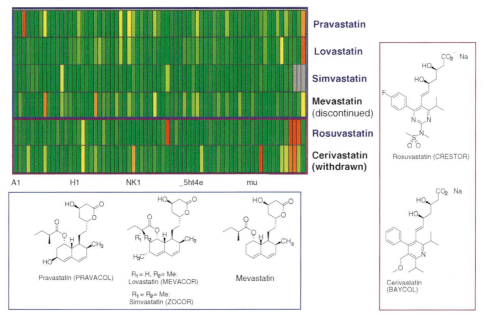

Figure 2.14 BioPrint® (partial) profile of some HMGCoA inhibitors. The two main structural classes are highlighted by a blue and purple box for both the activity fingerprint and the 2D structures.

would be obtained experimentally. The BioPrint® profiles enable an evaluation of different methods/descriptors. Figure 2.16 (left) shows the clustering using the Daylight fingerprints discussed in Section 2.3.2.1, together with the result using Scitegic FCFP6 circular fingerprints, from the reevaluation of these compounds using different descriptors. The impact of different descriptors on the results is evident and in this case (but not in all cases) the FCFP6 fingerprints were able to differentiate the cleanest compound from the BioPrint® biological profiling (shown with thick orange

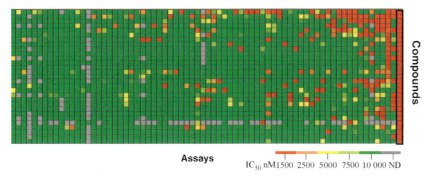

Figure 2.15 BioPrint® (partial) data for a set of compounds developed against the same enzyme target.

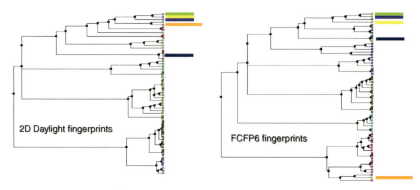

Figure 2.16 Comparison of the structure-based clustering of the lead compounds described in Section 2.3.2 using two different fingerprints. The color coding for the compounds is the same as for Figure 2.10. The impact of different descriptors on the results is evident, and in this case the Scitegic FCFP6 circular fingerprints were able to differentiate the cleanest compound from the BioPrint® biological profiling (shown with thick orange line).

line). The use of pharmacophore-based descriptors clearly has potential – as earlier studies discussed in Section 3.1 show 'fuzzy pharmacophoric' descriptors perform best for replicating biological similarities – and further work is required in this area. For a particular study, some form of validation of the relevance to the desired outcome of the descriptor and the clustering method to be used is recommended. Since the high confidence prediction of activity on many targets is not possible, ideally as much experimental data as possible should be obtained. Although profiles such as BioPrint® or a reduced version of it (e.g. 80–100 of the profiles with a higher hit rate) may appear expensive, they are actually cheap when compared to wasted resources on chemistry or other follow-up work, either on a suboptimal series or without knowledge of what the real selectivity targets are.

2.3.4
Predictive Models from Pharmacological Data

The BioPrint® data set provides a very consistent and complete data set for the generation of predictive models. A sufficient amount of data for actives exists to be able to develop at least simple binary models for about 50% of the assays using various modeling approaches, including Bayesian models and the FCPF6 circular fingerprint (Scitegic Pipeline Pilot software). These data were also combined [30] with published activity data (e.g. in *Journal of Medicinal Chemistry*) and in-house screening data from a large pharmaceutical company (e.g. Pfizer, that includes compounds/data from many other companies that have been incorporated), giving 617 694 data points on 238 655 compounds covering 698 targets. The set was used to generate Bayesian models and human polypharmacology interaction networks that represent relationships between proteins in chemical space (two proteins are deemed interacting in

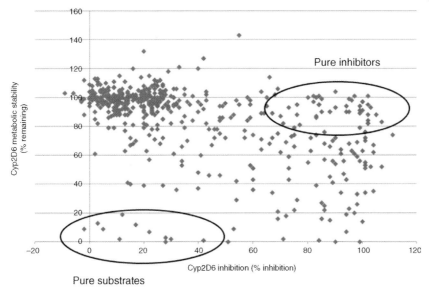

Figure 2.17 Percentage inhibition of CYP2D6 plotted on the x-axis against compound stability (percent remaining, y-axis) after incubation with the CYP2D6 enzyme.

chemical space, i.e. joined by an edge, if both bind one or more compounds within a defined difference in binding energy threshold).

2.3.5
Predictive Models from ADME Data – BioPrint® Learnings

An analysis of the large ADME data set allows for a greater understanding of how to interpret early *in vitro* ADME data and use it in predictive modeling. In Figure 2.17, for instance, percentage inhibition of CYP2D6 is plotted against compound stability (percentage remaining) after incubation with the CYP2D6 enzyme. The wide distribution of points in this plot highlights the fact that a compound can be a pure inhibitor of CYP2D6, solely a substrate of the enzyme or, as most often the case, both a substrate and a functional inhibitor of enzyme. This extra information enables a better use, and understanding of the limitations, of the inhibition data in the development of predictive models for CYP inhibition.

2.4
Chemogenomic Analysis – Target–Target Relationships

The full matrix nature of the BioPrint® database also enables an analysis of targets in drug chemical space. In this approach, a target is characterized by a fingerprint of the activities of a fixed set of compounds (the drug and reference compound set) against

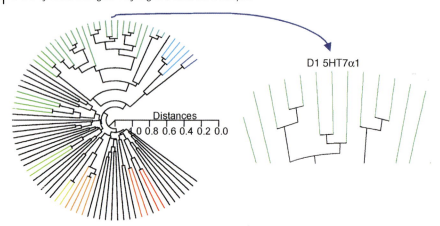

Figure 2.18 Chemogenomic analysis of targets: BioPrint® targets clustered in chemical space by fingerprint of activities against the consistent BioPrint® compound set.

that target. Thus, the target is probed by a relatively diverse set of compounds and the resultant binding affinities used to create a binary or numerical fingerprint. This can then be used in a hierarchical clustering, as illustrated in Figure 2.18, to enable the identification of targets that share more similar binding site properties. The expanded part of Figure 2.18 shows one of the clusters, in which D1, 5-HT_7 and α1 interact with similar compounds, highlighting key selectivity targets; such a clustering is not found in a phylogenetic analysis, but is indicated from binding site model comparisons and in practice when working on these targets. The clustering depends on the compounds used and can be done in a broader way by adding extra compounds that interact with new targets compared to the drug set; for example, peptidic GPCRs and kinases and newer 'chemotypes' used for project compounds. A new target can thus be analyzed to find where it is positioned in chemical space by screening the common set of compounds and clustering with all other targets that have screened these compounds (BioPrint® targets, in-house HTS targets, etc.).

2.5
In Vitro–In Vivo Relationships – Placing Drug Candidates in the Context of BioPrint®

BioPrint® is particularly useful in placing the new drug candidates in the context of drugs and related compounds that together make up the history of medicinal chemistry. New candidates are run on the same assays as the BioPrint® compounds and the resulting profile is analyzed. Profiles can be analyzed in two different ways; each individual hit can be analyzed and assessed for potential ADR liabilities or the entire profile can be used to identify compounds with similar profiles. Potential ADR liabilities are assessed based on those of the similar compounds identified. These two different approaches will be discussed in the following sections.

2.5.1
Analyzing Potential ADR Liabilities Based on Individual Hits

Hits on individual assays can be analyzed in a couple of different ways. First, drugs can be identified that have similar strength hits and then the ADR profiles of these drugs can be examined to identify ADRs that may be associated with these hits. Also, contained in BioPrint® are extensive collections of ADR associations [2], which have been identified by querying the database for statistically significant correlations between individual assays and individual ADRs. These ADRs are stored in the database and can be accessed by searching assay or the ADR. It is also useful to consult the pharmacokinetic data to confirm that the strength of the *in vitro* hit is consistent with *in vivo* exposure levels.

For example, in the case of compound XX, which among other activities has an IC_{50} of 60 nM on the calcium channel diltiazem site binding assay, a search for BioPrint® compounds with an activity in the similar range yields a series of compounds nearly all of which are known calcium channel blockers (see Table 2.1).

Thus, without even knowing the therapeutic area of the drug candidate in question, we can hypothesize that this compound is likely to share ADRs with the calcium channel blocker family of drugs. The ADRs for this family of drugs can be collected and tabulated to identify ADRs common to this family (see Table 2.2). For instance, the ADR 'tinnitus' is present in 9 out of 11 of the calcium channel blockers, whose ADRs are listed.

On searching the database among ADR associations we can retrieve a large number of associations for the calcium channel diltiazem site binding assay. Two ADR associations are shown in Figures 2.19 and 2.20. The ADR 'tinnitus' has a

Table 2.1 BioPrint® compounds with a similar strength hit on the calcium channel diltiazem site binding assay to compound XX.

Compound Name	Drug Status	Drug Class
Compound XX	Candidate	Test compound
Amlodipine maleate salt	Drug	Calcium channel blocker
BAYK8644	Nondrug	Calcium channel blocker
Benidipine HCl	Drug	Calcium channel blocker
Carpipramine diHCl	Drug	Antidepressant
Diltiazem HCl	Drug	Calcium channel blocker
Isradipine	Drug	Calcium channel blocker
Lercanidipine	Drug	Calcium channel blocker
levo-α-Acetylmethadol HCl	Prodrug	Narcotic analgesic
Nicardipine HCl	Drug	Calcium channel blocker
Nilvadipine	Drug	Calcium channel blocker
Nimodipine	Drug	Calcium channel blocker
Pimozide	Drug	Antipsychotic
Thonzonium bromide	Drug	Cationic detergent surfactant
Lomerizine diHCl	Drug	Calcium channel blocker

Table 2.2 ADRs for calcium channel blocker family of drugs.

ADR	Tinnitus	Asthenia	Palpitation	Rash	Dizziness	Nausea
Compounds	9 of 11	10 of 11	10 of 11	10 of 11	11 of 11	11 of 11
Amlodipine maleate	Infrequent (>0.1 and <1.0%)	Common (>3%)	Frequent (>1%)	Rare (<0.1%)	Common (>3%)	Frequent (>1%)
Nicardipine HCl	Frequency not stated	Common (>3%)	Common (>3%)	Infrequent (>0.1 and <1.0%)	Common (>3%)	Frequent (>1%)
Diltiazem HCl	Frequency not stated	Common (>3%)	Frequency not stated	Frequent (>1%)	Frequency not stated	Frequent (>1%)
Bepridil HCl	Infrequent (>0.1 and <1.0%)	Common (>3%)	Frequent (>1%)	Infrequent (>0.1 and <1.0%)	Common (>3%)	Common (>3%)
Isradipine		Common (>3%)	Common (>3%)		Common (>3%)	Frequent (>1%)
Nifedipine	Frequency not stated	Common (>3%)	Common (>3%)	Frequent (>1%)	Common (>3%)	Common (>3%)
Nimodipine			Frequency not stated	Infrequent (>0.1 and <1.0%)	Frequency not stated	Frequent (>1%)
Nitrendipine	Rare (<0.1%)	Rare (<0.1%)	Infrequent (>0.1 and <1.0%)	Infrequent (>0.1 and <1.0%)	Infrequent (>0.1 and <1.0%)	Infrequent (>0.1 and <1.0%)
Risedronate sodium	Common (>3%)	Common (>3%)		Common (>3%)	Common (>3%)	Common (>3%)
Verapamil HCl	Frequency not stated	Frequent (>1%)	Frequency not stated	Frequent (>1%)	Common (>3%)	Frequent (>1%)
Zonisamide	Frequent (>1%)	Frequent (>1%)	Infrequent (>0.1 and <1.0%)	Common (>3%)	Common (>3%)	Common (>3%)

2.5 In Vitro–In Vivo Relationships – Placing Drug Candidates in the Context of BioPrint

Selected assay	Body system	drug response	% of bioprint drugs listing ADR	Spearman correlation
Ca^{2+} channel (L-diltiazem site) (benzothiazepines)	Special senses	TINNITUS	19.69	0.96

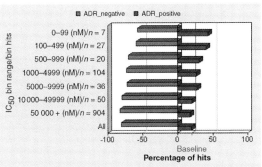

Figure 2.19 Tinnitus ADR association with Ca^{2+} channel (L-diltiazem site) activity.

baseline incidence among BioPrint® compounds of 19.69%. This is to say that 19.69% of the drug labels in BioPrint® list the ADR 'tinnitus'. Four out of seven (42.85%) of the drugs with a sub-100 nM hit on the calcium channel list 'tinnitus' on the label. Eleven out of 27 (41%) of the drugs with IC$_{50}$s between 100 and 500 nM list tinnitus on the label. A Spearman rank correlation is also performed to assess the dose response characteristics of the ADR association. In the case of the calcium channel/tinnitus ADR association, there is a strong Spearman rank correlation of 0.96. The association of the calcium channel with the ADR palpitations has an incidence rate of 57% in the highest IC$_{50}$ bin, well above the background rate of 20%. We can derive the thresholds for the IC$_{50}$ at which an *in vitro* side activity translates into an ADR. Only above a certain activity bin threshold the ADR rate becomes significantly higher than the background rate (e.g. at 500 nM in this case).

Selected assay	Body system	drug response	% of bioprint drugs listing ADR	Spearman correlation
Ca^{2+} channel (L-diltiazem site) (benzothiazepiness)	Cardiovascular	PALPITATION	23.69	0.88

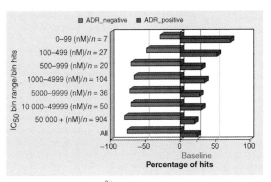

Figure 2.20 Palpitation ADR association with Ca^{2+} channel (L-diltiazem site) activity.

Table 2.3 Consensus profile neighbors for Compound XX.

Compound name	Drug class	MechTarget	Heat Map
Compound XX	unknown	unknown	
Nicardipine HCl	Drug	Ca L-channel antagonist	
Zolantidine Dimaleate	Non-drug	DA reuptake inhibitor, Sigma-2 antagonist	
Rimcazole DiHCl	Non-drug	H2 antagonist	
Fluoxetine HCl	Drug	5-HT reuptake inhibitor	
p-Chlorobenzhydrylpiperazine	Non-drug	5-HT 2C agonist	
Flunarizine diHCl	Drug	H1 antagonist, Ca channel antagonist	
GBR-12935 DiHCl	Non-drug	DA reuptake inhibitor	
SB-271046 HCl	Non-drug	5-HT6 receptor antagonist	
Raloxifene	Drug	estrogen receptor modulator	
Prenylamine Lactate Salt	Drug	Ca channel antagonist	
GBR 13069 DiHCl	Non-drug	DA reuptake inhibitor	
Mibefradil	Drug	Ca T-channel antagonist	
Pinaverium Bromide	Drug	Ca L-channel antagonist	

Heat maps representing pharmacological profiles of compound XX and its consensus profile neighbors: gray, not hit; red, 1–99 nM; orange, 100–999 nM; yellow, 1000–9999 nM; white, >10 000 nM.

Table 2.4 A selection of tabulated drug label ADRs for profile neighbors of compound XX.

ADR_Costartcode	Flunarizine diHCl	Fluoxetine HCl	Mibefradil	Nicardipine HCl	Pinaverium Bromide	Prenylamine Lactate Salt	Raloxifene
Nausea	Frequency not reported	Frequent (>1%)	Frequency not reported	Frequent (>1%)	Infrequent (>0.1 and <1.0%)	Frequency not reported	Common (>3%)
Dyspepsia	Frequency not reported	Common (>3%)			Infrequent (>0.1 and <1.0%)	Frequency not reported	Common (>3%)
Headache		Common (>3%)	Frequency not reported	Common (>3%)	Infrequent (>0.1 and <1.0%)		Frequent (>1%)
Rash	Frequency not reported	Infrequent (>0.1 and <1.0%)		Infrequent (>0.1 and <1.0%)		Frequency not reported	Common (>3%)
Asthenia	Frequency not reported	Common (>3%)	Frequency not reported	Common (>3%)			
Diarrhea		Infrequent (>0.1 and <1.0%)			Infrequent (>0.1 and <1.0%)	Frequency not reported	Frequent (>1%)
Dry mouth	Frequency not reported	Common (>3%)		Infrequent (>0.1 and <1.0%)	Infrequent (>0.1 and <1.0%)		
Edema peripheral		Infrequent (>0.1 and <1.0%)	Frequency not reported	Common (>3%)			Common (>3%)
Somnolence	Frequency not reported	Common (>3%)		Frequent (>1%)	Infrequent (>0.1 and <1.0%)		
Vasodilation		Common (>3%)	Frequency not reported	Frequency not reported			Common (>3%)
Vertigo		Infrequent (>0.1 and <1.0%)		Frequency not reported	Infrequent (>0.1 and <1.0%)		Frequent (>1%)
Vomiting		Frequent (>1%)		Infrequent (>0.1 and <1.0%)		Frequency not reported	Common (>3%)

(Continued)

Table 2.4 (Continued)

ADR_Costartcode	Flunarizine diHCl	Fluoxetine HCl	Mibefradil	Nicardipine HCl	Pinaverium Bromide	Prenylamine Lactate Salt	Raloxifene
Allergy reaction		Frequency not reported		Frequency not reported	Infrequent (>0.1 and <1.0%)		
Anxiety	Frequency not reported	Common (>3%)		Frequency not reported			
Depression	Frequency not reported						Common (>3%)
Extrapyr syndrome	Frequency not reported	Frequency not reported				Frequency not reported	
Insomnia	Frequency not reported	Common (>3%)					Common (>3%)
Pharyngitis		Common (>3%)		Frequency not reported			Common (>3%)
Sinusitis		Frequent (>1%)		Frequency not reported			Common (>3%)
Tremor	Frequency not reported	Common (>3%)					
Weighting	Frequency not reported	Frequent (>1%)					Common (>3%)

2.5.2
Analyzing Potential ADR Liabilities Based on Profile Similarity

Potential ADR liabilities can also be assessed using the whole BioPrint® profile. In this case, the profile is used to identify other BioPrint® compounds with similar profiles (see Table 2.3 for compound XX). We can identify compounds with similar profiles using standard statistical tools such as hierarchical clustering or correlations. BioPrint® also offers a profile similarity tool. This tool compares profiles of the seed compound with those in the BioPrint® database on an assay-by-assay basis calculating a penalty when the compared assay results do not fall within a set range. The most similar compounds have the lowest penalties. The BioPrint® similarity tool also has an added feature that allows the user to weight certain assays higher than other assays. This is particularly useful when compounds have very sparse profiles since similarity metrics will treat the compounds as similar because they all share the characteristic of having very few hits. Heavily weighting the hits makes it easier to identify meaningful neighbors.

Often, it is useful to identify profile neighbors using a variety of different techniques such that one can arrive at a consensus list of neighbors. Here too, as previously described for the compounds that shared strong hits on the diltiazem site calcium channel assay, the ADRs for the similar compounds can be collected and tabulated to aid in identifying ADRs that are shared by compounds with this type of profile. A selection of these ADRS that are common to a number of the profile neighbors can be found in Table 2.4.

The profile similarity approach was applied to Duloxetine (see Figure 2.8) before it was added to the database, and several potential ADRs were predicted based on its

ADR / Assay Association

Selected assay	Body system	Drug response	% of bioprint drugs listing ADR	Spearman correlation
_5ht7	Digestive	DRY MOUTH	30.48	0.91

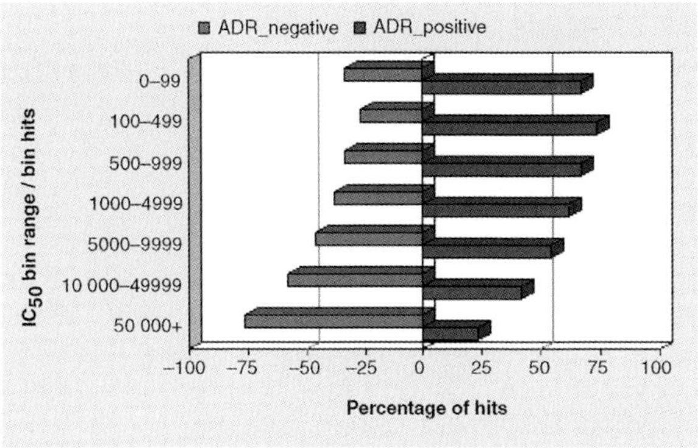

Figure 2.21 Association of 5-HT$_7$ activity with dry mouth as an ADR.

nearest neighbors. One was dry mouth that was reported during clinical trials that could be associated with 5-HT$_7$ activity (see Figure 2.21).

2.6
A Perspective for the Future

Biological fingerprints based mainly on *in vitro* binding assays are already showing great promise as a better way of describing and analyzing compounds targeted to biological systems. Now that functional assays providing an agonist and antagonist value for a compound for many GPCR targets are available at a cost/time similar to binding assays (e.g. from Cerep), more relevant data can be generated for drugs, candidates, attrited and project compounds. This should improve the ability to predict *in vivo* effects, including ADRs of compounds. As more *in vivo* data are generated for compounds, both human and animal, data sets such as BioPrint® enable a powerful analysis to seek the identification of any *in vitro* – *in vivo* associations.

Abbreviations

ADR	adverse drug reaction.
ADME(T)	absorption, distribution, metabolism, excretion (toxicity)
GPCR	G-protein-coupled receptor
NHR	nuclear hormone receptor
PK	pharmacokinetic
HTS	high-throughput screening

References

1 Jean, T. and Chapelain, B. (1999) Method of identification of leads or active compounds. CEREP, International Publication Number WO-09915894.

2 Krejsa, C.M., Horvath, D., Rogalski, S.L., Penzotti, J.E., Mao, B., Barbosa, F. and Migeon, J.C. (2003) Predicting ADME properties and side effects: the BioPrint® approach. *Current Opinion in Drug Discovery and Development*, **6**, 471–480.

3 Froloff, N., Hamon, V., Dupuis, P., Otto-Bruc, A., Mao, B., Merrick, S. and Migeon, J. (2006) Construction of a homogeneous and informative *in vitro* profiling database for anticipating the clinical effects of drugs. in *Chemogenomics: Knowledge-Based Approaches to Drug Discovery* (ed. E. Jacoby), Imperial College Press, London, pp. 175–206.

4 http://www.cerep.fr/cerep/users/pages/productsservices/bioprintservices.asp.

5 Hopkins, A.L. Mason, J.S. and Overington, J.P. (2006) Can we rationally design promiscuous drugs? *Current Opinion in Structural Biology*, **16**, 127–136.

6 Stanton, R. and Cao, Q. (2007) Biological fingerprints, in *Comprehensive Medicinal Chemistry II*, Volume 4 (eds-in-chief J.B. Taylor and D.J. Triggle) (ed J.S. Mason), Elsevier, Oxford, Chapter 32, pp. 807–18.

7 Fliri, A.F., Loging, W.T., Thadeio, P. and Volkmann, R.A. (2005) Biological spectra analysis: linking biological activity profiles to molecular structure. *Proceedings of the National Academy of Sciences of the United States of America*, **102**, 261–266.

8 Fliri, A.F., Loging, W.T., Thadeio, P. and Volkmann, R.A. (2005) Analysis of drug-induced effect patterns linking structure and side effects of medicines. *Nature Chemical Biology*, **1**, 389–397.

9 Fliri, A.F., Loging, W.T., Thadeio, P. and Volkmann, R.A. (2005) Biospectra analysis: model proteome characterizations for linking molecular structure and biological response. *Journal of Medicinal Chemistry*, **48**, 6918–6925.

10 Joost, P. and Methner, A. (2002) Phylogenic analysis of 277 human G-protein-coupled receptors as a tool for the prediction of orphan receptor ligands. *Genome Biology*, **3**, 63.1–63.16, http://genomebiology.com/2002/3/11/research/0063.

11 Manning, G., Whyte, D.B., Martinez, T., Hunter, T. and Sdarsanam, S. (2002) The protein kinase complement of the human genome. *Science*, **298**, 1912–1934.

12 Venter, J.C., Adams, M.D., Myers, E.W., Li, P.W., Mural, R.J., Sutton, G.G., Smith, H.O., Yandell, M. *et al.* (2001) The sequence of the human genome. *Science*, **291**, 1304–1351.

13 Hopkins, A.L. and Groom, C.R. (2002) The druggable genome. *Nature Reviews Drug discovery*, **1**, 727–730.

14 Drew, J. (2000) Drug discovery: a historical perspective. *Science*, **287**, 1960–1964.

15 Bleicher, K.H., Bohm, H.J., Muller, K. and Alanine, A.I. (2003) Hit and lead generation: beyond high-throughput screening. *Nature Reviews Drug Discovery*, **2**, 369–378.

16 Gozalbes, R., Rolland, C., Nicolaï, E., Paugam, M.-F., Coussy, L., Horvath, D., Barbosa, F., Mao, B., Revah, F. and Froloff, N. (2005) QSAR strategy and experimental validation for the development of a GPCR focused library. *QSAR and Combinatorial Science*, **24**, 508–516.

17 Cohen, P. (2002) Protein kinases – the major drug targets of the twenty-first century? *Nature Reviews Drug Discovery*, **1**, 309–315.

18 Mustelin, T. and Tasken, K. (2003) Positive and negative regulation of T-cell activation through kinases and phosphatases. *Biochemical Journal*, **371**, 15–27.

19 Sallantin, J., Dartnell, C. and Afshar, M. (2006) A pragmatic logic of scientific discovery, in *Discovery Science: Volume 4265 of Lecture Notes in Computer Science* (eds N. Lavrac, L. Todrovski and J.P., Jantke) Springer Verlag, Berlin, Heidelberg, pp.231–242.

20 Afshar, M., Lanoue, A. and Sallantin, J. (2007) Multiobjective/multicriteria optimization and decision support in drug discovery, in *Comprehensive Medicinal Chemistry II*, Volume 4 (eds-in-chief J.B. Taylor and D.J. Triggle) (ed J.S. Mason), Elsevier, Oxford, Chapter 30, pp 767–72.

21 UK QSAR 2006 http://www.iainm.demon.co.uk/ukqsar/meetings/2006-04-26.html.

22 Ariana Pharmaceuticals, Paris, www.arianapharma.com.

23 Martin, Y.C., Kofron, J.L. and Traphagen, L.M. (2002) Do structurally similar molecules have similar biological activity? *Journal of Medicinal Chemistry*, **45**, 4350–4358.

24 Daylight Fingerprints. Daylight Chemical Information Systems, Inc., www.daylight.com, Irvine, CA.

25 Horvath, D. and Jeandenans, C. (2003) Neighborhood behavior of *in silico* structural spaces with respect to *in vitro* activity spaces – a novel understanding of the molecular similarity principle in the context of multiple receptor binding

profiles. *Journal of Chemical Information and Computer Sciences*, **43**, 680–690.
26 Horvath, D. and Jeandenans, C. (2003) Neighborhood behavior of *in silico* structural spaces with respect to *in vitro* activity spaces – a benchmark for neighborhood behavior assessment of different *in silico* similarity metrics. *Journal of Chemical Information and Computer Sciences*, **43**, 691–698.
27 Horvath, D. (2001) High throughput conformational sampling and fuzzy similarity metrics: a novel approach to similarity searching and focused combinatorial library design and its role in the drug discovery laboratory, in *Principles, Software Tools and Applications* (ed. A. Ghose and V. Viswandhan) Marcel Dekker, New York, pp. 429–472.
28 Mason, J.S., Mills, J.E., Barker, C., Loesel, J., Yeap, K. and Snarey, M. (2003) Higher-throughput approaches to property and biological profiling, including the use of 3-D pharmacophore fingerprints and applications to virtual screening and target class-focused library design.Abstracts of Papers, 225th ACS National Meeting, New Orleans, LA, United States, March 23–27, COMP-343.
29 Mason, J.S. (2005) Understanding leads and chemotypes from a biological viewpoint: chemogenomic, biological profiling and data mining approaches. Book of Abstracts of First European Conference on Chemistry for Life Sciences: Understanding the Chemical Mechanisms of Life. October 4–8, Rimini, Italy.
30 Paolini, G.V., Shapland, R.H.B., van Hoorn, W.P., Mason, J.S. and Hopkins, A.L. (2006) Global mapping of pharmacological space. *Nature Biotechnology*, **24**, 805–815.

II
Antitargets: Ion Channels and GPCRs

3
Pharmacological and Regulatory Aspects of QT Prolongation
Fabrizio De Ponti[1]

3.1
Introduction

The long and growing list of non-antiarrhythmic drugs associated with the prolongation of the QT interval of the electrocardiogram has generated concern not only for regulatory interventions leading to drug withdrawal, but also for the unjustified view that QT prolongation is usually an intrinsic effect of a whole therapeutic class (e.g. antihistamines), whereas, in many cases, it is displayed only by some compounds within a given class of non-antiarrhythmic drugs because of an effect on cardiac repolarization.

This chapter provides insight into the strategies that should be followed during a drug development program when a drug is suspected of affecting the QT interval, with specific emphasis on the factors limiting the predictive value of preclinical and clinical studies (*in silico* studies are covered in other chapters). Mechanisms leading to QT prolongation (which are not limited to hERG blockade) are briefly discussed from a pharmacological point of view.

The chapter also highlights the requirements of clinical studies: although prolongation of the QT interval by non-antiarrhythmic drugs is not an unusual finding, potentially fatal arrhythmias such as *torsades de pointes* (TdP) are uncommon and are unlikely to occur during phase I–III clinical trials, when relatively few subjects are exposed to the investigational drug. Thus, QT prolongation has become a surrogate marker of cardiotoxicity and has received increasing regulatory attention. An important, unresolved problem is that there is no consensus on the degree of QT prolongation to be considered clinically significant and on the need of the so-called 'thorough QT study'.

---

[1] The author is a member of the Subcommittee for European Procedures at the Italian Medicines Agency. The opinions expressed herein are those of the author and do not necessarily reflect the views of the Italian Medicines Agency.

Table 3.1 Expression of (h)ERG K$^+$ channels in different systems and possible implications for drug development.

Localization/Expression	Function/Pathophysiology	Target (Desired Effect)	Antitarget (Side Effect)
Myocardium	Cardiac action potential Channelopathies: • Loss-of-function mutations of *hERG* are one of the possible causes of the congenital long QT syndrome • Gain-of-function mutations of *hERG* lead to the short QT syndrome [5]	Class III antiarrhythmics block hERG channels and prevent reentry; they may also be a second-line adjunct to therapy in patients with the short QT syndrome [2–4,6] hERG openers shorten QT interval (theoretically, this may be desirable in people with a loss-of-function hERG channelopathy) [7–9]	Many noncardiac drugs block hERG and prolong the QT interval (risk of *torsades de pointes*) as a side effect
Chromaffin cells	Modulate adrenaline release [10] Does it contribute to the sudden death phenotype associated with LQT2 syndrome?	Modulation of blood pressure?	—
Neurons	Several functions [11]: • Resting membrane potential • Neuritogenesis [12] • Spike-frequency adaptation [13]	Antiepileptic drugs?	Side effects of hERG blockers in the central nervous system?
Cancer	Properties of cancer cells (loss of cell-contact inhibition) [14–16]	hERG blockers as potential therapeutic agents? [17,18] Screening methods (hERG as a potential diagnostic marker) [19]	Effect on cell proliferation as a side effect of hERG blockers when used as antiarrhythmics?
Gut • Smooth muscle • Interstitial cells of Cajal	Motility and pacemaker activity [20–26]	hERG blockers to restore peristalsis? [27]	Potential side effects of hERG blockers at the moment precludes their development for potential gastrointestinal indications: • Intestinal spasms?

3.2
hERG: Target Versus Antitarget

The ether-a-go-go-related gene (ERG) K^+ channels and specifically those expressed in human cells (i.e. hERG K^+ channels) are fascinating to the medicinal chemist, pharmacologist and clinician because of their many possible implications both in drug development and in clinical practice.

To provide a general picture, it is important to recognize that hERG may indeed be not only an antitarget but also a desired target both in cardiology (development of antiarrhythmics: hERG blockade and QT prolongation are typical effects of class III agents such as dofetilide) and in other areas (e.g. oncology) that are starting to be more closely investigated. Table 3.1 summarizes the current knowledge on the localization of (h)ERG in different systems and its possible exploitation both as a target and an antitarget [1–27]. In addition, hERG K^+ channels may be responsible for a number of diseases ('channelopathies') based on loss-of-function or gain-of-function mutations (respectively, long and short QT syndromes) [2,28–30].

This chapter will focus on hERG K^+ channels as an antitarget during drug development: indeed, the blockade of hERG may be the unwanted effect of a large number of noncardiovascular drugs [31] and lead to prolongation of the QT interval of the electrocardiogram as a side effect. For both drug developers and regulators, the great interest on the QT prolonging potential of noncardiac drugs is due to a number of reasons.

First, drug-induced lengthening of the QT interval has been associated with the occurrence of ventricular tachyarrhythmias, namely TdP, a polymorphous ventricular arrhythmia that may cause syncope and degenerate into ventricular fibrillation and sudden death: although the incidence of TdP is a rare event (usually, less than 1 in 100 000) [32], even a low risk is not justified for drugs with uncertain benefits or drugs providing only symptomatic improvement of a mild disease.

Second, as many as 60% of new molecular entities developed as potential therapeutic agents, when assayed for hERG blocking liability, test positive and are thus abandoned early in development, although their true torsadogenic potential is unknown.

Third, when considering marketed drugs, there is a significant exposure to non-antiarrhythmic drugs with QT-prolonging potential in the general population: approximately 2–3% of all drug prescriptions in the United Kingdom and Italy involve medications that may unintentionally cause the long QT syndrome [33]. In an international drug utilization study, the total amount of QT-prolonging drugs dispensed through community pharmacies ranged from 13.1 to 19.6 defined daily doses/1000 inhabitants per day [34].

Finally, a recent review [35] reported that QT prolongation (with or without TdP), together with hepatotoxicity, was responsible for more than 60% of drug withdrawals over the last 16 years.

Thus, there is an urgent need for a multidisciplinary approach to assess the QT liability of new drugs [36].

3.3
Pharmacology of QT Prolongation

The QT interval (measured from the beginning of the Q wave to the end of the T wave of the surface electrocardiogram) reflects the duration of individual action potentials in cardiac myocytes (Figure 3.1): indeed, a prolongation of the action potential duration (APD) of myocytes will result in a prolonged QT interval.

Cardiac APD is controlled by a fine balance between inward and outward currents in the repolarization phase. Since outward K^+ currents, especially the delayed rectifier repolarizing current, I_K (which is the sum of two kinetically and pharmacologically distinct types of K^+ currents: a rapid, I_{Kr}, and a slow, I_{Ks}, component), play an important role during repolarization and in determining the configuration of the action potential, small changes in conductance can significantly alter the effective refractory period, hence the action potential duration.

Several studies support the notion that the basic mechanism by which many drugs prolong the QT interval is related to blockade of potassium currents. For instance, several antihistamines, antibacterial macrolides, fluoroquinolones and antipsychotics were shown to inhibit the rapid component of the delayed rectifier K^+ current (I_{Kr}) in electrophysiological studies and to block potassium channels encoded by $hERG$ [37–42].

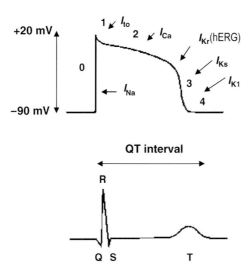

Figure 3.1 Upper panel: schematic representation of the action potential of a ventricular cell. Depolarizing currents include I_{Na} (inward Na^+ current) and I_{Ca} (inward Ca^{2+} current). Repolarizing currents (all K^+) include I_{to} (transient outward current), I_{Kr} (rapid component of the delayed rectifier current), I_{Ks} (slow component of the delayed rectifier current) and I_{K1} (inward rectifier current). Lower panel: electrical gradients in the myocardium can be detected on the body surface electrocardiogram: the diagram provides an illustrative example of the measurement of the QT interval (from the beginning of the Q wave to the end of the T wave).

3.3.1
Multiple Mechanisms Leading to QT Prolongation

There are a number of mechanisms leading to QT prolongation: this may be achieved not only by lengthening cardiac repolarization but also by widening the QRS complex (which corresponds to the ventricular depolarization phase). Thus, the duration of the QT interval may be affected by both the velocity of repolarization and ventricular conduction velocity. For instance, as a result of Na^+ channel blockade, class I antiarrhythmics and local anesthetics can reduce ventricular conduction velocity, cause widening of the QRS complex and therefore prolong the QT interval [43–45]. To further complicate the issue, recent data indicate that local anesthetic agents also interact with hERG channels [46], a property that may well contribute to the overall effect on the QT interval and explain part of their proarrhythmic action.

Cocaine is another example of a drug with a complex pharmacological profile responsible for different properties probably concurring due to QT prolongation. It has a local anesthetic action (and therefore shares the pharmacological properties mentioned above), but recent reports also indicate the blockade of hERG K^+ currents [47–49]. Thus, it is not unexpected that cocaine has been associated with QT prolongation and occurrence of TdP [50–53].

In any case, blockade of I_{Kr} is the key mechanism underlying the drug-induced QT prolongation, although, at least theoretically, actions on other potassium currents may also account for the prolongation of the action potential duration. These include the slow component of the delayed rectifier current I_{Ks}, the transient outward current I_{to}, the ultrarapidly activating delayed rectifier current I_{Kur} and the inward rectifier I_{K1} current [54–60]. The overall relevance of a given current may depend on the type of ion channels expressed in different parts of the heart (e.g. atrium versus ventricle), the species and the pathophysiological condition (low versus high heart rate; ischemic versus normal myocardium). A recent report [61] on antileishmanial antimonial drugs suggests that conversion of Sb(V) into active Sb(III) in patients produces a common mode of action for these agents, which increase cardiac risk not by a reduction of hERG/I_{Kr} currents, but by an increase in cardiac calcium currents. This aspect deserves further investigation.

A detailed discussion of the possible mechanisms leading to prolongation of the APD and QT interval is beyond the scope of this chapter, and the reader is referred to the reviews by Sheridan [62] and Malik and Camm [63].

3.3.2
hERG as the Key Mechanism for the Drug-Induced Long QT Syndrome

All agents so far identified as torsadogenic have been shown to block the hERG K^+ channels, which are undoubtedly the primary antitarget when exploring the 'QT liability' of a compound.

hERG channels have unique properties that make them an important, 'promiscuous' drug-binding site: this promiscuity is also indicated by a recent report showing that even dietary flavonoids [64], arachidonic acid and docosahexaenoic acid

[65] may impair hERG currents. The first feature is that, unlike most other potassium channels, hERG channels lack two specific proline residues that normally produce a kink limiting the volume of the inner cavity of the channel. Therefore, the larger hERG cavity volume allows relatively large molecules to enter and prevent potassium conduction. Second, they have two aromatic residues facing the inner cavity of the channel: results of alanine-scanning mutagenesis have led to the conclusion that these residues facilitate the interaction of hERG with aromatic groups on drugs in the inner cavity [66–68], again favoring drug-binding and channel blockade. Third, the unusual gating of hERG channels may also increase the chances of drugs being trapped within the inner cavity. Thus, unwanted clinical side effects from a wide spectrum of medications are mediated through blockade of I_{Kr} largely because of the unusual properties of the hERG pore-forming α-subunit.

However, it is important to recognize that (a) there are some hERG blockers that are unlikely to cause TdP; and (b) not only blockade, but also interference with the processing of mature hERG channels can lead to QT liability.

Concerning the first point, it is now widely recognized that the whole spectrum of the effects of a single agent must be considered. It is indeed possible that potent hERG blockers have a low torsadogenic potential because of additional pharmacological properties: two well-known examples are amiodarone and verapamil [63]. The latter is a typical 'false-positive' compound in the hERG liability test [69]. Antidepressants are also drugs with complex pharmacological actions on K^+, Na^+, and Ca^{2+} channels leading to variable effects on the QT interval *in vivo*, depending on the animal species and experimental model [70]. Therefore, it is important, within each therapeutic class, to investigate the whole pharmacological spectrum of each agent to identify those with a true torsadogenic potential [70–74].

Concerning the second aspect, pentamidine, a drug known to induce QT prolongation, has been shown to block I_{Kr}, but only at concentrations much higher than therapeutic levels [75]. Chronic treatment with pentamidine, however, results in disruption of KCNH2 channel protein trafficking and reduces surface membrane expression of otherwise functional channels [76]. Arsenic trioxide [77,78], cardiac glycosides [79], fluoxetine and norfluoxetine [80] are other examples of compounds interfering with hERG trafficking. This has important consequences and adds further complexity to the development of screening methods for QT liability [81] (see also [82]).

A loss of function by disruption of protein trafficking, rather than by production of channels reaching the cell membrane but unable to conduct current, has been recognized as a relatively common mechanism in the congenital long QT syndrome due to *hERG* (now termed *KCNH2*) mutations [83]. This has important pharmacogenetic implications, discussed in some recent papers [84–86]. Interestingly, Nakajima *et al.* [87] report that the K^+ channel regulatory protein 1 (KCR1) seems to protect hERG from pharmacological blockade.

3.3.3
Pharmacogenetic Aspects

The congenital form of long QT syndrome is characterized by a prolonged QT interval in the electrocardiogram and TdP [28–30]. Many patients with this syndrome suffer from severe cardiac events such as syncope and/or sudden cardiac death, most often during physical exercise or mental stress. However, cardiac events occasionally occur at rest, during sleep, or under specific circumstances with arousal. At least eight forms of congenital LQTS caused by mutations in genes of the potassium, sodium and calcium channels have been described. Mutations in KCNQ1 and KCNE1, the α- and β-subunits of the potassium channel gene, respectively, are responsible for defects (loss of function) in the slowly activating component of the delayed rectifier potassium current (I_{Ks}) underlying the LQT1 and LQT5 forms of LQTS. Mutations in KCNH2 and KCNE2 cause defects in the rapidly activating component of the delayed rectifier potassium current (I_{Kr}) responsible for the LQT2 and LQT6 forms. Mutations in SCN5A, the gene that encodes the α-subunit of the sodium channel, result in an increase (gain of function) in the late sodium current (I_{Na}) responsible for LQT3. The LQT1, LQT2 and LQT3 forms constitute more than 90% of genotyped patients with LQTS. For the sake of completeness, also gain-of-function mutations of hERG have been described. They are one of the possible underlying mechanisms leading to the short QT syndrome, another inherited disorder associated with familial atrial fibrillation and/or sudden death or syncope [2,3,6].

These pharmacogenetic aspects [88] pose a significant challenge to the drug developers, because in some subjects, single-point mutations may lead to hERG K^+ channels with increased affinity for the drug. There are a number of ion channel gene variants that predispose to I_{Kr}-associated drug-induced TdP. In particular, there are the so-called 'direct' mutations, which do not impair I_{Kr} at baseline but increase the sensitivity to drug blockade (see [89] for a classification of 'indirect', 'direct' and 'compound' mutations). One example is the T8A-MiRP1 single nucleotide polymorphism, which was identified in a patient who developed prolonged QTc after taking sulfamethoxazole/trimetoprim, probably as a consequence of the fourfold increase in the sensitivity of MiRP1/hERG to sulfamethoxazole [50]. This polymorphism destroys an N-glycosylation site of MiRP1, and it has been proposed that defective glycosylation does not affect the ability of the T8A mutant to form stable complexes with hERG but rather increases drug accessibility to the receptor because of the absence of the oligosaccharide groups [90].

3.4
Significance of Drug-Induced QT Prolongation

3.4.1
Prolonged QT/QTc and Occurrence of TdP

QT prolongation is a convenient end point to assess, but it is important to recognize that it is only a surrogate marker of cardiotoxicity and there is no consensus on the degree of QT prolongation that becomes clinically significant. Although the risk of developing TdP is proportional to the degree of QT prolongation (a QTc greater than 500 ms raises clear concerns about the potential to induce TdP) [91], actual occurrence of TdP and ventricular fibrillation depends on a number of concomitant risk factors that may be associated in a single patient (Table 3.2). Indeed, there are reported cases of TdP in patients with apparently normal QT interval [92].

Table 3.2 Risk factors for the occurrence of *torsades de pointes*.

Subject related	• Female gender • Congenital long QT syndrome • QTc > 440 ms in males; >460 ms in females (high risk with QTc >500 ms) • increased QT dispersion? • Clinically significant bradycardia (<50 beats per minute) • History of clinically significant heart disease (e.g. myocardial hypertrophy, heart failure) • Electrolyte imbalance (e.g. hypokalemia, hypomagnesemia, hypocalcemia) • Endocrine disorders (e.g. diabetes, hypothyroidism) • Impaired hepatic or renal function: these conditions may reduce the clearance of drugs or nutrients (e.g. flavonoids) with a potential to block hERG K^+ channels

Drug related	• Use of antiarrhythmic agents (class I or class III)
	• Use of drugs for which hERG is an antitarget (e.g. cisapride, erythromycin, terfenadine, thioridazine, etc.)
	• Drug interactions:
	  • Use of a hERG blocker in a patient also taking drugs inducing electrolyte imbalance (e.g. risk of hypokalemia with diuretics): this is a pharmacodynamic interaction
	  • Use of a hERG blocker in a patient also taking CYP3A4 inhibitors (e.g. antibacterial macrolides, azole antifungals, HIV protease inhibitors) or CYP2D6 inhibitors (quinidine, halofantrine, fluoxetine, paroxetine, thioridazine, terbinafine): the hERG blocker, if mostly metabolized by these CYP isoforms, may accumulate because of a pharmacokinetic interaction. In some cases, the interaction is both pharmacokinetic and pharmacodynamic because some of the aforementioned CYP2D6 inhibitors mentioned above also interfere with the hERG channel (blockade or inhibition of trafficking or both as in the case of fluoxetine [81])

To further complicate the issue, no international standard exists even for QT measurement itself. Several authors have already drawn the attention to the inherent difficulties involved in accurate measurement of the QT interval. These are summarized in Table 3.3.

The QT interval is a dynamic physiological variable depending on multiple factors such as cardiac cycle length (heart rate), autonomic nervous system activity, age, gender, plasma electrolyte concentrations, genetic variations in ion channels involved in cardiac repolarization. In addition, circadian and seasonal variations of the QT interval have been described [93].

The determinant that mostly influences the QT interval duration is cycle length (RR interval): the longer the RR interval, the longer the QT interval and vice versa. Therefore, a number of formulas (see [94] for a list) are used to normalize the QT interval for heart rate and obtain a corrected QT interval (QTc), a key issue especially

3.4 Significance of Drug-Induced QT Prolongation

Table 3.3 Preclinical evaluation of the proarrhythmic potential of QT prolonging drugs.

Model	Species (most used)	Advantages	Drawbacks
hERG K⁺ channels expressed in heterologous or human cells	Human embryonic kidney cells (HEK 293)	Mammalian cell lines are the ideal model for studying an effect on the current underlying I_{Kr}; they allow to use physiological temperatures (37 °C) for the human species HEK 293, CHO or L cells expressing hERG channels may be used as a primary test to study the pharmacological activity of a compound	Different laboratories may yield different IC_{50} values Activity of metabolites and enantiomers must be studied Compounds insoluble in water are difficult to assess Species vary in type and distribution of ion channel involved in cardiac repolarization Other repolarizing current and drug interactions with different hERG K⁺ channel subunits have to be characterized
	Chinese hamster ovary (CHO) cells		
	Oocytes of the amphibian *Xenopus laevis* (unreliable because poorly predictive)	Inhibition of [³H]-dofetilide binding to hERG channels has also been proposed, but this nonfunctional assay has important limitations	
Isolated intact heart (Langendorff preparation)	Rabbit, guinea pig	It allows screening of a large number of compounds The perfused heart of female rabbits is used in the SCREENIT system It is possible to induce experimental TdP	APD prolongation per se does not necessarily indicate proarrhythmia (the 'TRIaD' concept should be considered: see text)
Isolated tissues:	Dog, sheep, cat, rabbit, guinea pig	Isolated tissues allow screening of a large number of compounds and assessment of conditions that favor I_{Kr} block, such as low K⁺ concentrations and low stimulation rates	Not suitable for high-throughput screening

Table 3.3 (Continued)

Model	Species (most used)	Advantages	Drawbacks
Transmural wedge preparation of the left ventricle		The transmural wedge preparation allows to detect difference between M cells endo- and epicardial cells (it ensures that the extent of dispersion is explored)	High cost; not suitable for screening
Purkinje fibers Papillary muscle Isolated cardiac myocytes		Purkinje fibers are easily accessible	
ECG recording in conscious or anesthetized animals	Dog, pig, monkey (heart rate similar to human heart rate) Rabbit, rat, guinea pig, mouse (species with high baseline heart rate)	Ideal for studying the dose–response relationship for QT interval prolongation taking into account all the pharmacological properties of a compound The dog model is one of the most widely used; anesthetized rabbits (especially female rabbits) have also been proposed for high sensitivity It provides complementary information with respect to in vitro tests (activity of metabolites, measurement of plasma drug concentrations, calculation of the volume of distribution) Possibility to induce experimental TdP	In vivo models require a relatively large sample size to detect small differences in QTc When using anesthetized animals: use of anesthetics per se may affect the QTc interval Changes in heart rate require correction; different formulas may optimize correction in different species Definition of the end of the T wave is problematic in some species such as the dog, having a variable morphology of the T wave Data should be extrapolated to humans with caution

for those drugs that affect heart rate [63,95]. In these formulas, the reference heart rate is usually 60 beats/min (RR interval of 1 s or 1000 ms). The most used are Bazett's (square root) and Fredericia's (cubic root) formula, respectively: QTc = QT/$RR^{1/2}$ and QTc = QT/$RR^{1/3}$. However, these formulas (especially the Bazett's formula) are not ideal, since the correlation value between QTc and RR is significantly different from zero, showing that QTc is still dependent on underlying heart rate. Indeed, the Bazett's formula overcorrects at high heart rates and undercorrects at low heart rates [62,63]. Using this formula, an increase in QTc of 4–5 ms may depend on measurement bias, as shown for ebastine [94].

Derivation of the most appropriate QT correction formula should begin with the assessment of the QT/RR relationship in the population under study. Malik tried to obtain a formula for his specific data set using the generic equation QTc = QT/RR^{α}, where α was 0.314 in the pooled baseline data but could vary from 0.161 to 0.417 depending on the individual subject [94]. An important finding was that the QT/RR relationship had a high interindividual variability, but relative intraindividual stability, hence the need to examine the QT/RR relationship in each subject for an accurate assessment of drug effects on the QT interval, especially in phase I/II studies. From a regulatory point of view, correction formulas derived using individual subject data are considered most suitable for the 'thorough QT/QTc study' and early clinical studies, where it is possible to obtain many QT interval measurements for each study subject over a broad range of heart rates. On the contrary, a simple formula is needed by the clinician to make decisions in everyday clinical practice, in the light of the possible bias of automated analysis. In these circumstances, it is more important to understand the limitations of the correction formula rather than applying multiple formulas [94,96–99].

For the evaluation of potential clinical risks associated with QTc changes, individual QTc changes rather than mean values for study populations should be used. Identifying outliers when looking for drug-induced changes in the QT interval is an important issue. Their high ΔQTc value may be due to chance or they may be silent carriers of the long QT syndrome (subjects with normal QT interval, but carrying subtle genetic defects involving K^+ channels) [100–103]. A recent report proposes dynamic analysis of the QT interval in long QT1 syndrome patients with normal phenotype [104]. Thus, changes in T wave morphology and analysis of TU wave patterns may be more important than simple measurement of the QT interval.

A term that sometimes generates confusion is the definition of drug-induced TdPs as *idiosyncratic* adverse drug reactions. Roden, in a seminal paper [105], suggested that we should take the 'idio' out of the term 'idiosyncratic' and introduced the concept of *repolarization reserve*. He postulated that, in the normal ventricle, there is essentially no risk of developing TdP because the normal function of the repolarizing currents (mainly I_{Kr} and I_{Ks}) ensures a large repolarization reserve. However, the aforementioned risk factors, by reducing the repolarization reserve, greatly increase the likelihood of the occurrence of TdP. Thus, there are clinical circumstances in which TdP are not so unexpected, since a patient may have several associated risk factors [28]. For instance, a study reports that sympathetic activation by a low-salt diet increases the sensitivity to quinidine-induced QT prolongation [106]. The pharmacogenetic aspects discussed above in Section 3.3.3 raise particular concern, because patient with silent mutations will be particularly susceptible to drug-induced TdP.

3.4.2
Dose–Response Relationship for QT Prolongation

The statement sometimes found in the literature is that no clear-cut dose dependency can be observed for drug-induced QT prolongation or occurrence of TdP sometimes generates confusion. Actually, a recent study [107] confirms that QT prolongation by a wide dose range of dofetilide (a class III antiarrhythmic agent) is dose dependent. This has also been confirmed with sotalol in a pediatric population [108]. Thus, one can expect dose dependency for QT prolongation and likelihood of TdP with all hERG channel blockers, especially in case of drug interactions leading to very high plasma levels [31,105,109]. However, the fact that normal plasma levels may be associated with exaggerated increases in the QT interval and even occurrence of TdP led some to suggest a lack of dose dependency. Actually, several factors may reduce the repolarization reserve of a given subject, hence greatly increase the proarrhythmic potential of relatively low plasma levels to such an extent that establishing a dose–response relationship may be impossible, all the more so because drug-induced TdP are rare events.

In conclusion, QT prolongation by hERG channel blockers is per se dose dependent, whereas actual occurrence of TdP depends on the repolarization reserve, which is variable among subjects and over time. The exception to the rule are those hERG blockers with multiple, sometimes complementary electrophysiological actions, which lead to a bell-shaped concentration–response curve [109].

Concerning the higher sensitivity because of pharmacogenetic reasons, *in vitro* studies show that human K^+ channels with the same mutation detected in subjects with the long QT syndrome and expressed in *in vitro* systems are more sensitive than wild-type channels to blockade by certain drugs such as clarithromycin and sulfamethoxazole [50,110]. In silent carriers of the long QT syndrome, drug-induced TdPs may indeed be considered idiosyncratic and unpredictable with the current diagnostic standards, although knowledge of the underlying genetic defect would allow prediction of the possible occurrence of TdP.

A major problem in extrapolating results of *in vitro* electrophysiological studies (IC_{50} for inhibition of K^+ currents, IC_{50} for prolongation of APD, etc.) to the clinical setting is that the pharmacokinetic properties of the compound must be thoroughly studied to allow meaningful comparisons between *in vitro* and plasma concentrations. Plasma concentrations in humans should be considered along with the apparent volume of distribution and the metabolic pathways (metabolites may retain QT prolonging potential). The threshold concentration (or the IC_{50}) for hERG K^+ channel blockade *in vitro* may be higher than peak plasma concentrations achieved at therapeutic doses, but tissue concentrations (specifically, cardiac tissue concentrations) may exceed those found in plasma if the drug has a large volume of distribution. One example is provided by the comparison of pharmacodynamic and pharmacokinetic parameters of terfenadine and astemizole [31]. Terfenadine is readily metabolized to fexofenadine, which maintains good H_1-receptor blocking activity, but has no affect on the QT interval even at doses well above the therapeutic ones. Unmetabolized terfenadine plasma concentrations are usually below detection limits, but may become detectable in case of pharmacokinetic interactions with drugs known to inhibit the CYP3A4 isoenzyme, in case of overdosage, or concomitant hepatic disease. On the contrary, two of the main metabolites of astemizole (desmethylastemizole and norastemizole) retain the ability to block hERG K^+ currents at nanomolar concentrations [111]. In addition, the large volume of distribution of astemizole (indicating extensive tissue penetration: indeed, the concentration in cardiac muscle is estimated to be more than 100 times the plasma concentration [112,113]) and the long elimination half-life of desmethylastemizole (about 9.5 days) suggests a higher risk of potentially harmful effects on cardiac repolarization with astemizole than with terfenadine.

A recent study determined the myocardium to plasma concentration ratios of five antipsychotics and underscored the importance of interpreting hERG channel and electrophysiological data in conjunction with other pharmacokinetic parameters [114].

Finally, it is important to remember that, in case of racemic drugs, it is important to resort to stereoselective methods to detect possible differences in the actions of the test compounds at the desired target and hERG channels. For instance, a recent study detected differences between (*R*)- and (*S*)-methadone [115].

3.5
Regulatory Aspects of QT Prolongation

3.5.1
Regulatory Guidance Documents

In May 2005, the International Conference on Harmonization (ICH) adopted the final texts for clinical (ICH topic E14) [116] and nonclinical (ICH topic S7B) [117] strategies by which drugs should be investigated for their QT liability during their development. ICH is composed of representatives from regulatory authorities and industry associations in the United States, European Union and Japan (WHO, European Free Trade Area and Canada attended as observers (see http://www.ich.org). It should be acknowledged that the two documents represent a compromise among different positions held by regulatory agencies, as discussed in detail in recent reviews [118–123].

ICH E14 provides recommendations to sponsors concerning the design, conduct, analysis and interpretation of clinical studies to assess the potential of a drug to delay cardiac repolarization. Specifically, it calls for a clinical 'thorough QT/QTc study' (typically conducted in healthy volunteers), which is intended to determine whether a drug has a threshold pharmacological effect on cardiac repolarization, as detected by QT/QTc interval prolongation.

It is worth noting that, presently, *in silico* prediction of hERG liability is not considered in regulatory documents, although this is an area of intense investigation

(see Chapters 4 and 5) [36,124]. In brief, the main problems to be faced by *in silico* predictive models are as follows:
- hERG liability does indicate QT liability, but does not necessarily predict TdP liability;
- although hERG blockade is the key mechanism underlying QT prolongation, a single 'QT prolongation' pharmacophore does not exist because multiple mechanisms may be involved in cardiac repolarization (hERG blockade, impaired hERG trafficking, possible role of other ion channels);
- pharmacogenetic aspects: single-point mutations may increase drug affinity for hERG K^+ channels in a single patient, who will be particularly prone to develop arrhythmias.

3.5.2
Preclinical *In Vitro* and *In Vivo* Studies

The ICH S7B guideline [117] describes a nonclinical testing strategy to identify the potential of a test substance and its metabolites to delay ventricular repolarization and to relate the extent of delayed ventricular repolarization to the concentrations of a test substance and its metabolites.

Although several nonclinical approaches are referred to in the guideline, the core testing systems are *in vitro* assay of the rapid component of the delayed rectifier potassium current (I_{Kr}), that is the target of virtually all torsadogenic drugs, and *in vivo* study in laboratory animals (especially dog, monkey, swine, rabbit, ferret and guinea pig). Because no preclinical model has an absolute predictive value, the guideline proposes a concept of integrated risk assessment that can contribute to the design of clinical investigations and interpretation of their results. It also requires that *in vitro* I_{Kr} and *in vivo* QT assays, when performed for regulatory submission, should be conducted in compliance with good laboratory practice.

It is important to mention that the European Union (together with Japan) and the FDA have divergent views on the predictive value of nonclinical data: while the EU authorities do not regard the thorough QT/QTc clinical study as an infallible tool and acknowledge the predictive value of an integrated evaluation of nonclinical studies, the FDA tends to require a clinical study for all new drugs regardless of the nonclinical data, with few exceptions. The final version of the ICH E14 guideline has taken into account these differences by acknowledging that factors that *could reduce* the need for a thorough (and very expensive) QT/QTc study include the inability to conduct the study in healthy volunteers or patients, how the drug is studied and used (e.g. administered under continuous monitoring) and nonclinical data. It is evident that the wording 'could reduce' is subject to variable interpretations.

Several reviews are available on preclinical models [63,98,125–131], which will not be discussed in detail here. Table 3.3 provides a synopsis of some nonclinical models with advantages and drawbacks of each system.

The sensitivity of nonclinical tests (i.e. their ability to label as positive those drugs with a real risk of inducing QT prolongation in humans) is sufficiently good [132–137] but their specificity (i.e. their ability to label as negative those drugs carrying no risk) is not well established [120,138]. Verapamil is a notable example of a false positive: it blocks hERG K$^+$ channels, but is reported to have little potential to trigger TdP [139–141]. The existence of false 'nonclinical' negatives (i.e. compounds carrying clinical risk but resulting as negative in the integrated assessment of nonclinical tests) is a matter of debate, as discussed in a recent review [120].

It must be acknowledged that no single nonclinical model has an absolute predictive value or can be considered as a gold standard. Therefore, the use of several models facilitates accurate decision making and is recommended by most experts of the field [142]. There seems to be consensus, at least in Europe, that nonclinical data should guide not only the safety monitoring during early human tolerance studies but also the need for a thorough QT/QTc study.

Determining the IC$_{50}$ value for inhibition of K$^+$ conductance in native or cloned hERG channels was proposed as a primary test for screening purposes [125,140]; however, it is important to remember that hERG trafficking is emerging as a possible mechanism and that metabolites must be specifically tested in this *in vitro* test. Wible *et al.* [82] recently reported a novel high-throughput screen using the so-called hERG-Lite system, which correctly predicted hERG risk for all the 100 compounds tested with no false positives or negatives: the 50 hERG blockers were detected as drugs with hERG risk in the hERG-Lite assay and fell into two classes: B (for blocker) and C (for complex; block and trafficking inhibition).

In all safety studies, it is important to establish safety margins, and it should be noted that this aspect is not clearly addressed in the ICH S7B guideline. Redfern *et al.* [143] determined the relative value of nonclinical cardiac electrophysiological data (*in vitro* and *in vivo*) for predicting clinical risk of TdP. In addition to cardiac APD *in vitro* and QT prolongation *in vivo* in dogs, they used published data on hERG (or I_{Kr}) activity to compare QT effects with reports of TdP in humans. Their data on hERG or I_{Kr} IC50 (the concentration producing 50% inhibition of the channel) suggest that, in general, a 30-fold margin between hERG IC50 and peak free therapeutic plasma concentrations may be adequate to exclude a clinical effect on cardiac repolarization. However, the authors correctly emphasize that the acceptable safety margin should depend on the seriousness of the disease to be treated – ranging from 10-fold for a lethal disease to more than 100-fold for symptomatic treatment of a benign condition. Thus, if a safety margin is generally less than 30-fold, nonclinical data could be considered positive and require close monitoring in subsequent clinical development regardless of a negative thorough QT/QTc study.

When calculating this safety margin, there are three possible strategies. It can be the ratio of *in vitro* IC$_{50}$ for hERG inhibition to (a) EC50 (concentration required to elicit 50% response at the intended pharmacological target); (b) effective free plasma concentrations; or (c) effective myocardial concentration that is thus dependent on drug lipophilicity. Depending on the method of calculation, the three corresponding cardiac safety index values may considerably differ. For instance, Cavero and Crumb [140] elegantly discussed the case of terfenadine (selected as a drug that accumulates substantially in the myocardium: ratio of myocardium concentration of the drug and the concentration of the drug perfused at 1 µM for 2 h in guinea pig hearts: ∼260) compared to cetirizine (taken as a drug that has no particular affinity for the myocardium: ratio of myocardium concentration of the drug and the concentration of the drug perfused at 1 μM for 2 h in the guinea pig hearts: ~0.5). The results of their calculations indicate that the cardiac safety index can change greatly as a function of the parameter selected for its calculation. When comparative IC_{50} values are used, the cardiac safety index of terfenadine is less than 1, whereas that of cetirizine is 3. On the contrary, if the safety index is calculated by using its free plasma concentration (2 nM) achievable with therapeutic doses, terfenadine turns out to have an excellent safety margin (104 for terfenadine versus 28 for cetirizine). The latter result is misleading because it does not take into account the marked accumulation of terfenadine in the myocardium. When this factor (~260) is introduced in the calculation of the safety index, it clearly appears that terfenadine no longer possesses a reassuring cardiac safety index (0.4 versus 56 for cetirizine, which has a reasonable safety margin, because its cardiac concentration is slightly less than the plasma concentration: free plasma concentration 46 nM). However, for the sake of completeness, it should be pointed out that, with therapeutic doses, free plasma concentrations of terfenadine are usually lower than 2 nM, except in poor metabolizers and in case of metabolic inhibition (terfenadine is a CYP3A4 substrate and is metabolized into fexofenadine, having virtually no hERG liability). For a detailed discussion of these aspects, see a previous review [31].

The Langendorff-perfused female rabbit heart model measures action potential duration, conduction and also allows evaluation of the so-called TRIaD parameters (triangulation, reverse use dependence, instability and dispersion), as suggested by Hondeghem et al. [144–147], who provided important insights to understand why lengthening of the APD does not invariably correlate with a proarrhythmic effect (occurrence of EAD and TdP). The cardiac electrophysiological effects of 702 chemicals were studied [144] in the Langendorff preparation (rabbit perfused heart) and it was found that only those agents that caused lengthening of the APD associated with instability of APD, triangulation and reverse use dependence were proarrhythmic. Instability was an index of variability of APD and was defined as the difference (in ms) between the bottom and the top twenty-fifth percentile APD_{60}. It was elegantly visualized by creating Poincaré plots [APD of the nth action potential was plotted against the $(n-1)$th APD]: acetylsalicylic acid (a nontorsadogenic drug) did not change APD_{60} and thus did not show chaotic behavior in the Poincaré plot (identical action potentials project to a single point), whereas the reverse was true for haloperidol (a torsadogenic drug). Triangulation was an index of duration of fast repolarization and was the difference (in ms) between APD_{30} and APD_{90} (as triangulation increases, the action potential assumes a more triangular shape). Notably, prolongation of the action potential plateau without instability or triangulation was antiarrhythmic rather than proarrhythmic. These observations challenge the use of APD lengthening as a surrogate marker for proarrhythmia and strengthens the notion that the whole spectrum of pharmacological properties of a compound must be considered to draw conclusions on its proarrhythmic potential *in vivo*. Hondeghem and Hoffmann further refined the rabbit perfused rat model (SCREENIT system [145]) and proposed it as a tool in a preclinical proarrhythmia test battery. In a recent review criticizing the FDA policy focusing on QT interval measurement in the 'thorough study', Hondeghem [147] states that 'TRIaD appears to be a direct cause of torsades, as not a single agent is known to be torsasogenic without TRIaD. QT prolongation cannot make such claim'. To get some insight into the reason why action potential prolongation per se without triangulation is insufficient to cause TdP, Guo et al. [148], using rabbit ventricular myocardium in an elegant study, showed that action potential triangulation accelerates I_{Ca-L} recovery from inactivation, leading to instability of the cell membrane potential during repolarization, a condition capable of initiating TdP.

Since transmural dispersion of depolarization seems to be a good predictor of the torsadogenic potential of a compound, Antzelevitch's group [149,150] developed another proarrhythmia model based on the transmural heterogeneity of the expression of cardiac ion channels. Using a left ventricular wedge preparation of the canine heart, transmembrane action potentials from epicardial and M-regions were simultaneously measured. The data support the hypothesis that the risk for the development of TdP is related to the increase in transmural dispersion of repolarization rather than to prolongation of the QT interval. The usefulness of this approach is indicated by drugs such as pentobarbital, which reduce transmural dispersion of repolarization and can decrease the likelihood of TdP, despite their ability to prolong the QT interval [150].

Recently, Bottino et al. [151] developed mathematical models that use hERG IC_{50} data and APD results measured from dog Purkinje fibers to predict drug interaction with other cardiac ion currents and dispersion of repolarization in transmural ECG. As regards cardiac Purkinje cells, a recent study suggested that rabbit cells might be more suitable for screening purposes because they possess a lower repolarization reserve than canine cells [152].

In *in vivo* studies, an important issue is the quality of ECG recordings and the methods used for the analysis. A few years ago, a survey reported on the practice in the pharmaceutical industry to assess the potential for QT prolongation and concluded that 'the majority view in the industry is not necessarily best practice' [153]. After this publication, a lot of work has been done to improve the performance of *in vivo* models, although it should be acknowledged that their cost (together with some ethical considerations on the rational use of animals) makes them unsuitable for large-scale screening.

The dog is among the most popular species for *in vivo* studies [153]: this preference seems to be justified by the fact that the heart rate range is closer to that of humans (smaller animals have much higher baseline heart rates) and by the similarities in ionic determinants of Purkinje fiber and ventricular action potentials. However, it is important to remember that T wave morphology and RR intervals are highly variable in dogs and must be analyzed by an expert veterinarian. Investigators should be aware that the accuracy of the Bazett's or Fredericia's formulas to correct the QT interval for heart rate is most uncertain in animals because formulas used in humans cannot be readily applied to animals. So far, the dog is the species in which the problem of QT correction has been investigated in detail by several groups [154–158]. Although it is now recognized that the Bazett's formula is even more unsatisfactory in dogs, it should be noted that at present none of the proposed formulas can be endorsed as the ideal formula, because each has advantages and drawbacks (see [156]). If accuracy of the correction formula is a particular concern for the investigator, one could attempt to derive a heart rate correction algorithm for each animal from the QT/RR relationship.

Since the conscious dog model is relatively expensive and is not suitable for screening a large number of compounds, calculation of the sample size to have sufficient statistical power is an important issue. With the Van de Water's formula, it has been suggested that a sample size of five to eight dogs is sufficient to detect changes in QTc of 10% [159]. Finally, these studies should always include a positive control (i.e. a drug known to prolong the QT interval) [153] and a vehicle-injected control (to assess the intrinsic variability of the method used to correct the QT interval) [160].

3.5.3
Clinical Studies

Several specialized reviews on detection of QT liability in the clinical development phase have already been published and the reader is referred to these publications [63]. Guidelines of the International Society for Holter and Noninvasive Electrocardiology (ISHNE) for electrocardiographic evaluation of drug-related QT prolongation are also available [161]. The main issues related to measurement of the QT interval in clinical studies are summarized in Table 3.4. An important aspect is the calculation of sample size: usually 40–60 subjects per treatment arm are required, implying high cost [162–164].

The ICH E14 guideline [116] calls for a clinical 'thorough QT/QTc study' (typically conducted in healthy volunteers), which is intended to determine whether a drug has a threshold pharmacological effect on cardiac repolarization, as detected by QT/QTc interval prolongation. Its recommendations are generally applicable not only to new drugs that have systemic bioavailability, but also to approved drugs when one explores a new dose, route of administration or target population that may result in an increased risk. The guideline provides for exceptions when this study may not be required (e.g. products with highly localized distribution and those administered topically and not absorbed). Uncertainties remain on the application of the guideline to biotechnology products, vaccines, enzyme replacement therapy and blood or plasma products.

In general, it is difficult to determine whether there is an effect on the mean QT/QTc interval that is so small as to be of no clinical consequence. However, drugs that prolong the mean QT/QTc interval by around 5 ms or less do not appear to cause TdP. On that basis, the threshold level of regulatory concern is 5 ms.

The thorough QT/QTc study should be conducted early during clinical development, but should not be the first study, because it is necessary to have basic clinical data for its design and conduct, including tolerability, pharmacokinetics and activity of the metabolites. The thorough QT/QTc study should be well controlled and designed to deal with potential bias (use of randomization, appropriate blinding and concurrent placebo control group). Because of its critical role in determining the

Table 3.4 Problems encountered in conducting the 'thorough QT/QTc study' in adult healthy volunteers.

Subject variability	High intraindividual variability in QTc values (circadian and seasonal variation; law of regression to the mean) High interindividual variability in QTc values (males versus females) Unknown prevalence in the general population of subjects carrying silent mutations in the ion channels responsible for cardiac repolarization (these subjects have normal QTc value but reduced repolarization reserve) Variability in the individual metabolic capacity for a given drug
Study design	40–60 subjects per treatment arm (high cost) Need for a placebo control arm and a positive control group Need to study more than one dose level of the test compound to assess dose dependency and the safety margin of therapeutic doses
Measurement of QT interval	Definition of the end of the T wave. Changes in T wave morphology and occurrence of U waves (these may be important warning signs and precede the occurrence of TdP) Errors in manual measurement in QT interval Variability in the heart rate (need to correct the QT value for heart rate) Lack of reliable correlation between readings from Holter recordings and standard ECG Lack of standardization of automated ECG readings (computerized methods are often unreliable) Need for a central core laboratory to analyze data
Pharmacokinetics	Timing of ECG measurements with respect to peak/steady-state drug plasma concentrations; consider also volume of distribution and tissue concentrations Need to consider plasma concentrations of both parent drug and its metabolites (especially if they maintain the ability to block K^+ channels) Need for enantioselective methods to monitor plasma concentrations of racemic compounds
Data analysis and interpretation (i.e. extrapolation to the target patient population)	Different formulas are used to correct duration of the QT interval for heart rate; some formulas may overcorrect at high heart rates and undercorrect at low heart rates (e.g. Bazett's formula): consider that with some formulas (e.g. Bazett's) a QTc increase of 4–5 ms may result from measurement bias Need for an individualized correction formula

(Continued)

> Extrapolation of the result to the patient population (especially children or patients with hepatic or renal diseases affecting drug metabolism)
> The 5-ms threshold (and a confidence interval beyond 10 ms) is a conventional threshold established by the ICH E14 guideline: it calls for but does not necessarily imply a clinically significant risk of torsades de pointes or a negative risk/benefit balance
> A *mean* QTc value during drug treatment is not representative of those patients having important QTc increases (QTc > 500 ms or ΔQTc > 60 ms).
> These outliers must be carefully considered to determine whether their high QTc value is due to chance or they have clinically silent long QT syndrome

intensity of ECG data collection during later stages of development, this study should guarantee a high degree of confidence in its ability to detect differences of clinical significance. This can be greatly enhanced by the use of a concurrent positive control group to establish assay sensitivity. The positive control should have an effect on the mean QT/QTc interval of about 5 ms (i.e. the threshold level of regulatory concern). Detecting the positive control's effect will establish the ability of the study to detect such an effect of the study drug. If an investigational drug belongs to a chemical or pharmacological class that has been associated with QT/QTc prolongation, a positive control selected from other members of the same class should be considered to permit a comparison of effect sizes, preferably at equipotent therapeutic doses. In other cases, moxifloxacin is frequently used as a comparator [165].

Figure 3.2 provides exemplifications of possible results of a thorough QT/QTc study indicating point estimates (and one-sided 95% confidence intervals) of the placebo-corrected effect on the QTc interval of the test drug and the positive control. According to the ICH E14 guideline, the positive control should have an effect of about 5 ms, but no bounds are set for the one-sided 95% confidence intervals. For the test drug, a negative 'thorough QT/QTc study' is the one in which the upper bound of the 95% one-sided confidence interval for the largest time-matched mean effect of the drug on the QTc interval excludes 10 ms [116]. This definition is chosen to provide reasonable assurance that the mean effect of the study drug on the QT/QTc interval is not greater than around 5 ms. Three examples of negative drugs are provided in the figure.

When the largest time-matched difference exceeds the threshold, the study is termed 'positive'. A positive study influences the evaluations carried out during later stages of drug development but does not imply that the drug is proarrhythmic. As with other data, the presence of outliers should also be explored. A positive 'thorough QT/QTc study' will almost always call for an expanded ECG safety evaluation during later stages of drug development.

The guideline acknowledges that 'there could be very unusual cases in which the "thorough QT/QTc study" is negative but the available nonclinical data are strongly positive (e.g. hERG positive at low concentrations and *in vivo* animal model results

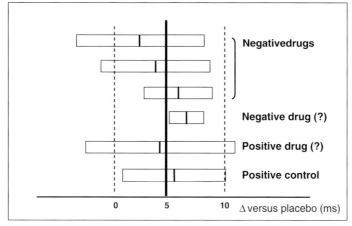

Figure 3.2 Exemplification of possible results of a thorough QT/QTc study indicating point estimates (and 95% confidence intervals) of the placebo-corrected effect on the QTc interval. According to the ICH E14 guideline, the positive control (frequent use of moxifloxacin) should have an affect on the mean QT/QTc interval of about 5 ms (i.e. the threshold of regulatory concern), but no bounds are set for the one-sided 95% confidence intervals. For the test drug, a negative 'thorough QT/QTc study' is one in which the upper bound of the 95% one-sided confidence interval for the largest time-matched mean effect of the drug on the QTc interval excludes 10 ms. Three examples are provided in the figure. However, the guideline leaves some questions unanswered, especially on the designation of a study as positive or negative when there is a disparity between the mean effect (above or below the 5-ms threshold) and the 95% upper confidence bound (below or above the 10-ms threshold, respectively). In the two examples with a question mark, one drug is defined as negative because the 95% upper confidence bound is below 10 s, although the mean effect is likely to be above the 5-ms threshold of concern. The other drug (and indeed this case may be more frequent) is defined as positive because the 95% upper confidence bound goes beyond 10 ms, although the mean effect is below 5 ms (in this case, the large dispersion leaves uncertainty on the true value).

that are strongly positive). If this discrepancy cannot be explained by other data, and the drug is in a class of pharmacological concern, expanded ECG safety evaluation during later stages of drug development might be appropriate'.

Although the definition of a negative thorough QT/QTc study result is based on changes in central tendency as described above, the guideline also clearly states that categorical analyses can provide relevant information on clinical risk assessment. Categorical analyses of QT/QTc interval data are based on the number and percentage of patients meeting or exceeding some predefined upper limit value. Clinically noteworthy QT/QTc interval changes might be defined in terms of absolute QT/QTc intervals or changes from baseline. There is no consensus concerning the choice of upper limit values for absolute QT/QTc interval and changes from baseline. While lower limits increase the false-positive rate, higher limits increase the risk of failure to detect a signal of concern. The guideline states that multiple analyses using different limits are a reasonable approach to this uncertainty (e.g. QTc interval >450, >480 or >500 ms; change from baseline in QTc interval >30 or >60 ms).

It should be noted that the guideline leaves some questions unanswered, especially on the designation of a study as positive or negative when there is a disparity between the mean effect (above or below the 5-ms threshold) and the 95% upper confidence bound (respectively, below or above the 10-ms threshold), as indicated in Figure 3.2.

Another unanswered question is the evaluation of the effect on cardiac repolarization of oncologic drugs, for which the thorough QT/QTc study in volunteers cannot be performed [166]. In these cases, not only central tendency (i.e. mean QTc increase) and proportion of outliers but also other findings such as syncope, ventricular tachyarrhythmias and other cardiac effects should be more closely defined.

Finally, the ICH E14 guideline does not address the possible consequences of QT/QTc shortening, which may be associated with significant arrhythmias [5,167]. Digitalis intoxication is a known example of drug-induced short QT syndrome associated with polymorphic ventricular tachycardia [168]. It should be acknowledged that this is an area of active research and no guidelines can be put forward at the moment.

3.6
Conclusions

The field of drug-induced QT prolongation is rapidly evolving, thanks to significant technological advances in the last 5 years. Although no straightforward and absolute rules can be put forward because the risk/benefit balance is unique for each drug, a number of principles have emerged and are summarized below.

- Early and reliable detection of QT liability (hERG liability) has become an important aspect in drug development and requires expertise from different areas.
- QT prolongation is a surrogate marker used in cardiac safety studies, but several lines of experimental evidence indicate that it is a poor surrogate of TdP: a number of drugs, because of their complex pharmacological profile, can prolong the QTc with a relatively low proarrhythmic risk (e.g. amiodarone).
- Compounds of the same therapeutic class may profoundly differ as to QT prolonging potential; in addition, the assessment of the proarrhythmic risk must take into account pharmacokinetic aspects including metabolism, since a very low baseline proarrhythmic risk may become clinically important in case of drug interactions leading to higher than expected plasma levels.
- An effect on QTc is an important argument to stop drug development, since even a QT prolongation of a few milliseconds may be a reason for not obtaining marketing authorization by regulatory agencies, unless unique benefits in a serious disease are proven.
- The final documents adopted by the International Conference on Harmonization for clinical (ICH topic E14) and nonclinical (ICH topic S7B) strategies to investigate

QT liability during drug development are important milestones but leave many questions unanswered: for example, do 'false nonclinical negatives' with a significant clinical torsadogenic risk exist? Are there reliable markers of the true torsadogenic potential of a new drug?
- The predictive value of an integrated evaluation of nonclinical studies should be acknowledged, all the more so because the thorough QT/QTc clinical study is not an infallible tool and relies on a surrogate marker of toxicity.

Abbreviations

APD	action potential duration
EMEA	European Medicines Agency
FDA	Food and Drug Administration
hERG	human ether-a-go-go-related gene
IC_{50}	concentration producing 50% inhibition
ICH	International Conference on Harmonization
ISHNE	International Society for Holter and Noninvasive Electrocardiology
LQTS	long QT syndrome
TdP	*torsades de pointes*
TRIaD	triangulation, reverse use dependence, instability and dispersion
WHO	World Health Organization

Acknowledgment

The author's research activity on this topic is supported by a grant from the University of Bologna.

References

1. Witchel, H.J. (2007) The hERG potassium channel as a therapeutic target. *Expert Opinion on Therapeutic Targets*, **11**, 321–336.
2. Wolpert, C., Schimpf, R., Veltmann, C., Giustetto, C., Gaita, F. and Borggrefe, M. (2005) Clinical characteristics and treatment of short QT syndrome. *Expert Review of Cardiovascular Therapy*, **3**, 611–617.
3. Wolpert, C., Schimpf, R., Giustetto, C., Antzelevitch, C., Cordeiro, J., Dumaine, R., Brugada, R., Hong, K., Bauersfeld, U., Gaita, F. and Borggrefe, M. (2005) Further insights into the effect of quinidine in short QT syndrome caused by a mutation in HERG. *Journal of Cardiovascular Electrophysiology*, **16**, 54–58.
4. Giustetto, C., Di Monte, F., Wolpert, C., Borggrefe, M., Schimpf, R., Sbragia, P., Leone, G., Maury, P., Anttonen, O., Haissaguerre, M. and Gaita, F. (2006) Short QT syndrome: clinical findings and diagnostic–therapeutic implications. *European Heart Journal*, **27**, 2440–2447.
5. Brugada, R., Hong, K., Dumaine, R., Cordeiro, J., Gaita, F., Borggrefe, M.,

Menendez, T.M., Brugada, J., Pollevick, G.D., Wolpert, C., Burashnikov, E., Matsuo, K., Wu, Y.S., Guerchicoff, A., Bianchi, F., Giustetto, C., Schimpf, R., Brugada, P. and Antzelevitch, C. (2004) Sudden death associated with short-QT syndrome linked to mutations in HERG. *Circulation*, **109**, 30–35.

6 Antzelevitch, C., Pollevick, G.D., Cordeiro, J.M., Casis, O., Sanguinetti, M.C., Aizawa, Y., Guerchicoff, A., Pfeiffer, R., Oliva, A., Wollnik, B., Gelber, P., Bonaros, E.P. Jr, Burashnikov, E., Wu, Y., Sargent, J.D., Schickel, S., Oberheiden, R., Bhatia, A., Hsu, L.F., Haissaguerre, M., Schimpf, R., Borggrefe, M. and Wolpert, C. (2007) Loss-of-function mutations in the cardiac calcium channel underlie a new clinical entity characterized by ST-segment elevation, short QT intervals, and sudden cardiac death. *Circulation*, **115**, 442–449.

7 Shivkumar, K. and Valderrabano, M. (2004) Use of potassium channel openers for pharmacologic modulation of cardiac excitability. *Journal of Cardiovascular Electrophysiology*, **15**, 821–823.

8 Malykhina, A.P., Shoeb, F. and Akbarali, H.I. (2002) Fenamate-induced enhancement of heterologously expressed HERG currents in Xenopus oocytes. *European Journal of Pharmacology*, **452**, 269–277.

9 Zeng, H., Lozinskaya, I.M., Lin, Z., Willette, R.N., Brooks, D.P. and Xu, X. (2006) Mallotoxin is a novel human ether-a-go-go-related gene (hERG) potassium channel activator. *The Journal of Pharmacology and Experimental Therapeutics*, **319**, 957–962.

10 Gullo, F., Ales, E., Rosati, B., Lecchi, M., Masi, A., Guasti, L., Cano-Abad, M.F., Arcangeli, A., Lopez, M.G. and Wanke, E. (2003) ERG K+ channel blockade enhances firing and epinephrine secretion in rat chromaffin cells: the missing link to LQT2-related sudden death? *FASEB Journal*, **17**, 330–332.

11 Furlan, F., Taccola, G., Grandolfo, M., Guasti, L., Arcangeli, A., Nistri, A. and Ballerini, L. (2007) ERG conductance expression modulates the excitability of ventral horn GABAergic interneurons that control rhythmic oscillations in the developing mouse spinal cord. *Journal of Neuroscience*, **27**, 919–928.

12 Furlan, F., Guasti, L., Avossa, D., Becchetti, A., Cilia, E., Ballerini, L. and Arcangeli, A. (2005) Interneurons transiently express the ERG K+ channels during development of mouse spinal networks *in vitro*. *Neuroscience*, **135**, 1179–1192.

13 Chiesa, N., Rosati, B., Arcangeli, A., Olivotto, M. and Wanke, E. (1997) A novel role for HERG K+ channels: spike-frequency adaptation. *The Journal of Physiology*, **501** (Pt 2), 313–318.

14 Conti, M. (2004) Targeting K+ channels for cancer therapy. *Journal of Experimental Therapeutics and Oncology*, **4**, 161–166.

15 Camacho, J. (2006) Ether a go-go potassium channels and cancer. *Cancer Letters*, **233**, 1–9.

16 Wang, H., Zhang, Y., Cao, L., Han, H., Wang, J., Yang, B., Nattel, S. and Wang, Z. (2002) HERG K+ channel, a regulator of tumor cell apoptosis and proliferation. *Cancer Research*, **62**, 4843–4848.

17 Chen, S.Z., Jiang, M. and Zhen, Y.S. (2005) HERG K+ channel expression-related chemosensitivity in cancer cells and its modulation by erythromycin. *Cancer Chemotherapy and Pharmacology*, **56**, 212–220.

18 Shao, X.D., Wu, K.C., Hao, Z.M., Hong, L., Zhang, J. and Fan, D.M. (2005) The potent inhibitory effects of cisapride, a specific blocker for human ether-a-go-go-related gene (HERG) channel, on gastric cancer cells. *Cancer Biology & Therapy*, **4**, 295–301.

19 Farias, L.M., Ocana, D.B., Diaz, L., Larrea, F., Avila-Chavez, E., Cadena, A., Hinojosa, L.M., Lara, G., Villanueva, L.A., Vargas, C., Hernandez-Gallegos, E.,

Camacho-Arroyo, I., Duenas-Gonzalez, A., Perez-Cardenas, E., Pardo, L.A., Morales, A., Taja-Chayeb, L., Escamilla, J., Sanchez-Pena, C. and Camacho, J. (2004) Ether a go-go potassium channels as human cervical cancer markers. *Cancer Research*, **64**, 6996–7001.

20 Zhu, Y., Golden, C.M., Ye, J., Wang, X.Y., Akbarali, H.I. and Huizinga, J.D. (2003) ERG K^+ currents regulate pacemaker activity in ICC. *The American Journal of Physiology*, **285**, G1249–G1258

21 McKay, C.M. and Huizinga, J.D. (2006) Muscarinic regulation of ether-a-go-go-related gene K+ currents in interstitial cells of Cajal. *The Journal of Pharmacology and Experimental Therapeutics*, **319**, 1112–1123.

22 McKay, C.M., Ye, J. and Huizinga, J.D. (2006) Characterization of depolarization-evoked ERG K currents in interstitial cells of Cajal. *Neurogastroenterology and Motility*, **18**, 324–333.

23 Farrelly, A.M., Ro, S., Callaghan, B.P., Khoyi, M.A., Fleming, N., Horowitz, B., Sanders, K.M. and Keef, K.D. (2003) Expression and function of KCNH2 (HERG) in the human jejunum. *The American Journal of Physiology*, **284**, G883–G895

24 Shoeb, F., Malykhina, A.P. and Akbarali, H.I. (2003) Cloning and functional characterization of the smooth muscle ether-a-go-go-related gene K+ channel. Potential role of a conserved amino acid substitution in the S4 region. *The Journal of Biological Chemistry*, **278**, 2503–2514.

25 Parr, E., Pozo, M.J., Horowitz, B., Nelson, M.T. and Mawe, G.M. (2003) ERG K+ channels modulate the electrical and contractile activities of gallbladder smooth muscle. *The American Journal of Physiology*, G398–G492

26 Ohya, S., Asakura, K., Muraki, K., Watanabe, M. and Imaizumi, Y. (2002) Molecular and functional characterization of ERG, KCNQ, and KCNE subtypes in rat stomach smooth muscle. *The American Journal of Physiology*, **282**, G277–G287

27 Mohammad, S., Zhou, Z., Gong, Q. and January, C.T. (1997) Blockage of the HERG human cardiac K^+ channel by the gastrointestinal prokinetic agent cisapride. *The American Journal of Physiology*, **273**, H2534–H2538

28 Roden, D.M. (2006) Long QT syndrome: reduced repolarization reserve and the genetic link. *Journal of Internal Medicine*, **259**, 59–69.

29 Schwartz, P.J. (2006) The congenital long QT syndromes from genotype to phenotype: clinical implications. *Journal of Internal Medicine*, **259**, 39–47.

30 Shimizu, W. (2005) The long QT syndrome: therapeutic implications of a genetic diagnosis. *Cardiovascular Research*, **67**, 347–356.

31 De Ponti, F., Poluzzi, E. and Montanaro, N. (2000) QT-interval prolongation by non-cardiac drugs: lessons to be learned from recent experience. *European Journal of Clinical Pharmacology*, **56**, 1–18.

32 Darpo, B. (2001) Spectrum of drugs prolonging QT interval and the incidence of torsades de pointes. *European Heart Journal Supplements*, **3**, K70–K80

33 De Ponti, F., Poluzzi, E., Montanaro, N. and Ferguson, J. (2000) QTc and psychotropic drugs. *Lancet*, **356**, 75–76.

34 De Ponti, F., Poluzzi, E., Vaccheri, A., Bergman, U., Bjerrum, L., Ferguson, J., Frenz, K.J., McManus, P., Schubert, I., Selke, G., Terzis-Vaslamatzis, G. and Montanaro, N. (2002) Non-antiarrhythmic drugs prolonging the QT interval: considerable use in seven countries. *British Journal of Clinical Pharmacology*, **54**, 171–177.

35 Shah, R.R. (2006) Can pharmacogenetics help rescue drugs withdrawn from the market? *Pharmacogenomics*, **7**, 889–908.

36 Recanatini, M., Poluzzi, E., Masetti, M., Cavalli, A. and De Ponti, F. (2005) QT

prolongation through hERG K$^+$ channel blockade: current knowledge and strategies for the early prediction during drug development. *Medicinal Research Reviews*, **25**, 133–166.

37 Suessbrich, H., Waldegger, S., Lang, F. and Busch, A.E. (1996) Blockade of HERG channels expressed in Xenopus oocytes by the histamine receptor antagonists terfenadine and astemizole. *FEBS Letters*, **385**, 77–80.

38 Taglialatela, M., Pannaccione, A., Castaldo, P., Giorgio, G., Zhou, Z., January, C.T., Genovese, A., Marone, G. and Annunziato, L. (1998) Molecular basis for the lack of HERG K+ channel block-related cardiotoxicity by the H1 receptor blocker cetirizine compared with other second-generation antihistamines. *Molecular Pharmacology*, **54**, 113–121.

39 Bischoff, U., Schmidt, C., Netzer, R. and Pongs, O. (2000) Effects of fluoroquinolones on HERG currents. *European Journal of Pharmacology*, **406**, 341–343.

40 Kang, J., Wang, L., Chen, X.L., Triggle, D.J. and Rampe, D. (2001) Interactions of a series of fluoroquinolone antibacterial drugs with the human cardiac K$^+$ channel HERG. *Molecular Pharmacology*, **59**, 122–126.

41 Tie, H., Walker, B.D., Valenzuela, S.M., Breit, S.N. and Campbell, T.J. (2000) The heart of psychotropic drug therapy. *Lancet*, **355**, 1825.

42 Suessbrich, H., Schonherr, R., Heinemann, S.H., Attali, B., Lang, F. and Busch, A.E. (1997) The inhibitory effect of the antipsychotic drug haloperidol on HERG potassium channels expressed in Xenopus oocytes. *British Journal of Pharmacology*, **120**, 968–974.

43 Kennerdy, A., Thomas, P. and Sheridan, D.J. (1989) Generalized seizures as the presentation of flecainide toxicity. *European Heart Journal*, **10**, 950–954.

44 Platia, E.V., Estes, M., Heine, D.L., Griffith, L.S., Garan, H., Ruskin, J.N. and Reid, P.R. (1985) Flecainide: electrophysiologic and antiarrhythmic properties in refractory ventricular tachycardia. *The American Journal of Cardiology*, **55**, 956–962.

45 Bruelle, P., LeFrant, J.Y., de La Coussaye, J.E., Peray, P.A., Desch, G., Sassine, A. and Eledjam, J.J. (1996) Comparative electrophysiologic and hemodynamic effects of several amide local anesthetic drugs in anesthetized dogs. *Anesthesia and Analgesia*, **82**, 648–656.

46 Siebrands, C.C., Schmitt, N. and Friederich, P. (2005) Local anesthetic interaction with human ether-a-go-go-related gene (HERG) channels: role of aromatic amino acids Y652 and F656. *Anesthesiology*, **103**, 102–112.

47 O'Leary, M.E. (2001) Inhibition of human ether-a-go-go potassium channels by cocaine. *Molecular Pharmacology*, **59**, 269–277.

48 Zhang, S., Rajamani, S., Chen, Y., Gong, Q., Rong, Y., Zhou, Z., Ruoho, A. and January, C.T. (2001) Cocaine blocks HERG, but not KvLQT1+minK, potassium channels. *Molecular Pharmacology*, **59**, 1069–1076.

49 Guo, J., Gang, H. and Zhang, S. (2006) Molecular determinants of cocaine block of human ether-a-go-go-related gene potassium channels. *The Journal of Pharmacology and Experimental Therapeutics*, **317**, 865–874.

50 Sesti, F., Abbott, G.W., Wei, J., Murray, K.T., Saksena, S., Schwartz, P.J., Priori, S.G., Roden, D.M., George, A.L.J. and Goldstein, S.A. (2000) A common polymorphism associated with antibiotic-induced cardiac arrhythmia. *Proceedings of the National Academy of Sciences of the United States of America*, **97**, 10613–10618.

51 Gamouras, G.A., Monir, G., Plunkitt, K., Gursoy, S. and Dreifus, L.S. (2000) Cocaine abuse: repolarization abnormalities and ventricular arrhythmias. *The American Journal of the Medical Sciences*, **320**, 9–12.

52. Parker, R.B., Perry, G.Y., Horan, L.G. and Flowers, N.C. (1999) Comparative effects of sodium bicarbonate and sodium chloride on reversing cocaine-induced changes in the electrocardiogram. *Journal of Cardiovascular Pharmacology*, **34**, 864–869.
53. Wang, R.Y. (1999) pH-dependent cocaine-induced cardiotoxicity. *American Journal of Emergency Medicine*, **17**, 364–369.
54. Fedida, D., Chen, F.S. and Zhang, X. (1998) The 1997 Stevenson Award Lecture. Cardiac K^+ channel gating: cloned delayed rectifier mechanisms and drug modulation. *Canadian Journal of Physiology and Pharmacology*, **76**, 77–89.
55. Nattel, S. (1999) The molecular and ionic specificity of antiarrhythmic drug actions. *Journal of Cardiovascular Electrophysiology*, **10**, 272–282.
56. Vandenberg, J.I., Walker, B.D. and Campbell, T.J. (2001) HERG K^+ channels: friend and foe. *Trends in Pharmacological Sciences*, **22**, 240–246.
57. Shieh, C.C., Coghlan, M., Sullivan, J.P. and Gopalakrishnan, M. (2000) Potassium channels: molecular defects, diseases, and therapeutic opportunities. *Pharmacological Reviews*, **52**, 557–594.
58. Towbin, J.A. and Vatta, M. (2001) Molecular biology and the prolonged QT syndromes. *The American Journal of Medicine*, **110**, 385–398.
59. Tseng, G.N. (2001) I_{Kr}: the hERG channel. *Journal of Molecular and Cellular Cardiology*, **33**, 835–849.
60. Roden, D.M. and George, A.L.J. (1997) Structure and function of cardiac sodium and potassium channels. *The American Journal of Physiology*, **273**, H511–H525.
61. Kuryshev, Y.A., Wang, L., Wible, B.A., Wan, X. and Ficker, E. (2006) Antimony-based antileishmanial compounds prolong the cardiac action potential by an increase in cardiac calcium currents. *Molecular Pharmacology*, **69**, 1216–1225.
62. Sheridan, D.J. (2000) Drug-induced proarrhythmic effects: assessment of changes in QT interval. *British Journal of Clinical Pharmacology*, **50**, 297–302.
63. Malik, M. and Camm, A.J. (2001) Evaluation of drug-induced QT interval prolongation: implications for drug approval and labelling. *Drug Safety*, **24**, 323–351.
64. Zitron, E., Scholz, E., Owen, R.W., Luck, S., Kiesecker, C., Thomas, D., Kathofer, S., Niroomand, F., Kiehn, J., Kreye, V.A., Katus, H.A., Schoels, W. and Karle, C.A. (2005) QTc prolongation by grapefruit juice and its potential pharmacological basis: HERG channel blockade by flavonoids. *Circulation*, **111**, 835–838.
65. Guizy, M., Arias, C., David, M., Gonzalez, T. and Valenzuela, C. (2005) {Omega}-3 and {omega}-6 polyunsaturated fatty acids block HERG channels. *The American Journal of Physiology*, **289**, C1251–C1260.
66. Mitcheson, J.S., Chen, J., Lin, M., Culberson, C. and Sanguinetti, M.C. (2000) A structural basis for drug-induced long QT syndrome. *Proceedings of the National Academy of Sciences of the United States of America*, **97**, 12329–12333.
67. Mitcheson, J.S. (2003) Drug binding to HERG channels: evidence for a 'non-aromatic' binding site for fluvoxamine. *British Journal of Pharmacology*, **139**, 883–884.
68. Stansfeld, P.J., Sutcliffe, M.J. and Mitcheson, J.S. (2006) Molecular mechanisms for drug interactions with hERG that cause long QT syndrome. *Expert Opinion on Drug Metabolism and Toxicology*, **2**, 81–94.
69. Karagueuzian, H.S. (2000) Acquired long QT syndromes and the risk of proarrhythmia. *Journal of Cardiovascular Electrophysiology*, **11**, 1298.
70. Pacher, P., Ungvari, Z., Nanasi, P.P., Furst, S. and Kecskemeti, V. (1999) Speculations on difference between tricyclic and selective serotonin reuptake inhibitor antidepressants on their

cardiac effects. Is there any? *Current Medicinal Chemistry*, **6**, 469–480.

71 Nau, C., Seaver, M., Wang, S.Y. and Wang, G.K. (2000) Block of human heart hH1 sodium channels by amitriptyline. *The Journal of Pharmacology and Experimental Therapeutics*, **292**, 1015–1023.

72 Isenberg, G. and Tamargo, J. (1985) Effect of imipramine on calcium and potassium currents in isolated bovine ventricular myocytes. *European Journal of Pharmacology*, **108**, 121–131.

73 Delpon, E., Tamargo, J. and Sanchez-Chapula, J. (1991) Further characterization of the effects of imipramine on plateau membrane currents in guinea-pig ventricular myocytes. *Naunyn-Schmiedeberg's Archives of Pharmacology*, **344**, 645–652.

74 Reich, M.R., Ohad, D.G., Overall, K.L. and Dunham, A.E. (2000) Electrocardiographic assessment of antianxiety medication in dogs and correlation with serum drug concentration [published erratum appears in Journal of American Veterinary Medical Association 2000;216 (12):1936]. *Journal of the American Veterinary Medical Association*, **216**, 1571–1575.

75 Cordes, J.S., Sun, Z., Lloyd, D.B., Bradley, J.A., Opsahl, A.C., Tengowski, M.W., Chen, X. and Zhou, J. (2005) Pentamidine reduces hERG expression to prolong the QT interval. *British Journal of Pharmacology*, **145**, 15–23.

76 Kuryshev, Y.A., Ficker, E., Wang, L., Hawryluk, P., Dennis, A.T., Wible, B.A., Brown, A.M., Kang, J., Chen, X.L., Sawamura, K., Reynolds, W. and Rampe, D. (2005) Pentamidine-induced long QT syndrome and block of hERG trafficking. *The Journal of Pharmacology and Experimental Therapeutics*, **312**, 316–323.

77 Ficker, E., Kuryshev, Y.A., Dennis, A.T., Obejero-Paz, C., Wang, L., Hawryluk, P., Wible, B.A. and Brown, A.M. (2004) Mechanisms of arsenic-induced prolongation of cardiac repolarization. *Molecular Pharmacology*, **66**, 33–44.

78 Katchman, A.N., Koerner, J., Tosaka, T., Woosley, R.L. and Ebert, S.N. (2006) Comparative evaluation of HERG currents and QT intervals following challenge with suspected torsadogenic and nontorsadogenic drugs. *The Journal of Pharmacology and Experimental Therapeutics*, **316**, 1098–1106.

79 Wang, L., Wible, B.A., Wan, X. and Ficker, E. (2007) Cardiac glycosides as novel inhibitors of human ether-a-go-go-related gene channel trafficking. *The Journal of Pharmacology and Experimental Therapeutics*, **320**, 525–534.

80 Rajamani, S., Eckhardt, L.L., Valdivia, C.R., Klemens, C.A., Gillman, B.M., Anderson, C.L., Holzem, K.M., Delisle, B.P., Anson, B.D., Makielski, J.C. and January, C.T. (2006) Drug-induced long QT syndrome: hERG K^+ channel block and disruption of protein trafficking by fluoxetine and norfluoxetine. *British Journal of Pharmacology*, **149**, 481–489.

81 Hancox, J.C. and Mitcheson, J.S. (2006) Combined hERG channel inhibition and disruption of trafficking in drug-induced long QT syndrome by fluoxetine: a case-study in cardiac safety pharmacology. *British Journal of Pharmacology*, **149**, 457–459.

82 Wible, B.A., Hawryluk, P., Ficker, E., Kuryshev, Y.A., Kirsch, G. and Brown, A.M. (2005) HERG-Lite: a novel comprehensive high-throughput screen for drug-induced hERG risk. *Journal of Pharmacological and Toxicological Methods*, **52**, 136–145.

83 Anderson, C.L., Delisle, B.P., Anson, B.D., Kilby, J.A., Will, M.L., Tester, D.J., Gong, Q., Zhou, Z., Ackerman, M.J. and January, C.T. (2006) Most LQT2 mutations reduce Kv11.1 (hERG) current by a class 2 (trafficking-deficient) mechanism. *Circulation*, **113**, 365–373.

84 Rajamani, S., Anderson, C.L., Anson, B.D. and January, C.T. (2002) Pharmacological rescue of human K(+)

channel long-QT2 mutations: human ether-a-go-go-related gene rescue without block. *Circulation*, **105**, 2830–2835.
85. Liu, K., Yang, T., Viswanathan, P.C. and Roden, D.M. (2005) New mechanism contributing to drug-induced arrhythmia: rescue of a misprocessed LQT3 mutant. *Circulation*, **112**, 3239–3246.
86. Kannankeril, P.J. and Roden, D.M. (2007) Drug-induced long QT and torsade de pointes: recent advances. *Current Opinion in Cardiology*, **22**, 39–43.
87. Nakajima, T., Hayashi, K., Viswanathan, P.C., Kim, M.Y., Anghelescu, M., Barksdale, K.A., Shuai, W., Balser, J.R. and Kupershmidt, S. (2007) HERG is protected from pharmacological block by {alpha}-1,2-glucosyltransferase function. *The Journal of Biological Chemistry*, **282**, 5506–5513.
88. Judson, R.S., Salisbury, B.A., Reed, C.R. and Ackerman, M.J. (2006) Pharmacogenetic issues in thorough QT trials. *Molecular Diagnosis and Therapy*, **10**, 153–162.
89. Anantharam, A., Markowitz, S.M. and Abbott, G.W. (2003) Pharmacogenetic considerations in diseases of cardiac ion channels. *The Journal of Pharmacology and Experimental Therapeutics*, **307**, 831–838.
90. Park, K.H., Kwok, S.M., Sharon, C., Berga, R. and Sesti, F. (2003) N-Glycosylation-dependent block is a novel mechanism for drug-induced cardiac arrhythmia. *FASEB Journal*, **17**, 2308–2309.
91. Morganroth, J. (2001) Focus on issues in measuring and interpreting changes in the QTc interval duration. *European Heart Journal Supplements*, **3**, K105–K111.
92. Paltoo, B., O'Donoghue, S. and Mousavi, M.S. (2001) Levofloxacin induced polymorphic ventricular tachycardia with normal QT interval. *Pace-Pacing and Clinical Electrophysiology*, **24**, 895–897.
93. Beyerbach, D.M., Kovacs, R.J., Dmitrienko, A.A., Rebhun, D.M. and Zipes, D.P. (2007) Heart rate-corrected QT interval in men increases during winter months. *Heart Rhythm*, **4**, 277–281.
94. Malik, M. (2001) Problems of heart rate correction in assessment of drug-induced QT interval prolongation. *Journal of Cardiovascular Electrophysiology*, **12**, 411–420.
95. Bednar, M.M., Harrigan, E.P., Anziano, R.J., Camm, A.J. and Ruskin, J.N. (2001) The QT interval. *Progress in Cardiovascular Diseases*, **43**, 1–45.
96. Funck-Brentano, C. and Jaillon, P. (1993) Rate-corrected QT interval: techniques and limitations. *The American Journal of Cardiology*, **72**, 17B–22B.
97. Aytemir, K., Maarouf, N., Gallagher, M.M., Yap, Y.G., Waktare, J.E. and Malik, M. (1999) Comparison of formulae for heart rate correction of QT interval in exercise electrocardiograms. *Pace-Pacing and Clinical Electrophysiology*, **22**, 1397–1401.
98. Cavero, I., Mestre, M., Guillon, J.M. and Crumb, W. (2000) Drugs that prolong QT interval as an unwanted effect: assessing their likelihood of inducing hazardous cardiac dysrhythmias. *Expert Opinion on Pharmacotherapy*, **1**, 947–973.
99. Puddu, P.E., Jouve, R., Mariotti, S., Giampaoli, S., Lanti, M., Reale, A. and Menotti, A. (1988) Evaluation of 10 QT prediction formulas in 881 middle-aged men from the seven countries study: emphasis on the cubic root Fridericia's equation. *Journal of Electrocardiology*, **21**, 219–229.
100. Donger, C., Denjoy, I., Berthet, M., Neyroud, N., Cruaud, C., Bennaceur, M., Chivoret, G., Schwartz, K., Coumel, P. and Guicheney, P. (1997) KVLQT1 C-terminal missense mutation causes a forme fruste long-QT syndrome. *Circulation*, **96**, 2778–2781.
101. Priori, S.G., Napolitano, C. and Schwartz, P.J. (1999) Low penetrance in the long-QT syndrome: clinical impact. *Circulation*, **99**, 529–533.

102 Chiang, C.E. and Roden, D.M. (2000) The long QT syndromes: genetic basis and clinical implications. *Journal of the American College of Cardiology*, **36**, 1–12.

103 Napolitano, C., Schwartz, P.J., Brown, A.M., Ronchetti, E., Bianchi, L., Pinnavaia, A., Acquaro, G. and Priori, S.G. (2000) Evidence for a cardiac ion channel mutation underlying drug-induced QT prolongation and life-threatening arrhythmias. *Journal of Cardiovascular Electrophysiology*, **11**, 691–696.

104 Lande, G., Kyndt, F., Baro, I., Chabannes, D., Boisseau, P., Pony, J.C., Escande, D. and Le Marec, H. (2001) Dynamic analysis of the QT interval in long QT1 syndrome patients with a normal phenotype. *European Heart Journal*, **22**, 410–422.

105 Roden, D.M. (1998) Taking the "idio" out of "idiosyncratic": predicting torsades de pointes. *Pace-Pacing and Clinical Electrophysiology*, **21**, 1029–1034.

106 Darbar, D., Fromm, M.F., Dellorto, S. and Roden, D.M. (2001) Sympathetic activation enhances QT prolongation by quinidine. *Journal of Cardiovascular Electrophysiology*, **12**, 9–14.

107 Allen, M.J., Nichols, D.J. and Oliver, S.D. (2000) The pharmacokinetics and pharmacodynamics of oral dofetilide after twice daily and three times daily dosing. *British Journal of Clinical Pharmacology*, **50**, 247–253.

108 Saul, J.P., Ross, B., Schaffer, M.S., Beerman, L., Melikian, A.P., Shi, J., Williams, J., Barbey, J.T., Jin, J. and Hinderling, P.H. (2001) Pharmacokinetics and pharmacodynamics of sotalol in a pediatric population with supraventricular and ventricular tachyarrhythmia. *Clinical Pharmacology and Therapeutics*, **69**, 145–157.

109 Roden, D.M. (2000) Acquired long QT syndromes and the risk of proarrhythmia. *Journal of Cardiovascular Electrophysiology*, **11**, 938–940.

110 Abbott, G.W., Sesti, F., Splawski, I., Buck, M.E., Lehmann, M.H., Timothy, K. W., Keating, M.T. and Goldstein, S.A. (1999) MiRP1 forms IKr potassium channels with HERG and is associated with cardiac arrhythmia. *Cell*, **97**, 175–187.

111 Zhou, Z., Vorperian, V.R., Gong, Q., Zhang, S. and January, C.T. (1999) Block of HERG potassium channels by the antihistamine astemizole and its metabolites desmethylastemizole and norastemizole. *Journal of Cardiovascular Electrophysiology*, **10**, 836–843.

112 Michiels, M., Van Peer, A., Woestenborghs, R. and Heykants, J. (1986) Pharmacokinetics and tissue distribution of astemizole in the dog. *Drug Development Research*, **8**, 53–62.

113 Sugiyama, A., Aye, N.N., Katahira, S., Saitoh, M., Hagihara, A., Matsubara, Y. and Hashimoto, K. (1997) Effects of nonsedating antihistamine, astemizole, on the *in situ* canine heart assessed by cardiohemodynamic and monophasic action potential monitoring. *Toxicology and Applied Pharmacology*, **143**, 89–95.

114 Titier, K., Canal, M., Deridet, E., Abouelfath, A., Gromb, S., Molimard, M. and Moore, N. (2004) Determination of myocardium to plasma concentration ratios of five antipsychotic drugs: comparison with their ability to induce arrhythmia and sudden death in clinical practice. *Toxicology and Applied Pharmacology*, **199**, 52–60.

115 Eap, C.B., Crettol, S., Rougier, J.S., Schlapfer, J., Sintra, G.L., Deglon, J.J., Besson, J., Croquette-Krokar, M., Carrupt, P.A. and Abriel, H. (2007) Stereoselective block of hERG channel by (S)-methadone and QT interval prolongation in CYP2B6 slow metabolizers. *Clinical Pharmacology and Therapeutics*, **81**, 719–728.

116 ICH Topic E14 – The Clinical Evaluation of QT/QTc Interval Prolongation and Proarrhythmic Potential for Non-Antiarrhythmic Drugs. EMEA.(2005) http://emea.eu.int/pdfs/human/ich/000204en.pdf. (last accessed 2/25/2007).

117 ICH Topic S7B – The Nonclinical Evaluation of the Potential for delayed Ventricular Repolarization QT Interval Prolongation) by Human Pharmaceuticals. EMEA.(2005)http://www.emea.europa.eu/pdfs/human/ich/042302en.pdf. (last accessed 2/25/2007).

118 Cavero, I. and Crumb, W. (2006) Moving towards better predictors of drug-induced Torsade de Pointes *Expert Opinion on Drug Safety*, **5**, 335–340.

119 Cavero, I. and Crumb, W. (2006) Safety Pharmacology Society: 5th Annual Meeting. *Expert Opinion on Drug Safety*, **5**, 181–185.

120 Cavero, I. and Crumb, W. (2005) The use of electrocardiograms in clinical trials: a public discussion of the proposed ICH E14 regulatory guidance *Expert Opinion on Drug Safety*, **4**, 795–799.

121 Cavero, I. and Crumb, W. (2005) ICH S7B draft guideline on the non-clinical strategy for testing delayed cardiac repolarisation risk of drugs: a critical analysis. *Expert Opinion on Drug Safety*, **4**, 509–530.

122 Shah, R.R. (2005) Drugs, QTc interval prolongation and final ICH E14 guideline: an important milestone with challenges ahead. *Drug Safety*, **28**, 1009–1028.

123 Shah, R.R. (2005) Drugs, QT interval prolongation and ICH E14: the need to get it right. *Drug Safety*, **28**, 115–125.

124 Cavalli, A., Poluzzi, E., De Ponti, F. and Recanatini, M. (2002) Toward a pharmacophore for drugs inducing the long QT syndrome: insights from a CoMFA study of HERG K^+ channel blockers. *Journal of Medicinal Chemistry*, **45**, 3844–3853.

125 Netzer, R., Ebneth, A., Bischoff, U. and Pongs, O. (2001) Screening lead compounds for QT interval prolongation. *Drug Discovery Today*, **6**, 78–84.

126 Haverkamp, W., Breithardt, G., Camm, A.J., Janse, M.J., Rosen, M.R., Antzelevitch, C., Escande, D., Franz, M., Malik, M., Moss, A. and Shah, R. (2000) The potential for QT prolongation and proarrhythmia by non-antiarrhythmic drugs: clinical and regulatory implications. Report on a policy conference of the European Society of Cardiology. *European Heart Journal*, **21**, 1216–1231.

127 Champeroux, P., Martel, E., Vannier, C., Blanc, V., Leguennec, J.Y., Fowler, J. and Richard, S. (2000) The preclinical assessment of the risk for QT interval prolongation. *Therapie*, **55**, 101–109.

128 Finlayson, K., Witchel, H.J., McCulloch, J. and Sharkey, J. (2004) Acquired QT interval prolongation and HERG: implications for drug discovery and development. *European Journal of Pharmacology*, **500**, 129–142.

129 Hoffmann, P. and Warner, B. (2006) Are hERG channel inhibition and QT interval prolongation all there is in drug-induced torsadogenesis? A review of emerging trends. *Journal of Pharmacological and Toxicological Methods*, **53**, 87–105.

130 Carlsson, L. (2006) *in vitro* and *in vivo* models for testing arrhythmogenesis in drugs. *Journal of Internal Medicine*, **259**, 70–80.

131 Thomsen, M.B., Matz, J., Volders, P.G. and Vos, M.A. (2006) Assessing the proarrhythmic potential of drugs: current status of models and surrogate parameters of torsades de pointes arrhythmias. *Pharmacology & Therapeutics*, **112**, 150–170.

132 Davis, A.S. (1998) The pre-clinical assessment of QT interval prolongation: a comparison of *in vitro* and *in vivo* methods. *Human & Experimental Toxicology*, **17**, 677–680.

133 Carlsson, L. (2001) Drug-induced torsade de pointes: the perspectives of industry. *European Heart Journal Supplements*, **3**, K114–K120.

134 Gintant, G.A., Limberis, J.T., McDermott, J.S., Wegner, C.D. and Cox, B.F. (2001) The canine Purkinje fiber: an *in vitro* model system for acquired long

QT syndrome and drug-induced arrhythmogenesis. *Journal of Cardiovascular Pharmacology*, 37, 607–618.

135 Hamlin, R.L., Cruze, C.A., Mittelstadt, S.W., Kijtawornrat, A., Keene, B.W., Roche, B.M., Nakayama, T., Nakayama, H., Hamlin, D.M. and Arnold, T. (2004) Sensitivity and specificity of isolated perfused guinea pig heart to test for drug-induced lengthening of QTc. *Journal of Pharmacological and Toxicological Methods*, 49, 15–23.

136 Aubert, M., Osterwalder, R., Wagner, B., Parrilla, I., Cavero, I., Doessegger, L. and Ertel, E.A. (2006) Evaluation of the rabbit Purkinje fibre assay as an *in vitro* tool for assessing the risk of drug-induced torsades de pointes in humans. *Drug Safety*, 29, 237–254.

137 Lu, H.R., Vlaminckx, E., Van de, W.A., Rohrbacher, J., Hermans, A. and Gallacher, D.J. (2006) In-vitro experimental models for the risk assessment of antibiotic-induced QT prolongation. *European Journal of Pharmacology*, 553, 229–239.

138 Cavero, I., Yan, G.X., Lux, R. and Steinhoff, U. (2007) Safety pharmacology and prolongation of the QT interval. *Journal of Electrocardiology*, 40S58–S61.

139 Zhang, S., Zhou, Z., Gong, Q., Makielski, J.C. and January, C.T. (1999) Mechanism of block and identification of the verapamil binding domain to HERG potassium channels. *Circulation Research*, 84, 989–998.

140 Cavero, I. and Crumb, W.J. (2001) Native and cloned ion channels from human heart: laboratory models for evaluating the cardiac safety of new drugs. *European Heart Journal Supplements*, 3, K53–K63.

141 Lacerda, A.E., Kramer, J., Shen, K.Z., Thomas, D. and Brown, A.M. (2001) Comparison of block among cloned cardiac potassium channels by non-antiarrhythmic drugs. *European Heart Journal Supplements*, 3, K23–K30.

142 Hanson, L.A., Bass, A.S., Gintant, G., Mittelstadt, S., Rampe, D. and Thomas, K. (2006) ILSI-HESI cardiovascular safety subcommittee initiative: evaluation of three non-clinical models of QT prolongation. *Journal of Pharmacological and Toxicological Methods*, 54, 116–129.

143 Redfern, W.S., Carlsson, L., Davis, A.S., Lynch, W.G., MacKenzie, I., Palethorpe, S., Siegl, P.K., Strang, I., Sullivan, A.T., Wallis, R., Camm, A.J. and Hammond, T.G. (2003) Relationships between preclinical cardiac electrophysiology, clinical QT interval prolongation and torsade de pointes for a broad range of drugs: evidence for a provisional safety margin in drug development. *Cardiovascular Research*, 58, 32–45.

144 Hondeghem, L.M., Carlsson, L. and Duker, G. (2001) Instability and triangulation of the action potential predict serious proarrhythmia, but action potential duration prolongation is antiarrhythmic. *Circulation*, 103, 2004–2013.

145 Hondeghem, L.M. and Hoffmann, P. (2003) Blinded test in isolated female rabbit heart reliably identifies action potential duration prolongation and proarrhythmic drugs: importance of triangulation, reverse use dependence, and instability. *Journal of Cardiovascular Pharmacology*, 41, 14–24.

146 Shah, R.R. and Hondeghem, L.M. (2005) Refining detection of drug-induced proarrhythmia: QT interval and TRIaD. *Heart Rhythm*, 2, 758–772.

147 Hondeghem, L.M. (2006) Thorough QT/QTc not so thorough: removes torsadogenic predictors from the T-wave, incriminates safe drugs, and misses profibrillatory drugs. *Journal of Cardiovascular Electrophysiology*, 17, 337–340.

148 Guo, D., Zhao, X., Wu, Y., Liu, T., Kowey, P.R. and Yan, G.X. (2007) L-type calcium current reactivation contributes to arrhythmogenesis associated with action

potential triangulation. *Journal of Cardiovascular Electrophysiology*, **18**, 196–203.
149 Weissenburger, J., Nesterenko, V.V. and Antzelevitch, C. (2000) Transmural heterogeneity of ventricular repolarization under baseline and long QT conditions in the canine heart *in vivo*: torsades de pointes develops with halothane but not pentobarbital anesthesia. *Journal of Cardiovascular Electrophysiology*, **11**, 290–304.
150 Antzelevitch, C. (2004) Arrhythmogenic mechanisms of QT prolonging drugs: is QT prolongation really the problem? *Journal of Electrocardiology*, **37**, (Suppl.),15–24.
151 Bottino, D., Penland, R.C., stamps, A., Traebert, M., Dumotier, B., Georgiva, A., Helmlinger, G. and Lett, G.S. (2006) Preclinical cardiac safety assessment of pharmaceutical compounds using an integrated systems-based computer model of the heart. *Progress in Biophysics and Molecular Biology*, **90**, 414–443.
152 Dumaine, R. and Cordeiro, J.M. (2007) Comparison of K^+ currents in cardiac Purkinje cells isolated from rabbit and dog. *Journal of Molecular and Cellular Cardiology*, **42**, 378–389.
153 Hammond, T.G., Carlsson, L., Davis, A. S., Lynch, W.G., MacKenzie, I., Redfern, W.S., Sullivan, A.T. and Camm, A.J. (2001) Methods of collecting and evaluating non-clinical cardiac electrophysiology data in the pharmaceutical industry: results of an international survey. *Cardiovascular Research*, **49**, 741–750.
154 Van de Water, A., Verheyen, J., Xhonneux, R. and Reneman, R.S. (1989) An improved method to correct the QT interval of the electrocardiogram for changes in heart rate. *Journal of Pharmacological Methods*, **22**, 207–217.
155 Matsunaga, T., Mitsui, T., Harada, T., Inokuma, M., Murano, H. and Shibutani, Y. (1997) QT corrected for heart rate and relation between QT and RR intervals in beagle dogs. *Journal of Pharmacological and Toxicological Methods*, **38**, 201–209.
156 Spence, S., Soper, K., Hoe, C.M. and Coleman, J. (1998) The heart rate-corrected QT interval of conscious beagle dogs: a formula based on analysis of covariance. *Toxicological Sciences*, **45**, 247–258.
157 Classen, W., Altmann, B., Gretener, P., Souppart, C., Skelton-Stroud, P. and Krinke, G. (1999) Differential effects of orally versus parenterally administered qinghaosu derivative artemether in dogs. *Experimental and Toxicologic Pathology*, **51**, 507–516.
158 Davis, A.S. and Middleton, B.J. (1999) Relationship between QT interval and heart rate in Alderley Park beagles. *The Veterinary Record*, **145**, 248–250.
159 Gralinski, M.R. (2000) The assessment of potential for QT interval prolongation with new pharmaceuticals: impact on drug development. *Journal of Pharmacological and Toxicological Methods*, **43**, 91–99.
160 Escande, D. (2001) Inhibition of repolarizing ionic currents by drugs. *European Heart Journal Supplements*, **3**, K17–K22
161 Moss, A.J., Zareba, W., Benhorin, J., Couderc, J.P., Kennedy, H., Locati-Heilbron, E. and Maison-Blanche, P. (2001) ISHNE Guidelines for Electrocardiographic Evaluation of Drug-related QT Prolongation and Other Alterations in Ventricular Repolarization: Task Force Summary. A Report of the Task Force of the International Society for Holter and Noninvasive Electrocardiology (ISHNE), Committee on Ventricular Repolarization. *Annals of Noninvasive Electrocardiology*, **6**, 333–341.
162 Malik, M., Hnatkova, K., Batchvarov, V., Gang, Y., Smetana, P. and Camm, A.J. (2004) Sample size, power calculations, and their implications for the cost of thorough studies of drug induced QT

interval prolongation. *Pace-Pacing and Clinical Electrophysiology*, **27**, 1659–1669.

163 Darpo, B., Nebout, T. and Sager, P.T. (2006) Clinical evaluation of QT/QTc prolongation and proarrhythmic potential for nonantiarrhythmic drugs: the International Conference on Harmonization of Technical Requirements for Registration of Pharmaceuticals for Human Use E14 guideline. *Journal of Clinical Pharmacology*, **46**, 498–507.

164 Morganroth, J. (2007) Cardiac repolarization and the safety of new drugs defined by electrocardiography. *Clinical Pharmacology and Therapeutics*, **81**, 108–113.

165 Malhotra, B.K., Glue, P., Sweeney, K., Anziano, R., Mancuso, J. and Wicker, P. (2007) Thorough QT study with recommended and supratherapeutic doses of tolterodine. *Clinical Pharmacology and Therapeutics*, **81**, 377–385.

166 Morganroth, J. (2007) *Evaluation of the effect on cardiac repolarization (QTc interval) of oncologic drugs*. Ernst Schering Research Foundation Workshop, Springer-Verlag, Berlin, 171–184.

167 Gaita, F., Giustetto, C., Bianchi, F., Wolpert, C., Schimpf, R., Riccardi, R., Grossi, S., Richiardi, E. and Borggrefe, M. (2003) Short QT Syndrome: a familial cause of sudden death. *Circulation*, **108**, 965–970.

168 Garberoglio, L., Giustetto, C., Wolpert, C. and Gaita, F. (2007) Is acquired short QT due to digitalis intoxication responsible for malignant ventricular arrhythmias? *Journal of Electrocardiology*, **40**, 43–46.

4
hERG Channel Physiology and Drug-Binding Structure–Activity Relationships

Sarah Dalibalta, John S. Mitcheson

4.1
Introduction

Potassium channels, a structurally diverse group of membrane proteins, are found in almost all cells and involved in a multitude of physiological functions. The six transmembrane helix voltage-gated (Kv) channels are a family of K^+ channels that respond to changes in the electrical field by changing conformation and selectively allowing K^+ ions to pass through the membrane. This property allows them to play a critical role in shaping action potentials and modulating the electrical activity of excitable cells [1]. The hERG (human ether-a-go-go-related gene) potassium channel plays a critical role in the repolarization of the cardiac action potential. In the past decade or more, it has become apparent that not only are hERG channels unusually susceptible to pharmacological block, but also block has the potential to cause cardiac arrhythmias and sudden death. Determining if new compounds have the potential to block hERG as well as finding ways to overcome this problem has become an important goal in drug discovery and development.

The hERG channel (also referred to as human *erg1*, KCNH2 or Kv11.1) belongs to the EAG family of voltage-gated potassium channels. In 1995, it was identified by two separate research groups as the pore-forming α-subunit of channels that conduct the rapid delayed rectifier K^+ current (I_{Kr}), an important current for normal cardiac electrical activity [2,3]. Later, two other genes, *erg2* and *erg3*, were shown to be expressed in both rat and human nervous systems and to contribute to the firing properties of neurons [4]. Since then, hERG channels have been found in multiple tissues including neuronal, smooth muscle, neuroendocrine glands and some cancerous tissues [5,6]. Their physiological roles in these tissues are yet to be fully elucidated. In the heart, where its role has been best characterized, hERG is important for ventricular and atrial myocyte action potential repolarization, and it also helps in conducting current during the diastolic depolarization in pacemaker cells [7,8]. In this chapter, we focus on its role in the ventricular action potential, the adverse consequences for cardiac function that arise from blocking the channel and recent insights into the molecular basis for drug interactions with hERG.

4.2
hERG Channel Structure

hERG is a tetramer formed by the coassembly of four identical subunits, each comprises six transmembrane α-helical segments denoted S1–S6 (Figure 4.1a). Segments S1–S4 form the voltage-sensing domain while S5 and S6 comprise the pore-forming domain. A peptide linker between S5 and S6 forms the outer mouth of the channel and the selectivity filter. Channel subunits also include large intracellular amino (NH_2) and carboxy (COOH) termini that are likely to play a major role in regulating channel function [9–12]. A crystal structure for the hERG channel has not yet been determined. However, crystal structures of the bacterial channels KcsA, MthK, KvAP, KirBac1.1 [13–16] and more recently that of the mammalian channel Kv1.2 [17] have provided dramatic insight into the structural basis of K^+ channel function.

The pore of K^+ channels has four identical subunits that coassemble to form an inverted 'teepee' structure [13] with the selectivity filter on the extracellular side (Figure 4.1b). Although the extracellular side of the channel is dedicated to K^+ selectivity, the intracellular side, which is lined by the inner helices, can alter its conformation to regulate ion movement through the pore. In the closed state, the inner helices are straight and form a bundle at the cytoplasmic entrance to the pore that is impermeable to K^+ ions. The crystal structure (at 3.3 Å resolution) of a calcium-gated potassium channel MthK from the archeon *Methanobacterium thermoautotrophicum* was the first open K^+ channel structure to be solved. Instead of a bundle crossing, the inner helices are splayed open creating a wide opening. Figure 4.1b shows homology models of hERG based on the KcsA and MthK structures. Comparison of the KcsA and MthK structures shows that the inner helices of MthK bend at approximately 30° at a hinge point located just below the selectivity filter. The gating hinge corresponds to Gly83 in MthK and Gly99 in KcsA (Gly648 in hERG). The structure of a bacterial Kv channel homologue, KvAP, was solved soon afterward [15]. The open pore of the KvAP channel is nearly as wide as the MthK channel, and its inner helices seem to bend at the conserved glycine hinge, suggesting that the activation gating movements are similar in ligand- and voltage-gated channels. The crystal structure of KvAP includes the voltage sensors, although these domains are distorted by the crystallization process. More recently, the crystal structure of a mammalian Kv channel, Kv1.2, was determined in the open state [17]. Kv1.2 is a mammalian homologue of *Shaker*, the first K^+ channel to be cloned and the most intensively studied. A feature of the Kv1–4 family of channels is the presence of a highly conserved Pro-X-Pro motif in the S6 inner helices. This motif is thought to create a kink in S6 forming a flexible hinge region for channel gating [18]. Mutating these prolines to an alanine or a glycine has resulted in either nonfunctional channels or channels with drastically altered gating properties, hence indicating their importance in normal channel opening [18].

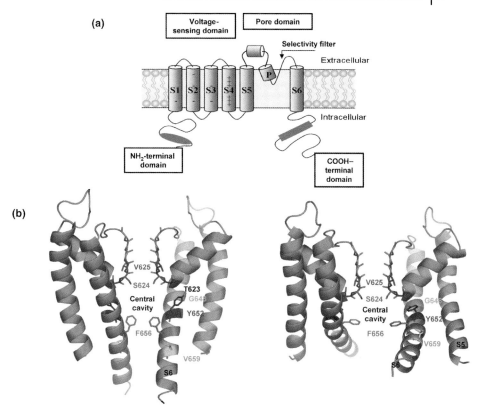

Figure 4.1 hERG channel secondary structure (a) and homology models of the pore (b). (a) hERG channels are tetramers consisting of four identical subunits with six transmembrane spanning helical segments labeled S1–S6. S1–S4 forms the voltage-sensing domain, while S5–S6 forms the pore domain of the channel. The majority of drugs that directly modify hERG function bind to the channel pore. Between S5 and S6 is a peptide loop that contains the pore helix and the selectivity filter. N- and C-terminal domains are on the cytoplasmic side of the channel. (b) Homology models of the pore of hERG based on KcsA (left panel) and KvAP (right panel). Only two of the four subunits are shown. The KcsA-based model represents the closed state, whereas the KvAP-based structure is an open-state model. A kink at Gly648 in S6 allows the inner helices to be splayed open. The residues lining the pore helix and central cavity involved in drug binding are labeled and illustrated by sticks. These residues are Val625, Ser624, Gly648, Tyr652, Phe656 and Val659.

4.3
hERG Potassium Channels and the Cardiac Action Potential

The ventricular cardiac action potential is characterized by five phases and is shaped by the complex interplay of a variety of Na^+, Ca^{2+} and K^+ currents (Figure 4.2). The distinct voltage-dependent properties of hERG channels [19,20] govern the time course of I_{Kr} and the manner in which it contributes to the outward K^+ current during the repolarization phase of the cardiac action potential. The opening and closing of

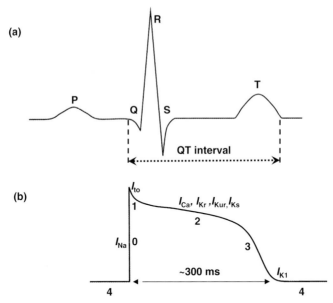

Figure 4.2 Cartoon representation of an ECG trace and ventricular cardiac action potential. (a) A representation of an ECG trace with its five typical deflections (PQRST) arising from the spread of electrical activity through the heart. The QRS wave denotes the ventricular depolarization, while the T wave represents ventricular repolarization. The QT interval therefore estimates the duration of a ventricular action potential. (b) Schematic of the five phases of a ventricular action potential. Phase 0 is the rapid depolarization phase due to a large influx of Na^+ ions (I_{Na}). Phase 1 occurs with the inactivation of Na^+ channels and the onset of transient outward (repolarizing) currents (I_{to}) forming a 'notch' in the action potential. These two phases correspond to the QRS wave of the ventricular action potential. Phase 2 is the plateau phase and corresponds to the ST segment of the ECG. In this phase, there is a balance between the inward Ca^{2+} current (I_{Ca}) and the outward delayed rectifier K^+ currents (I_{Kur}, I_{Ks} and I_{Kr}). hERG is the pore-forming subunit of I_{Kr} channels. During phase 3, the K^+ currents during phase 2 plus the inward rectifier I_{K1} repolarize the membrane potential. The spread of action potential repolarization produces the T wave on the ECG. Subsequently, phase 4 is a return to the resting membrane potential.

ion channels is referred to as gating. hERG channels exhibit two types of gating processes involving specific channel structures: activation gating that leads to opening of the channel at the cytoplasmic entrance to the pore and C-type inactivation gating that leads to a collapse of the selectivity filter and loss of pore conductance. These gating processes are coupled to the movement of the voltage-sensing domain and strongly depend on voltage and time. K^+ ion movement across the pore may occur only when both gates are open. At positive potentials, during the early stages of the plateau phase of a cardiac action potential, hERG currents are small. This is because the activation gates open with a relatively slow time course, while the inactivation gates close with a faster time course. However, during the later stages of the plateau phase, when membrane potentials return to more negative values, the inactivation gates reopen (recovery from inactivation) allowing the outward current to flow through the channels. The resulting delayed increase in K$^+$ current is critical for regulating the duration of the action potential and for the normal electrical and contractile function of the heart. Loss of I_{Kr} function, by mutations in the hERG channel or by drug blockade, prolongs the action potential duration, which can lead to Ca^{2+} overload and early depolarizations, which may trigger lethal cardiac arrhythmias [21].

There is evidence that hERG is associated with auxiliary subunits that may modify its function. In particular, a class of subunits with a single transmembrane spanning helix topology, minK (KCNE1) and MiRP1 (KCNE2), are reported to coassemble with hERG in heterologous systems [22]. Coassembly of *MiRP1* with hERG was initially reported to accelerate I_{Kr} deactivation and the rate of onset of drug block [23]. However, the level of expression of MiRP1 in atrial and ventricular muscle appears to be low [24,25], and recent studies report minor effects of MiRP1 expression on hERG biophysical and pharmacological properties [26]. Although MiRP1 may contribute to hERG channel complexes within specialized cells of the cardiac conduction system (e.g. Purkinje cells [25]), it is unlikely to be a critical component in most cells of the heart [6].

4.4
Mutations in hERG Are Associated with Cardiac Arrhythmias

Long QT syndrome (LQTS) is a collection of diseases that are diagnosed on the basis of prolongation of the QT interval on a body surface electrocardiogram (Figure 4.2) and are associated with a greatly increased risk of cardiac arrhythmias. The QT interval gives a crude measure of the duration of the ventricular action potential. Genetic defects in hERG cause LQT2, an inherited form of LQTS [27]. The most characteristic arrhythmia in LQTS patients is *torsades de pointes* (TdP), a distinctive polymorphic ventricular tachycardia characterized by QRS waves that oscillate in amplitude about the isoelectric line on the ECG. TdP can degenerate into ventricular fibrillation and cause sudden cardiac death.

Over 200 mutations in hERG have been shown to be associated with LQTS and most of them cluster around the pore and the surrounding membrane spanning regions. These mutations lead to a loss of channel function by altered gating or a dominant negative suppression of hERG function when mutant and wild-type (WT) subunits are coexpressed [28,29]. The majority of these mutations are missense mutations that appear to disrupt protein folding and trafficking to the cell surface; hence, the protein is retained in endoplasmic reticulum (ER) as core-glycosylated protein [30]. These trafficking-defective mutations have been identified throughout much of the hERG protein, including the pore, transmembrane domains, N- and C-termini, and the cyclic nucleotide-binding domain. A number of these mutations have been rescued by reducing the temperature or by pharmacological block, which is thought to stabilize intermediate states of the protein, thus promoting native protein folding and its export to the surface. Interestingly, from a potential therapeutic perspective, fexofenadine, a compound, which is closely related to the hERG blocker terfenadine but has no associated risks of arrhythmias, appears to rescue certain mutations (G601S and N470D) and permits protein trafficking to the membrane [31]. However, other compounds appear to disrupt channel trafficking, providing yet another mechanism for acquired LQTS [32,33].

The net result of most of these LQT-associated mutations is loss of hERG channel function leading to a reduction in current magnitude and a prolongation of cardiac action potential duration with varying severity. On the contrary, a gain-of-function mutation in the S5-P loop region of hERG (N588K) has been identified and shown to reduce inactivation and increase the magnitude of repolarizing current, thus accelerating cardiac repolarization [34]. This short QT syndrome (SQTS) can also result in life-threatening atrial and ventricular arrhythmias. Thus, mutations in hERG that cause gain or loss of channel function are associated with cardiac dysfunction.

4.5
Acquired Long QT Syndrome

Acquired LQTS is more common than the inherited form and most frequently results from the action of medications that block the pore of the channel or disrupt channel trafficking to the cell surface. hERG is blocked by many antiarrhythmic agents as well as a wide range of noncardiovascular drugs. These compounds share the potential to cause drug-induced QT prolongation and TdP [35]. Drug-induced arrhythmias are rare. However, even a low risk is deemed unacceptable by drug regulatory authorities, particularly for medications against non-life-threatening diseases. In the 1990s, noncardiac drugs such as terfenadine and astemizole (antihistamines), cisapride (gastrointestinal agent), sertindole (antipsychotic) and grepafloxacin (antibiotic), to name a few, were taken off the market or their use was restricted because of their association with arrhythmias. The diverse nature of QT-prolonging drugs, both structurally and therapeutically, has made it difficult to use conventional drug design approaches to evade QT prolongation. The risk of this side effect is recognized as a significant obstacle for the pharmaceutical industry as it strives to develop safer drugs. Drugs are now tested for hERG blockade at much earlier stages of the development program. High-throughput, automated electrophysiological devices have made the screening of new compounds for hERG liability feasible. Nevertheless, as many as 40–70% of compounds may be discarded early on in the drug development process because of the blockade of hERG [36]. It is important to note that these compounds may have minimal arrhythmogenic risks, and therefore current methodologies may discard many therapeutically valuable compounds on this basis. An attractive alternative approach would be to use *in silico* approaches to identify compounds with the propensity to block hERG [37]. In recent years, considerable progress has been made in defining the structural determinants of hERG channel block and ligand structure–activity relationships [37,38].

4.6
Drug-Binding Site of hERG

Early experiments by Armstrong showed that intracellularly applied quaternary amines (QA) blocked voltage-gated K^+ channels in the squid axon only when the channel was opened by a depolarizing pulse [39]. Furthermore, the larger QA blockers apparently slowed deactivation by preventing K^+ channels from closing upon repolarization. This phenomenon is referred to as 'foot in the door' type block because the drug behaves like a wedge, preventing channel closure until the drug has unbound and exited the channel. These observations provided early evidence that drugs bind within a space located behind the activation gate (known as the inner cavity) where access is gained only when the channel opens [40]. In contrast, small QA compounds such as TEA block the channel without changing the deactivation rate, suggesting that TEA is trapped within the inner cavity by the closure of the activation gate [41]. Numerous studies on hERG channel blockers have revealed that compounds block in the open state [42] and from the intracellular side of the membrane [43]. Recovery from block for many of these compounds is also extremely slow and virtually nonexistent at more negative voltages [44,45]. This finding suggests that closure of the activation gate prevents the drug from exiting the channel pore, consistent with a drug-trapping mechanism. The drug-trapping hypothesis was confirmed by studying the pharmacological properties of a highly unusual mutant, D540K hERG [46]. Unlike wild-type hERG, D540K channels can be opened by very negative potentials and this permits recovery from block. Evidently, the negative potential provides the driving force for positively charged compounds to exit the open channel. These experiments provided direct evidence that the mechanism of slow recovery from block is due to trapping within the inner cavity by closure of the activation gate.

These observations highlight the ability of hERG to trap relatively large and bulky drug molecules. As other Kv channels can trap only much smaller compounds, it has been suggested that the inner cavity of the hERG channel should be larger than that of other Kv channels [47]. This has been attributed to the absence of a highly conserved Pro-X-Pro motif in Kv channels, found near the C-terminal end of S6 (Figure 4.3) that may cause a kink in S6 and limit the size of the inner vestibule [48]. Importantly, the drug-trapping studies identified the inner cavity as the primary site of drug blockade and made it possible to use mutagenesis approaches to determine the residues that form the binding site.

4.7
Structural Basis for hERG Block

Lees-Miller et al. [49] were the first to use mutagenesis approaches to demonstrate the importance of an inner cavity residue in high-affinity binding. The mutation F656V substantially increased the IC_{50} for dofetilide and quinidine, without altering the potency of extracellularly applied TEA [49]. The binding site was more

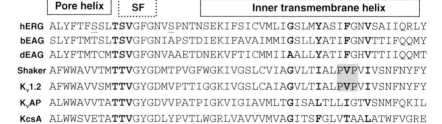

```
                Pore helix           SF              Inner transmembrane helix
hERG    ALYFTFSSLTSVGFGNVSPNTNSEKIFSICVMLIGSLMYASIFGNVSAIIQRLY
bEAG    SLYFTMTSLTSVGFGNIAPSTDIEKIFAVAIMMIGSLLYATIFGNVTTIFQQMY
dEAG    ALYFTMTCMTSVGFGNVAAETDNEKVFTICMMIIAALLYATIFGHVTTIIQQMT
Shaker  AFWWAVVTMTTVGYGDMTPVGFWGKIVGSLCVIAGVLTIALPVPVIVSNFNYFY
Kv1.2   AFWWAVVSMTTVGYGDMVPTTIGGKIVGSLCAIAGVLTIALPVPVIVSNFNYFY
KvAP    ALWWAVVTATTVGYGDVVPATPIGKVIGIAVMLTGISALTLLIGTVSNMFQKIL
KcsA    ALWWSVETATTVGYGDLYPVTLWGRLVAVVVMVAGITSFGLVTAALATWFVGRE
```

Figure 4.3 Sequence alignment of hERG and other representative K$^+$ channels: alignments of the pore helix, selectivity filter and inner helices. The inner helix of KcsA corresponds with S6 of the other K$^+$ channels shown here, which are all voltage-gated channels. hERG residues close to the selectivity filter and on S6 implicated in drug binding are shown in bold. Bovine EAG (bEAG) and *Drosophila* EAG (dEAG) are very closely related to hERG. The eukaryotic channels Shaker and Kv1.2 contain a Pro-X-Pro motif (in gray) not found in the EAG family of K channels, KvAP or KcsA. KvAP and KcsA are prokaryotic channels for which crystal structures are available [13, 15].

comprehensively characterized by an Ala-scanning mutagenesis approach to identify S6 and pore helix residues of hERG that interact with drugs [47]. Residues were individually substituted to Ala and the impact on the IC$_{50}$ of MK-499, cisapride and terfenadine was assessed. Mutation of Phe656 and Tyr652 decreased the affinity of all three compounds. These aromatic residues, located on S6, have since been shown to be important sites of interaction for a variety of compounds such as propafenone [50] and ibutilide [44,45]. Further mutagenesis at these positions indicated that hydrophobicity was important for Phe656, whereas the aromatic properties of Tyr652 were required for high-affinity block [51]. This suggests that electrostatic interactions involving the π-electrons of Tyr652 with cations such as the positively charged amines of drugs or π–π interactions with phenyl rings are likely to be important. Most Kv channels have aliphatic isoleucine or valine residues in positions equivalent to Tyr652 and Phe656. The presence of eight aromatic residues per channel means hERG is able to interact with compounds in a variety of different ways, providing it with unique structural features that help explain its pharmacological promiscuity.

Other inner cavity residues contribute to the binding site. Mutation of the pore helix residues Thr623, Ser624 and Val625 has variable effects on compound affinity. These residues are conserved in K$^+$ channels. Although not unique to hERG, they nevertheless form an important part of the binding site for many high-affinity blockers. The double mutant T623A:S624A can increase the IC$_{50}$ of compounds by more than 1000-fold [45]. A recent study on close analogues of ibutilide and clofilium indicated that small differences in the *para*-substituent on the phenyl ring of these compounds were sufficient to alter potency by 200-fold and probably via differential interactions with Thr623 and Ser624. The hydroxyl groups of these residues are likely to be suitably positioned to form hydrogen bonds with compounds and may function together to coordinate binding [44,45,52]. Modeling studies suggest the side chain of Val625 is shielded from the inner cavity and thus unlikely to directly interact with drugs. Therefore, the V625A mutation is likely to have an allosteric effect on drug binding – probably by disrupting drug interactions with Thr623 and Ser624 [52].

Mutation of Gly648 and Val659, located in the S6 domain of hERG, also has an impact on the potency of some drugs. V659A hERG channels exhibit extremely slow deactivation kinetics. Slow closure may permit recovery from block and account for the observed reduced drug sensitivity. Further studies are required to distinguish between effects on channel gating and direct interactions with compounds. G648A significantly reduces MK-499 and ibutilide binding but has little effect on cisapride, terfenadine and propafenone [44,45,47,52]. Gly648 in hERG also corresponds to the proposed glycine hinge point in potassium channels [53]. This glycine residue is highly conserved in K^+ channels and is thought to facilitate bending of the inner helices during activation gating. Mutating this glycine in hERG would be expected to have substantial effects on its activation gating, similar to what has been reported for other K^+ channels [54,55]. In fact, the voltage dependence of activation of G648A channels is not remarkably different from wild-type hERG [47], suggesting that G648A does not prevent compounds from gaining access to the binding site. Instead, the Ala substitution at Gly648 is likely to have effects on drug binding by occluding access to a binding pocket or hampering interactions with other neighboring residues such as Thr623 or Tyr652.

4.8
Alternative Mechanisms of Block

For most drugs tested to date, the S6 residues Phe656 and Tyr652 appear to be a requirement for high-affinity hERG block. However, for a small selection of drugs, one or both of these residues may not be an absolute prerequisite. Low-affinity hERG blockers such as vesnarinone [56], propafenone [50] and bepridil [57] as well as the high-affinity blockers dronedarone [58] and thioridizine [59] appear to preferentially interact with the Phe656 side chains, with minor interactions with Tyr652. The potency of the well-characterized hERG blocker quinidine is also strongly reduced by the F656A mutation (125-fold increase in IC_{50}) as compared to the Y652A mutation (3-fold increase in IC_{50}) [60]. Fluvoxamine appears to be relatively insensitive to both F656A and Y652A mutations [61], raising the possibility of an alternative site of block.

One such site may be the outer mouth of the channel, where a number of peptide toxins bind. Several potent and selective scorpion toxins have been identified that block hERG and native I_{Kr} channels [62–64]. Sites for toxin binding have been mapped to the S5-P linker and P-S6 linkers and are therefore used to gain further insights into the channel structure. Be-Km1, a scorpion toxin shown to interact with specific residues in the outer mouth of hERG, was used to create a model of the hERG S5-P linker [65]. This linker in hERG is unusually long and has no sequence similarity to that of other K^+ channels. Molecular dynamics simulations suggest that the S5-P linker forms an amphipathic α-helix that is inherently unstable and can interact with other channel subunits as well as with residues at the outer entrance of the selectivity filter, influencing both inactivation and channel selectivity (Stansfeld *et al.*, submitted). A study on another peptide toxin, CnErg1 [66], aims to further understand the mechanism of its binding to hERG. A model is proposed by which the toxin binds to the amphipathic α-helix without occluding the ion conduction pathway. This is followed by conformational changes in the toxin–channel complex, permitting occlusion of the pore. The latter conformation is thought to be weak, explaining the incomplete blockade of hERG current by CnErg1. This highlights the possibility of various means of hERG blockade by toxins consistent with the dynamic nature of the outer mouth of the channel.

Recently, specific activators of hERG channels have been reported, such as RPR260243, PD-118057 and NS1643 [67–69]. These compounds enhance hERG current by various different mechanisms but share the ability to shorten the action potential duration. RPR260243 causes a significant slowing of hERG channel deactivation, leading to almost a 20% increase in hERG current magnitude through a binding site distinct from that of dofetilide [67]. In contrast, the agonist effects of NS1643 appear to be mediated through effects on the rate of onset and voltage dependence of inactivation, rather than deactivation, and are facilitated by Phe656 mutations [69]. Thus, this drug acts as a partial agonist, which may form π–π interactions with the F656 residue in hERG. These hERG activators are an interesting new development and understanding their distinct modes of action should provide further insight into hERG channel structure–function relations. It remains to be seen if they will be useful for treating LQTS.

4.9
Role of Inactivation in hERG Block

EAG channels have a high level of sequence homology with hERG, and all the previously described hERG drug-binding residues are retained. Despite this, EAG channels are relatively insensitive to most hERG blockers, with the exception of clofilium [70]. An important difference between EAG and hERG is that EAG channels do not inactivate. Intriguingly, mutations that introduce hERG-like C-type inactivation into EAG can increase its drug sensitivity [71]. Many mutations that disrupt inactivation have been shown to reduce the high-affinity drug binding of methanesulfonanilides to hERG [72,73]. Serine residues in positions 620 and 631, in the outer mouth of the hERG channel, when mutated to Thr and Ala, respectively, result in a loss of C-type inactivation, in turn reducing the potency of dofetilide, a high-affinity hERG blocker [72]. Thus, the fast inactivation exhibited by hERG appears important for high-affinity drug binding.

However, inactivation does not seem to be necessary for some low-affinity blockers such as disopyramide [74] and fluvoxamine, which was only partially attenuated by the S631A mutant [61]. Furthermore, MK-499 blocks G628C:S631C mutant hERG channels, which lack inactivation, with an IC_{50} comparable to that of wild-type hERG channels [47]. Other mutations that increase inactivation compared to WT hERG channels (G648A, T623A and F656A) show greatly reduced drug sensitivity. This indicates that there may be other factors required to account for the apparent inactivation-induced increases in drug potency. Another hypothesis supposes that the Phe656 and Tyr652 residues are optimally positioned for drug binding in hERG by conformational changes associated with C-type inactivation. Repositioning the S6 aromatic residues of hERG by one position either toward the N-terminus or toward the C-terminus reduced the sensitivity of hERG channels to cisapride block [75]. In contrast, repositioning the equivalent EAG residues, Tyr481 and Phe485, enhanced cisapride sensitivity. Thus, the alignment of key residues may differ in EAG and hERG channels. So, inactivation gating may change the conformation of the inner cavity and influence drug-binding affinities. Studies on Shaker channels also support a model in which conformational changes alter the accessibility of residues within the inner cavity during inactivation gating [76].

4.10
Inhibition of hERG Trafficking by Pharmacological Agents

As discussed earlier, the familial form of LQTS can be a result of defective hERG protein trafficking to the cell membrane, leading to a loss of normal channel function. Some mutations can be pharmacologically rescued by drugs that bind to the inner cavity of the channel [77]. In contrast, some compounds appear to reduce hERG surface expression, producing a novel form of acquired LQTS. Arsenic trioxide (As_2O_3), used for the treatment of acute promyelocytic leukemia, was the first documented case of a drug that inhibits hERG channel trafficking [78]. The reduction in hERG surface expression after long-term exposure to As_2O_3 is a direct consequence of the interference of As_2O_3 with the binding of hERG to the cytosolic chaperones, Hsp70 and Hsp90. These chaperones play an important role in promoting proper channel folding and assembly and appear to directly interact with WT hERG channels during maturation in the ER [79].

Geldanamycin, a benzoquinoid antibiotic, also inhibits the function of Hsp90, reducing hERG surface expression by increasing proteasomal degradation [79,80]. Its derivatives are also in clinical studies for the treatment of many cancers with no reports yet of adverse cardiac effects [81]. Similarly pentamidine, an antiprotozoal agent, inhibits hERG channel processing and trafficking to the cell surface [82]. This can lead to a prolongation of cardiac action potential at therapeutic concentrations. These drugs show an alternative, indirect mechanism for I_{Kr} reduction raising the question as to whether the current screening methodologies for hERG blockade should also incorporate the study of indirect effects of drugs on hERG protein trafficking [33].

4.11
Computational Approaches to Predict hERG K^+ Channel Block

The evolving structural insights into the hERG drug-binding site show that this channel has a relatively larger inner cavity, with many aromatic and polar residues intricately positioned to interact with drugs. Utilization of *in silico* techniques, refined by growing databases of hERG blocking ligands and further insights of structural determinants of the hERG-binding site, offers the potential for effective tools for identifying hERG blockers and drug design strategies to circumvent QT liability [37].

Quantitative structure–activity relationship (QSAR) methodologies have been used to determine the physicochemical features of hERG blockers. These studies are the subject of the following chapter and are covered in far greater detail there. Within the context of the present chapter, it will be useful to describe briefly the common pharmacophoric features using the model by Cavalli et al. [83] as an example. Typically, there is a central protonated nitrogen (N) linked to a phenyl ring (C0) and potentially two other hydrophobic groups (C1 and C2) that may also be aromatic. The Cavalli model is shown docked into a homology model of hERG in Figure 4.4.

A number of homology models of hERG have been generated based on available crystallographic structures of the pore domains of K^+ channels. These models have mainly used the crystal structures of KcsA and MthK as templates of hERG in the closed and open states, respectively. These studies aim to explain structurally the interaction between drugs and key amino acid residues identified by mutagenesis studies or to relate ligand-based models with the hERG-binding site [45,47,50,84–87]. Although these models do not possess a predictive value, they aid in the visualization of results, complementing experimental data, analysis of potential binding modes and, when combined with pharmacophore models of hERG blockers, make it possible to interpret the role of key amino acid residues in the channel (Figure 4.4). Common features are that the protonated nitrogen atom found in most hERG blockers may form cation–π interactions with aromatic residues in the pore [88]. The phenyl ring of compounds may promote π-stacking interactions with the Tyr652 residue, and since it often has a halogen or a polar group attached to its *para*-position, it may then interact with the Thr623 and Ser624 residues. Mutagenesis studies indicate that Phe656 forms predominantly hydrophobic interactions with drugs [51], consistent with interactions with the C2 feature, although inverted poses in which C1 and/or C2 interact with Phe656 have also been described [86].

The results from mutagenesis studies are not necessarily well reproduced by homology models of hERG. In particular, detailed inspection of the models often predicts interactions of drugs with two serines (Ser649 and Ser660) and yet mutation of these residues has no impact on drug potency. Furthermore, key residues such as Tyr652 and Phe656 are not optimally positioned in the models for favorable interactions with ligands. In an attempt to overcome these shortcomings, a recent hERG homology model by Stansfeld et al. [52] incorporates a clockwise rotation of the S6 helices toward the C-terminus. This rotation is supported by the previously discussed studies on hERG channel gating that infer a repositioning of inner cavity residues during inactivation gating [75]. Initially, the Phe656 residues were closely packed together, hindering the binding of larger ligands. Therefore, channel opening was simulated using the inherent flexibility of the inner helix at the glycine hinge. The rotated–hinged model created by these modifications retains residues implicated in drug interactions within the binding site and displaces Ser649 and Ser660. Drug dockings in this model are far more consistent with available mutagenesis data and ligand-based *in silico* studies [52].

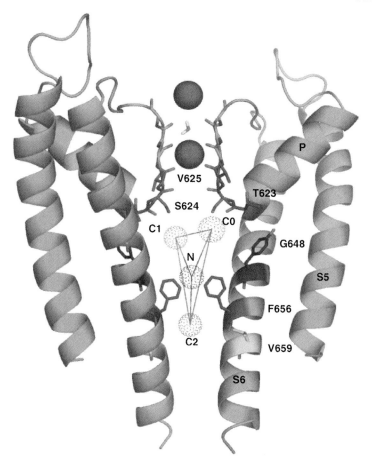

Figure 4.4 Placement of the hERG blocker pharmacophore in the inner cavity of the channel. The Cavalli pharmacophore [83] is placed within the pore illustrating its main features: N (positively charged central nitrogen), C0 (aromatic group), C1 and C2 (hydrophobic groups). The N feature is thought to form cation–π interactions with Tyr652. C1 possibly interacts with Tyr652 through π-stacking or hydrophobic interactions. C2 is predicted to bind to Phe656, and C0 is thought to interact with Ser624 and/or Thr623. The solid spheres in the selectivity filter represent K^+ ion sites.

4.12
Conclusions

The potential of drugs to cause QT prolongation and the associated risk of cardiotoxicity remains an important issue in drug development. Although the advent of automated electrophysiological systems and other preclinical screens has helped in early identification of hERG channel blockers, the outcome can be highly unsatisfactory. The process is wasteful and expensive and the development of potentially valuable compounds is halted. The wealth of information continually uncovered from *in vitro*, *in vivo* and *in silico* approaches means that we now have a much better understanding of the structural basis of hERG channel block and of the molecular features of hERG channel blockers. This offers the prospect of rationally designing out hERG channel liability and improving *in silico* methods to predict compounds with the potential to cause cardiac arrhythmias.

Abbreviations

ER	endoplasmic reticulum
hERG	human ether-a-go-go-related gene
I_{Kr}	delayed rectifier potassium current
Kv	voltage-gated potassium channel
LQTS	long QT syndrome
QA	quaternary amines
QSAR	quantitative structure–activity relationship
SQTS	short QT syndrome
TdP	*torsades de pointes*

Acknowledgments

We would like to thank all recent members of the Mitcheson lab for their help and support. In particular, we are grateful to Dr Phill Stansfeld for his modeling expertise and assistance with the figures, and to Professor Mike Sutcliffe (University of Manchester) who supervised the modeling work and has been a pleasure to work with. We are also grateful to Dr Cavalli and Professor Recanatini for providing the coordinates of their pharmacophore model.

References

1 Hille, B. (2001) *Ion Channels of Excitable Membranes*, 3rd edn, Sinauer Associates, Sunderland, MA.

2 Sanguinetti, M.C., Jiang, C., Curran, M.E. and Keating, M.T. (1995) A mechanistic link between an inherited and an acquired cardiac arrhythmia: *HERG* encodes the I_{Kr} potassium channel. *Cell*, **81**, 299–307.

3 Trudeau, M.C., Warmke, J.W., Ganetzky, B. and Robertson, G.A. (1995) hERG, a human inward rectifier in the voltage-gated potassium channel family. *Science*, **269**, 92–95.

4 Shi, W., Wymore, R.S., Wang, H.S., Pan, Z., Cohen, I.S., McKinnon, D. and Dixon, J.E. (1997) Identification of two nervous system-specific members of the *erg* potassium channel gene family. *Journal of Neuroscience*, **17**, 9423–9432.

5 Vandenberg, J.I., Torres, A.M., Campbell, T.J. and Kuchel, P.W. (2004) The HERG K^+ channel: progress in understanding the molecular basis of its unusual gating kinetics. *European Biophysics Journal*, **33**, 89–97.

6. Sanguinetti, M.C. and Tristani-Firouzi, M. (2006) hERG potassium channels and cardiac arrhythmia. *Nature*, **440**, 463–469.
7. Ono, K. and Ito, H. (1995) Role of rapidly activating delayed rectifier K^+ current in sinoatrial node pacemaker activity. *The American Journal of Physiology*, **269** (2 Pt 2), H453–H462.
8. Mitcheson, J.S. and Hancox, J.C. (1999) An investigation of the role played by the E-4031 sensitive (rapid delayed rectifier) potassium current in isolated rabbit atrioventricular nodal and ventricular myocytes. *Pflugers Archiv: European Journal of Physiology*, **438**, 843–850.
9. Wang, J., Trudeau, M.C., Zappia, A.M. and Robertson, G.A. (1998) Regulation of deactivation by an amino terminal domain in human ether-a-go-go-related gene potassium channels. *The Journal of General Physiology*, **112**, 637–647.
10. Wang, J., Myers, C.D. and Robertson, G.A. (2000) Dynamic control of deactivation gating by a soluble amino-terminal domain in hERG K^+ channels. *The Journal of General Physiology*, **115**, 749–758.
11. Cui, J., Kagan, A., Qin, D., Mathew, J., Melman, Y.F. and McDonald, T.V. (2001) Analysis of the cyclic nucleotide binding domain of the HERG potassium channel and interactions with KCNE2. *The Journal of Biological Chemistry*, **276**, 17244–17251.
12. Cockerill, S.L., Tobin, A.B., Torrecilla, I., Willars, G.B., Standen, N.B. and Mitcheson, J.S. (2007) Modulation of hERG potassium currents in HEK-293 cells by protein kinase C. Evidence for direct phosphorylation of pore forming subunits. *The Journal of Physiology*, **581**, 479–493.
13. Doyle, D.A., Morais Cabral, J.H., Pfuetzner, R.A., Kuo, A., Gulbis, J.M., Cohen, S.L., Chait, B.T. and MacKinnon, R. (1998) The structure of the potassium channel: molecular basis of K^+ conduction and selectivity. *Science*, **280**, 69–77.
14. Jiang, Y., Lee, A., Chen, J., Cadene, M., Chait, B.T. and MacKinnon, R. (2002) Crystal structure and mechanism of a calcium-gated potassium channel. *Nature*, **417**, 515–522.
15. Jiang, Y., Lee, A., Chen, J.Y., Cadene, M., Chait, B.T. and MacKinnon, R. (2003) X-ray structure of a voltage dependent K^+ channel. *Nature*, **423**, 33–41.
16. Kuo, A., Gulbis, J.M., Antcliff, J.F., Rahman, T., Lowe, E.D., Zimmer, J., Cuthbertson, J., Ashcroft, F.M., Ezaki, T. and Doyle, D.A. (2003) Crystal structure of the potassium channel KirBac1.1 in the closed state. *Science*, **300**, 1922–1926.
17. Long, S.B., Campbell, E.B. and MacKinnon, R. (2005) Crystal structure of a mammalian voltage-dependent *Shaker* family K^+ channel. *Science*, **309**, 897–903.
18. Labro, A.J., Raes, A.L., Bellens, I., Ottschytsch, N. and Snyders, D.J. (2003) Gating of *Shaker*-type channels requires the flexibility of S6 caused by prolines. *The Journal of Biological Chemistry*, **278**, 50724–50731.
19. Hancox, J.C., Levi, A.J. and Witchel, H.J. (1998) Time course and voltage dependence of expressed hERG current compared with native "rapid" delayed rectifier K^+ current during the cardiac ventricular action potential. *Pflugers Archiv: European Journal of Physiology*, **436**, 843–853.
20. Zhou, Z., Gong, Q., Ye, B., Fan, Z., Makielski, J.C., Robertson, G.A. and January, C.T. (1998) Properties of HERG channels stably expressed in HEK 293 cells studied at physiological temperature. *Biophysical Journal*, **74**, 230–241.
21. Roden, D.M., Balser, J.R., George, A.L. and Anderson, M.E. (2002) Cardiac ion channels. *Annual Review of Physiology*, **64**, 431–475.
22. Anantharam, A. and Abbott, G.W. (2005) Does hERG coassemble with a beta subunit? Evidence for roles of MinK and MiRP1. *Novartis Foundation Symposium*, **266**, 100–112.

23 Abbott, G.W., Sesti, F., Splawski, I., Buck, M.E., Lehmann, M.H., Timothy, K.W., Keating, M.T. and Goldstein, S.A. (1999) MiRP1 forms I_{Kr} potassium channels with hERG and is associated with cardiac arrhythmia. *Cell*, **97**, 175–187.

24 Lundquist, A.L., Manderfield, L.J., Vanoye, C.G., Rogers, C.S., Donahue, B.S., Chang, P.A., Drinkwater, D.C., Murray, K.T. and George, A.L. (2005) Expression of multiple KCNE genes in human heart may enable variable modulation of I(Ks). *Journal of Molecular and Cellular Cardiology*, **38**, 277–287.

25 Pourrier, M., Schram, G. and Nattel, S. (2003) Properties, expression and potential roles of cardiac K^+ channel accessory subunits: MinK, MiRPs, KChIP, and KChAP. *The Journal of Membrane Biology*, **194**, 141–152.

26 Weerapura, M., Nattel, S., Chartier, D., Caballero, R. and Herbert, T.E. (2002) A comparison of currents carried by HERG, with and without coexpression of MiRP1, and the native rapid delayed rectifier current. Is MiRP1 the missing link? *The Journal of Physiology*, **540** (Pt 1), 15–27.

27 Splawski, I., Shen, J., Timothy, K.W., Lehmann, M.H., Priori, S., Robinson, J.L., Moss, A.J., Schwartz, P.J., Towbin, J.A., Vincent, G.M. and Keating, M.T. (2000) Spectrum of mutations in long QT syndrome genes. KvLQT1, HERG, SCN5A, KCNE1, KCNE2. *Circulation*, **102**, 1178–1185.

28 Sanguinetti, M.C., Curran, M.E., Spector, P.S. and Keating, M.T. (1996) Spectrum of HERG K^+ channel dysfunction in an inherited cardiac arrhythmia. *Proceedings of the National Academy of Sciences of the United States of America*, **93**, 2208–2212.

29 Keating, M.T. and Sanguinetti, M.C. (2001) Molecular and cellular mechanisms of cardiac arrhythmias. *Cell*, **104**, 569–580.

30 Delisle, B.P., Anson, B.D., Rajamani, S. and January, C.T. (2004) Biology of cardiac arrhythmias: ion channel protein trafficking. *Circulation Research*, **94**, 1418–1428.

31 Rajamani, S., Anderson, C.L., Anson, B.D. and January, C.T. (2002) Pharmacological rescue of human K^+ channel long-QT2 mutations: human ether-a-go-go-related gene rescue without block. *Circulation*, **105**, 2830–2835.

32 Rajamani, S., Eckhardt, L.L., Valdivia, C.R., Klemens, C.A., Gillman, B.M., Anderson, C.L., Holzem, K.M., Delisle, B.P., Anson, B.D., Makielski, J.C. and January, C.T. (2006) Drug-induced long QT syndrome: hERG K^+ channel block and disruption of protein trafficking by fluoxetine and norfluoxetine. *British Journal of Pharmacology*, **149**, 481–489.

33 Wible, B.A., Hawryluk, P., Ficker, E., Kuryshev, Y.A., Kirsch, G. and Brown, A.M. (2005) HERG-Lite: a novel comprehensive high-throughput screen for drug-induced hERG risk. *Journal of Pharmacological and Toxicological Methods*, **52**, 136–145.

34 Brugada, R., Hong, K., Dumaine, R., Cordeiro, J., Gaita, F., Borggrefe, M., Menendez, T.M., Brugada, J. Pollerick, G.D., Wolpert, C., Burashnikov, E., Matsuo, K., Wu, Y.S., Guerchicoff, A., Biarchi, F., Giustetto, C., Schimpf, R., Brugada, P. and Antzelevitch, C. (2004) Sudden death associated with short-QT syndrome linked to mutations in HERG. *Circulation*, **109**, 30–35.

35 Yap, Y.G. and Camm, A.J. (2003) Drug-induced QT prolongation and *torsades de pointes*. *Heart*, **89**, 1363–1372.

36 Witchel, H.J. (2007) The hERG potassium channel as a therapeutic target. *Expert Opinion on Therapeutic Targets*, **11**, 321–336.

37 Sanguinetti, M.C. and Mitcheson, J.S. (2005) Predicting drug–hERG channel interactions that cause acquired long QT syndrome. *Trends in Pharmacological Sciences*, **26**, 119–124.

38 Recanatini, M., Poluzzi, E., Masetti, M., Cavalli, A. and De Ponti, F. (2005) QT prolongation through hERG K^+ channel blockade: current knowledge and strategies for the early prediction during drug development. *Medicinal Research Reviews*, **25**, 133–166.

39 Armstrong, C.M. (1969) Inactivation of the potassium conductance and related phenomena caused by quaternary ammonium ion injection in squid axons. *The Journal of General Physiology*, **54**, 553–575.

40 Armstrong, C.M. (1971) Interaction of tetraethylammonium ion derivatives with the potassium channels of giant axons. *The Journal of General Physiology*, **58**, 413–437.

41 Yellen, G. (1998) The moving parts of voltage-gated ion channels. *Quarterly Reviews of Biophysics*, **31**, 239–295.

42 Spector, P.S., Curran, M.E., Keating, M.T. and Sanguinetti, M.C. (1996) Class III antiarrhythmic drugs block hERG, a human cardiac delayed rectifier K^+ channel. Open channel block by methanesulfonalides. *Circulation Research*, **78**, 499–503.

43 Zou, A., Curran, M.E., Keating, M.T. and Sanguinetti, M.C. (1997) Single HERG delayed rectifier K^+ channels expressed in *Xenopus* oocytes. *The American Journal of Physiology*, **272** (3 Part 2), H1309–H1314.

44 Perry, M., De Groot, M.J., Helliwell, R., Leishman, D., Tristani-Firouzi, M., Sanguinetti, M.C. and Mitcheson, J.S. (2004) Structural determinants of hERG channel block by clofilium and ibutilide. *Molecular Pharmacology*, **66**, 240–249.

45 Perry, M., Stansfeld, P.J., Leaney, J., Wood, C., De Groot, M.J., Leishman, D., Sutcliffe, M.J. and Mitcheson, J.S. (2006) Drug binding interactions in the inner cavity of HERG channels: molecular insights from structure–activity relationships of clofilium and ibutilide analogs. *Molecular Pharmacology*, **69**, 509–519.

46 Mitcheson, J.S., Chen, J. and Sanguinetti, M.C. (2000) Trapping of a methanesulfonanilide by closure of the hERG potassium channel activation gate. *The Journal of General Physiology*, **115**, 229–240.

47 Mitcheson, J.S., Chen, J., Lin, M., Culberson, C. and Sanguinetti, M.C. (2000) A structural basis for drug-induced long QT syndrome. *Proceedings of the National Academy of Sciences of the United States of America*, **97**, 12329–12333.

48 del Camino, D., Holmgren, M., Liu, H. and Yellen, G. (2000) Blocker protection in the pore of a voltage-gated K^+ channel and its structural implications. *Nature*, **403**, 321–325.

49 Lees-Miller, J.P., Duan, Y., Teng, G.Q. and Duff, H.J. (2000) Molecular determinant of high affinity dofetilide binding to hERG1 expressed in *Xenopus* oocytes: involvement of S6 sites. *Molecular Pharmacology*, **57**, 367–374.

50 Witchel, H.J., Dempsey, C.E., Sessions, R.B., Perry, M., Milnes, J.T., Hancox, J.C. and Mitcheson, J.S. (2004) The low-potency, voltage-dependent HERG blocker propafenone-molecular determinants and drug trapping. *Molecular Pharmacology*, **66**, 1201–1212.

51 Fernandez, D., Ghanta, A., Kauffman, G.W. and Sanguinetti, M.C. (2004) Physicochemical features of the hERG channel drug binding site. *The Journal of Biological Chemistry*, **279**, 10120–10127.

52 Stansfeld, P.J., Gedeck, P., Gosling, M., Cox, B., Mitcheson, J.S. and Sutcliffe, M.J. (2007) Drug block of the hERG potassium channel: insight from modeling. *Proteins*, **68**, 568–580.

53 Jiang, Y., Lee, A., Chen, J.Y., Cadene, M., Chait, B.T. and MacKinnon, R. (2002) The open pore conformation of potassium channels. *Nature*, **417**, 523–526.

54 Magidovich, E. and Yifrach, O. (2004) Conserved gating hinge in ligand- and

voltage-dependent K$^+$ channels. *Biochemistry*, **43**, 13242–13247.

55 Ding, S., Ingleby, L., Ahern, C.A. and Horn, R. (2005) Investigating the putative glycine hinge in Shaker potassium channel. *The Journal of General Physiology*, **126**, 213–226.

56 Kamiya, K., Mitcheson, J.S., Yasui, K., Kodama, I. and Sanguinetti, M.C. (2001) Open channel block of HERG K$^+$ channels by vesnarinone. *Molecular Pharmacology*, **60**, 244–253.

57 Kamiya, K., Niwa, R., Mitcheson, J.S. and Sanguinetti, M.C. (2006) Molecular determinants of HERG channel block. *Molecular Pharmacology*, **69**, 1709–1716.

58 Ridley, J.M., Milnes, J.T., Witchel, H.J. and Hancox, J.C. (2004) High affinity HERG K$^+$ channel blockade by the antiarrhythmic agent dronedarone: resistance to mutations of the S6 residues Y652 and F656. *Biochemical and Biophysical Research Communications*, **352**, 883–891.

59 Milnes, J.T., Witchel, H.J., Leaney, J., Leishman, D. and Hancox, J.C. (2006) hERG K$^+$ channel blockade by the antipsychotic drug thioridazine: an obligatory role for the S6 helix residue F656. *Biochemical and Biophysical Research Communications*, **351**, 273–280.

60 Sanchez-Chapula, J.A., Ferrer, T., Navarro-Polanco, R.A. and Sanguinetti, M.C. (2003) Voltage-dependent profile of human ether-a-go-go-related gene channel block is influenced by a single residue in the S6 transmembrane domain. *Molecular Pharmacology*, **63**, 1051–1058.

61 Milnes, J.T., Crociani, O., Arcangeli, A., Hancox, J.C. and Witchel, H.J. (2003) Blockade of hERG potassium currents by fluvoxamine: incomplete attenuation by S6 mutations at F656 or Y652. *British Journal of Pharmacology*, **139**, 887–898.

62 Gurrola, G., Rosati, B., Rocchetti, M., Pimienta, G., Zaza, A., Arcangeli, A., Olivotto, M., Possani, L.D. and Wanke, E. (1999) A toxin to nervous, cardiac, and endocrine ERG K$^+$ channels isolated from *Centruroides noxius* scorpion venom. *FASEB Journal*, **13**, 953–962.

63 Pardo-Lopez, L., Zhang, M., Liu, J., Jiang, M., Possani, L.D. and Tseng, G.N. (2002) Mapping the binding site of a human ether-a-go-go-related gene-specific peptide toxin (ErgTx) to the channel's outer vestibule. *The Journal of Biological Chemistry*, **277**, 16403–16411.

64 Milnes, J.T., Dempsey, C.E., Ridley, J.M., Crociani, O., Arcangeli, A., Hancox, J.C. and Witchel, H.J. (2003) Preferential closed channel blockade of hERG potassium currents by chemically synthesised BeKm-1 scorpion toxin. *FEBS Letters*, **547**, 20–26.

65 Tseng, G.N., Sonawane, K.D., Korolkova, Y.V., Zhang, M., Liu, J., Grishin, E.V. and Guy, H.R. (2007) Probing the outer mouth structure of the hERG channel with peptide toxin footprinting and molecular modeling. *Biophysical Journal*, **92**, 3524–3540.

66 Hill, A.P., Sunde, M., Campbell, T.J. and Vandenberg, J.I. (2007) Mechanism of block of the hERG K$^+$ channel by the scorpion toxin CnErg1. *Biophysical Journal*, **92**, 3915–3929.

67 Kang, J., Chen, X.L., Wang, H., Ji, J., Cheng, H., Incardona, J., Reynolds, W., Viviani, F., Tabart, M. and Rampe, D. (2005) Discovery of a small molecule activator of the human ether-a-go-go-related gene (HERG) cardiac K$^+$ channel. *Molecular Pharmacology*, **67**, 827–836.

68 Zhou, J., Augelli-Szaran, C.E., Bradley, J.A., Chen, X., Koci, B.J., Volberg, W.A., Sun, Z. and Cordes, J.S. (2005) Novel potent human ether-a-go-go-related gene (hERG) potassium channel enhancers and their *in vitro* antiarrhythmic activity. *Molecular Pharmacology*, **68**, 876–884.

69 Casis, O., Olesen, S.P. and Sanguinetti, M.C. (2006) Mechanism of action of a novel human ether-a-go-go-related gene activator. *Molecular Pharmacology*, **69**, 658–665.

70 Gessner, G. and Heinemann, S.H. (2003) Inhibition of hEAG1 and hERG1 potassium channels by clofilium and its tertiary analogue LY97241. *British Journal of Pharmacology*, **138**, 161–171.

71 Ficker, E., Jarolimek, W. and Brown, A.M. (2001) Molecular determinants of inactivation and dofetilide block in ether-a-go-go (EAG) channels and EAG-related K^+ channels. *Molecular Pharmacology*, **60**, 1343–1348.

72 Ficker, E., Jarolimek, W., Kiehn, J., Baumann, A. and Brown, A.M. (1998) Molecular determinants of dofetilide block of hERG K^+ channels. *Circulation Research*, **82**, 386–395.

73 Wang, S., Morales, M.J., Liu, S., Strauss, H.C. and Ramusson, R.L. (1997) Modulation of HERG affinity for E-4031 by $[K^+]_o$ and C-type inactivation. *FEBS Letters*, **417**, 43–47.

74 Paul, A.A., Witchel, H.J. and Hancox, J.C. (2001) Inhibition of HERG potassium channel current by the class 1a antiarrhythmic agent disopyramide. *Biochemical and Biophysical Research Communications*, **280**, 1243–1250.

75 Chen, J., Seebohm, G. and Sanguinetti, M.C. (2002) Position of aromatic residues in the S6 domain, not inactivation, dictates cisapride sensitivity of hERG and eag potassium channels. *Proceedings of the National Academy of Sciences of the United States of America*, **99**, 12461–12466.

76 Panyi, G. and Deutsch, C. (2007) Probing the cavity of the slow inactivated conformation of Shaker potassium channels. *The Journal of General Physiology*, **129**, 403–418.

77 Ficker, E., Obejero-Paz, C.A., Zhao, S. and Brown, A.M. (2002) The binding site for channel blockers that rescue misprocessed human long QT syndrome type 2 ether-a-go-go-related gene (HERG) mutations. *The Journal of Biological Chemistry*, **277**, 4989–4998.

78 Ficker, E., Kuryshev, Y.A., Dennis, A.T., Obejero-Paz, C.A., Wang, L., Hawryluk, P., Wible, B.A. and Brown, A.M. (2004) Mechanisms of arsenic-induced prolongation of cardiac repolarisation. *Molecular Pharmacology*, **66**, 33–44.

79 Ficker, E., Dennis, A.T., Wang, L. and Brown, A.M. (2003) Role of the cytosolic chaperones Hsp70 and Hsp90 in maturation of the cardiac potassium channel HERG. *Circulation Research*, **92**, e87–100.

80 Roe, S.M., Prodromou, C., O'Brien, R., Ladbury, J.E., Piper, P.W. and Pearl, L.H. (1999) Structural basis for inhibition of the Hsp90 molecular chaperone by the antitumour antibiotics radicicol and geldanamycin. *Journal of Medicinal Chemistry*, **42**, 260–266.

81 Powers, M.V. and Workman, P. (2006) Targeting of multiple signalling pathways by heat shock protein 90 molecular chaperone inhibitors. *Endocrine-Related Cancer*, **13** (Suppl. 1), S125–S135.

82 Kuryshev, Y.A., Ficker, E., Wang, L., Hawryluk, P., Dennis, A.T., Wible, B.A., Brown, A.M., Kang, J., Chen, X.L., Sawamura, K., Reynolds, W. and Rampe, D. (2005) Pentamidine-induced long QT syndrome and block of hERG trafficking. *The Journal of Pharmacology and Experimental Therapeutics*, **312**, 316–323.

83 Cavalli, A., Poluzzi, E., De Ponti, F. and Recanatini, M. (2002) Toward a pharmacophore for drugs inducing the long QT syndrome: insights from a CoMFA study of hERG K^+ channel blockers. *Journal of Medicinal Chemistry*, **45**, 3844–3853.

84 Sanchez-Chapula, J.A., Navarro-Polanco, R.A., Culberson, C., Chen, J. and Sanguinetti, M.C. (2002) Molecular determinants of voltage dependent human ether-a-go-go related gene (hERG) K^+ channel block. *The Journal of Biological Chemistry*, **277**, 23587–23595.

85 Gessner, G., Zacharias, M., Bechstedt, S., Schonherr, R. and Heinemann, S.H. (2004) Molecular determinants for high affinity block of human EAG

potassium channels by antiarrhythmic agents. *Molecular Pharmacology*, **65**, 1120–1129.

86 Pearlstein, R.A., Vaz, R.J., Kang, J., Chen, X.L., Preobrazhenskaya, M., Shchekotikhin, A.E., Korolev, A.M., Lysenkova, L.N., Miroshnikova, O.V., Hendrix, J. and Rampe, D. (2003) Characterisation of HERG potassium channel inhibition using CoMSiA 3D QSAR and homology modeling approaches. *Bioorganic & Medicinal Chemistry Letters*, **13**, 1829–1835.

87 Pearlstein, R.A., Vaz, R.J. and Rampe, D. (2003) Understanding the structure–activity relationship of the human ether-a-go-go-related gene cardiac K^+ channel. A model for bad behaviour. *Journal of Medicinal Chemistry*, **46**, 2017–2022.

88 Stansfeld, P.J., Sutcliffe, M.J. and Mitcheson, J.S. (2006) Molecular mechanisms for drug interactions with hERG that cause long QT syndrome. *Expert Opinion on Drug Metabolism and Toxicology*, **2**, 81–94.

5
QSAR and Pharmacophores for Drugs Involved in hERG Blockage

Maurizio Recanatini, Andrea Cavalli

5.1
Introduction

The blockade of the I_{Kr} current is widely recognized as the main mechanism at the base of the QT prolongation, which may eventually result in the induction of a potentially lethal form of ventricular arrhythmia (*torsades de pointes*, TdP) [1]. The hERG potassium channel is the molecular system responsible for conducting this current, and the blockade of its activity by drugs can thus be associated with the risk of developing the drug-induced form of long QT syndrome (LQTS) [2]. The detection of the ability of pharmaceutical agents to elicit such an unwanted side effect is a crucial issue in the contemporary drug discovery process [3], and methods for the early prediction of the hERG blockade potential of candidate drugs and for monitoring the LQTS liability of marketed drugs are actively pursued [4,5]. At present, different preclinical models are available to investigate the risk of QT interval prolongation by drugs (*in vitro* and *in vivo*), but none per se is sufficiently predictive and the costs are also high [6]. In addition, during clinical trials, very limited number of patients are usually enrolled to obtain information on the actual risk of inducing LQTS.

The availability of validated *in silico* methods to predict interactions with the hERG K^+ channel in the early phase of drug development would certainly increase the screening rate and also lower the costs compared to experimental assay methods. Moreover, both the exploration of the features of hERG-blocking drugs and modeling of the hERG/drug interactions can contribute to shed light on the molecular mechanism at the base of the hERG channel blockade with a possible useful impact on drug design practices [7,8]. Therefore, it is not surprising to find in the literature an increasing number of papers dealing with modeling aspects of hERG and hERG blockers. The findings of such studies have already constituted an initial body of knowledge that is worth examining and discussing in the hope that useful indications can be obtained from them and exploited in view of a faster and accurate identification of safer medicines.

Antitargets. Edited by R. J. Vaz and T. Klabunde
Copyright © 2008 WILEY-VCH Verlag GmbH & Co. KGaA, Weinheim
ISBN: 978-3-527-31821-6

Computational modeling studies in drug research are carried out following two main kinds of strategies, depending on the starting experimental information available. If data exist only for the molecules displaying the biological property of interest, a so-called ligand-based approach is taken; if information (mostly, resolved 3D structures) is available on the macromolecular target(s) of these compounds, target-based studies can be performed. Recently, a third approach that might be designated as system based is gaining importance. In this case, the knowledge about the cellular/physiological system is taken into account to build simulation models that are able to predict the behavior of the system in normal conditions and even in the presence of perturbations (e.g. mutations or administration of a drug) [9]. Examples regarding LQTS have been presented in some recent papers [10,11], and the perspective impact of this kind of simulations in the field was briefly discussed in a review article [7].

Given the topic of this chapter, the focus will be on ligand-based studies that have been made possible by the great amount of available experimental evidence about the action of many pharmacological agents toward the hERG channel. Among these *in silico* works, we will review 2D and 3D quantitative structure–activity relationship (QSAR) models that, besides allowing one to predict the hERG liability of molecules, are also aimed at identifying physicochemical properties and other molecular features putatively responsible for the blockage of the channel by drugs. Also, we will present results of classification procedures that not only are best suited to sort or classify compounds as active or inactive against hERG but can also provide information on the molecular determinants of the effect. The main achievements of all these studies in terms of features characterizing the ability of binding to and blocking of hERG will be summarized. Some works that take into consideration docking models of drugs into the channel inner cavity will also be illustrated and discussed in the light of their relevance toward ligand-based studies. Finally, open questions and perspectives on the use of these *in silico* approaches will be briefly outlined.

5.2
Ligand-Based Models for hERG-Blocking Activity

Typical output of studies that take into consideration the pharmacological action of series of compounds is the determination of structure–activity relationships (SAR) that correlate the variations of activity to structural differences among molecules. The SAR can be embodied in a conceptual framework defined as the pharmacophore [12] for that biological action (in the present case, the hERG blockade). Early pharmacophores were simply built as schematic diagrams indicating the features responsible for the activity and their spatial localization, while more recent approaches tend to join the qualitative scheme to a statistical model that allows one to calculate predictions of the biological activity associated to the pharmacophore (predictive or quantitative pharmacophores). This evolution of the concept of pharmacophore applied to the field of pro-arrhythmic drugs is well exemplified by the works of Morgan and Sullivan [13] on one side and Ekins *et al.* [14] and Cavalli *et al.* [15] on the other. In the first case, the pharmacophore comprises a *para*-substituted phenyl ring linked to a basic nitrogen through a one- to four-atom linking chain and bears substituents on the nitrogen atom, one (or both) of which includes an aromatic ring located close (one to three atoms) to the basic center. The pharmacophores proposed in 2002 by Ekins *et al.* and by Cavalli *et al.* show substantially similar molecular features as that shown by Morgan and Sullivan, but, in addition, they provide the possibility of calculating through a QSAR model the hERG-blocking potency of the molecules that are fitted to them. Thus, quantitative pharmacophores have the dual property of allowing both prediction of biological activity and identification of the main features determining it.

It is evident how models of this kind, if accurate and reliable, could be useful in the challenge posed by the need of avoiding hERG liability of drugs. However, given the variety of different methodological tools available and the consequent different emphasis cast on some aspects of the models or of their derivation (e.g. consideration of molecular or physicochemical features, 2D or 3D description of molecules, assessment of single compounds or classification in groups of activity, etc.), several other kinds of QSAR analyses on hERG blockers have been performed and recently published. To take into account their peculiarities, we decided to group the studies into three categories, which will be illustrated separately: 3D QSAR models, 2D QSAR models and classification models. The main results within each approach will be presented in the following sections in an attempt to highlight common aspects and peculiarities.

5.2.1
3D QSAR

The first 3D QSAR model of hERG-blocking drugs was published in 2002 by Ekins *et al.* [14], followed by the work of Cavalli *et al.* [15] in the same year. Both groups, working independently, applied the same approach to the analysis of a limited set of compounds and developed a quantitative pharmacophore using the Catalyst (Accelrys Inc., San Diego, USA) or the CoMFA (Tripos Inc., St Louis, USA) software, respectively. The molecules of the training sets were different in the two studies, but both groups used electrophysiological IC_{50} data derived from the literature. Both models have been subsequently updated [7,16], and a critical comparison between them has been reported [7]. In particular, with regard to the pharmacophoric features, in both cases, the structural scheme is centered around an ionizable basic nitrogen function bearing some hydrophobic groups located at similar distances (Figure 5.1). On the contrary, the two pharmacophoric frames seem to differ in the position of the central function with respect to the surrounding hydrophobic moieties. In fact, considering the pharmacophore like a pyramid with the aminic moiety on top, the Ekins' structure seems to have a smaller base and a greater height with respect to the Cavalli's one, which in turn is flatter and less high. However, the different shapes of the two pyramidal structures representing the pharmacophoric schemes probably reflect the shape of the conformers that were selected to build the frames, and it is a consequence of the conformation selection procedure applied in each study. It has

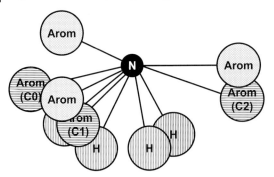

Figure 5.1 Superimposition of the pharmacophoric schemes derived from the 3D QSAR models published by Ekins et al. [14] (vertical lines filling), Cavalli et al. [15] (horizontal lines filling) and Pearlstein et al. [17] (dot filling). The central black circle represents the protonatable nitrogen atom common to all compounds used to build the pharmacophores. The indications of the pharmacophoric features linked to the central nitrogen correspond to those originally indicated by the authors.

been suggested that the two pharmacophores probably represent the same molecular characteristics either in folded or in extended conformation [7].

Considering the limited numbers and the kind of drug molecules used in the above studies to derive the 3D QSAR models, it is not surprising that the resulting pharmacophores looked like a general stereoelectronic picture of the molecular requirements associated with the ability to bind the hERG K^+ channel. To proceed further in the definition of the hERG pharmacophore, one clearly needs to identify other features that could complement the general picture and reveal more details about the specific molecular characteristics of hERG blockers. A way toward this end might be to develop 3D QSAR models that include as many compounds as possible, such as to provide the maximum coverage of the chemical space allowed for that activity. However, to avoid the intrinsic structural redundancy typical of large collections of molecules, the building of local sets of carefully varied analogues of selected parent structures might represent an appropriate starting point in this direction.

Vaz and coworkers showed the feasibility of this approach by reporting the development of 3D QSAR models that, still in agreement with the general view proposed by the previous pharmacophores (Figure 5.1), highlighted some features that had remained undisclosed in the earlier models. In the first paper [17] based on a CoMSiA (Tripos Inc., St Louis, USA) analysis of 28 molecules, mostly sertindole analogues, Pearlstein et al. were able to identify a sterically forbidden region in the middle part of the pharmacophoric frame corresponding to the presence of unfavorable (for the hERG blockade) substituents on the indole phenyl and N-phenyl of sertindole, and, quite interestingly, to the bent conformation of fexofenadine. A further analysis [18] performed on a large set of 882 compounds with experimentally measured IC_{50} values for hERG channel inhibition and comprising both basic and nonbasic molecules (note that the presence of a basic nitrogen atom has long been considered a crucial requisite for binding to hERG) allowed this group to add further details to the pharmacophoric scheme of hERG-blocking agents. In fact, by carrying out a PLS analysis on the two subsets of compounds using GRIND descriptors, Cianchetta et al. [17] not only confirmed some results of the previous CoMSiA model, such as the relevance of the charged nitrogen and hydrophobic groups for the hERG-blocking activity, but also identified two new pharmacophoric features: a H-bond donor function at different distances from a hydrophobic area in the compounds that carry a protonated nitrogen and in those that do not, and a H-bond acceptor function in both subsets.

Indeed, considering the latter 3D QSAR model, the features that make a molecule suitable to bind to the hERG channel start delineating in a chemically interpretable manner, but, it is rather clear how these kinds of models emphasize mostly the 3D steric aspects of molecules, depending mainly on factors such as the conformation (or the conformational analysis protocol) or the alignment of the molecules. To obtain a description of the characteristics of hERG-blocking molecules in terms of measurable (computable) properties in a way that the physicochemical determinants of the activity can be identified, the classical 2D QSAR approach is well suited.

5.2.2
2D QSAR

The first 2D QSAR model accounting for the hERG-blocking ability of drugs published by Aptula and Cronin in 2004 [19] was based on a limited set of 19 compounds, which, however, were consistently tested in a single laboratory and covered a range of potency as wide as six orders of magnitude. Quite a large number of descriptors were taken into consideration in the statistical analysis, comprising not only classical physicochemical parameters such as log P, log D and pK_a but also electronic, steric and topological descriptors. The best model obtained through a stepwise multiple regression analysis was the following:

$$pIC_{50} = 0.58(\pm 0.09)\log D + 0.30(\pm 0.06) D_{max} - 0.36(\pm 1.04)$$
$$n = 19, \quad r^2 = 0.87, \quad q^2 = 0.81, \quad s = 0.73, \quad F = 54, \quad (5.1)$$

where log D is the partition coefficient corrected for the ionization at physiological pH 7.4 (distribution coefficient), and D_{max} is the maximum diameter of molecules, an estimate of their maximum length. In Equation 5.1 (and in subsequent Equation 5.2), the numbers in parentheses are the 95% confidence intervals, and the statistics reported to assess the statistical value of the model are n (the number of compounds), r^2 (the squared correlation coefficient), q^2 (the leave-one-out cross-validated correlation coefficient), s (the standard error of the regression) and F (the value of the Fisher's test). This simple and elegant model points to two clearly identifiable characteristics as the main determinants of the hERG-blocking potential: lipophilicity of ionizable compounds and length. The authors found that most (but not all) of the potent ($pIC_{50} >$ 6.0) hERG blockers have log $D > 2.5$ and $D_{max} > 18$ Å.

In an independent study, Yoshida and Niwa [20] analyzed a larger and more diverse set of molecules (104 compounds) and developed a 2D QSAR model, which gave results similar to that of Cronin [19] but added some more details with regard to the physicochemical properties involved in the hERG blockade by drugs. Equation 5.2 represents the best model:

$$\begin{aligned} \text{pIC}_{50} = {} & 0.221(\pm 0.097)\, C\log P - 0.017(\pm 0.005)\text{TPSA} \\ & + 0.231(\pm 0.055)\, \text{diameter} + 0.037(\pm 0.016)\, \text{PEOE\_VSA-4} \\ & - 0.793(\pm 0.330)\, \text{cell} + 2.993(\pm 0.671) \\ & n = 104 \quad r^2 = 0.704 \quad q^2 = 0.671 \quad s = 0.763 \quad F = 46.6. \end{aligned} \qquad (5.2)$$

Despite the large structural variation of the training set, only four variables were sufficient to provide a QSAR model of acceptable statistical significance. Quite interestingly, the two most important variables were almost the same as those of Equation 5.1, that is, those accounting for lipophilicity ($C\log P$, the calculated partition coefficient for neutral molecules, instead of $\log D$) and length (diameter, a graph-based descriptor corresponding to the distance between a vertex and the most remote point in the graph) of the molecules. Other properties explaining the variation of hERG-blocking potency in the model of Equation 5.2 are those described by the topological variables TPSA (polar surface area of H-bond donor or acceptor fragments) and PEOE_VSA-4 (sum of the surface areas of atoms with partial charges from -0.25 to -0.20). Cell is an indicator variable introduced to account for the different biological systems employed to measure the activity, that is, HEK and CHO cells: it is equal to 0 for compounds tested in HEK cells expressing hERG channels and 1 for compounds tested in CHO cells. Besides confirming the key role played by lipophilicity and dimensions of the molecules in the hERG blockade, the model described by Equation 5.2 identifies some factors related to the polarity of molecules as a further (even if minor) determinant of the activity.

The works presented so far focused mostly on the identification of the properties involved in the hERG blockade and explicitly aimed at deriving 'simple, transparent and interpretable QSAR models' [19] 'by using descriptors that are familiar to medicinal chemists and that have clear physicochemical meanings' [20]. However, it has been shown by other authors that, through the use of rather sophisticated statistical tools and extended sets of molecular descriptors, it is possible to develop models of increasing complexity capable of capturing further details of the determinants of the hERG blockade, which eventually might remain hidden in the more general models. As an example of such an approach, Seierstad and Agrafiotis [21] analyzed a set of 439 compounds tested experimentally for their ability to block the hERG-mediated K^+ currents in mammalian cells scanning about 650 descriptors through six statistical feature selection methods. The result of this extended QSAR work was a robust predictive model ($q^2 = 0.76$) involving 20 descriptors, for all of which the significance in predicting hERG activity was estimated by calculating a model using each descriptor alone. It showed that the most important parameters were those related to hydrophobicity, but again new details were added to the general picture by revealing that the number of fluorine atoms connected to sp^2 carbons (fluoroaromatic groups) and the count of nonring methylene units (as an indicator of flexibility) were important. Remarkably, the presence of hydroxyl groups was also deemed important, but in a clear inverse correlation with the hERG activity.

The relevance of size-related properties of hERG-blocking molecules was also detected in a 2D QSAR model developed by Coi et al. [22] after the analysis of 82 compounds through the CODESSA method. These authors developed two multiparameter models with strong predictive properties, from which, besides the involvement of hydrophobic features, the importance of linearity as opposed to globularity of the hERG blockers emerged.

Following an approach similar to that of Seierstad and Agrafiotis [21], Song and Clark [23] developed a QSAR model based on the use of 2D fragments as descriptors of the molecular features. The training set consisted of 71 compounds, and three statistical methods were used for feature selection and model calculation. The best model ($r^2 = 0.912$, $q^2 = 0.636$, $s = 0.440$) reported by the authors included 45 descriptors (fragments) and was obtained through a method called support vector machine. From the analysis of the fragments involved in the hERG-blocking activity, it was shown that lipophilic as well as fluorine or methane sulfonamide substituted fragments and methylpiperidine groups were positively correlated, while hydrophilic fragments such as carboxylic acids, ketones, hydroxyls and primary aliphatic amines were negatively correlated to the activity.

In Table 5.1, we present a list of the main physicochemical and structural properties associated with the descriptors included in the 2D QSAR models discussed above. Of course, we did some generalizations in an attempt to refer different parameters and descriptors to the same property, but the effort was devoted at identifying the smallest number of significant features positively or negatively correlated to the hERG blockade by small molecules. Examining the properties

Table 5.1 Molecular properties identified as relevant in 2D QSAR models of hERG blockade by small molecules.

Property of hERG Blockers	Sign of Correlation with hERG Blockade	References
Lipophilicity	+	[17,19–23]
Length	+	[19,20,22]
Topological polar surface area (H-bond donor/acceptor atoms)	−	[20]
Surface area of atoms with partial charges $-0.20 > q > -0.25$	+	[20]
Flexibility	+	[21]
Presence of F-aromatic groups	+	[21,23]
Presence of hydrophilic groups (OH, carboxyl)	−	[21,23]
Presence of N-methylpiperidine	+	[23]
Presence of methane sulfonamide	+	[23]

reported in the first column of Table 5.1 and the QSAR models from which they emerged, it is seen that the parameters most relevantly linked to the hERG inhibition are lipophilicity and length of the molecules, which, together with flexibility, outline a sort of basic skeleton of hERG blockers. Other features that can alert one with respect to hERG liability are related to the presence of fragments, among which the fluoroaromatic and methane sulfonamidic ones seem to be the most significant. Interestingly, it appears that some (but not all) of the features allowing for increased hydrophilicity are negatively (inversely) correlated with the hERG inhibitory activity.

5.2.3
Classification Models

The aim of classification methods can generally be considered as that of assigning the members of more or less large collections of compounds to one of the two (or more) classes defined according to activity values taken as thresholds. For instance, for hERG blockers, an IC_{50} value of 1 µM might separate 'active' ($IC_{50} < 1$ µM) compounds from the 'inactive' ($IC_{50} > 1$ µM) ones. To obtain this result, statistical methods are applied that work on a set of variables, whose role is to take into account the molecular features of the different compounds. Similar to the QSAR applications, the output of the classification methods is twofold, that is, a model predicting the hERG liability of compounds and a set of descriptors that identify the main properties or molecular features affecting the hERG-blocking ability.

We have recently reviewed the early classification studies by Roche et al. [24] and Keserü [25] on QT [7]. The former was particularly interesting as it was the first study carried out on an extended set of hERG-blocking compounds (472, even though the best results were obtained on a nonredundant subset of 244 derivatives) using a large number of descriptors (1258). The predictive tool developed by the Hoffmann–La Roche group was a nonlinear model obtained through a supervised neural network system able to discriminate between compounds with low (<1 µM) and high (>10 µM) IC_{50} values. Interestingly, among the descriptors, the greatest difference between high- and low-potency hERG blockers was found for $C \log P$, the classical lipophilicity parameter. More recently, several other classification models have been published, whose main achievements are briefly summarized below.

In studies by Tobita et al. [26], Dubus et al. [27], Bains et al. [28] and Sun [29], classification models each obtained with a different classification technique such as support vector machine (SVM), recursive partitioning (RP), genetic programming and naive Bayes classifier, respectively, were reported. In every case, the model development was based on different numbers of compounds (training sets of 73, 203, 124 and 1979 compounds, respectively), and the common goal was to obtain an efficient and validated classifier capable of recognizing hERG blockers, given predefined activity values (thresholds) to separate the classes of potency. Two models were developed by Tobita et al. [26] with IC_{50} threshold values of 1 and 40 µM; Dubus et al. [27] defined three classes of activity characterized by $IC_{50} < 1$ µM, 1 µM $< IC_{50} <$ 10 µM and $IC_{50} > 10$ µM; Bains et al. [28] and Sun [29] set the cutoff at 1 and 3 µM, respectively. The validation procedures to which these models were submitted showed the reliability of each of them as a classifying tool, but in addition, from the general point of view of the identification of the molecular determinants of the hERG liability, their contributions were differentiated, though complementary and consistent with the picture emerging from the other ligand-based studies presented above. From the models by Tobita et al. [26] and Dubus et al. [27], some descriptors emerged as the most important in the classification, and they were related to lipophilicity, the presence of polar zones on the molecular surface, and eventually to H-bonding capability and flexibility. In the study by Tobita et al. [26], some molecular motifs that reflected the properties accounted for by the descriptors were sketched, and these structures were consistent with the features identified by Bains et al. [28] as most relevant to the hERG blockade prediction. The latter authors reported a rather detailed list of structural fragments that gave medicinal chemists a clear perception of the features of a hERG-binding molecule. The study by Sun [29] went on to detail such features by listing atom types associated with promoting or decreasing hERG affinity. Given the extended training set on which it is based (1979 compounds), the model by Sun can be regarded as very sound, and the structural indications inferred from it are as strong as general statements: acidic groups abolish hERG liability, cyclic basic groups are warning signals, and a positively charged nitrogen is essential for hERG blockers in the nanomolar range but not for weaker blockers (micromolar range); the number of aromatic rings might be a good predictor of hERG-binding capability.

Two classification models [30,31] were developed with an explicit focus on the prediction of TdP-inducing potential instead of 'simple' hERG liability, and both were based on almost the same training set comprising three classes of TdP-causing compounds (agents with risk of TdP, agents with possible risk of TdP and agents to be avoided by congenital long QT patients) and one subset of non-TdP-causing drugs. The statistical methods employed were SVM and a decision tree based on RP; the LSER descriptors were used in the former, while several topological and physicochemical descriptors were used in the latter. From their study, Yap et al. [30] concluded that SVM gave the highest accuracy in the prediction as compared with three other classification methods, and that the few LSER descriptors were equally useful for the prediction of TdP compared with a more extended and diverse set of descriptors. On the contrary, Gepp and Hutter [31] identified a structural fragment as the most relevant descriptor to discriminate between TdP-causing and non-TdP-causing compounds. Such fragment comprising a tertiary nitrogen and a phenyl ring at variable distance is expressed by the SMARTS string N([H,C])(C)(C[!O]*~*~c) and corresponds to the following general structure:

A different strategy from that of the studies illustrated above was followed by Ekins et al. [32], who used three different techniques to generate classification models based on a set of 99 hERG blockers. He compared the performance of the different methods and explored the suitability of a consensus approach for the prediction of a test set of 35 compounds. RP and two mapping methods (Sammon nonlinear maps and Kohonen self-organizing maps) were used to obtain a classification in three classes, 0: $IC_{50} < 1\,\mu M$, 1: $1\,\mu M < IC_{50} < 10\,\mu M$ and 2: $IC_{50} > 10\,\mu M$; consensus scores were calculated for the compounds belonging to the extreme classes 0 and 2. The conclusions of this study were supported by rigorous data and arguments and can somehow be taken as a general assessment of the potential of classification methods as filtering schemes for hERG liable compounds. It was concluded that compounds belonging to the intermediate class 1 were more frequently mispredicted (see also [27]), while the state-of-the-art statistical methods seemed well suited for an acceptable classification of unknown compounds, given an internally consistent training set. The consensus approach performed slightly better than some individual methods; however, its accuracy was lower than the Sammon mapping.

Finally, the work of Aronov on hERG blockers is worth noting because of the interesting approach followed in his studies, which integrate classification schemes and pharmacophoric models in an original and efficient manner. In the first paper [33], the combination of pharmacophoric (3D) and topology-based (2D) methods was explored through the use of some in-house programs that allowed the development of both 3D and 2D classification schemes, which were subsequently combined in a holistic model (called VPharm-FP/TOPO/AP from the three combined procedures) in a 'veto' format (a molecule flagged by either method is considered a hERG blocker). On a training set of 85 'actives' and 329 'inactives' (threshold IC_{50} value $40\,\mu M$), the combined classification model outperformed the single methods. Moreover, three of the pharmacophoric hypotheses generated by the 3D classification model and associated with the highest discriminating content were in good agreement with previously developed 3D QSAR models. In a more recent work [34], by analyzing a training set of 194 potent ($IC_{50} < 10\,\mu M$) uncharged hERG blockers, Aronov was able to identify a six-point pharmacophore as an effective tool for filtering out hERG channel blockers that are uncharged at physiological pH. This was the first study completely devoted to uncharged compounds in the field of potential QT prolongation inducers and contributed to elucidate molecular features associated with the hERG-binding capability of this kind of molecules, eventually different from those characterizing charged ligands. Besides the common role of lipophilicity, H-bond accepting ability was proposed to play a role in the hERG blockade by uncharged drugs.

Different from 2D QSAR, classification models reported in the papers cited above do not always allow the identification of descriptors related to the hERG activity; however, in some cases, descriptors or molecular features crucial for the assignment to either of the classes were explicitly indicated. This allowed us to tentatively collect them in Table 5.2, which when compared with Table 5.1 provides (as expected) a very similar picture of the molecular properties involved in the blockade of hERG by drugs. Even though the properties listed in Table 5.2 are not associated with a positive or a negative sign (they can only be indicated as relevant for the classification), they

Table 5.2 Molecular properties identified as relevant in classification models of hERG blockers.

Property of hERG Blockers	References
Lipophilicity	[24,26,27]
Topological polar surface area (H-bond donor/acceptor atoms)	[26,27]
Flexibility	[26]
Presence of carboxyl group	[29]
Presence of cyclic basic amines	[29]
Number of aromatic rings	[29]

are in some cases the same as those of Table 5.1 (lipophilicity, flexibility, TPSA), and in other cases, they are clearly consistent with the features reported in Table 5.1.

5.3
Ligand-Derived Models in the Light of the hERG Channel Structure

A long sought goal of modelers working in the drug research field has been the integration of ligand- and target-based models [35]. Recently, the development of powerful docking algorithms has given impetus to the building of ligand/target complex models, whose characteristics can thus be compared with the outcomes of QSAR or classification models in the search for consistency between the results of the two different approaches. In the present section, we will consider some of these integrated works but will not review articles dealing with the hERG channel structure and drug docking recently appearing in the literature. Despite the great interest generated by such works (see, for example the paper by Farid *et al.* [36]), their presentation would entail a discussion on the modeling of the hERG channel (for a short account on this topic, see [37]), which is out of the scope of this chapter and will not be considered here.

In the context of hERG studies, the comparative approach was first suggested by Cavalli *et al.* [15], who discussed their pharmacophore in the light of the interaction model of MK499 with the hERG channel developed by Mitcheson *et al.* [38]. A more in-depth comparison of a ligand 3D QSAR model with a channel structure model was reported by Pearlstein *et al.* [17], who built their CoMSiA model (see above) to describe the 3D features of sertindole derivatives and then compared it with the 3D characteristics of the channel cavity as revealed by a homology model of hERG in the open state. On this basis, the authors proposed a hypothesis of hERG/blockers interaction [17,39], suggesting that the inhibitor molecules penetrate into the channel pore from the intracellular side orienting themselves with the long 'tail' pointing toward the selectivity filter of the channel and the hydrophobic head ('handle') to block the intracellular entrance of the pore. Steric and electrostatic characteristics of the CoMSiA model matched ideally with the position of some critical residues of the channel proteins, such as Tyr652 and Phe656 localized at the inner mouth of the pore. Interestingly, there was agreement among the interaction hypotheses presented so far

[29] on the possibility of a cation–π interaction between the protonated nitrogen of hERG blockers and Tyr652 side chains as one important aspect of the inhibitors' binding mode. It is worth noting that the presence of this interaction has been further investigated through extensive experimental works (see the previous chapter), whose results are consistent with its role in binding blocker molecules. Nevertheless, some recent reports of docking analysis put the role of the cation–π interaction in a rather different perspective than a simple anchorage point for hERG ligands [36].

Along the line of comparing different kind of models, Yoshida and Niwa [20] investigated the correspondence between their 2D QSAR model for hERG blockers (see above) and a model of the complex between the channel in its open state and a representative blocker (pimozide). The main characteristics of the channel pore region as revealed by the hERG model matched well the requirements of the ligands pointed out by the QSAR model represented by Equation 5.2. In particular, the length of the molecule (descriptor: diameter) and the presence of atoms with partial negative charge (descriptor: PEOE_VSA-4) are compatible with the width of the pore region and the presence therein of H-bond donors groups such as the hydroxyls of Thr623, Ser624, Ser649 and Tyr652. Moreover, the abundance of hydrophobic residues in the drug-binding cavity contributes to provide a region where ligands with higher log P values can interact more favorably.

In the studies mentioned above, the correspondence between the features of the ligands coming from a QSAR model and the structural characteristics of the drug-binding region within the channel was investigated after the independent development of the two models. However, considering an inverse approach, one would like to be able to derive a model suitable for predicting the channel affinity of ligands directly from the 3D models of the drug/channel complexes. Calculating accurate and reliable binding affinities of small molecules to protein targets is one of the toughest challenges for computational (bio)chemists, but methods based on different approaches are available, all of which rely on molecular mechanics simulations and aim at obtaining an estimate of the free energy of binding [40]. In the field of hERG blockers, as well as in many other therapeutic or toxicological contexts, the ability to predict a ligand affinity from a virtual docking experiment would be highly desirable, but there are still several theoretical and methodological obstacles that hamper a quick solution of the problem. In regard to hERG, besides the computational demand that can become prohibitive for a large series of compounds, the main problem is the lack of a reliable description of the 3D structure of the tetrameric protein complex that can be obtained from X-ray crystallography. Nevertheless, to compute the free energy of binding attempts have been recently reported to calculate the binding affinities of a small series of compounds to hERG using homology models of the channel, standard docking methods and the linear interaction energy (LIE) method [41].

Österberg and Åqvist [42] built an homology model of hERG based on the crystallographic structure of the archeobacterial KvAP channel and used the AutoDock 3.0 software to obtain reliable poses for six sertindole analogues. Molecular dynamics (MD) simulations were carried out for each compound, and relative binding free energies were then estimated from the MD simulations with the LIE approach. The authors reported a plot showing a rather good correlation between experimental and predicted binding affinities, for which we calculated an r^2 value of 0.972. Following a similar approach, Rajamani et al. [43] calculated the binding affinities for 32 hERG blockers and compared them to the experimental values. The drugs were docked to hERG models that supposedly represent different gating states of the channel obtained by moving the S6 helices that form the pore in such a way as to derive a number of intermediate states between the open and the closed ones. The authors then developed models for the correlation of the binding affinity with electrostatic and van der Waals energy parameters combining the data from the docking to two states (open and partially open): the best model was obtained for 27 compounds (omitting 5 outliers), and the reported statistics are $r^2 = 0.82$ and $s = 0.56$.

Clearly, the latter studies are suggestive of a promising approach toward the prediction of hERG liability based on a direct estimation of the drug affinity for the antitarget. However, they also show that it is early to assume such methods as standard tools (even provided that the computational problems can be solved in a short time) given some uncertainties that are inherent in the procedure (first of all, the choice and the reliability of both the channel model and the docking poses), and cast some doubts on the general applicability of the results.

5.4
Conclusions

The ultimate goal of computational methods applied to drug research-related problems is to help us speed up the process of discovering and developing new safe and therapeutically useful chemical entities. In this context, studies on hERG and hERG blockers are certainly aimed at the early identification of potential channel blockers and inducers of LQTS. The ligand-based studies reviewed in this chapter have contributed mainly to two important aspects concerning the design of candidate drugs devoid of hERG liability: the identification of the structural and physicochemical determinants of the hERG-binding ability and the construction of statistical models capable of predicting the hERG inhibitory activity of untested molecules.

As regards the first aspect, the properties reported in Tables 5.1 and 5.2 have to be considered. Combining the right structural and physicochemical elements should result in appropriate suggestions on how to avoid the possibility that a new drug binds to and blocks the hERG channel. Only the practical application of such suggestions by medicinal chemists will allow to assess the correctness of the indications, but several examples collected by Jamieson et al. [44] already exist and seem to be consistent with the output of the theoretical models listed in Tables 5.1 and 5.2 (see also Chapter 17).

On the pharmacophoric and statistical models side, we have seen that several QSAR and classification models have been published recently, and that quite high levels of accuracy have been reached in their development. It is worth noting that both extension of training and test sets and internal consistency of the data sets have been substantially improved since the first models, relying mostly on a relatively few literature data. Even though the statistical significance did not increase much, the use of large amounts of experimental data produced by a single laboratory has certainly contributed to the overall reliability of the models. As regards the statistical performances, it is interesting to realize how the models of the three 'categories' examined (2D QSAR, 3D QSAR and classification methods) based on the largest training sets of basic compounds (544 [18], 439 [21] and 1979 [29] compounds, respectively) showed very similar statistics expressed as q^2 for the QSARs (0.74 and 0.76 for the 2D [18] and 3D [21] models, respectively), and as receiver operating characteristic (ROC) accuracy for the naive Bayes classifier (0.87 [29]). The predictive power of the latter model is best appreciated by considering its accuracy in classifying the test set of 66 compounds that reached 87.9%.

Considering the achievements mentioned above in terms of features associated with the hERG blockade and methods for predicting the hERG liability, it can be stated that a significant progress has been made toward the understanding of the molecular reasons at the base of the hERG blockade by drugs and the development of methods to predict such blockade. However, some things remain to be done, besides improving and refining the methodological approaches in view of the building of better and more predictive models. In particular, we think that it might be worth exploring the possibility of using quantitative pharmacophores to directly design molecules possessing the features required to prevent the binding to hERG. Also, the potential of the docking technique combined with other simulation procedures such as molecular dynamics has not been fully exploited yet. In the perspective of obtaining the experimentally determined 3D structure of the channel complex, the development of accurate docking protocols might open the way to integrated ligand-/target-based applications such as virtual screening or target-based 3D QSAR and classification.

Abbreviations

CHO	Chinese Hamster Ovary
$C \log P$	calculated logarithm of the partition coefficient
CODESSA	comprehensive descriptors for structural and statistical analysis
CoMFA	comparative molecular field analysis
CoMSiA	comparative molecular similarity analysis
D_{max}	maximum diameter
F	Fisher's test
GRIND	grid-independent descriptors
HEK	human embryonic kidney
hERG	human *ether-a-go-go*-related gene
I_{Kr}	rapid component of the delayed rectifier repolarizing current
LIE	linear interaction energy
$\log D$	logarithm of the distribution coefficient
$\log P$	logarithm of the partition coefficient
LQTS	long QT syndrome

LSER	linear solvation energy relationships
MD	molecular dynamics
PEOE_VSA	sum of the Van der Waals surface area of atoms where the PEOE (partial equalization of orbital electronegativities) atomic partial charges are in a predefined range
PLS	partial least squares
q^2	cross-validated correlation coefficient
QSAR	quantitative structure–activity relationships
QT	the time interval between initial depolarization and final repolarization of the ventricles measured by electrocardiogram
r^2	squared correlation coefficient
ROC	receiver operating characteristic
RP	recursive partitioning
s	standard error of the regression
SAR	structure–activity relationships
SMARTS	SMILES (simplified molecular input line entry specification) arbitrary target specification
SVM	support vector machine
TdP	*torsades de pointes*
TPSA	topological polar surface area
VPharm-FP/TOPO/AP	see [33]

References

1 Keating, M.T. and Sanguinetti, M.C. (2001) Molecular and cellular mechanisms of cardiac arrhythmias. *Cell*, **104**, 569–580.

2 Sanguinetti, M.C. and Tristani-Firouzi, M. (2006) hERG potassium channels and cardiac arrhythmia. *Nature*, **440**, 463–469.

3 Fermini, B. and Fossa, A.A. (2003) The impact of drug-induced QT interval prolongation on drug discovery and development. *Nature Reviews. Drug Discovery*, **2**, 439–447.

4 De Ponti, F., Poluzzi, E. and Montanaro, N. (2001) Organising evidence on QT prolongation and occurrence of Torsades de Pointes with non-antiarrhythmic drugs: a call for consensus. *European Journal of Clinical Pharmacology*, **57**, 185–209.

5 Fenichel, R.R., Malik, M., Antzelevitch, C., Sanguinetti, M., Roden, D.M., Priori, S.G., Ruskin, J.N., Lipicky, R.J. and Cantilena, L.R. (2004) Drug-induced torsades de pointes and implications for drug development. *Journal of Cardiovascular Electrophysiology*, **15**, 475–495.

6 De Ponti, F., Poluzzi, E., Cavalli, A., Recanatini, M. and Montanaro, N. (2002) Safety of non-antiarrhythmic drugs that prolong the QT interval or induce torsades de pointes: an overview. *Drug Safety*, **25**, 263–286.

7 Recanatini, M., Poluzzi, E., Masetti, M., Cavalli, A. and De Ponti, F. (2005) QT prolongation through hERG K(+) channel blockade: current knowledge and strategies for the early prediction during drug development. *Medicinal Research Reviews*, **25**, 133–166.

8 Aronov, A.M. (2005) Predictive *in silico* modeling for hERG channel blockers. *Drug Discovery Today*, **10**, 149–155.

9 Noble, D. (2006) Systems biology and the heart. *Biosystems*, **83**, 75–80.

10 Rudy, Y. (2006) Modelling and imaging cardiac repolarization abnormalities. *Journal of Internal Medicine*, **259**, 91–106.

11 Bottino, D., Penland, R.C., Stamps, A., Traebert, M., Dumotier, B., Georgiva, A., Helmlinger, G. and Lett, G.S. (2006) Preclinical cardiac safety assessment of pharmaceutical compounds using an integrated systems-based computer model of the heart. *Progress in Biophysics and Molecular Biology*, **90**, 414–443.

12 Guner, O.F. (2002) History and evolution of the pharmacophore concept in computer-aided drug design. *Current Topics in Medicinal Chemistry*, **2**, 1321–1332.

13 Morgan, T.K., Jr and Sullivan, M.E. (1992) An overview of class III electrophysiological agents: a new generation of antiarrhythmic therapy. *Progress in Medicinal Chemistry*, **29**, 65–108.

14 Ekins, S., Crumb, W.J., Sarazan, R.D., Wikel, J.H. and Wrighton, S.A. (2002) Three-dimensional quantitative structure–activity relationship for inhibition of human ether-a-go-go-related gene potassium channel. *The Journal of Pharmacology and Experimental Therapeutics*, **301**, 427–434.

15 Cavalli, A., Poluzzi, E., De Ponti, F. and Recanatini, M. (2002) Toward a pharmacophore for drugs inducing the long QT syndrome: insights from a CoMFA study of HERG K(+) channel blockers. *Journal of Medicinal Chemistry*, **45**, 3844–3853.

16 Ekins, S. (2003) *In silico* approaches to predicting drug metabolism, toxicology and beyond. *Biochemical Society Transactions*, **31**, 611–614.

17 Pearlstein, R.A., Vaz, R.J., Kang, J., Chen, X.L., Preobrazhenskaya, M., Shchekotikhin, A.E., Korolev, A.M., Lysenkova, L.N., Miroshnikova, O.V., Hendrix, J. and Rampe, D. (2003) Characterization of HERG potassium channel inhibition using CoMSiA 3D QSAR and homology modeling approaches. *Bioorganic & Medicinal Chemistry Letters*, **13**, 1829–1835.

18 Cianchetta, G., Li, Y., Kang, J., Rampe, D., Fravolini, A., Cruciani, G. and Vaz, R.J. (2005) Predictive models for hERG potassium channel blockers. *Bioorganic & Medicinal Chemistry Letters*, **15**, 3637–3642.

19 Aptula, A.O. and Cronin, M.T. (2004) Prediction of hERG K^+ blocking potency: application of structural knowledge. *SAR and QSAR in Environmental Research*, **15**, 399–411.

20 Yoshida, K. and Niwa, T. (2006) Quantitative structure–activity relationship studies on inhibition of HERG potassium channels. *Journal of Chemical Information and Modeling*, **46**, 1371–1378.

21 Seierstad, M. and Agrafiotis, D.K. (2006) A QSAR model of HERG binding using a large, diverse, and internally consistent training set. *Chemical Biology and Drug Design*, **67**, 284–296.

22 Coi, A., Massarelli, I., Murgia, L., Saraceno, M., Calderone, V. and Bianucci, A.M. (2006) Prediction of hERG potassium channel affinity by the CODESSA approach. *Bioorganic & Medicinal Chemistry Letters*, **14**, 3153–3159.

23 Song, M. and Clark, M. (2006) Development and evaluation of an *in silico* model for hERG binding. *Journal of Chemical Information and Modeling*, **46**, 392–400.

24 Roche, O., Trube, G., Zuegge, J., Pflimlin, P., Alanine, A. and Schneider, G. (2002) A virtual screening method for prediction of the HERG potassium

channel liability of compound libraries. *Chembiochem*, **3**, 455–459.

25 Keserü G.M. (2003) Prediction of hERG potassium channel affinity by traditional and hologram QSAR methods. *Bioorganic & Medicinal Chemistry Letters*, **13**, 2773–2775.

26 Tobita, M., Nishikawa, T. and Nagashima, R. (2005) A discriminant model constructed by the support vector machine method for HERG potassium channel inhibitors. *Bioorganic & Medicinal Chemistry Letters*, **15**, 2886–2890.

27 Dubus, E., Ijjaali, I., Petitet, F. and Michel, A. (2006) In silico classification of HERG channel blockers: a knowledge-based strategy. *ChemMedChem*, **1**, 622–630.

28 Bains, W., Basman, A. and White, C. (2004) HERG binding specificity and binding site structure: evidence from a fragment-based evolutionary computing SAR study. *Progress in Biophysics and Molecular Biology*, **86**, 205–233.

29 Sun, H. (2006) An accurate and interpretable Bayesian classification model for prediction of HERG liability. *ChemMedChem*, **1**, 315–322.

30 Yap, C.W., Cai, C.Z., Xue, Y. and Chen, Y.Z. (2004) Prediction of torsade-causing potential of drugs by support vector machine approach. *Toxicological Sciences*, **79**, 170–177.

31 Gepp, M.M. and Hutter, M.C. (2006) Determination of hERG channel blockers using a decision tree. *Bioorganic & Medicinal Chemistry Letters*, **14**, 5325–5332.

32 Ekins, S., Balakin, K.V., Savchuk, N. and Ivanenkov, Y. (2006) Insights for human ether-a-go-go-related gene potassium channel inhibition using recursive partitioning and Kohonen and Sammon mapping techniques. *Journal of Medicinal Chemistry*, **49**, 5059–5071.

33 Aronov, A.M. and Goldman, B.B. (2004) A model for identifying HERG K^+ channel blockers. *Bioorganic & Medicinal Chemistry Letters*, **12**, 2307–2315.

34 Aronov, A.M. (2006) Common pharmacophores for uncharged human ether-a-go-go-related gene (hERG) blockers. *Journal of Medicinal Chemistry*, **49**, 6917–6921.

35 Hansch, C. and Klein, T.E. (1986) Molecular graphics and QSAR in the study of enzyme–ligand interactions. On the definition of bioreceptors. *Accounts of Chemical Research*, **19**, 392–400.

36 Farid, R., Day, T., Friesner, R.A. and Pearlstein, R.A. (2006) New insights about HERG blockade obtained from protein modeling, potential energy mapping, and docking studies. *Bioorganic & Medicinal Chemistry Letters*, **14**, 3160–3173.

37 Recanatini, M., Cavalli, A. and Masetti, M. (2005) In silico modelling-pharmacophores and hERG channel models. *Novartis Foundation Symposium*, **266** 171–181, discussion 181–175.

38 Mitcheson, J.S., Chen, J., Lin, M., Culberson, C. and Sanguinetti, M.C. (2000) A structural basis for drug-induced long QT syndrome. *Proceedings of the National Academy of Sciences of the United States of America*, **97**, 12329–12333.

39 Pearlstein, R., Vaz, R. and Rampe, D. (2003) Understanding the structure–activity relationship of the human ether-a-go-go-related gene cardiac K^+ channel. A model for bad behavior. *Journal of Medicinal Chemistry*, **46**, 2017–2022.

40 Huang, N., Kalyanaraman, C., Bernacki, K. and Jacobson, M.P. (2006) Molecular mechanics methods for predicting protein–ligand binding. *Physical Chemistry Chemical Physics*, **8**, 5166–5177.

41 Åqvist, J., Medina, C. and Samuelsson, J.E. (1994) A new method for predicting binding affinity in computer-aided drug design. *Protein Engineering*, **7**, 385–391.

42 Österberg, F. and Åqvist, J. (2005) Exploring blocker binding to a homology

model of the open hERG K$^+$ channel using docking and molecular dynamics methods. *FEBS Letters*, **579**, 2939–2944.

43 Rajamani, R., Tounge, B.A., Li, J. and Reynolds, C.H. (2005) A two-state homology model of the hERG K$^+$ channel: application to ligand binding. *Bioorganic & Medicinal Chemistry Letters*, **15**, 1737–1741.

44 Jamieson, C., Moir, E.M., Rankovic, Z. and Wishart, G. (2006) Medicinal chemistry of hERG optimizations: highlights and hang-ups. *Journal of Medicinal Chemistry*, **49**, 5029–5046.

6
GPCR *Antitarget* Modeling: Pharmacophore Models to Avoid GPCR-Mediated Side Effects

Thomas Klabunde, Andreas Evers

6.1
Introduction: GPCRs as *Antitargets*

In recent years, the term *antitarget* has been coined within the drug discovery community. Several enzymes, receptors or channels were identified as molecular basis for several severe side effects observed for development candidates (or even for marketed drugs) and were thus termed antitargets. Antitarget-mediated side effects can put the further development of promising clinical candidates at risk. Thus, several pharmaceutical companies have now started to implement appropriate *in vitro* assays in the early phase of the drug discovery chain. In addition, structural information on these antitargets, their ligands and structure–activity relationships is compiled and *in silico* tools are developed for antitarget modeling. These computational tools can guide the chemical optimization of novel lead series toward clinical candidates lacking antitarget-mediated side effects. Probably, the best known example of an antitarget is the K^+ channel encoded by the human ether-a-go-go-related gene (hERG) (Figure 6.1a) [1]. In this book, Chapters 3–5 are dedicated to this ion channel. In short, the hERG K^+ channel plays a crucial role for normal action potential repolarization in the heart. In recent years, several noncardiac drugs have been found to inhibit the hERG K^+ channel, resulting in a drug-induced long QT syndrome and sudden cardiac death. Also terfenadine (Figure 6.1b), a drug released for the treatment of seasonal rhinitis, was found to inhibit the hERG channel with an IC_{50} of 56 nM [2]. It caused significant QT prolongation and had to be withdrawn from the market. Interestingly, fexofenadine, a close analogue of terfenadine, does not inhibit the hERG channel and is free from any cardiac-related side effects (Figure 6.1c). Now, pharmacophore [3], structure-based [4–6], 3D quantitative structure–activity relationship (QSAR) [3] and neural network models [7] have been developed for the hERG channel to support the chemical optimization of novel drug candidates toward molecules with no hERG-mediated cardiac side effects.

Historically, the discovery of drugs acting at G-protein-coupled receptors (GPCRs) has been extremely successful with 50% of all recently launched drugs targeting

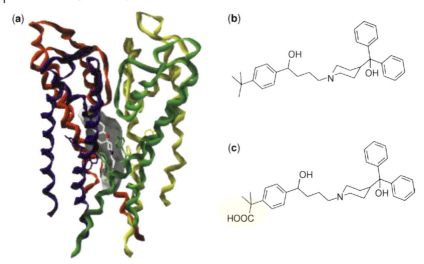

Figure 6.1 The potassium hERG channel as antitarget within drug discovery: (a) homology model of hERG channel with compound MK-499 bound; (b) histamine H1 antagonist terfenadine, which was found to be a strong blocker of the hERG channel inducing the long QT syndrome; (c) fexofenadine, a close analogue of terfenadine, reveals no hERG channel affinity and is successfully marketed for seasonal rhinitis (trade name Allegra).

against GPCRs [8], which form a large protein family that plays an important role in many physiological and pathophysiological processes. Especially, the subfamily of biogenic amine binding GPCRs has provided excellent drug targets (given in parentheses) for the treatment of numerous diseases: schizophrenia (mixed D2/D1/5-HT$_2$ antagonists), psychosis (mixed D2/5-HT$_{2A}$ antagonists), depression (5-HT$_1$ agonists), migraine (5-HT$_1$ agonists), allergies (H1 antagonists), asthma (β2 agonists, M1 antagonists), ulcers (H2 antagonist) or hypertension (α1 antagonist, β1 antagonist). Although representing excellent therapeutic targets, the central role that many biogenic amine binding GPCRs play in cell signaling also poses a risk to new drug candidates that reveal side affinities toward these receptor sites. These candidates have the potential to interfere with the physiological signaling process and to cause undesired effects in preclinical or clinical studies. To obtain a clean clinical profile for novel development candidates, strong molecular interactions with dopamine and serotonin receptors (like 5-HT$_{2A}$ and D2 receptors), representing the molecular targets for many antipsychotics (e.g. olanzapine or risperidone), need to be avoided. Furthermore, the α$_{1A}$ adrenergic receptor modulates the relaxation of the vascular muscle tone and is thus important for blood pressure regulation. It has been suggested as an antitarget that mediates cardiovascular side effects of many drug candidates causing orthostatic hypotension, dizziness and fainting spells [9,10].

The role of the 5-HT$_{2B}$ receptor as GPCR antitarget was underlined only recently by two independent studies reinforcing the causal association between 5-HT$_{2B}$ agonism and valvular heart disease [11–13]. Drug-induced valvular heart disease led to the

Table 6.1 Biogenic amine binding GPCR antitargets and side effects mediated by high-affinity agonists and antagonists of the respective receptor.

Receptor	Associated Side Effects
Adrenergic α_{1a} antagonist	Orthostatic hypotension, dizziness and fainting spells
Dopaminergic D2 antagonist	Extrapyramidal syndrome (EPS), tardive dyskinesia
Serotonin 5-HT$_{2C}$ antagonist	Weight gain, obesity
Serotonin 5-HT$_{2B}$ agonist	Valvular heart disease
Muscarinic M1 antagonist	Attention deficits, hallucinations and memory deficits

withdrawal of the appetite suppressant fenfluramine (Pondimin) in 1997 [14]. Only several years later, it was found that norfenfluramine, the active metabolite of fenfluramine, is a potent agonist of the 5-HT$_{2B}$ receptor suggesting that activation of the cardiac 5-HT$_{2B}$ receptor causes the toxic effects of Pondimin [15]. Two recent studies now show that pergolide and cabergoline (used for the treatment of Parkinson's disease), two dopamine agonists that are also potent 5-HT$_{2B}$ agonists, significantly increase the risk of valvular heart disease [11,12]. These findings stress the importance of avoiding side affinities to the 5-HT$_{2B}$ receptor when developing novel drugs to avoid drug-induced valvular heart disease. Table 6.1 lists the currently known GPCR antitargets and the potential side effects mediated by high-affinity antagonists or agonists of these receptors.

6.2
In Silico Tools for GPCR Antitarget Modeling

To monitor affinity profiles of new drug candidates and to predict undesired GPCR-mediated side effects, we have established a panel of biogenic amine receptor binding assays. Profiling of several hundred compounds within this panel showed that several lead compounds entering the chemical optimization phase reveal affinities toward several members of the biogenic amine antitarget panel. Reliable *in silico* tools to identify compounds with strong GPCR antitarget affinity and computational models to guide the chemical optimization toward compounds with a more favorable GPCR affinity profile thus appear to be of great value for the design and development of new drug candidates.

The challenge in the generation of these pharmacophores for antitarget modeling is the requirement that these models need to describe the receptor interaction points not only for a single chemical series but also for several different compound classes. In addition, these *cross-chemotype* pharmacophores need to capture sufficient pharmacophore points to describe all relevant receptor–ligand interactions. 3D pharmacophore models rationalizing the affinity of several different chemical series have been recently described for the α_{1A}, the 5-HT$_{2A}$ and the D2 receptor [16]. This chapter is focused on the antitarget pharmacophore models of the α_{1A} adrenergic receptor, the prototype of a GPCR antitarget. Using the α_{1A} receptor as an example, the generation and validation of cross-chemotype pharmacophore models as well as first applications of these antitarget models will be described. It will be shown here how these antitarget pharmacophore models are capable to rationalize the strong antitarget affinity of the novel lead series and how they can guide the chemical optimization toward development candidates with a superior safety index.

6.3
GPCR Antitarget Pharmacophore Modeling: The α_{1a} Adrenergic Receptor

Pharmacophore models for the α_{1A} adrenergic receptor have also been described by others [17,18] and are reviewed in the following chapters of this book. Barbaro et al. [17] have used a series of pyridiazionone derivatives based on biological data on the rat receptor as a training set for pharmacophore generation. The model appears to be well suited for a quantitative prediction of the biological activity of the training set molecules and for a chemically closely related series (Figure 6.2a). However, it does not represent a cross-chemotype model suitable to map a diverse set of different α_{1A} chemical series. The model generated by Bremner et al. [18,19], on the contrary, has been derived from a diverse set of 38 compounds. However, it comprises only three pharmacophoric features and is thus quite generic and cannot be expected to be very specific for the α_{1A} receptor (Figure 6.2b). As both available models appeared to be unsuitable for the purpose of antitarget modeling, cross-chemotype pharmacophore models for the human α_{1A} adrenergic receptor have been generated [16].

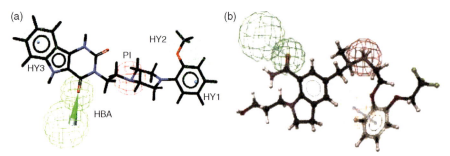

Figure 6.2 α_{1A} adrenergic receptor pharmacophore models (a) by Barbaro et al. [17] using a series of pyridiazionone derivatives and biological data from rat receptor, Reprinted with permission from [17], copyright (2001) American Chemical Society; (b) by Bremner et al. derived from a diverse set of 38 compounds [18,19]. Reprinted from [19], copyright 2000, with permission from Elsevier.

6.3.1
Generation of Cross-Chemotype Pharmacophore Models

The common-features hypothesis generation module of Catalyst 4.7 [20] (termed *HipHop*) was used for the generation of two cross-chemotype 3D pharmacophores describing α_{1A} antagonists. The common-features hypothesis generation module is designed specifically to find chemical features shared by a set of compounds belonging to different chemical classes. It provides the compounds relative alignments with the hypothesis expressing these common features. The training set used for the generation of the α_{1a} adrenergic pharmacophore model has been extracted from the Aureus database, a structure–activity database for GPCR ligands compiled and maintained by Aureus Pharma [21]. The database covers all biological data published on GPCRs and provides chemical structural information, references to the original publication or patent, as well as detailed information on the experimental conditions (e.g. assay type, cell line or radioligand used). Adrenergic α_{1A} receptor antagonists with a K_i value of smaller than 100 nM tested against the recombinant human wild-type receptor were extracted from the Aureus database. The structural analysis of the compounds reveals that they can be grouped into two classes, probably binding overlapping but not identical binding sites within the receptor. Thus, two diverse training sets covering chemotype examples of both classes were selected: (i) class II antagonists are represented by six compounds revealing two aromatic rings and a positively ionizable group positioned at two to four bond lengths from the aromatic features, (ii) 14 representatives of class I antagonists, revealing a positively ionizable group that is separated from the first aromatic ring by two to three bond lengths and by six to seven bond lengths from the second aromatic ring. Table 6.2 and Scheme 6.1 show the chemical structures of both sets of compounds used for generation of the pharmacophores together with the reported binding affinities.

6.3.2
Description of Cross-Chemotype Pharmacophore Models

The two common-feature pharmacophores are depicted in Figure 6.3a and b showing the mapping onto a reference molecule representative for both classes of high-affinity α_{1A} antagonists, respectively [16]. The models describe the key chemical features required for binding of structurally diverse ligands to this adrenergic receptor subtype: the class I pharmacophore (Figure 6.3a) represents a five-point pharmacophore, which is composed of three hydrophobic (HY) moieties connected though a positively ionizable (PI) group (matched by the N2 group of the quinazoline ring) and a hydrogen bond acceptor (HBA) group (mapped by the amide group of the shown compound prazosin). The class II pharmacophore (Figure 6.3b) describes the four main pharmacophoric points of the smaller class of α_{1A} ligands lacking the hydrogen bond acceptor group: two ring aromatic (RA) features, one hydrophobic feature and one positively ionizable feature. The similarity of the right part of both pharmacophores (class I: positively ionizable, hydrophobic, hydrophoic; class II: positively ionizable, hydrophobic, ring aromatic) indicates that the 'head' groups of class I and

Table 6.2 Training set molecules for α_{1A} adrenergic receptor and their affinity measured in a radioligand displacement assay.

Class	Compound	K_i (nM) [16]
I	1	0.2
I	2	0.2
I	Prazosin	0.3
I	NAN 190	0.4
I	RS 17053	0.5
I	3	0.5
I	Doxazosin	0.8
I	4	1.0
I	5	2.8
I	6	4.6
I	Cyclazosin	12.3
I	7	27.1
I	8	44
I	9	72.4
II	YM 617	0.04
II	WB 4104	0.1
II	ARC 239	0.4
II	BE 2254	0.4
II	Spiperone	25.1
II	10	28.2

class II ligands mapping this part of the pharmacophore interact with the same site at the adrenergic receptor (see Section 6.3.4). However, both pharmacophores also reflect the differences between the two different classes of α_{1A} receptor antagonists found in the left part of the molecules. Class I ligands appear to share an acceptor group and a second hydrophobic group separated from the central positive charge by 9.5 Å (five to six bond lengths). The shorter class II ligands, however, reveal an aromatic group connected by only 7.2 Å (two to four bond lengths) to the positively charged nitrogen.

6.3.3
Validation of Antitarget Pharmacophore Models

6.3.3.1 Virtual Screening: Hit Rates and Yields

The purpose of the antitarget pharmacophores is to recognize and rationalize antitarget side affinities within chemotypes different from those used in the training set. Thus, it is crucial to validate the pharmacophore hypotheses using external test set molecules, which have not been used for generation of the pharmacophore. The predictive power of both α_{1A} pharmacophore models has been evaluated by virtual screening using a test database of 50 known α_{1A} antagonists embedded into 1000 drug-like molecules (active and inactive sets were taken from the MDL MDDR database) [16]. To mark the predictive power, hit rates [hit rate = (number of α_{1A} antagonists in hit list)/(total number of compounds in hit list) × 100] and yields [yield = (number of α_{1A} antagonists in hit list)/(number of α_{1A} antagonists in full

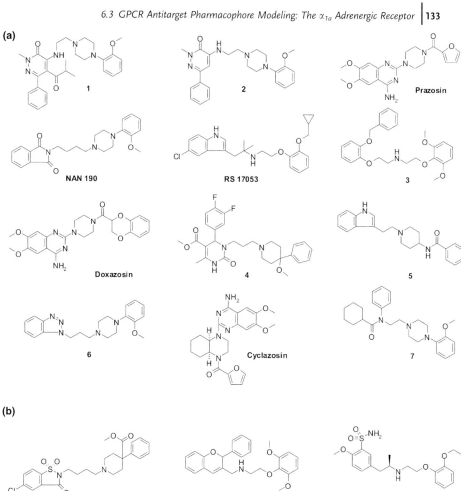

Scheme 6.1 Chemical structures of adrenergic receptor antagonists for generation of antitarget pharmacophores: class I (a) and class II (b).

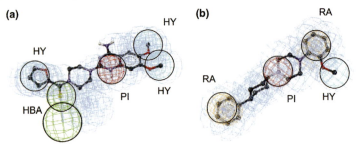

Figure 6.3 Common-feature pharmacophores of α_{1a} adrenergic receptor antagonists [16]. Onto each pharmacophore the reference has been mapped. (a) Class I pharmacophore model aligned to prazosin, (b) class II pharmacophore model aligned to compound **10**. Pharmacophoric features are red for positively ionizable, green for hydrogen bond acceptors, light blue for hydrophobic and orange for ring aromatic. Shape restraints are shown in light blue.

database) × 100] were calculated. The results are depicted in Table 6.3. For both pharmacophores a hit rate of approximately 30% was obtained, which is sixfold higher than a random selection. In addition, with a yield of 84%, the class II pharmacophore was able to identify most of the α_{1A} antagonists within the test set still revealing excellent specificity reflected by a good hit rate. The virtual screening thus suggests that especially the less stringent class II four-point pharmacophore is suitable to recognize most of the known α_{1A} antagonists and to provide mappings of compounds having significant α_{1A} affinity. Taken together both pharmacophores are able to recognize 90% of the α_{1A} antagonists embedded into the test data set.

6.3.3.2 Virtual Screening: Fit Values and Enrichment Factors

In many cases, the performance of a pharmacophore-based virtual screening can be improved when the quality of the mapping to the respective pharmacophore is considered. Fit values of all test set molecules onto both pharmacophores can be calculated and the compounds can be sorted using the fit value of their mappings. The resulting enrichment graphs for both α_{1A} pharmacophores are shown in Figure 6.4 Both enrichment curves show a steep beginning, going almost parallel to the ideal curve (black line). The flattening of the curves toward the right can be explained by the fact that some α_{1A} compounds of the MDDR database cannot be

Table 6.3 Hit rate and yield from virtual screen of MDDR test set database using both α_{1A} pharmacophores as selection filter [16].

Class	Number of Virtual Hits	Number of Identified α_{1A} Antagonists	Hit Rate[a] (%)	Yield[b] (%)
I	82	26	32	52
II	146	42	29	84
I or II	168	45	27	90

[a]Hit rate = (number of true actives in hit list)/(number of compounds in hit list) × 100.
[b]Yield = (number of true actives in hit list)/(number of true actives in full database) × 100.

Table 6.4 Hit rate, yield and enrichment factors for the top 5% scorers from virtual screen of MDDR test set database using both α_{1A} pharmacophores as selection filters.

Class	Number of α_{1A} Antagonists Among Top 5% of Database	Hit Rate^a (%)	Yield^b (%)	Enrichment Factor
I	22	42	44	8.8
II	25	48	50	10

All compounds have been scored based on their fit value onto the respective pharmacophore [16].
^{a}Hit rate = (number of true actives in hit list)/(number of compounds in hit list) × 100.
^{b}Yield = (number of true actives in hit list)/(number of true actives in full database) × 100.

mapped by the pharmacophore and thus obtain fit values of 0. The steepness of the enrichment curve on the left, however, reflects that among the top ranked compounds of the database (e.g. 1–10%) a high percentage of true α_{1A} ligands can be identified by these antitarget pharmacophores (e.g. among the top 10 scored virtual hits using class II pharmacophore six are α_{1A} antagonists indicating a hit rate of 60% among the top 1% of the virtual hits). Hit values, yield and enrichment factor (hit rate found versus random selection) are listed for both pharmacophores for the top 5% scorers in Table 6.4: (i) both pharmacophores provide an excellent yield with, respectively, 44 and 50% of the α_{1A} antagonists being found among the top 5% scorers of the ranked database, (ii) enrichment factors between 9 and 10 were obtained comparing the hit rate of the pharmacophore-based selection to a random selection. The excellent hit rate and yield generated cannot be explained by the structural similarity of MDDR test set molecules to the Aureus training set molecules. Figure 6.4 reveals that the yield and hit rate obtained by ranking the database compounds on the basis of their maximal Tanimoto similarity (calculation is based on 2D fingerprints: green curve) to one of the six Aureus class II training set molecules are not significantly higher than those obtained by a random selection.

The excellent performance of both pharmacophores in terms of yield and enrichment factor suggests that they can also be useful filters for virtual screening to identify α_{1A} antagonists within large compound repositories. Indeed, both pharmacophores have been successfully applied in a virtual screening approach combining pharmacophore-based and homology-model-based virtual screening [22]. Using this combined approach novel α_{1A} antagonists with nanomolar affinity could be identified from the corporate compound collection.

6.3.4
Mapping of Pharmacophore Models into Receptor Site

Numerous site-directed mutagenesis studies have provided a conclusive picture for the molecular interactions between the receptor-activating biogenic amines (e.g. serotonin, epinephrine, dopamine) and their receptors [23–27]: a highly conserved aspartate residue in transmembrane (TM) helix TM3 (Asp 3.32 according to the Ballosteros–Weinstein nomenclature) [28], conserved serine residues in TM5 (e.g.

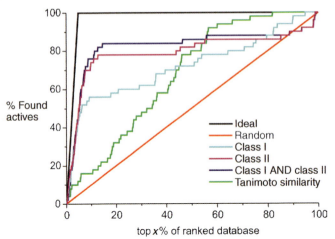

Figure 6.4 Enrichment graph for virtual screening of α_{1A} antagonists embedded into a random MDDR library comprising 1048 compounds [16]. The curve shows the relative ranking of the 50 α_{1A} antagonists. Database compounds are ranked along the x-axis based on the fit value calculated for the mapping on the respective pharmacophore. Cyan: class I pharmacophore; magenta: class II pharmacophore; blue: sum of class I and class II; green: maximal Tanimoto similarity to reference compounds in the training set; red: random; black: ideal. The enrichment by the class II pharmacophore (magenta) at a yield of 50% is 10-fold better than by a random selection.

Ser 5.42 and Ser 5.46 for α_{1A}) as well as hydrophobic phenylalanine residues from TM6 have been identified to be important for agonist binding. In addition, through mutational studies and comparative affinity determinations based on ligand binding, the essential amino acids involved in recognizing antagonists could be identified for the α_{1A} receptor [27,29,30]. According to these studies, the binding pocket of the prototype biogenic amine receptor antagonist stretches from the agonist binding site formed by TM3, TM5 and TM6 – interacting with the antagonists 'head' group – toward the transmembrane helices TM1, TM2 and TM7, which have been suggested to harbor the lipophilic 'tail' moiety of several antagonists. On the basis of these experimental data a topographical interaction model for the α_{1A} receptor has been generated as shown in Figure 6.5 [11]. The two pharmacophore models have been

Figure 6.5 Topographical interaction model of α_{1A} adrenergic receptor generated on the basis of public site-directed mutagenesis. Both pharmacophore models have been mapped into the topographical model of the receptor. The model reveals putative receptor interaction sites for most of the pharmacophore features observed within each antagonist class. (a) Class I pharmacophore with prazosin as reference compound, (b) class II pharmacophore with compound **10** as reference. Pharmacophoric features are red for positively ionizable, green for hydrogen bond acceptors, light blue for hydrophobic and orange for ring aromatic. Shape restraints are shown in light blue. Arrows indicate putative molecular interactions between the pharmacophoric points and receptor sites. Color codes indicate the type of molecular interaction. Light blue: hydrophobic; red: salt bridge to negative ionizable group from receptor; orange: aromatic stacking interaction.

6.3 *GPCR Antitarget Pharmacophore Modeling: The α₁ₐ Adrenergic Receptor* | 137

(a)

(b)

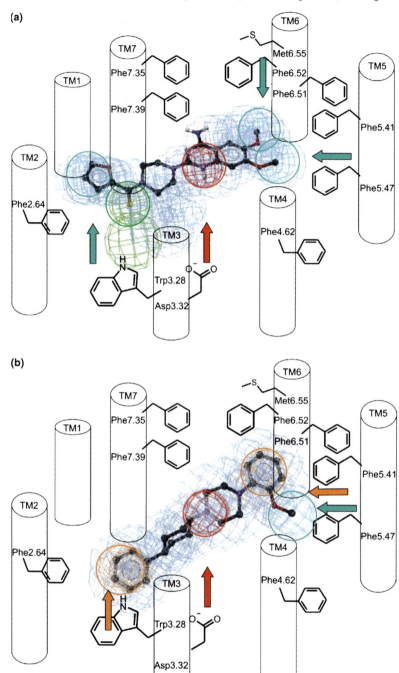

mapped into the topographical interaction model to indicate the putative interaction points of each pharmacophoric feature with its receptor: (i) the positive ionizable pharmacophoric feature is thought to be anchored via a salt bridge to the conserved aspartate residue in TM3, (ii) the hydrophobic and aromatic features of the 'head' moieties are harbored within hydrophobic microdomains formed by aromatic and aliphatic side chains of TM4, TM5 and TM6. The 'floor' of this hydrophobic microdomain is formed by several conserved aromatic amino acids (Phe 6.44, Trp 6.48, Phe 5.47), which are conserved among the family of biogenic amine GPCRs, (iii) the hydrophobic or ring aromatic feature observed within the 'tail' moiety of almost all α_{1A} antagonists is likely to be directed toward aromatic and hydrophobic residues within TM helices TM3 (Trp 3.28) and TM2 (Phe 2.64).

6.3.5
Guidance of Chemical Optimization to Avoid GPCR-Mediated Side Effects

Our company has established a panel of biogenic amine receptor binding assays to monitor affinity profiles of novel drug candidates. Up to now several hundreds of compounds coming from GPCR-directed libraries have been profiled against this panel revealing that approximately 14% of the profiled compounds have moderate α_{1A} affinity in the submicromolar range and 3.5% of all tested compounds reveal strong α_{1A} binding with affinities of smaller than 100 nM. These results once more show the need to optimize the side affinity profile of several compounds to support further development of these drug candidates.

The main application of the generated antitarget pharmacophore hypotheses is to rationalize these experimental findings by providing pharmacophore-based

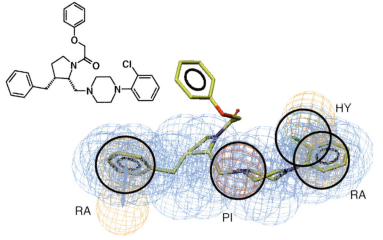

Figure 6.6 Mapping of high-affinity α_{1A} binder onto class II adrenergic α_{1A} pharmacophore model. All pharmacophoric points are mapped. The alignment suggests that removal of the chlorine substituent within the 4-phenyl piperidine will reduce the unfavorable side affinity on α_{1A}.

mappings. Recognition of the key chemical features, responsible for the side affinities of a chemical series, could then provide guidance for the chemical optimization of the series toward compounds with a more favorable side affinity profile. Most important, 80% of all experimentally identified α_{1A} binders could be mapped onto the class II α_{1A} pharmacophore, fulfilling all four pharmacophoric points. Mapping of one of these compounds onto the α_{1A} class II pharmacophore is shown in Figure 6.6. The mapping directly suggests the chemical features that are mediating the strong affinity toward this subtype of the adrenergic receptor: these are the positive charge of the piperazine moiety, the *ortho*-substituted phenyl ring at the 4-position of the piperazine as well as the aromatic ring of the phenoxy chain. The mapping thus provides direct guidance for the chemical optimization of the respective series to avoid the undesired α_{1A} affinity (e.g. removal of the chlorine substituent within the 4-phenyl piperazine).

6.4
Summary

As shown in this chapter, cross-chemotype pharmacophore models, which describe the key pharmacophoric features required for receptor binding, can be generated from training sets covering chemically diverse ligands. When site-directed mutagenesis data are available, the models can be mapped into the receptor recognition site linking each pharmacophoric point to its interaction site within the receptor. These 3D pharmacophores, when applied as filters within the virtual screening, offer acceptable levels of predictivity as revealed by good yields and enrichment factors. Furthermore, these *in silico* tools can be directly applied to guide the chemical optimization of novel GPCR drug candidates toward clinical candidates with less α_{1A}-mediated side effects (orthostatic hypotension, dizziness, fainting spells).

Abbreviations

5-HT	serotonin receptor
α	alpha-adrenergic receptor
β	beta-adrenergic receptor
D	dopaminergic receptor
H	histamine receptor
EPS	extrapyramidal syndrome
GPCR	G-protein-coupled receptor
hERG	human ether-a-go-go-related gene
K_i	inhibition constant
M	muscarinic receptor
QSAR	quantitative structure–activity relationship
TM	transmembrane

References

1 Keating, M.T. and Sanguinetti, M.C. (2001) Molecular and cellular mechanisms of cardiac arrhythmias. *Cell*, **104**, 569–580.

2 Pearlstein, R.A., Vaz, R.J., Kang, J., Chen, X.L., Preobrazhenskaya, M., Shchekotikhin, A.E., Korolev, A.M., Lysenkova, L.N., Miroshnikova, O.V., Hendrix, J. and Rampe, D. (2003) Characterization of HERG potassium channel inhibition using CoMSiA 3D QSAR and homology modeling approaches. *Bioorganic & Medicinal Chemistry Letters*, **13**, 1829–1835.

3 Cavalli, A., Poluzzi, E., De, P.F. and Recanatini, M. (2002) Toward a pharmacophore for drugs inducing the long QT syndrome: insights from a CoMFA study of HERG K(+) channel blockers. *Journal of Medicinal Chemistry*, **45**, 3844–3853.

4 Pearlstein, R., Vaz, R. and Rampe, D. (2003) Understanding the structure–activity relationship of the human ether-a-go-go-related gene cardiac K^+ channel. A model for bad behavior. *Journal of Medicinal Chemistry*, **46**, 2017–2022.

5 Mitcheson, J.S., Chen, J. and Sanguinetti, M.C. (2000) Trapping of a methanesulfonanilide by closure of the HERG potassium channel activation gate. *The Journal of General Physiology*, **115**, 229–240.

6 Mitcheson, J.S., Chen, J., Lin, M., Culberson, C. and Sanguinetti, M.C. (2000) A structural basis for drug-induced long QT syndrome. *Proceedings of the National Academy of Sciences of the United States of America*, **97**, 12329–12333.

7 Roche, O., Trube, G., Zuegge, J., Pflimlin, P., Alanine, A. and Schneider, G. (2002) A virtual screening method for prediction of the HERG potassium channel liability of compound libraries. *Chembiochem: A European Journal of Chemical Biology*, **3**, 455–459.

8 Klabunde, T. and Hessler, G. (2002) Drug design strategies for targeting G-protein-coupled receptors. *Chembiochem: A European Journal of Chemical Biology*, **3**, 928–944.

9 Kehne, J.H., Baron, B.M., Carr, A.A., Chaney, S.F., Elands, J., Feldman, D.J., Frank, R.A., van Giersbergen, P.L., McCloskey, T.C., Johnson, M.P., McCarty, D.R., Poirot, M., Senyah, Y., Siegel, B.W. and Widmaier, C. (1996) Preclinical characterization of the potential of the putative atypical antipsychotic MDL 100,907 as a potent 5-HT2A antagonist with a favorable CNS safety profile. *The Journal of Pharmacology and Experimental Therapeutics*, **277**, 968–981.

10 Peroutka, S.J., U'Prichard, D.C., Greenberg, D.A. and Snyder, S.H. (1977) Neuroleptic drug interactions with norepinephrine alpha receptor binding sites in rat brain. *Neuropharmacology*, **16**, 549–556.

11 Schade, R., Andersohn, F., Suissa, S., Haverkamp, W. and Garbe, E. (2007) Dopamine agonists and the risk of cardiac-valve regurgitation. *The New England Journal of Medicine*, **356**, 29–38.

12 Zanetti, R., Antonini, A., Gatto, G., Gentile, R., Tesei, S. and Pezzoli, G. (2007) Valvular heart disease and the use of dopamine agonists for Parkinson's disease. *The New England Journal of Medicine*, **356**, 39–46.

13 Roth, B.L. (2007) Drugs and valvular heart disease. *The New England Journal of Medicine*, **356**, 6–9.

14 Conelly, H.M., Crary, J.L., McGoon, M.D. Hensrud, D.D., Edwards, B.S., Edwards, W.D., Schaff, H.V. (1997) Valvular disease associated with fenfluramine–phentermine. *The New England Journal of Medicine*, **337**, 581–588.

15 Rothman, R.B., Baumann, M.H., Savage, J.E., Rauser, L., McBride, A., Hufeisen, S. J. and Roth, B.L. (2000) Evidence for

possible involvement of 5-HT2B receptors in the cardiac valvulopathy associated with fenfluramine and other serotonergic medications. *Circulation*, 102, 2836–2841.

16 Klabunde, T. and Evers, A. (2005) GPCR anti-target modeling: pharmacophore models for biogenic amine binding GPCRs to avoid GPCR-mediated side effects. *Chembiochem: A European Journal of Chemical Biology*, 6, 876–889.

17 Barbaro, R., Betti, L., Botta, M., Corelli, F., Giannaccini, G., Maccari, L., Manetti, F., Strappaghetti, G. and Corsano, S. (2001) Synthesis biological evaluation, and pharmacophore generation of new pyridazinone derivatives with affinity toward alpha(1)- and alpha(2)-adrenoceptors. *Journal of Medicinal Chemistry*, 44, 2118–2132.

18 Bremner, J.B., Griffith, R. and Coban, B. (2001) Ligand design for alpha(1) adrenoceptors. *Current Medicinal Chemistry*, 8, 607–620.

19 Bremner, J.B., Coban, B., Griffith, R., Groenewoud, K.M. and Yates, B.F. (2000) Ligand design for alpha1 adrenoceptor subtype selective antagonists. *Bioorganic & Medicinal Chemistry*, 8, 201–214.

20 Catalyst, version 4.7, Accelrys Inc, San Diego, CA, 2002. 2004.

21 Aureus Pharma. (2004) www.aureus-pharma.com.

22 Evers, A. and Klabunde, T. (2005) Structure-based drug discovery using GPCR homology modelling: successful virtual screening for antagonists of the alpha1a receptor. *Journal of Medicinal Chemistry*, 48, 1088–1097.

23 Chambers, J.J. and Nichols, D.E. (2002) A homology-based model of the human 5-HT2A receptor derived from an *in silico* activated G-protein coupled receptor. *Journal of Computer-Aided Molecular Design*, 16, 511–520.

24 Hwa, J., Graham, R.M. and Perez, D.M. (1995) Identification of critical determinants of alpha 1-adrenergic receptor subtype selective agonist binding. *The Journal of Biological Chemistry*, 270, 23189–23195.

25 Ishiguro, M. (2004) Ligand-binding modes in cationic biogenic amine receptors. *Chembiochem: A European Journal of Chemical Biology*, 5, 1210.

26 Pollock, N.J., Manelli, A.M., Hutchins, C.W., Steffey, M.E., MacKenzie, R.G. and Frail, D.E. (1992) Serine mutations in transmembrane V of the dopamine D1 receptor affect ligand interactions and receptor activation. *The Journal of Biological Chemistry*, 267, 17780–17786.

27 Shi, L. and Javitch, J.A. (2002) The binding site of aminergic G protein-coupled receptors: the transmembrane segments and second extracellular loop. *Annual Review of Pharmacology and Toxicology*, 42, 437–467.

28 Ballosteros, J.A. and Weinstein, H. (1995) Integrated methods for the construction of three-dimensional models and computational probing of structure–function relations of G protein-coupled receptors. *Methods in Neurosciences*, 25, 366–428.

29 Hamaguchi, N., True, T.A., Saussy, D.L., Jr and Jeffs, P.W. (1996) Phenylalanine in the second membrane-spanning domain of alpha 1A-adrenergic receptor determines subtype selectivity of dihydropyridine antagonists. *Biochemistry*, 35, 14312–14317.

30 Hamaguchi, N., True, T.A., Goetz, A.S., Stouffer, M.J., Lybrand, T.P. and Jeffs, P.W. (1998) Alpha 1-adrenergic receptor subtype determinants for 4-piperidyl oxazole antagonists. *Biochemistry*, 37, 5730–5737.

7
The Emergence of Serotonin 5-HT$_{2B}$ Receptors as Drug Antitargets
Vincent Setola, Bryan L. Roth

Pharmacotherapies contribute greatly to human health and longevity. As such, researchers in the public and private sectors make an enormous investment in drug design and refinement and preclinical and clinical testing. Such efforts often result in a marketed compound that expands the options for the treatment and prevention of human diseases. However, on rare occasions, medications associated with adverse effects – ranging from undesirable to potentially fatal – that greatly reduce or even eliminate their therapeutic utility reach the marketplace.

The prediction and prevention of adverse effects are a major focus in drug design efforts. Receptorome screening is a powerful approach that has proven useful in deciphering the mechanisms of the salutary as well as adverse effects of compounds targeting the central nervous system. In this chapter, we present a brief introduction of receptorome screening, highlighting some of its successes, the most significant of which was the identification of the serotonin 5-HT$_{2B}$ receptor as the likely culprit for the valvulopathogenicity of the now-banned amphetamine derivative fenfluramine (the anorectic component of the Fen-Phen diet supplement). We will discuss this in detail, as well as integrate findings from later studies that corroborated the results of the receptorome screening pertinent to fenfluramine-induced valvular heart disease (VHD) and extended them to drug-induced pulmonary hypertension (PH), another side effect of fenfluramine use. Next, we will discuss how studies of fenfluramine and other VHD- and PH-associated drugs resulted in the identification of a handful of structural classes – all exhibiting high-to-moderate affinity for, and agonist potency at, 5-HT$_{2B}$ receptors – with a propensity for inducing cardiopulmonary complications. After presenting ample evidence to conclude that 5-HT$_{2B}$ receptors represent a dangerous 'antitarget' for drugs destined for human use, we will suggest methodologies applicable during the early stages of drug design and testing that might permit the identification of potentially fatal cardiopulmonary side effects.

Antitargets. Edited by R. J. Vaz and T. Klabunde
Copyright © 2008 WILEY-VCH Verlag GmbH & Co. KGaA, Weinheim
ISBN: 978-3-527-31821-6

7.1
Receptorome Screening to Identify Drug Targets and Antitargets

Drugs exert their desired biological actions by affecting the activities of molecules involved in biological processes. For instance, antipsychotics alleviate psychotic symptoms primarily by blocking the activation of D_2 and serotonin 5-HT$_{2A}$ receptors [1], which are involved in perception; antidepressants ameliorate depression, in part, by prolonging serotonergic and noradrenergic neurotransmission via inhibition of serotonin transporter (SERT)- and noradrenaline transporter (NET)-mediated reuptake into presynaptic terminals [1]. The D_2 and 5-HT$_{2A}$ receptors, in the aforementioned examples might be considered 'targets' for antipsychotic medications; SERT and NET might be thought of as 'targets' for antidepressant drugs. For the purposes of this chapter, we consider a 'target' to be a receptor whose activity is affected by a drug to achieve a desired effect.

Receptorome screening involves measuring drug affinity for and/or activity at a large number of recombinant neurotransmitter receptors and transporters [1,2]. A recent success of the receptorome screening approach to determine drug targets was its application to identify κ-opioid receptors as the site of action of salvinorin A, a hallucinogen found in the mint *Salvia divinorum* [3,4]. Interestingly, the chemical structure of salvinorin A (a *trans*-neoclerodane diterpenoid) is unique among hallucinogens, both natural and synthetic, in that it lacks an amine group. Functionally, salvinorin A is distinct from other hallucinogens as it exhibits no affinity for or activity at 5-HT$_{2A}$ receptors [4], the target for other hallucinogenic compounds [5]. Thus, salvinorin A was identified as a novel hallucinogen both chemically and mechanistically. Receptorome screening allowed a discovery-based approach to identifying the target of salvinorin A in the absence of any insights into its mechanism of action.

Receptorome screening has also been successfully applied to gain mechanistic insights into the adverse or undesirable effects of drugs. One such example involves the receptorome screening of a large panel of antipsychotic medications, conducted in part to elucidate novel targets for antipsychotic efficacy [1]. In an attempt to understand why some antipsychotic medications caused serious weight gain while others induced less or none, Kroeze *et al.* [6] compared the results of the antipsychotic receptorome screening with clinical studies of antipsychotic-induced weight gain. Statistical analyses of the two data sets revealed a positive association between histamine H1 receptor affinity and antipsychotic-induced weight gain, suggesting that interactions between antipsychotics and H1 receptors may deleteriously alter feeding behavior and/or metabolism. In other words, the H1 receptor represents an 'antitarget' for antipsychotic medications, that is, a receptor for which a drug has appreciable affinity that might be or is involved in unanticipated, undesirable effects. Recently, Snyder's group corroborated these findings and identified AMP kinase as a downstream effector in this pathway [7].

Perhaps, the most significant success of receptorome screening was the implication of the serotonin 5-HT$_{2B}$ receptor in the VHD associated with fenfluramine use. The anorexigen fenfluramine was widely used in the 1990s as an adjunct to diet and exercise, with great success [8]. Originally, racemic fenfluramine was used with phentermine, the mild stimulant properties of the latter helping to offset the drowsiness induced by the former; the combination came to be known as Fen-Phen. Later, the (+)-rotamer of fenfluramine (dexfenfluramine, Pondimin) was identified as being a more potent appetite suppressant, allowing lower dosing and better tolerance vis-à-vis uncomfortable side effects. Despite its efficacy in weight-loss regimens, fenfluramine was withdrawn from the US market voluntarily by its manufacturer due to its association with VHD and PH [9].

In 2000, in collaboration with Richard Rothman (National Institute of Drug Abuse), we performed a receptorome screening of fenfluramine and other VHD-associated drugs (e.g. the antimigraine drugs methysergide and dihydroergotamine) with the hope of identifying neurotransmitter receptor(s) and/or transporter(s) possibly involved in fenfluramine-induced VHD [10]. We found that drugs associated with VHD (e.g. fenfluramine and methysergide) and their metabolites (e.g. norfenfluramine and methylergonovine) (see Figure 7.1) all shared moderate-to-high affinity for and agonist potency – with the metabolites being more potent and efficacious – at 5-HT$_{2B}$ receptors [10], which are expressed in heart valves [11,12], the site of the characteristic proliferative lesions in VHD. In addition, around the same time, a group from DuPont Pharmaceuticals demonstrated that norfenfluramine stimulated mitogen-activated protein kinase (MAPK) activity via activation of recombinant 5-HT$_{2B}$ receptors expressed in HEK-293 cells [12]. Thus, based on the results of the receptorome screening of VHD-associated drugs, we predicted that drugs and/or their metabolites that are 5-HT$_{2B}$ receptor agonists might be associated with increased risk for VHD in humans [10].

7.2
Post-Receptorome Screening Data Implicate 5-HT$_{2B}$ Receptors in Drug-Induced VHD and PH

Further evidence implicating 5-HT$_{2B}$ receptors in drug-associated VHD came in the years following our predictions. In 2003, we used primary cultures of human heart valve interstitial cells (hVICs) to assess whether fenfluramine and/or norfenfluramine induced cell proliferation – an *in vitro* approximation of the proliferative lesions found in the valves of VHD patients – in a 5-HT$_{2B}$ receptor-dependent manner [11]. Treatment of primary hVIC cultures with either fenfluramine or norfenfluramine increased biochemical indices of cell proliferation, an effect that was blocked by cotreatment with a 5-HT$_{2B}$ receptor-selective antagonist [11].

In early 2000, a receptorome screening of 3,4-methylenedioxymethamphetamine (MDMA, 'ecstasy') had been performed and showed that MDMA had moderate affinity for 5-HT$_{2B}$ receptors (unpublished data). We confirmed that MDMA and its N-demethylated metabolite, 3,4-methylenedioxyamphetamine (MDA) – also a drug of abuse – had moderate affinity for and agonist potency at 5-HT$_{2B}$ receptors, with MDA being a more efficacious agonist [11]. Next, we sought to test the predictions of Rothman *et al. in vitro* using MDMA and MDA, whose potential to induce VHD was unknown. Consistent with the 5-HT$_{2B}$ receptor/VHD hypothesis, MDMA and MDA

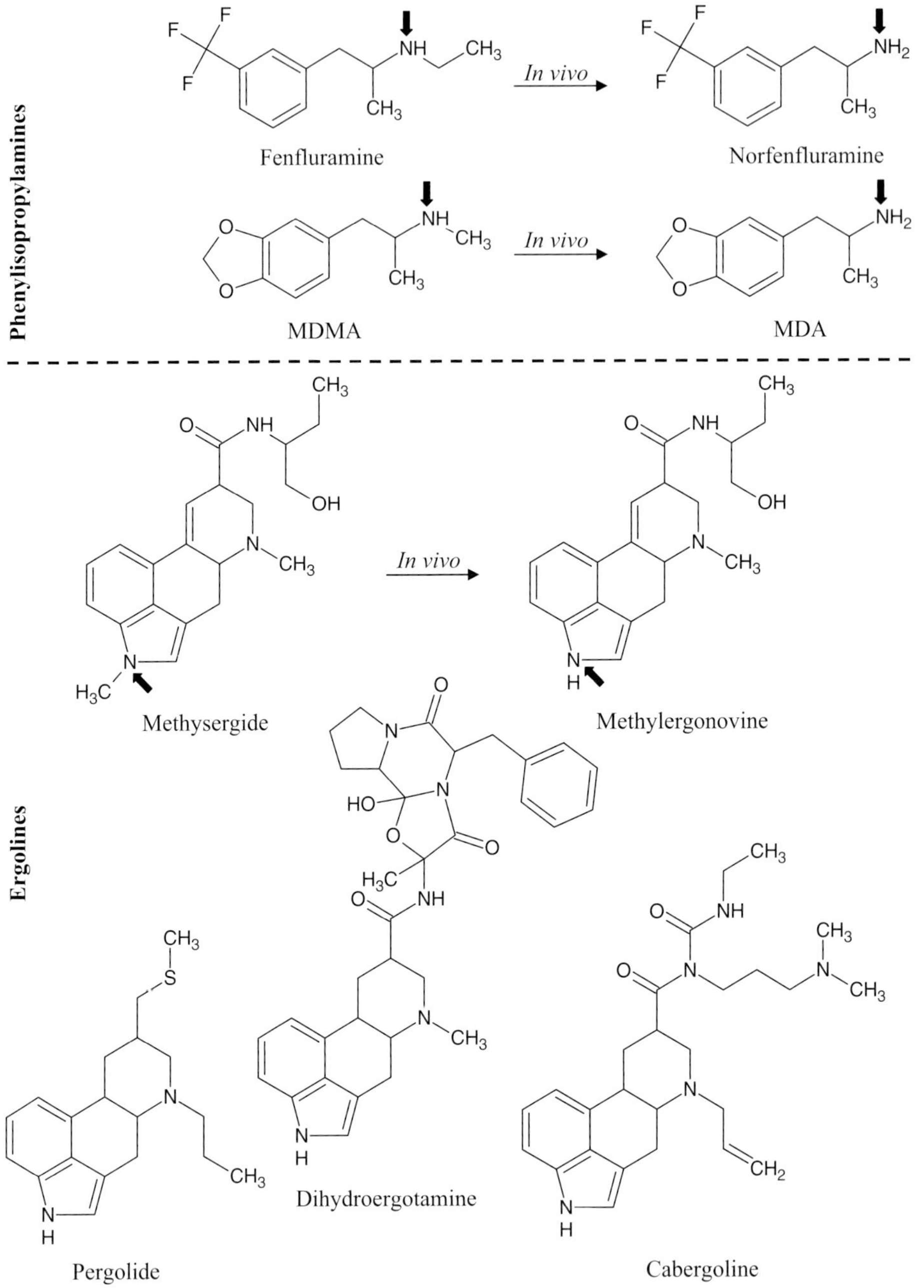

Figure 7.1 Chemical structures of VHD/PH-associated drugs that are also 5-HT$_{2B}$ receptor agonists. The chemical structures of all currently known VHD/PH-inducing drugs – all exhibit high-to-moderate 5-HT$_{2B}$ receptor agonist potency – are shown. MDMA and MDA are included as they behave similarly to VHD-associated drugs *in vitro*. For fenfluramine, MDMA and methysergide, *in vivo* oxidation gives rise to a more potent 5-HT$_{2B}$ receptor agonist than the parent compound. The oxidized nitrogen is indicated by a solid arrow. Note that all drugs fall into one of the two structural classes: phenylisopropylamines (top) and ergolines (bottom).

each increased biochemical measures of cell proliferation in primary cultures of hVICs in a 5-HT$_{2B}$ receptor-dependent manner [11].

In 2002, a group at the Mayo Clinic reported VHD in patients taking the anti-Parkinsonian pergolide (Permax) [13]. Simultaneously, we assessed whether pergolide exhibited agonist properties at 5-HT$_{2B}$ receptors and published our results in 2003: pergolide did indeed exhibit high-affinity binding and potent, efficacious agonist activity at recombinant 5-HT$_{2B}$ receptors [11]. Since the initial 2002 report, many cases of pergolide-induced VHD have been identified [14,15]. In addition, another anti-Parkinsonian – cabergoline – has been associated with VHD [15] and shown to be a 5-HT$_{2B}$ receptor agonist [16].

In addition to its association with VHD, fenfluramine use has also been linked to PH in humans [9]. Insights into the mechanism of fenfluramine-induced PH have been gleaned using a chronic hypoxic mouse model of PH. While fenfluramine treatment *per se* does not induce PH in mice, exposure to chronic hypoxia does elicit the condition and treatment with fenfluramine exacerbates chronic hypoxia-induced PH [17]. In 2002, Launay *et al.* reported that 5-HT$_{2B}^{-/-}$ mice, as well as wild-type mice treated with a potent and selective 5-HT$_{2B}$ receptor antagonist, were completely resistant to PH induced by chronic hypoxia, irrespective of fenfluramine treatment, as measured by right ventricular systolic pressure (RVSP) [17]. Chronic hypoxia also induces – both in mice and in humans – an increase in pulmonary artery smooth muscle cell proliferation, an effect worsened by fenfluramine treatment. Furthermore, pulmonary artery smooth muscle tissue from humans and mice with PH expresses significantly higher levels of 5-HT$_{2B}$ receptor than normal pulmonary artery smooth muscle tissue [17]; given the importance of pulmonary smooth muscle cell proliferation on PH development and the mitogenic activity of 5-HT$_{2B}$ receptors [11,12,18,19], the increased expression of the receptor could generate a positive feedback loop with endogenous 5-HT and/or exogenous drugs that are themselves and/or have metabolites that are 5-HT$_{2B}$ receptor agonists. Furthermore, as predicted by RVSP measurements, genetic or pharmacologic blockade of 5-HT$_{2B}$ function completely blocks the smooth muscle cell proliferation induced by either chronic hypoxia or chronic hypoxia-plus-fenfluramine [17].

In considering the aforementioned mouse studies linking 5-HT$_{2B}$ receptors to PH, one must be judicious when extrapolating the data to humans. In the case of PH, there are similarities between mice and humans regarding 5-HT$_{2B}$ receptors (i.e. both are upregulated in pulmonary artery smooth muscle tissue from mice and humans with PH), 5-HT (i.e. plasma 5-HT is increased in both mouse and human PH) and in vascular remodeling, all of which might legitimize comparisons between a mouse model of PH and the human disease. Significantly, data from a human study of PH that link 5-HT$_{2B}$ receptor activity with the disease exist. In 2003, Blanpain *et al.* identified the R393X single nucleotide polymorphism (SNP) in 1 out of 10 PH patients who had taken fenfluramine; the SNP was not found in 18 PH fenfluramine-naïve controls, nor was it observed in 80 non-PH control subjects [20]. The R393X SNP encodes a 5-HT$_{2B}$ receptor that lacks a C-terminal tail and, when expressed in cell lines, exhibits neither any coupling to inositol trisphosphate production nor any increase in intracellular calcium levels, as does the wild-type receptor [20]. Thus,

Blanpain et al. suggested that a lack of 5-HT$_{2B}$ receptor activity is associated with fenfluramine-induced PH. A few years later, Deraet et al. probed further into the intracellular signaling of the 5-HT$_{2B}$R393X variant, measuring other second messengers and biological processes that are known to be modulated by 5-HT$_{2B}$ receptors in murine LMTK fibroblasts [19]. As Blanpain et al. had reported, Deraet et al. found that the R393X SNP did eliminate receptor-dependent inositol trisphosphate production, an effect that they concluded was due to reduced receptor-Gα_q coupling [19]. In addition, the R393X variant 5-HT$_{2B}$ receptor exhibited a partial loss of coupling to nitric oxide synthase [19], a known effector of the 5-HT$_{2B}$ receptor [21]. However, the R393X SNP increased the 5-HT$_{2B}$ receptor-mediated MAPK activation and [^{3}H]thymidine incorporation via coupling to Gα_{13} and eliminated agonist-induced internalization, revealing that the C-terminal truncation represents a gain-of-function mutation with respect to cell proliferation. Thus, combining the results of Blanpain et al. and Deraet et al. with those showing that 5-HT$_{2B}$ receptors, which markedly stimulate mitogenic processes [11,12,18,19], are upregulated in pulmonary artery smooth muscle cells from mice and humans with PH [17] results in a plausible link between 5-HT$_{2B}$ receptor activity and the pulmonary artery smooth muscle cell proliferation that mark the initial stages of PH.

Although humans and mice with PH display marked increases in plasma 5-HT [17,22], the effect in human primary PH is 50-fold [22]. Recently, Callebert et al. measured plasma 5-HT in wild type and 5-HT$_{2B}$$^{-/-}$ exposed to chronic hypoxia [23]. As had been shown before [17], chronic hypoxia induced PH in wild-type mice but not in mice either lacking 5-HT$_{2B}$ receptors or wild-type mice treated with a 5-HT$_{2B}$ receptor-selective antagonist; in addition, plasma 5-HT levels were elevated only in mice with intact 5-HT$_{2B}$ receptor function [23]. Notably, PH (as well as VHD) is a complication found among patients with serotonin-secreting carcinoid tumors and in whom plasma 5-HT levels are greatly elevated [24]. Thus, one possible mechanism by which 5-HT$_{2B}$ receptor activity contributes to PH development is modulating plasma 5-HT concentration. Indeed, Launay et al. recently described 5-HT-induced, 5-HT$_{2B}$ receptor-mediated decrease in 5-HT uptake by the plasma membrane SERT in 1C11 neuroectodermal progenitor cells and also in primary cultures of raphe neurons [25]. In addition, acute administration of a 5-HT$_{2B}$ receptor-selective agonist to mice causes a transient increase in plasma 5-HT that is abrogated by SERT blockade [23].

The preceding discussion provides substantial evidence to implicate 5-HT$_{2B}$ receptors in drug-induced VHD and PH. To summarize, (1) all known VHD-associated drugs and/or their metabolites are potent, efficacious 5-HT$_{2B}$ receptor agonists; (2) VHD-associated drugs elicit 5-HT$_{2B}$ receptor-dependent mitogenic responses from primary cultures of VICs, an *in vitro* model of the proliferative lesions seen in the valves of VHD patients; (3) functional 5-HT$_{2B}$ receptors are absolutely required for the development of PH in a mouse model of the disease and fenfluramine exacerbates experimentally induced PH in wild-type mice; and (4) administration of a 5-HT$_{2B}$ receptor agonist causes an increase in plasma 5-HT levels in mice, a condition that in both mice and humans (e.g. carcinoid syndrome) is associated with VHD and PH. On the basis of evidence presented here, our original suggestion [10] 'that all clinically available medications with serotonergic activity and their active metabolites be screened for agonist activity at 5-HT$_{2B}$ receptors and that clinicians should consider suspending their use of medications with significant activity at 5-HT$_{2B}$ receptors' – that is, the 5-HT$_{2B}$ receptor is an antitarget for drugs used in humans – bears, 7 years later, considerably more importance.

The importance we ascribed to assessing 5-HT$_{2B}$ activity of drug metabolites [10] merits some elaboration. As mentioned above, the N-deethylated metabolite of fenfluramine, norfenfluramine, is a 10-fold more potent and 2.5-fold more efficacious 5-HT$_{2B}$ receptor agonist [10,11]. In their PH study, Launay et al. measured plasma fenfluramine and norfenfluramine levels in mice treated with a 'therapeutic dose' of fenfluramine (2.5 mg/kg/day) for 5 weeks; they found that plasma levels of fenfluramine were in the 5 nM range [17] – about 80 times lower than the EC$_{50}$ [10,11]; and those of norfenfluramine were in the 500 nM range [17] – about 25 times greater than the EC$_{50}$ [10,11]. Assuming similar pharmacodynamics in humans, and based on the relative efficacies of fenfluramine (about 10% compared to the E_{max} of 5-HT [10,11]) and norfenfluramine (70–100% compared to the E_{max} of 5-HT [10,11]), one would expect essentially maximal 5-HT$_{2B}$ receptor activation by plasma norfenfluramine and no antagonism of the response by the weak partial agonist fenfluramine. Indeed, if one accepts the 5-HT$_{2B}$ model of fenfluramine-induced VHD and PH, one must also accept, given the pharmacodynamics, that norfenfluramine is the actual culprit, not fenfluramine. Another example of metabolism altering the potency and efficacy of a drug is trazodone (Desyrel), which is an antagonist at 5-HT$_2$ receptors [10]. In vivo, trazodone is metabolized to meta-chlorophenylpiperazine (mCPP) [26,27], which is a 5-HT$_{2B/2C}$ receptor agonist [10,28]. Serendipitously, plasma levels of the parent compound trazodone are about 10-fold greater than those of the metabolite [26,27]; thus, nonmetabolized trazodone would be expected to block any activation of 5-HT$_{2B/2C}$ receptors by mCPP and might explain why trazodone use has not been associated with VHD or PH [10]. These are just two examples highlighting the importance of characterizing the major metabolites of drugs and to screen them and their parent compounds for agonist activity at the 5-HT$_{2B}$ receptor antitarget, before they are used in humans, to identify potential VHD/PH-associated drugs.

7.3
Drug Structural Classes and VHD/PH

To date, all but one of the VHD/PH-associated drugs (dihydroergotamine, methysergide, pergolide and cabergoline) are ergoline derivatives (see Figure 7.1). Notably, a study of VHD incidence in ergoline-treated and non-ergoline-treated Parkinson's disease patients versus controls revealed no association between non-ergoline treatments and VHD [15]. Thus, it is not likely that dopamine receptor agonism participates in ergoline-induced VHD. The sole non-ergoline VHD-associated drug is fenfluramine, which is a phenylisopropylamine (see Figure 7.1). MDMA and MDA are also phenylisopropylamines (see Figure 7.1) and while their abuse in humans has not been linked to VHD, our in vitro findings that both compounds induce

VIC proliferation are just cause for suspicion. Thus, it seems that ergolines and phenylisopropylamines are both structural classes likely to contain VHD-inducing members. Furthermore, since all VHD-associated drugs are 5-HT$_{2B}$ receptor agonists, the ergoline and phenylisopropylamine structural classes are also likely to contain compounds that are themselves and/or have metabolites that are 5-HT$_{2B}$ receptor agonists. As such, using ergolines and phenylisopropylamines as pharmaceuticals should be avoided, unless it is established that neither the parent compound nor its metabolites are 5-HT$_{2B}$ receptor agonists.

It is interesting to contemplate whether ergolines such as methysergide and pergolide and phenylisopropylamines such as fenfluramine can be modified in such a way as to eliminate agonist actions at 5-HT$_{2B}$ receptors, as the drugs have been effective tools in treating common, and in some cases, debilitating diseases. We recently applied a molecular modeling and receptor-and-ligand mutagenesis approach to determine whether it is possible to decrease the affinity of norfenfluramine for its 5-HT$_{2B}$ receptor antitarget. Our combined modeling and mutagenesis efforts suggested that the α-methyl group of (+)-norfenfluramine contributes more to the compound's affinity for and potency at the 5-HT$_{2B}$ receptor antitarget than the 5-HT$_{2C}$ receptor target [29]. Consistent with our predictions, the affinity of the 5-HT$_{2B}$ receptor antitarget for α-desmethylnorfenfluramine [i.e. (+)-norfenfluramine lacking an α-methyl group] was reduced three- to fourfold compared with (+)-norfenfluramine, while the affinities of the 5-HT$_{2C}$ receptor target for α-desmethylnorfenfluramine and for (+)-norfenfluramine were not significantly different [29]. However, the affinities of the 5-HT$_{2B}$ receptor antitarget and the 5-HT$_{2C}$ receptor target for α-desmethylnorfenfluramine were similar (55 nM versus 69 nM, respectively [29]), implying that therapeutic ranges would still activate the 5-HT$_{2B}$ receptor antitarget. With additional chemical modifications, however, it might be possible to further and selectively reduce 5-HT$_{2B}$ receptor antitarget binding affinity, and our data suggest that combined molecular modeling and receptor-and-ligand mutagenesis approach might be useful in such attempts.

Ergolines might represent a more reasonable class of compounds for rational drug optimization, as their target receptors are not members of the same subtype as their antitarget. For instance, the antimigraine efficacy of dihydroergotamine and methysergide, while not completely understood, is believed to involve agonist activity at 5-HT$_1$ receptor subtypes [30,31]; the anti-Parkinsonian effects of pergolide and cabergoline are accepted to be due to dopamine receptor agonism [32]. Thus, unlike norfenfluramine, whose target and antitarget have quite similar pharmacological profiles, the ergolines by virtue of their targets and antitarget being more pharmacologically distinct might lend themselves better to reducing antitarget affinity without affecting target affinity.

7.4
Conclusions

Our studies clearly implicate 5-HT$_{2B}$ receptor agonism as causes of serious and potentially fatal drug side effects in humans – VHD and PH. Given the ease with which drugs can be screened *in vitro* to identify this potential activity [33], it is imperative that all currently approved drugs, their major metabolites and drugs in development (along with their identified metabolites) be screened for 5-HT$_{2B}$ agonist activity. Candidate medications with potent agonist activities should not be advanced to clinical trials unless the benefits greatly outweigh potential risks associated with VHD and PH. Medications, already in use, with potent 5-HT$_{2B}$ agonism need to be used with caution.

Abbreviations

EC$_{50}$	effective concentration giving 50% of maximal biological response
5-HT	5-hydroxytryptamine, serotonin
hVICs	human heart valve interstitial cells
LMTK	thymidine kinase-deficient mouse L cells (fibroblasts)
MDA	3,4-methylenedioxyamphetamine
MDMA	3,4-methylenedioxymethamphetamine
mCPP	*meta*-chlorophenylpiperazine
NET	noradrenaline transporter
PH	pulmonary hypertension
RVSP	right ventricular systolic pressure
SERT	serotonin transporter
SNP	single nucleotide polymorphism
VHD	valvular heart disease
VICs	(heart) valve interstitial cells

Acknowledgments

VS is supported by a European Union Marie Curie International Fellowship; BLR is supported by the NIMH PDSP as well as grants from NIMH and NIDA.

References

1 Roth, B.L., Sheffler, D.J. and Kroeze, W.K. (2004) Magic shotguns versus magic bullets: selectively non-selective drugs for mood disorders and schizophrenia. *Nature Reviews. Drug Discovery*, **3** (4), 353–359.

2 Armbruster, B.N. and Roth, B.L. (2005) Mining the receptorome. *The Journal of Biological Chemistry*, **280** (7), 5129–5132.

3 Sheffler, D.J. and Roth, B.L. (2003) Salvinorin A: the "magic mint" hallucinogen finds a molecular target in the kappa opioid receptor. *Trends in Pharmacological Sciences*, **24** (3), 107–109.

4 Roth, B.L., Baner, K., Westkaemper, R., Siebert, D., Rice, K.C., Steinberg, S., Ernsberger, P. and Rothman, R.B. (2002) Salvinorin A: a potent naturally occurring nonnitrogenous kappa opioid selective

agonist. *Proceedings of the National Academy of Sciences of the United States of America*, **99** (18), 11934–11939.

5 Nichols, D.E. (2004) Hallucinogens. *Pharmacology & Therapeutics*, **101** (2), 131–181.

6 Kroeze, W.K., Hufeisen, S.J., Popadak, B.A., Renock, S.M., Steinberg, S., Ernsberger, P., Jayathilake, K., Meltzer, H.Y. and Roth, B.L. (2003) H1-histamine receptor affinity predicts short-term weight gain for typical and atypical antipsychotic drugs. *Neuropsychopharmacology*, **28** (3), 519–526.

7 Kim, S.F., Huang, A.S., Snowman, A.M., Teuscher, C. and Snyder, S.H. (2007) From the cover: Antipsychotic drug-induced weight gain mediated by histamine H1 receptor-linked activation of hypothalamic AMP-kinase. *Proceedings of the National Academy of Sciences of the United States of America*, **104** (9), 3456–3459.

8 Weintraub, M., Sundaresan, P.R., Madan, M., Schuster, B., Balder, A., Lasagna, L. and Cox, C. (1992) Long-term weight control study. I (weeks 0 to 34). The enhancement of behavior modification, caloric restriction, and exercise by fenfluramine plus phentermine versus placebo. *Clinical Pharmacology and Therapeutics*, **51** (5), 586–594.

9 Connolly, H.M., Crary, J.L., McGoon, M.D., Hensrud, D.D., Edwards, B.S., Edwards, W.D. and Schaff, H.V. (1997) Valvular heart disease associated with fenfluramine–phentermine. *The New England Journal of Medicine*, **337** (9), 581–588.

10 Rothman, R.B., Baumann, M.H., Savage, J.E., Rauser, L., McBride, A., Hufeisen, S.J. and Roth, B.L. (2000) Evidence for possible involvement of 5-HT(2B) receptors in the cardiac valvulopathy associated with fenfluramine and other serotonergic medications. *Circulation*, **102** (23), 2836–2841.

11 Setola, V., Hufeisen, S.J., Grande-Allen, K.J., Vesely, I., Glennon, R.A., Blough, B., Rothman, R.B. and Roth, B.L. (2003) 3,4-Methylenedioxymethamphetamine (MDMA, "Ecstasy") induces fenfluramine-like proliferative actions on human cardiac valvular interstitial cells *in vitro*. *Molecular Pharmacology*, **63** (6), 1223–1229.

12 Fitzgerald, L.W., Burn, T.C., Brown, B.S., Patterson, J.P., Corjay, M.H., Valentine, P.A., Sun, J.H., Link, J.R., Abbaszade, I., Hollis, J.M., Largent, B.L., Hartig, P.R., Hollis, G.F., Meunier, P.C., Robichaud, A.J. and Robertson, D.W. (2000) Possible role of valvular serotonin 5-HT(2B) receptors in the cardiopathy associated with fenfluramine. *Molecular Pharmacology*, **57** (1), 75–81.

13 Pritchett, A.M., Morrison, J.F., Edwards, W.D., Schaff, H.V., Connolly, H.M. and Espinosa, R.E. (2002) Valvular heart disease in patients taking pergolide. *Mayo Clinic Proceedings*, **77** (12), 1280–1286.

14 Van Camp, G., Flamez, A., Cosyns, B., Weytjens, C., Muyldermans, L., Van Zandijcke, M., De Sutter, J., Santens, P., Decoodt, P., Moerman, C. and Schoors, D. (2004) Treatment of Parkinson's disease with pergolide and relation to restrictive valvular heart disease. *Lancet*, **363** (9416), 1179–1183.

15 Zanettini, R., Antonini, A., Gatto, G., Gentile, R., Tesei, S. and Pezzoli, G. (2007) Valvular heart disease and the use of dopamine agonists for Parkinson's disease. *The New England Journal of Medicine*, **356** (1), 39–46.

16 Newman-Tancredi, A., Cussac, D., Quentric, Y., Touzard, M., Verriele, L., Carpentier, N. and Millan, M.J. (2002) Differential actions of antiparkinson agents at multiple classes of monoaminergic receptor. III. Agonist and antagonist properties at serotonin, 5-HT

(1) and 5-HT(2), receptor subtypes. *The Journal of Pharmacology and Experimental Therapeutics*, **303** (2), 815–822.

17 Launay, J.M., Herve, P., Peoc'h, K., Tournois, C., Callebert, J., Nebigil, C.G., Etienne, N., Drouet, L., Humbert, M., Simonneau, G. and Maroteaux, L. (2002) Function of the serotonin 5-hydroxytryptamine 2B receptor in pulmonary hypertension. *Nature Medicine*, **8** (10), 1129–1135.

18 Nebigil, C.G., Launay, J.M., Hickel, P., Tournois, C. and Maroteaux, L. (2000) 5-Hydroxytryptamine 2B receptor regulates cell-cycle progression: cross-talk with tyrosine kinase pathways. *Proceedings of the National Academy of Sciences of the United States of America*, **97** (6), 2591–2596.

19 Deraet, M., Manivet, P., Janoshazi, A., Callebert, J., Guenther, S., Drouet, L., Launay, J.M. and Maroteaux, L. (2005) The natural mutation encoding a C terminus-truncated 5-hydroxytryptamine 2B receptor is a gain of proliferative functions. *Molecular Pharmacology*, **67** (4), 983–991.

20 Blanpain, C., Le Poul, E., Parma, J., Knoop, C., Detheux, M., Parmentier, M., Vassart, G. and Abramowicz, M.J. (2003) Serotonin 5-HT(2B) receptor loss of function mutation in a patient with fenfluramine-associated primary pulmonary hypertension. *Cardiovascular Research*, **60** (3), 518–528.

21 Manivet, P., Mouillet-Richard, S., Callebert, J., Nebigil, C.G., Maroteaux, L., Hosoda, S., Kellermann, O. and Launay, J.M. (2000) PDZ-dependent activation of nitric-oxide synthases by the serotonin 2B receptor. *The Journal of Biological Chemistry*, **275** (13), 9324–9331.

22 Herve, P., Launay, J.M., Scrobohaci, M.L., Brenot, F., Simonneau, G., Petitpretz, P., Poubeau, P., Cerrina, J., Duroux, P. and Drouet, L. (1995) Increased plasma serotonin in primary pulmonary hypertension. *The American Journal of Medicine*, **99** (3), 249–254.

23 Callebert, J., Esteve, J.M., Herve, P., Peoc'h, K., Tournois, C., Drouet, L., Launay, J.M. and Maroteaux, L. (2006) Evidence for a control of plasma serotonin levels by 5-hydroxytryptamine(2B) receptors in mice. *The Journal of Pharmacology and Experimental Therapeutics*, **317** (2), 724–731.

24 Egermayer, P., Town, G.I. and Peacock, A.J. (1999) Role of serotonin in the pathogenesis of acute and chronic pulmonary hypertension. *Thorax*, **54** (2), 161–168.

25 Launay, J.M., Schneider, B., Loric, S., Da Prada, M. and Kellermann, O. (2006) Serotonin transport and serotonin transporter-mediated antidepressant recognition are controlled by 5-HT2B receptor signaling in serotonergic neuronal cells. *FASEB Journal*, **20** (11), 1843–1854.

26 Mihara, K., Otani, K., Suzuki, A., Yasui, N., Nakano, H., Meng, X., Ohkubo, T., Nagasaki, T., Kaneko, S., Tsuchida, S., Sugawara, K. and Gonzalez, F.J. (1997) Relationship between the CYP2D6 genotype and the steady-state plasma concentrations of trazodone and its active metabolite *m*-chlorophenylpiperazine. *Psychopharmacology (Berl)*, **133** (1), 95–98.

27 Ishida, M., Otani, K., Kaneko, S., Ohkubo, T., Osanai, T., Yasui, N., Mihara, K., Higuchi, H. and Sugawara, K. (1995) Effects of various factors on steady state plasma concentrations of trazodone and its active metabolite *m*-chlorophenylpiperazine. *International Clinical Psychopharmacology*, **10** (3), 143–146.

28 Porter, R.H., Malcolm, C.S., Allen, N.H., Lamb, H., Revell, D.F. and Sheardown, M.J. (2001) Agonist-induced functional desensitization of recombinant human 5-HT2 receptors expressed in CHO-K1 cells. *Biochemical Pharmacology*, **62** (4), 431–438.

29 Setola, V., Dukat, M., Glennon, R.A. and Roth, B.L. (2005) Molecular determinants for the interaction of the valvulopathic

anorexigen norfenfluramine with the 5-HT2B receptor. *Molecular Pharmacology*, **68** (1), 20–33.

30 Newman-Tancredi, A., Conte, C., Chaput, C., Verriele, L., Audinot-Bouchez, V., Lochon, S., Lavielle, G. and Millan, M.J. (1997) Agonist activity of antimigraine drugs at recombinant human 5-HT1A receptors: potential implications for prophylactic and acute therapy. *Naunyn-Schmiedeberg's Archives of Pharmacology*, **355** (6), 682–688.

31 Saxena, P.R. and Den Boer, M.O. (1991) Pharmacology of antimigraine drugs. *Journal of Neurology*, **238** (Suppl. 1), S28–S35.

32 Factor, S.A. (1999) Dopamine agonists. *Medical Clinics of North America*, **83** (2), 415–443, vi–vii.

33 O'Connor, K.A. and Roth, B.L. (2005) Finding new tricks for old drugs: an efficient route for public-sector drug discovery. *Nature Reviews. Drug Discovery*, **4** (12), 1005–1014.

8
Computational Modeling of Selective Pharmacophores at the α_1-Adrenergic Receptors

Francesca Fanelli, Pier G. De Benedetti

8.1
Introduction

G-protein-coupled receptors (GPCRs) form the largest family of signal transduction membrane proteins and are the most privileged targets for currently marketed drugs and for the wealth of drug candidates that high-throughput methods promise to deliver in the immediate future [1–3].

Although varying considerably in molecular size, any GPCR polypeptide sequence contains seven hydrophobic α-helices that span the lipid bilayer and dictate the typical macromolecule architecture. Seven transmembrane domains bundled up to form a polar internal tunnel and expose the N-terminus and three interconnecting loops, to the exterior, and the C-terminus with a matching number of loops, to the interior of the cell [1–3]. This structural information was recently confirmed by the resolution of the crystal structure of rhodopsin [4,5].

Until 1995, GPCR ligands were classified as antagonists, full or partial agonists, depending on whether they produced null, full or partial stimulus upon binding to the receptor [6]. However, recent advances in receptor theory and experimental technology for ligand screening have led to the discovery of many additional types of receptor ligands. Some of these are (a) inverse agonists, which inhibit constitutively active GPCRs (i.e. agonist-independent spontaneous activity that emanates from the system itself); (b) allosteric agonists, which function as agonists by interacting with a site distinct from that of the endogenous agonist (usually a nonpeptide ligand for a peptide receptor); (c) allosteric modulators (antagonists), which block receptor function but do not necessarily interfere with ligand–receptor interaction (receptor occupancy); and (d) allosteric enhancers, which potentiate the agonist effect on the receptor (reviewed in [6]). These aspects cannot be ignored when dealing with the challenging issue of drug selectivity toward different target receptors. Prediction and prevention of side effects are further complicated by the continuous identification of new receptor subfamilies and subtypes within the same subfamily. This process imposes a frequent updating of the pharmacological profile and the related

Antitargets. Edited by R. J. Vaz and T. Klabunde
Copyright © 2008 WILEY-VCH Verlag GmbH & Co. KGaA, Weinheim
ISBN: 978-3-527-31821-6

pharmacophore and (quantitative) structure–activity relationship (QSAR) models of receptor ligands. In this respect, α_1-adrenergic receptors (α_1-ARs) constitute a representative example [7–9]. The α_1-ARs, which belong to class A of GPCRs, are constituted by three subtypes, α_{1a}, α_{1b} and α_{1d}, as established by molecular biology, radioligand binding and functional studies [7,8,10].

Their endogenous ligands are the catecholamine neurotransmitters epinephrine and norepinephrine, characterized by a β-hydroxyethyl chain that separates a protonated nitrogen atom from a catecholic ring (Figure 8.1). The structural requirements for the α_1-AR agonism are (a) a protonated nitrogen atom that could interact with a negatively charged residue (aspartate or glutamate) of the receptor; (b) the hydroxy groups at positions *meta* and *para* of the aromatic ring that could recognize H-bonding donor/acceptor groups of the receptor; and (c) the aromatic ring itself that could interact with aromatic or hydrophobic residues of the receptor. Mutational analysis on the α_{1b}-AR suggests that (a) an aspartate residue conserved in all the members of the amine GPCR subfamily [11–15], D3.32 (this numbering follows the scheme by Ballesteros and Weinstein [16]), recognizes the protonated nitrogen atom of agonists and antagonists; (b) S5.42 and S5.46, with prominence for the former, recognize the catechol hydroxy groups of norepinephrine; (c) F6.51 of an aromatic cluster in helix 6 recognizes the aromatic ring of the natural agonist (Figure 8.1) [17]. Most of these features are also present in the α_{1a}-AR [18]. Moreover, the α_{1a}-AR shows a considerable binding site conformational plasticity, which allows for high affinity binding of wide, structurally heterogeneous series of compounds. This situation makes the α_{1a}-AR both an interesting target and an intriguing antitarget.

The α_1-ARs are present in many tissues including brain, heart, blood vessels, liver, kidney, prostate and spleen, where they mediate a variety of physiological effects such as neurotransmission, vasoconstriction, cardiac inotropy and chronotropy and

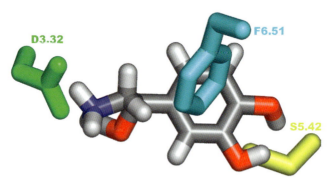

Figure 8.1 Details of the intermolecular interactions from a minimized complex between norepinephrine and a rhodopsin-based model of the α_{1b}-AR are shown. As for the amino acid residues (D3.32, S5.42 and F6.51), only the polar hydrogen atoms are shown. Norepinephrine is colored by atom type, whereas the amino acid residues are colored according to their helix location.

glycogenolysis [19]. Moreover, various studies showed that other signaling pathways can be activated by different α_1-AR subtypes [19]. Thu, drug affinity toward α_1-AR subtypes can mediate undesired side effects. Subtype selectivity toward the three α_1-ARs must also be interpreted in terms of efficacy. *In vitro* studies on α_{1a}- and α_{1b}-AR subtypes indeed revealed subtype-dependent inverse agonist or neutral antagonist features for a set of classical α_1-AR antagonists [20]. Whereas various ligands displayed variable degrees of negative efficacies (i.e. resulting in either neutral antagonists or inverse agonists) at the α_{1a}-AR, most of the tested ligands behaved as inverse agonists at the α_{1b}-AR, independent of their structural features.

The three α_1-AR subtypes share 50% identical amino acid sequences and 70% similarity, whereas the membrane-spanning regions share even higher homology. Since ligands for these receptors presumably recognize the same anionic site in the receptor (i.e. D3.32), subtype selectivity should consist in subtle sequence differences in the amino acids that surround such an anionic site. Similar arguments are also applicable to other biogenic amine receptors characterized by high sequence similarity to the α_1-ARs such as 5-HT$_{1A}$-receptor (5-HT$_{1A}$-R), α_2-AR, D$_2$-dopamine receptor (D$_2$-R) and so on [11–15]. Thus, while on the one hand, α_1-ARs are very interesting and challenging targets for selective pharmacophore modeling and drug design, they are, on the other hand, important antitargets that are thought to mediate relevant side effects. This scenario, which depicts a static situation, becomes more complicated when we consider the three-dimensional (3D) structure and the associated dynamic nature of ligands, receptors and their molecular interactions. In this respect, the most relevant dynamic structural features of ligands and receptors, connected with their mutual and selective molecular recognition mechanisms, are the well-known conformational equilibria (bioactive and functional conformation/s) and prototropic equilibria (acid–base and tautomeric forms) [21]. GPCR ligands are flexible molecules with acidic and/or basic functions containing heteroatoms and heterocycles that may induce tautomeric equilibria. Consequently, there are complex and interdependent prototropic equilibria that generate different prototropic molecular forms with different conformations and stereoelectronic features. Although the relevance of prototropic equilibria and ionization on the pharmacokinetic and pharmacodynamic properties of drugs is well known [22,23], their explicit evaluation in both pharmacophore modeling and QSAR studies is not yet a common practice in many ligand/drug design approaches [24] and target/antitarget virtual screening.

Furthermore, there is emerging evidence that GPCRs exist as homo- and heterodimers/oligomers [25–34] and α_1-ARs are no exception to this rule [35]. Thus, regulated protein–protein interactions are key features of many aspects of GPCR function, and there is now increasing evidence for GPCRs acting as part of multicomponent units comprising a variety of signaling and scaffolding molecules [1,2]. The concept that GPCR heterodimerization could play a role in pharmacological diversity was first indicated by studies on the δ and κ-opioid receptors [36]. If this is a general phenomenon, heterodimerization among pharmacologically distinct receptors could underlie a level of pharmacological diversity that would have far-reaching implications for drug development. In particular, it could provide new opportunities for the development of more selective compounds that would target specific heterodimers without affecting the individual protomers [28].

GPCRs are allosterically modulated proteins, which exist as complex statistical conformational ensembles [37–39]. They hold regions of high stability (i.e. low flexibility) and regions of low stability (i.e. high flexibility) that communicate with each other, even if they are distal. The functional properties of a GPCR are related to the distribution of states within the native ensemble [38,39]. This distribution is modulated by ligands, interacting proteins, amino acid mutations and/or prototropic equilibria (acid–base and tautomeric) [21,38,39]. Of course, the different oligomeric states of a GPCR may contribute to change the distribution of the receptor states.

However, in spite of the outstanding molecular and functional complexity of GPCRs, thanks to the development of information and computer technologies, different procedures for rational and specific target and antitarget-oriented drug design have been developed. These approaches can be schematically classified as ligand-based (pharmacophore and QSAR) modeling and receptor structure-based modeling and drug design [21].

In this work, we will try to illustrate, among the large variety of approaches and procedures proposed at different levels of complexity and sophistication, the most relevant methods and results obtained from different authors on computational modeling of both selective antagonism at α_1-ARs and their side affinity propensity (antitarget modeling and screening).

This chapter consists of seven main sections. Section 8.2 summarizes the basic concepts and approaches to pharmacophore and QSAR modeling. Section 8.3 describes modeling approaches to the generic α_1-AR affinity, with particular focus on three main classes of compounds: quinazolines, 1,4-benzodioxanes and arylpiperazines. It also deals with 5-HT$_{1A}$/generic α_1-AR target and antitarget modeling. Section 8.4 shows examples of ligand-based and receptor-based pharmacophore and QSAR modeling of α_1-AR subtype selectivity. Section 8.5 delineates on antitarget modeling of biogenic-amine GPCRs. Finally, Sections 8.6 and 8.7 provide conclusions and perspectives, respectively.

8.2
Ligand-Based and Receptor-Based Pharmacophore Modeling and QSAR Analysis

Two very similar molecules are two different physical objects [40]. Hence, chemical/structural comparison of similarity is a subtle and relative concept that acquires significance in a well-defined reference physical context. In other words, it is necessary to define in what respect and to what extent two different molecules are similar. Molecular series of specific and selective ligands interacting, *in vitro* and in equilibrium conditions, with a specific receptor constitute a sophisticated example of chemical similarity–diversity classification. This classification is based on the experimental binding affinity ($\Delta G°$) values that quantify a particular molecular recognition phenomenon, which is, essentially, a noncovalent process [41]. This implies, according to the Fisher's lock and key model, that structural complementarity (static and, even more, dynamic) between the interacting partners is the most important feature of the process, raising the problem of its chemical-structural description at a (sub)molecular level [42,43]. Many modeling strategies are currently used to design new drugs. However, they can be broadly classified into 'direct' and 'indirect' methods, according to the known (direct or target structure-based) or hypothetical (indirect or ligand-based) 3D structure of the target binding site [21,41].

The quantitative comparison of the optimized 3D structure of a selected set of ligands allows the development of their minimal 3D structural requirements for the recognition and activation of the biological target, that is, the pharmacophore hypothesis, and gives a sound 3D rationale to the available SARs [21]. A more complete and mechanistically relevant approach to the development of the 3D pharmacophore consists in its translation into a numerical molecular descriptor that quantifies the molecular–pharmacophore similarity–diversity for computational QSAR modeling [21,41].

The methods of computational chemistry are being increasingly applied in computer-aided drug design procedures and QSAR studies [21,44]. QSAR analysis, which is a fundamental prerequisite to rational drug design methods, addresses the problem of generating statistics-based correlative models between experimentally determined biological/pharmacological data for a set of compounds and their experimental and/or computational physicochemical/structural parameters [45,46]. The aim is twofold: to explain the biological activity, at the (sub)molecular level, in terms of physicochemical/structural properties of the ligands and to predict similar biological activity for new analogues. This implies that the variation in the biological activity of the studied compound set (molecular series = statistical sample of ligands) is reduced and translated into a chemical formalism (molecular descriptors), with great benefit to the physical interpretability of the QSAR model produced. When structural information on the biological target is unknown, the indirect ligand design approach and associated indirect QSAR modeling procedures are applicable. Computational indirect methods of QSAR analysis, including 3D QSAR, constitute the natural evolution of classical 2D QSAR or 'Hansch approach' [45], with one or more added dimension and the employment of computational descriptors instead of empirical parameters [47].

An essential aspect of the 3D pharmacophoric generation is the search for different and specific prototropic bioactive molecular forms, and their corresponding conformations, present in the biological test solution [24]. Accordingly, the computational descriptors should be defined and computed, in principle, on all the conformationally optimized prototropic forms, energetically accessible by the ligands. The selection of the most physically and statistically informative descriptors will then proceed according to different (but consistent) roles of these prototropic forms (and their associated conformations) on the hypothetical mechanism of interaction with the receptor binding site, according to the ligand 3D pharmacophoric similarity/target receptor complementarity paradigm. Many different chemometric techniques are available for descriptor–predictor selections and 3D QSAR modeling [46].

The advent of accessible computing facilities such as the ease of computer data generation, acquisition and their elaboration by multivariate statistical techniques has stimulated the use of high-dimensional descriptors for QSAR modeling and ligand design procedures [48–50]. Many extensions and variants of 3D QSAR modeling have been proposed mainly to improve the predictive aspects of different approaches [51,52]. After generation of the 3D QSAR model following the indirect approach, the next step consists in ligand design by making use of the most suitable computational descriptors and the derived pharmacophore hypothesis, for both database search and compound design. The biological potency of a collection of virtual molecules can be predicted by interpolation of the QSAR model and/or appropriate scoring functions (virtual screening). Then, the most promising molecules will be synthesized and biologically tested and the pharmacophore and QSAR models checked for their quality (model validation). The same procedure is followed by the direct ligand design approach by using ligand–receptor interaction-based pharmacophore and QSAR models and the relative computed intermolecular descriptors. Hence, indirect and direct computational approaches to ligand design (and antitarget screening) differ in a way that the second one generates QSAR models, which explicitly consider the 3D structure of the target binding site.

The direct or structure-based approach is a well-established procedure and constitutes the ideal situation, starting from the experimental structure of the biological target [42,44], and, even better, from the experimental structure of the ligand–target complex. Once the 3D structure of the biological target has been energetically optimized and the ligand-binding site identified, ligand–receptor docking simulations and energy minimization of the ligand–target complex allow the explicit optimization of the ligand–target complementarity in terms of intermolecular forces (electrostatic, vdW, hydrogen bonds), computed at different levels of accuracy. This provides a detailed picture of the receptors' amino acids involved in the interaction with the different ligand moieties as well as the extraction of the corresponding intermolecular descriptors. On the basis of these considerations, QSAR equations can be modeled and used for both ligand design purposes and biological potency prediction. A good and probably the first study that compares direct and indirect computational QSAR modeling approaches was conducted on a large series of sulfonamide carbonic anhydrase inhibitors [53,54]. Interestingly, the results obtained revealed that the information necessary for the biological response that the inhibitor elicits from the interaction is largely encoded in the molecular structure of the drugs and in their appropriate computational descriptors. The computational direct QSAR modeling approach developed on the carbonic anhydrase-sulfonamide inhibitor complexes was subsequently generalized and extended to the biological target with known primary sequence but unknown 3D structure. This approach, named heuristic-direct (target-based) and applicable when the atomic resolved structure of the biological target is unknown but predictable, is able to handle the heterogeneous experimental information on the ligands and their receptors and combine and translate them into QSAR models [55–61]. This procedure, although computationally quite intensive, gives a detailed and explicit picture of the interacting atoms or groups and their energetics in the ligand–receptor complex, taking advantage from its strong complementarity with the indirect (ligand-based) QSAR approaches. Hence, indirect QSAR modeling approaches to selective ligand design are prerequisites to the heuristic-direct QSAR modeling protocol.

8.3
The General α_1-AR Pharmacophore

8.3.1
Ligand-Based Pharmacophore and Virtual Screening

When dealing with the rationalization of biological responses (binding affinity, selectivity and efficacy) of GPCR ligands, the crucial problem of the large molecular diversity of the receptors and their subtypes must be faced [21]. On the one hand, there exists a general mode of interaction (privileged fragments or structures) of biogenic amine ligands with their receptors, which has been strictly conserved through receptor evolution; that is, all ligands have a basic nitrogen function protonated at physiological pH. Therefore, long-range electrostatic interaction should constitute the first recognition step. On the other hand, a large variation in the details of molecular interactions between the chemical signaling molecules (transmitters) and the receptors exists, and few molecular determinants are responsible for discriminating compounds among receptor classes, subtypes or variants. Thus, the ability to obtain a good quantitative rationalization of the selective binding properties of highly active and structurally different ligands depends primarily on the availability of descriptors to be able to capture the strict ligand–receptor complementarity criteria, which determine the biological property of interest. A relevant aspect of chemical research is to predict the behavior of new molecules from their structure, prior to synthesis [62]. Most often, structurally similar molecules have a comparable range of physicochemical properties and biological activities. On the contrary, structurally dissimilar molecules differ in their properties both qualitatively and quantitatively. Correlation analysis and, hence, QSAR and QSPR (quantitative structure–property relationship) are implicitly based on this simple fact. In the quantitative description of molecular shape, and its application in the context of molecular similarity comparisons, correlation analysis of chemical properties (QSPR) and QSAR modeling has been the objective of a number of methodological studies [46,62]. However, in our opinion, all these approaches have been mainly developed for the optimization of the predictive aspects of QSAR analysis. In fact, these approaches use, in many cases, sophisticated descriptors of properties related to molecular shape that, while improving, in some cases, the statistics of the QSAR models (i.e. they are good predictors for similar molecules of the test set), however, lose lead design capability. In other words, if we recall the well-known dualistic aspect of QSAR models, that is, the interpretative and predictive ones, a sort of 'uncertainty principle' governs their performance: if we improve the predictive aspect, we obscure the interpretative one and vice versa. Hence, simplifying, we can state that when molecular descriptors are good predictors, they are, concomitantly, poor 'designers', that is, they have poor propensity to suggest directly new leads.

Recently, we have shown that the supermolecule approach and the derived ad hoc size and shape descriptors defined on the ligand bioactive prototropic molecular form have been successful in deriving simple QSAR models for the molecular series of structurally very heterogeneous ligands of biogenic amine neurotransmitter GPCRs [24,41,63–71]. According to the ligand 3D pharmacophore, the supermolecule approach assumes that the volume obtained by superimposing the pharmacophoric elements of structurally the most different ligands (supermolecule) that show the highest affinities for the same receptor might reflect the overall shape and conformational flexibility (induced fit and accessory binding area) of the high-affinity state of the receptor binding site. In this approach, size and shape descriptors are computed by comparing the vdW volume of the minimized structure of each ligand with the vdW volume of a supermolecule chosen as a template. For each subset of analogues, the ligand showing the highest affinity for a given receptor is chosen as a component of the respective reference supermolecule. The energy minimized ligand/s chosen for the construction of each supermolecule are superimposed by a topologically rigid body fit procedure based on given pharmacophoric criteria that, for the α_1-ARs and the 5-HT$_{1A}$-R, are essentially (a) the hydrogen of the protonated nitrogen atom and (b) the aromatic rings closest to and farthest from the protonated nitrogen. To compute molecular descriptors relative to the supermolecule, all other energy optimized compounds constituting the training set are, hence, rigidly superimposed on the appropriate supermolecule with each ligand being superimposed on the analogue compound present in the supermolecule or on its structurally closest compound. The following size and shape descriptors have been considered: V_{in} and V_{out}, which are, respectively, the intersection and the outer vdW volume of the ligand considered with respect to the volume of the reference supermolecule, and V_{dif}, which is computed according to the following formula: $V_{dif} = (V_{in} - V_{out})/V_{sup}$, where V_{sup} is the molecular volume of the reference supermolecule. According to the definition of these size and shape descriptors, higher affinities are realized by maximizing V_{in} and by minimizing V_{out}. For its formulation, V_{dif} is a normalized size and shape descriptor, which takes into account the information content encoded by both V_{in} and V_{out}. Finally, V_{in}/V_{mol} is simply V_{in} normalized with respect to the vdW molecular volume (V_{mol}) of the ligand. By definition, ligands constituting the supermolecules show the following singularities: $V_{mol} = V_{in}$, $V_{out} = 0$ and $V_{in}/V_{mol} = 1$. Therefore, size and shape descriptors can be defined ad hoc (i.e. on a specific molecular series and in connection with a specific biological activity) with respect to the supermolecule, and this constitutes the main advantage over molecular descriptors defined and performed for a single structure and a single conformation. The good performance of the supermolecule approach was shown in a number of QSAR modeling studies [24,41,63–71]. Indeed, it is demonstrated to be a flexible tool to describe both congeneric and noncongeneric series of compounds in an extended (global) chemical space, which can be continuously and easily upgraded with new experimental data. QSAR models based on ad hoc defined supermolecules and the corresponding size–shape descriptors have been employed for deciphering the molecular features responsible for the affinity and selectivity in a wide series of congeneric and noncongeneric α_1-AR antagonists [41,63–66,68,70,71]. In this respect, one of the first attempts to obtain pharmacophore and QSAR models involved a series of prazosin analogues (quinazolines, quinolines and isoquinolines), 1,4-benzodioxan (WB-4101) related compounds and arylpiperazines. These models were generated using generic α_1-AR binding affinities because of the lack of in vitro measurements on the cloned α_1-AR subtypes. As such they did not aim to describe α_1-AR selectivity. The following paragraphs summarize the main achievements of the QSAR modeling of generic α_1-AR affinity and of 5-HT$_{1A}$-R versus generic α_1-AR selectivity.

8.3.1.1 Prazosin Analogues (2,4-Diamino-6,7-dimethoxyquinazoline Derivatives)

Among the compounds classified as antagonists of the α_1-ARs, the derivatives of 2,4-diamino-6,7-dimethoxyquinazoline proved to be the first to show potent and selective activity [7,8,72–75]. In Scheme 8.1 the structure of the earliest example, prazosin (**1**), is shown, together with those of other potent α_1-ARs antagonists (**2–6**) of this class. SAR studies on the wide series of 2,4-diamino-6,7-dimethoxyquinazoline [9,72–79] and -quinoline [80] derivatives have qualitatively rationalized the affinity and selectivity of these compounds for the α_1-ARs. It has been proposed that the 2,4-diamino-6,7-dimethoxyquinazoline scaffold acts as conformationally restricted bioisosteric replacement of norepinephrine with the N_1-protonated form (i.e. the nitrogen atom marked by an asterisk in Scheme 8.1) being particularly suited for effective charge-reinforced hydrogen bonding with the carboxylate counterion of the conserved aspartate in the α_1-AR binding site. This is corroborated by the lack of significant affinity and antihypertensive activity of the structurally closely related isoquinolines, where the bioisosteric substitution (C—H instead of N_1) abolishes any biological activity [81]. Further support comes from a study on molecular modeling and QSAR analysis of a heterogeneous series of quinazoline, quinoline and isoquinoline derivatives [82]. In this study, the QSAR models obtained by correlating computational molecular descriptors with both the experimental acidity constants and the α_1-AR binding affinity data values clearly depicted the fundamental role of the protonated quinazoline nucleus for a productive interaction with the receptor. These early inferences find support in a recent study by Kinsella et al. [83], in which the proton affinity of a large family of α_1-AR agonists and antagonists was computed by quantum mechanical methods at the B3LYP/6–31G* level of theory. A good correlation was found between the proton affinity values corresponding to the first protonation in the gas phase of some antagonists and their corresponding experimental binding affinity constants for the α_{1a}-AR. These results stress the importance of the prototropic molecular forms in the design of new potential drugs [24]. In fact, drugs interact with their receptors in an aqueous environment and at physiological pH. Therefore, the protonation state of the ligand, in such conditions, will play an essential role in ligand–receptor interactions. Thus, conformational and protonation analyses of the considered ligands are essential steps in the study of ligand–receptor interactions and in the design of new drugs [24,83].

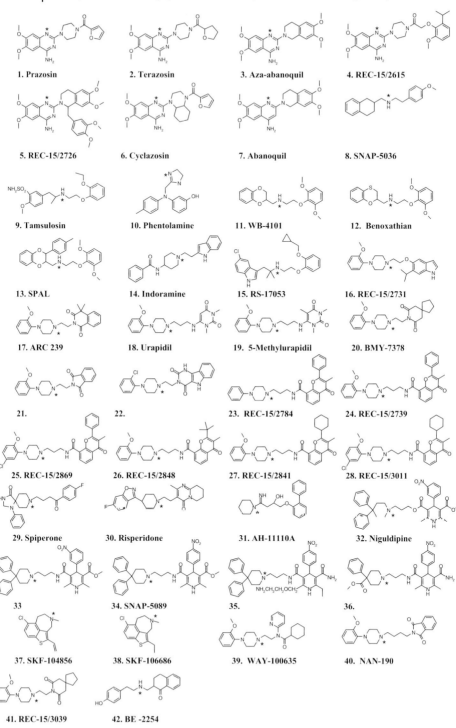

Scheme 8.1

Subsequent studies [63] pointed out that, once the electronic requirements of the common quinazoline moiety are satisfied, the binding affinities are modulated by the molecular shape of the quinazoline 2-substituent, through the optimization of both dispersive and steric interactions. In this respect, ad hoc derived size and shape descriptors defined on the ligand bioactive molecular form proved to be very successful in deriving QSAR models for several molecular series of GPCR ligands [65,66]. To validate on this basis the previously proposed computational QSAR models, novel piperazine and nonpiperazine derivatives of 2,4-diamino-6,7-dimethoxyquinazoline were designed, synthesized and tested for their binding affinity for the α_1-AR in rat cortex [70]. Selectivity toward other GPCRs, namely, α_2-AR, D_2-R and 5-HT$_{1A}$-R sites, was also investigated [70].

Successful correlation equations were obtained for the α_1-AR binding affinity by means of computational descriptors derived on a single structure (i.e. molecular orbital indexes and charged partial surface area descriptors) and based on size and shape descriptors obtained through the supermolecule approach. The descriptors involved in the rationalization of the α_1-AR binding affinity generated different models of almost similar quality and emphasized (a) the role of the quinazoline N_1-nitrogen atom as a hydrogen bonding donor (Equations 8.1 and 8.2), (b) the dispersive and steric interactions realized by the quinazoline substituents (Equations 8.2–8.4) and (c) the hydrogen bonding acceptor propensity of the ligands (Equations 8.4 and 8.5) [70]:

$$pK_{i\alpha 1} = -971.8\,(\pm 132.2)\,Vf(H) + 905.4\,(\pm 122.1),$$
$$n = 22,\; r = 0.85,\; r_{cv} = 0.81,\; s = 0.50,\; F = 54.05, \tag{8.1}$$

$$pK_{i\alpha 1} = 1410.7(\pm 244.8)\,SL(N) - 15.08\,(\pm 3.94),$$
$$n = 23,\; r = 0.78,\; r_{cv} = 0.73,\; s = 0.66,\; F = 33.20, \tag{8.2}$$

$$pK_{i\alpha 1} = 8.47\,(\pm 1.03)\,V_{dif} + 3.23\,(\pm 0.53),$$
$$n = 26,\; r = 0.86,\; r_{cv} = 0.81,\; s = 0.57,\; F = 66.96, \tag{8.3}$$

$$pK_{i\alpha 1} = 7.44\,(\pm 0.97)\,V_{dif} - 0.90\,(\pm 0.31)\,HACA{-}1 + 5.99\,(\pm 1.06),$$
$$n = 26,\; r = 0.90,\; r_{cv} = 0.85,\; s = 0.50,\; F = 48.04, \tag{8.4}$$

$$pK_{i\alpha 1} = -0.02(\pm 0.003)\,V_{out} - 891.1(\pm 160.2)\,HACA - 1/TMSA$$
$$+ 11.39\,(\pm 0.59),\; n = 26,\; r = 0.85,\; r_{cv} = 0.80,\; s = 0.59,\; F = 30.29, \tag{8.5}$$

where n is the number of compounds, r is the correlation coefficient, r_{cv} is the cross-validated r, s is the standard deviation and F is the value of the Fisher's ratio. The numbers in parentheses are the 95% confidence intervals of the regression coefficient and the intercept.

The crucial role of the quinazoline protonated N_1-nitrogen atom is well described by the valency of its hydrogen atom Vf(H). In fact, the negative value of the regression coefficient indicates that ligands with lower values for minimum valency give stronger hydrogen bonds (Equation 8.1).

The role of a long-range electrostatic interaction as a preliminary recognition step in the ligand–receptor complex formation is further emphasized by the calculated nucleophilic superdelocalizability of the protonated nitrogen atom ($SL(N)$) (Equation 8.2)). This index, which provides a numerical differentiation of the capability of the ligand to donate a proton, explains 60% of the variation in the binding affinity. Equations 8.3 and 8.4 involve the ad hoc derived size and shape descriptor V_{dif} (see above for its definition), implicating that, once the main docking has been accomplished, the binding affinity might be modulated by the optimization of the short-range intermolecular interactions and dispersion contributions of the quinazoline substituents.

A significant improvement in the stability of the regression involving V_{dif} is obtained in Equation 8.4 by including the HACA-1 index (HACA-1 $= q_A S_A$, where q_A is the charge on the hydrogen bonding acceptor atom and S_A is its surface area), as underlined by the cross-validated correlation coefficient ($r_{cv} = 0.85$). It is worth noting that this descriptor contributes to the regression with a negative sign; therefore, proliferation of hydrogen bond acceptor centers in the ligands does not appear to be essential for the modulation of the binding affinity. The HACA-1 index is also involved, together with V_{out}, in Equation 8.5, where it is normalized for the total molecular surface area (TMSA) of the ligand [70].

8.3.1.2 1,4-Benzodioxan (WB-4101) Related Compounds

It is well known that the identification of the pharmacophoric moiety (i.e. the essential atoms and molecular fragments of the bioactive compound, which selectively recognize the receptor eliciting the observed specific pharmacological effect) constitutes an important aspect of drug design procedures and SAR studies [84–86].

However, when dealing with receptor blockers such as the α_1-AR antagonists, that is, prazosin and WB-4101 (compound **11** in Scheme 8.1) analogues, the search for a common pharmacophoric pattern based on topological and topographical comparisons of their apparently unrelated chemical structures casts little light on the nature of the α_1-AR binding sites and hence on drug design.

In spite of the above considerations, the QSAR results obtained for a wide and very heterogeneous series of WB-4101-related compounds confirmed quantitatively and in terms of computational reactivity and shape descriptors previous SAR conclusions on the prazosin analogues [84–86]. Indeed, the protonated amine function, like for the prazosin analogues, plays a crucial role in the affinity for α_1-ARs via a charge-reinforced hydrogen bond with a primary nucleophilic site of the receptor. In fact, the more electrophilic (high $SL(N)$ and low E_{LUMO}) the protonated nitrogen atom, the stronger is the charge-reinforced hydrogen bond with the receptor and the higher is the α_1-ARs' binding affinity. The vdW volume of the most active derivative WB-4101 (in its protonated form and extended conformation) was taken as a reference shape and it was assumed to represent, for this class of compounds, the best complementary shape of the receptor. In other words, the energetically optimized three-dimensional structure of WB-4101 provided the shape that best optimizes the short-range intermolecular interactions in the drug–receptor complex. The more similar the three-dimensional shape of the antagonists to the shape of the reference molecule is, the more active the analogues as α_1-AR blockers are.

By making use of both the electrophilic reactivity (E_{LUMO}) and ad hoc shape (V_{dif}) descriptors, the following equation was obtained, which accounts for about 77% of the variance in the pharmacological data.

$$pK_i = 0.40(\pm 0.08)\, E_{LUMO} + 5(\pm 2)\, V_{dif} + 28(\pm 7),$$
$$n = 23,\; r = 0.88,\; s = 0.82,\; F = 38.0. \quad (8.6)$$

This equation shows that the most relevant computational information on the drug–receptor complex is encoded both in reactivity (hydrogen bond donor propensity) and in shape descriptors.

8.3.1.3 Arylpiperazine Derivatives

Aryl and heteroarylpiperazine and piperidine moieties are structural features present in many drugs (privileged fragments or structures) acting at the adrenergic, serotoninergic and dopaminergic receptors (see compounds **16–29** in Scheme 8.1). Several studies describe the conformational analysis of a wide series of arylpiperazines and the influence of the conformation on their binding affinity toward the 5-HT$_{1A}$-R [87,88]. However, in these reports, the compounds were considered in their neutral forms only, in spite of the recognized bioactive role of the N4-protonated form (i.e. the atom marked by an asterisk in compounds **16–29** of Scheme 8.1). In fact, also in this case, Coulombic attraction between the protonated amine of the ligands and D3.32 in the 5-HT$_{1A}$-R is probably the first step in the binding site recognition process followed by the formation of a charge reinforced hydrogen bond. The latter presumably shifts the conformational behavior of the protonated form toward that of the neutral one. The dependence of the conformational behavior of substituted arylpiperazines on their protonation state was explored by Cocchi *et al.* with the aim of building computational QSAR models able to rationalize the 3D structural features that modulate the binding affinity and selectivity at 5-HT$_{1A}$-R and α_1-AR. The study was based on 32 5-HT$_{1A}$-R ligands including arylpiperazines as well as the 5-HT$_{1A}$-R agonist serotonin and 8-OH-DPAT [67]. Conformational analyses highlighted higher conformational freedom in the neutral forms compared with the protonated ones. It was, therefore, speculated that conformationally the more constrained protonated form after desolvation recognizes via long-range attractive Coulombic interaction the protophilic residue D3.32 in the receptor binding site, consistent with *in vitro* evidence [89]. Following the formation of a charge reinforced hydrogen bond between the protonated ligand and D3.32, a conformational relaxation of the more flexible 'neutral form' is instrumental in the optimization of short-range (size/shape-dependent) intermolecular interactions (cooperative sequential interactions).

The supermolecule approach was able to rationalize changes in the ligand-binding affinities for the 5-HT$_{1A}$-R. The good linear correlations obtained through the V_{in} and V_{dif} indexes

$$pK_{5HT1A} = 0.04(\pm 0.005)\, V_{in} - 6.60(\pm 0.61),$$
$$n = 31,\; r = 0.83,\; s = 0.65,\; F = 66.20, \quad (8.7)$$

$$pK_{5HT1A} = 6.60(\pm 0.81)\, V_{dif} - 3.82(\pm 0.30),$$
$$n = 30,\; r = 0.83,\; s = 0.67,\; F = 62.67 \quad (8.8)$$

strengthened, also for the 5-HT$_{1A}$-R, the role of size and shape descriptors in the modulation of the ligand-binding affinity [67].

8.3.1.4 Target and Antitarget Pharmacophore Modeling

The supermolecule approach was also employed to model the 5-HT$_{1A}$-R target versus α_1-AR antitarget selectivity of arylpiperazine antagonists [66]. The approach consisted in building for each receptor a quantitative model based on the same descriptor (i.e. V_{dif}) and then combining the two models to account for selectivity. In detail, the correlative models describing the binding affinity for each receptor (pK_i) are represented by Equations 8.9 and 8.10

$$pK_{i(5HT1A)} = 8.0(\pm 0.1)\, V_{dif(5HT1A)} + 4.5(\pm 0.4), \\ n=18,\ r=0.90,\ s=0.72,\ F=65.27, \tag{8.9}$$

$$pK_{i(\alpha 1)} = 13.7(\pm 2.2)\, V_{dif(\alpha 1)} - 5.3(\pm 0.6), \\ n=13,\ r=0.89,\ s=0.68,\ F=37.92. \tag{8.10}$$

The QSAR model to describe 5-HT$_{1A}$-R/α_1 target/antitarget selectivity was obtained by correlating selectivity, expressed as the difference between the cologarithmic forms of the 5-HT$_{1A}$-R and α_1-AR binding affinities (i.e. $\Delta pK_i = pK_{i(5\text{-}HT1A)} - pK_{i(\alpha 1)}$), with the ΔV_{dif}, where ΔV_{dif} was the difference between the V_{dif} indexes computed for each receptor (i.e. $\Delta V_{dif} = V_{dif(5\text{-}HT1A)} - V_{dif(\alpha 1)}$). The linear correlation equation obtained thereof is reported as follows:

$$\Delta pK_i = 8.1(\pm 1.2)\, \Delta V_{dif} - 0.7(\pm 0.3), \\ n=13,\ r=0.91,\ s=1.01,\ F=46.92. \tag{8.11}$$

An extensive computation of size/shape and MO descriptors was successively carried out to model the 5-HT$_{1A}$-R and generic α_1-AR affinity and selectivity of a wide series of N4-substituted arylpiperazines [90]. Computations of MO descriptors and correlative analyses were carried out by means of the CODESSA3 (comprehensive descriptors for structural and statistical analysis) program [62], leading to a number of QSAR models. Collectively, the probed descriptors included (a) ad hoc size and shape descriptors defined with respect to a supermolecule of high affinity ligands (as already discussed above in more detail) and (b) descriptors derived on a single structure (i.e. MO indexes and charged partial surface area descriptors).

Size and shape descriptors confirmed their effectiveness in correlating with the binding affinities. In fact, the most significant QSAR models obtained involved V_{dif}, in the case of the 5-HT$_{1A}$R, and V_{out} or $V_{dif,}$ for the α_1-AR:

$$pK_{i(5HT1A)} = 3.64(\pm 0.45)\, V_{dif(5HT1A)} + 7.12(\pm 0.21), \\ n=29,\ r=0.84,\ s=0.24,\ F=64.46, \tag{8.12}$$

$$pK_{i(\alpha 1)} = -0.015(\pm 0.0024)\, V_{out(\alpha 1)} + 8.44(\pm 0.21), \\ n=11,\ r=0.91,\ s=0.37,\ F=42.82, \tag{8.13}$$

$$pK_{i(\alpha 1)} = 3.66(\pm 0.68)\, V_{dif(\alpha 1)} + 5.85(\pm 0.29), \\ n=11,\ r=0.88,\ s=0.43,\ F=29.28. \tag{8.14}$$

A significant improvement in the correlations was obtained by combining size/shape descriptors with MO indexes in multilinear regression models:

$$pK_{i(5HT1A)} = 3.18(\pm 0.39) V_{dif(5HT1A)} - 0.57(\pm 0.15) \mu h_{(5HT1A)} \\ + 7.95(\pm 0.27), \quad n = 29, r = 0.90, s = 0.19, F = 55.6, \tag{8.15}$$

$$pK_{i(\alpha 1)} = 0.03(\pm 0.003) f-\text{PNSA}-1 - 146(\pm 15.9) q_{tot}(H) + 39.9(\pm 3.78), \\ n = 11, r = 0.97, s = 0.21, F = 75.3, \tag{8.16}$$

where μh is the total hybridization component of the molecular dipole, f-PNSA-1 is the fractional partial negative surface area [Semi-MO] of the bicyclic fragment, $q_{tot}(H)$ is the net atomic charge of the proton of the N–H$^+$ group.

Finally, a large variety of QSAR models of 5-HT$_{1A}$-R/α_1-AR target/antitarget selectivity were also obtained. The following equation shows the model with best statistics:

$$\Delta pK_i = -3.88(\pm 0.52) E_{st}(N) - 0.015(\pm 0.003) \text{ PNSA}-1 + 711(\pm 94.4), \\ n = 11, r = 0.96, s = 0.22, F = 46.3, \tag{8.17}$$

where $E_{st}(N)$ is the minimum atomic state energy for the protonated nitrogen atom, and PNSA-1 is the partial negative surface area.

It is worth noting that the descriptors that led to the best statistics also held a mechanistic/functional meaning. In fact, MO descriptors localized on the protonated amine function of the ligand reflected the relevance of the electrophilic site of the ligand in recognizing a primary nucleophilic site on the receptor via Coulombic interactions. However, size/shape and MO descriptors extended to the whole molecule properly accounted for short-range attractive (polar and dispersive forces) and repulsive intermolecular interactions, which are essentially responsible for the modulation of the binding affinities and, hence, of the target/antitarget selectivity. It was also inferred that the ad hoc defined size and shape descriptors, being derived to account for the strict complementarity for the receptor binding site, are more suitable to be used in connection with pharmacological data referring to single population of target receptor subtypes.

A classical Hansch approach and an artificial neural networks approach were applied to a training set of 32 substituted phenylpiperazines characterized by their affinity for the 5-HT$_{1A}$-R and the generic α_1-AR [91]. The study was aimed at evaluating the structural requirements for the 5-HT$_{1A}/\alpha_1$ selectivity. Each chemical structure was described by six physicochemical parameters and three indicator variables. As electronic descriptors, the field and resonance constants of Swain and Lupton were used. Furthermore, the vdW volumes were employed as steric parameters. The hydrophobic effects exerted by the *ortho*- and *meta*-substituents were measured by using the Hansch π-*ortho* and π-*meta* constants [91]. The resulting models provided a significant correlation of electronic, steric and hydrophobic parameters with the biological affinities. Moreover, it was inferred that the

5-HT$_{1A}$ receptor is able to accommodate bulky substituents in its active site, whereas the steric requirements of the α$_1$-AR at this position are more restricted [91].

Another attempt to model the 5-HT$_{1A}$-R/α$_1$-AR target/antitarget selectivity is exemplified by a CoMFA study on a series of 48 *ortho-*, *meta-* and *para-*substituted bicyclohydantoin-phenylpiperazines with affinity for 5-HT$_{1A}$ and α$_1$-AR [92]. Good quantitative models (i.e. with high cross-validation correlations and predictive power) were achieved. The resulting 3D QSAR models rationalize steric and electrostatic factors, which modulate the binding to 5-HT$_{1A}$-R and α$_1$-AR and, consequently, the 5-HT$_{1A}$/α$_1$ selectivity.

A series of new pyridazin-3(2*H*)-one derivatives were evaluated by Barbaro *et al.*, for their *in vitro* affinity toward both α$_1$- and α$_2$-adrenoceptors [93]. All target compounds showed good affinities for the α$_1$-AR, with K_i values in the low nanomolar range. To gain insight into the structural features required for α$_1$-AR antagonist activity, the pyridazinone derivatives were submitted to a pharmacophoric generation procedure by using the software catalyst. The resulting pharmacophore model showed a high correlation of the experimental and calculated affinity values and a good predictive power.

The genetic function approximation algorithm was used by Maccari *et al.* [94] to derive a three-term QSAR equation that is able to correlate the structural properties of 119 arylpiperazine derivatives with their affinity toward the α$_1$-AR. The number of rotatable bonds (Rotbonds), the hydrogen bond properties (H bond-acceptor) and a variable belonging to a topological family of descriptors (CHI) showed significant roles in the binding process toward the α$_1$-AR:

$$pKi = 7.61 - 0.56(\pm 0.08) \text{ Hbond-acceptor} + 0.18(\pm 0.02) \text{ Rotbonds}$$
$$+ 1.15(\pm 0.13)\text{CHI}, \quad n = 119, r = 0.93, F = 70.8. \quad (8.18)$$

The presented model was also compared with a previous pharmacophore for α$_1$-AR antagonists and a QSAR model for α$_2$-AR antagonists with the aim of finding common or different key determinants influencing α$_1$-/α$_2$-AR target/antitarget affinity and selectivity. Collectively, the following conclusions were drawn: (a) the length of the alkyl spacer correlates positively with α$_1$- and α$_2$-AR affinities (i.e. longer chains lead to higher pK_i values toward both the ARs); (b) a further introduction of alkyl substituents onto the arylpiperazinyl scaffold has a detrimental effect on the α$_2$-AR affinity, while having no consequences on the α$_1$-AR affinity; (c) the molecular connectivity index correlates positively with the α$_1$-AR but not with the α$_2$-AR affinity (i.e. an increase in the index is associated with an increase in the α$_1$-AR affinity but not in the α$_2$-AR affinity) and (d) a low number of hydrogen bond acceptor features seem to be related to high α$_1$-AR affinity and low α$_2$-AR affinity.

8.4
Modeling the α$_1$-AR Subtype Selectivities of Different Classes of Antagonists

A relevant aspect in QSAR studies and pharmacophore modeling is the choice of the most appropriate molecular descriptors with respect to both the molecular series considered and the molecular mechanism of ligand–target/antitarget interaction. The selection of compounds for QSAR studies is based on a presumed congenericity, that is, compounds that can be derived from the same parent structure and act through the same mechanism. Receptor blockers such as the α_1-AR antagonists show very heterogeneous chemical structures and congenericity. The latter, intended as three-dimensional homology of the various molecular fragments of the considered compound, is apparently not satisfied. However, QSAR and pharmacophore modeling require more than strict congenericity comparison between 'homogeneous' data, that is, the functional relation between the variations of the structural properties (X) and the activity (Y) remains the same throughout the investigated structural domain. This homogeneity does depend both on the level of the molecular description and on the definition of the ad hoc molecular descriptors.

The QSAR analysis described above on a wide series of prazosin analogues, 1,4-benzodioxan (WB-4101) derivatives and arylpiperazine α_1-AR antagonists has definitely shown the crucial role of the protonated nitrogen atom on α_1-ARs antagonistic potency [63,64,66]. Moreover, once the hydrogen bond requirements are satisfied, the α_1-ARs' binding affinity and target/antitarget selectivity are modulated by the molecular size and shape of the compounds through the optimization of short-range intermolecular interactions and the hydrophobic contribution. These results also suggest, in the context of the static complementarity between ligands and receptor, some important features of both the architecture and the binding site of the receptor, as inferred from the three molecular classes of α_1-AR antagonists considered.

The following sections summarize the pharmacophore and QSAR modeling of the α_1-AR subtype affinity and selectivity through the employment of noncongeneric series of compounds.

8.4.1
Supermolecule-Based Subtype Pharmacophore and QSAR Models

The availability of *in vitro* binding affinity data on the cloned α_1-ARs allowed for the building of subtype specific QSAR models based on noncongeneric series of ligands. Early studies were based on 13 noncongeneric molecular series of α_{1A}-AR antagonists [65]. The considered series included compounds **1**, **3**, **9–11**, **14**, **19** and **32** in Scheme 8.1 and compounds **43** and **45** in Scheme 8.2. Compounds **9** and **32** were considered in both the enantiomeric forms, whereas compound **45**, corynanthine, was considered also in its diastereomeric form yohimbine [55]. The natural agonist, norepinephrine, was also considered, for comparative analysis [55].

In this noncongeneric molecular series, variations in the native α_{1A}-adrenoceptor binding affinity values did not correspond to any significant and/or organized variation in the values of the geometric and the computed MO reactivity descriptors.

On a quantitative basis, an accurate correlation analysis was carried out and some significant linear QSAR models were obtained, based on ad hoc defined size–shape descriptors. One of the best linear correlations based on the V_{in} index indicated that the higher the volume shared by a generic antagonist and the supermolecule (obtained by superimposing abanoquil and WB4101, Scheme 8.1), the higher is

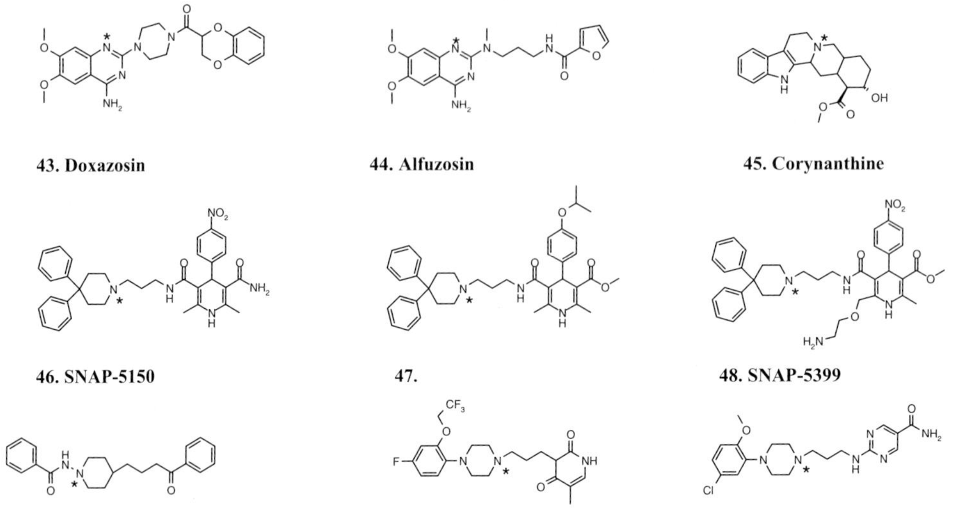

43. Doxazosin   **44. Alfuzosin**   **45. Corynanthine**

46. SNAP-5150   **47.**   **48. SNAP-5399**

49. SNAP-1069   **50. RS-100,975**   **51. SL-89-0591**

Scheme 8.2

the α_{1A}-AR binding affinity:

$$pK_{i\alpha 1A} = 0.022(\pm 0.004)V_{in} - 5.97(\pm 0.90),$$
$$n = 14, \ r = 0.85, \ s = 0.80, \ F = 31.62. \quad (8.19)$$

Similar results were obtained for the molecular descriptor V_{dif}. It is worth noting that the agonist norepinephrine participates in the QSAR model, being at the lowest extreme of the above linear regression. This result nicely illustrates, on a quantitative basis and for a very heterogeneous molecular series, the concept of accessory surface area of the receptors. The latter is postulated to be exquisitely suited to accommodate the usually exceeding molecular surface area or volume of the antagonists with respect to the agonists, which also usually show lower binding affinity values.

On the basis of the above results, a simple three-point pharmacophore model for the α_{1A}-AR antagonists was proposed. This model was characterized by two hydrophobic phenyl rings at an average distance of 1.43 and 6.41 Å, respectively, from the essential protonated nitrogen atom. It was also inferred that, for the conformationally restricted compounds such as **3** (Scheme 8.1) and **45** (Scheme 8.2), the coplanarity of the protonated nitrogen atom and its substituents could contribute to the high affinity toward the α_{1A}-AR [65].

The supermolecule-based QSAR modeling approach and the derived selective ligand design protocol were, then, extended to a larger set of structurally heterogeneous antagonists of the three cloned α_1-AR subtypes (i.e. compounds **1–38** in Scheme 8.1) [68]. The choice of the considered ligands depended on the availability of homogeneous *in vitro* binding affinity data on the three cloned α_1-ARs. In this respect, three supermolecules, one for each α_1-AR subtype, were built, which could represent the best mean 3D size and shape complementarity for their respective α_1-ARs [68]. In this respect, the numerical comparison of the vdW volumes of the

α_{1a}, α_{1b} and α_{1d} supermolecules suggested higher capacity and flexibility of the α_{1a}-AR binding site than the other two subtypes, which also differed in size and shape. These inferences were consistent with a computational structural/dynamics analysis of the binding site of the isolated receptor [95]. This α_{1a}-AR peculiarity makes this receptor subtype very important and interesting both for drug discovery and antitarget screening.

The supermolecules for the three α_1-ARs, which were obtained by superimposing the most structurally different ligands showing the highest affinities for the same receptor, hold the structural features differentiating the three α_1-AR subtypes although ligand selectivity was not explicitly taken into account while selecting the components of the three supermolecules. Ligand superimposition followed a topologic rigid body fit procedure based on a few essential pharmacophoric features: (a) the hydrogen of the protonated nitrogen atom; and (b) the aromatic rings closest to and farthest from the protonated nitrogen [68]. Thus, the structural peculiarities of the affinity-based supermolecules were able to account for the subtype-selectivity of a test set of compounds. In general, many structurally different sets of ligands show high affinity for the α_{1a}-AR, whereas only few classes of compounds are able to realize high affinity for the α_{1b}- and α_{1d}-AR subtypes, reflecting a more constrained binding mode for these two subtypes than the α_{1a}-AR. This information is captured by the three supermolecules. In fact, the number of structurally different compounds forming the α_{1a}-AR supermolecule is higher than that forming the α_{1b} AR and α_{1d}-AR supermolecules (Scheme 8.1 and Figure 8.2). In this respect, the α_{1a}-AR supermolecule involves four different types of protonated nitrogen atoms (i.e. quinazolinic, compound **3**; alkylaminic, compounds **9**, **11** and **15**; piperazinic, compound **25** and piperidinic, compound **32**, Scheme 8.1 and Figure 8.2). Fewer compounds build the supermolecule for the α_{1b} AR and the α_{1d}-AR subtypes (i.e. compounds **1**, **4**, **7** and **22** for the α_{1b}, and compounds **1**, **7**, **17**, **22** and **38** for α_{1d}, Scheme 8.1 and Figure 8.2). It is interesting to note that the supermolecules for the α_{1b} AR and α_{1d}-ARs share three ligands: compounds **1**, **7** and **22** (Scheme 8.1). However, the comparative visual inspection of the supermolecules for the three α_1-AR subtypes and of their vdW volumes shows significant size and shape diversity (Figure 8.2). This reflects the differences in the high-affinity binding site of the ensemble of ligands that constitute each supermolecule. In this respect, the vdW volume of the α_{1a}-AR supermolecule is larger than that of the other two supermolecules ($Vsup_{\alpha 1a} = 1016.00$ Å^3, $Vsup_{\alpha 1b} = 680.75$ Å^3 and $Vsup_{\alpha 1d} = 754.13$ Å^3; Figure 8.2) reflecting larger dimensions and/or higher flexibility of the α_{1a}-AR binding site than the other two subtypes. These inferences are consistent with a computational structural/dynamics analysis of the binding site of the isolated receptor [95].

The three supermolecules are characterized by aromatic and hydrophobic features on the two sides of a central and constrained part, consisting of the protonated nitrogen atom (Figure 8.2). Although the two aromatic rings and the protonated nitrogen atom are common features to the three supermolecules, their spatial arrangement is different in the three supermolecules. In fact, the average distances between the protonated nitrogen atom, on the one hand, and the center of the two

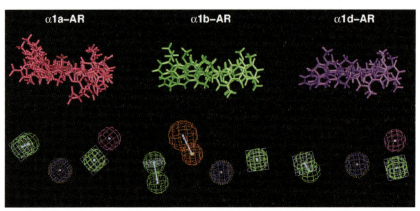

Figure 8.2 Top: frontal views of the supermolecules for the three α_1-AR subtypes are shown. The α_{1a}-AR supermolecule (colored in pink) is constituted by compounds **3**, **9**, **11**, **15**, **25** and **32**; the α_{1b}-AR supermolecule (colored in green) is constituted by compounds **1**, **4**, **7** and **22**; and the α_{1d}-AR supermolecule (colored in violet) is constituted by compounds **1**, **7**, **17**, **22** and **38**. The volume of the α_{1a}-AR supermolecule (1016.00 Å^3) resulted larger than those of the α_{1b}- and α_{1d}-AR supermolecules (680.75 Å^3 and 754.13 Å^3, respectively). Bottom: pharmacophore hypotheses for the α_{1a}- (left), α_{1b}- (middle) and α_{1d}-ARs (right) derived from the members of the three supermolecules. Mesh spheres represent the location constraints for the feature on the ligand (or receptor) and are colored according to the feature. In detail, positive ionizable, aromatic rings, hydrophobic groups and H-bonding acceptors are colored in blue, green, magenta and orange, respectively. The light blue squares indicate the position of the plane of the aromatic ring system of the ligand. For the aromatic and H-bonding acceptor features, the second larger sphere relates to the positions of the interacting aromatic or H-bonding donor groups, respectively, on the receptor.

aromatic rings closest to and farthest from the protonated nitrogen, on the other, are respectively 5.18 and 6.77 Å for the α_{1a}-AR, 3.53 and 8.68 Å for the α_{1b}-AR, as well as 4.25 and 7.68 Å for the α_{1d}-AR supermolecule. These average values indicate that, in the α_{1a}-AR supermolecule, the protonated nitrogen atom of the ligand is almost equidistant from the two aromatic rings, whereas, in the other two supermolecules (particularly in the α_{1b}-AR supermolecule), it is closer to one aromatic ring than to the other. This arrangement is mainly due to both the presence of the alkylamines in the α_{1a}-AR supermolecule and the high contribution of the quinazoline compounds in the α_{1b}-AR and α_{1d}-AR supermolecules (Scheme 8.1 and Figure 8.2).

The peculiar features of the three α_1-AR supermolecules were translated into pharmacophore hypotheses by means of Catalyst software (Figure 8.2, unpublished results). Catalyst treats molecular structures as templates placing chemical functions in 3D space to interact with the receptor. Molecular flexibility is taken into account by considering each compound as a collection of conformers representing different areas of the molecules conformational space within a given energy range.

The hypotheses were automatically generated on the ensembles of ligands constituting the α_{1a}-, the α_{1b}- and the α_{1d}-AR supermolecules (compound **38** was excluded). The ligands were taken in the same conformation as they have in the supermolecule. The following molecular features were considered for hypothesis generation: hydrophobic, H-bonding acceptor, H-bonding donor, positive charge and aromatic ring. For each α_1-AR subtype, various hypotheses were generated. The final pharmacophore hypothesis was selected on the basis of the highest scores and closer to the average geometric features displayed by the corresponding supermolecule (Figure 8.2). In particular, the pharmacophore hypothesis of the α_{1a}-AR is constituted by two aromatic features almost equidistant from the positively charged nitrogen (i.e. the distances between the center of the sphere representing the positively charged nitrogen and that of the sphere representing the closer and farther aromatic rings are 5.2 and 6.3 Å, respectively) and by a hydrophobic feature closer to one of the two aromatic rings. This model agrees with the class II α_{1a}-AR pharmacophore model proposed very recently by Klabunde and Evers [96]. In the α_{1b}-AR hypothesis, one aromatic feature is closer to the positive charge than the other (i.e. the distances between the centers of the relative spheres are 5.6 and 9.8 Å, respectively). Moreover, one H-bonding acceptor is present in between the positive charge and the farthest aromatic ring. The α_{1d}-AR pharmacophore is similar to the α_{1b}-AR one as the two aromatic features are not equidistant from the positive charge (i.e. the distance between the centers of the relative spheres is 5.6 and 7.6 Å, respectively). Furthermore, one hydrophobic feature can be found near the aromatic feature closest to the positive charge (Figure 8.2) or, in some other hypotheses, an H-bonding acceptor can be found in between the positive charge and the farthest aromatic ring.

The QSAR models obtained by the three α_1-AR supermolecules strengthened the concept that once the electronic/electrostatic requirements of the protonated nitrogen atom for the long-range recognition of the protophilic residue in the receptor are satisfied, the binding affinities are modulated mainly by short-range intermolecular interactions. These interactions depend on the molecular size and shape, defined and described with respect to the ad hoc reference supermolecules. For each subtype, the best linear regression equations involved the size/shape descriptors V_{in}/V_{mol} and V_{dif}:

$$pK_{1a} = 6.24(\pm 0.84) V_{in}/V_{mol} + 3.16(\pm 0.76),$$
$$n = 34, \ r = 0.79, \ s = 0.48, \ F = 55.19, \tag{8.20}$$

$$pK_{1a} = 5.18(\pm 0.83) V_{dif} + 7.19(\pm 0.27),$$
$$n = 34, \ r = 0.74, \ s = 0.53, \ F = 39.01, \tag{8.21}$$

$$pK_{1b} = 5.87(\pm 0.69) V_{in}/V_{mol} + 3.44(\pm 0.53),$$
$$n = 30, \ r = 0.84, \ s = 0.49, \ F = 71.50, \tag{8.22}$$

$$pK_{1b} = 5.87(\pm 0.72) V_{dif} + 6.26(\pm 0.22),$$
$$n = 30, \ r = 0.83, \ s = 0.51, \ F = 66.08, \tag{8.23}$$

$$pK_{1d} = 7.14(\pm 0.84) V_{in}/V_{mol} + 2.18(\pm 0.67),$$
$$n = 31, \ r = 0.84, \ s = 0.58, \ F = 71.99, \tag{8.24}$$

$$pK_{1d} = 9.74(\pm 1.14) V_{dif} + 5.17(\pm 0.33),$$
$$n = 31, \ r = 0.84, \ s = 0.57, \ F = 72.86. \tag{8.25}$$

Validation of the above-mentioned QSAR models was carried out by using the test set reported in Scheme 8.2. The agreement between predicted and measured data values was satisfactory [68]. These QSAR models were subsequently successfully challenged in their ability to predict the α_1-AR subtype affinity of other noncongeneric compounds (i.e. 28 compounds) [41]. These models were also used for designing and estimating the binding affinities and selectivities of new compounds [71]. Recalling that the simple supermolecule approach is mainly used for ligand design purposes and is based on QSAR models generated on highly noncongeneric molecular series, the agreement between the predicted and measured binding affinity and selectivity data values was considered satisfactory. In fact, these QSAR models were considered as useful ligand design and decision-making tools. In this case, these tools showed low predictive resolution (about 0.5 for α_{1a}, 1 for α_{1b} and 1.5 for α_{1d}, on a logarithmic scale). Nevertheless, they allowed for selection among the many designed ligands, by estimating their affinities for the α_1-ARs according to three levels: (a) low affinity, pK_i 7–8; (b) high affinity, pK_i 8–9 and (c) very high affinity, $pK_i > 9$. On this basis, a few wrong estimations occurred. Similarly, the three α_1-AR supermolecules can be used for *in silico* antitarget screening.

Interestingly, the majority of the compounds proposed in this work showed low affinities for the antitarget 5-HT$_{1A}$-R. This was mainly due to the presence of a chlorine substituent in position 5′ of the phenyl ring and/or to the elongation of the alkyl spacer. The latter, however, needed the presence of a phenyl ring in position 6 of the pyridazinone moiety to result in a drop in the 5-HT$_{1A}$-R affinity. Collectively, this study indicates that substitutions in positions 5 and/or 6 of the pyridazinone ring as well as in positions 2′ and 5′ of the phenyl ring together with the length of the linker are responsible for modulating the ligand α_1-AR binding affinities and α_1-AR/5-HT$_{1A}$ target/antitarget selectivities [71].

For modeling selectivity, a reasonable cutoff value of 1 was set. Thus, selective ligands were those characterized by pK_i difference values greater than 1, whereas nonselective ligands were characterized by pK_i difference values lower than 1. Selectivity predictions according to these cutoffs gave satisfactory results for the three α_1-AR subtypes [68]. Similar considerations are also applicable in the context of antitarget screening.

There are several problems connected with the attempt to rationalize, in quantitative terms, the pharmacological data for GPCRs. Cells expressing cloned receptors constitute useful tools for drug screening and facilitate the establishment of precise structure–affinity relationships. However, the ultimate goal is to generalize the information obtained on a pure population of receptors to the native receptors to design tissue-selective drugs. Given the high flexibility of its structure, the same receptor can accommodate very diverse ligands raising the problem of side-effect propensity and the need of antitarget modeling and screening. On the contrary, subtle molecular determinants might be responsible for the selectivity toward different receptor subtypes, and apparently small chemical modifications of a ligand can change its pharmacological phenotype dramatically. Therefore, the rule in this field is to deal with the series of noncongeneric compounds, showing a large variability in their structural features. In this framework, the task of deciphering the information relevant for ligand–receptor interaction can be accomplished only with key molecular descriptors, which are able to perceive nonintuitive differences among ligands. The heuristic statistical treatment implemented in the CODESSA program [62] is a valuable tool to select, from a large pool, the computational molecular descriptors that give powerful predictive and interpretative QSAR models for different types of pharmacological data.

A further quantitative rationalization of the pharmacological properties and selectivity of the same wide series of noncongeneric α_1-AR antagonists was attempted by combining the size and shape descriptors with a large variety of global and local computational molecular descriptors [97]. These descriptors were employed to compare their performance in rationalizing high-quality binding affinity data values and to extend previous QSAR models of *in vitro* binding affinities to models of *in vitro* and *in vivo* functional activities. The heuristic statistical treatment implemented in the CODESSA program was used to select a limited set of descriptors able to describe the molecular determinants for binding affinity of native (α_{1A} and α_{1B}) and cloned (α_{1a}, α_{1b} and α_{1d}) ARs, functional activity of vascular and lower urinary tract tissues and *in vitro* and *in vivo* subtype selectivities [97].

The most relevant bilinear regression models obtained for the binding affinity of the α_1-AR subtypes are the following:

$$pK_{1A} = 0.02(\pm 0.003)\ V_{\text{out}(1A)} - 8.50(\pm 3.16)\ \text{ABIC2} + 15.8(\pm 2.44),$$
$$n = 23,\ r = 0.88,\ s = 0.44,\ F = 33.17, \quad (8.26)$$

$$pK_{1B} = 5.81(\pm 0.48)\ V_{\text{dif}(1B)} - 23.7(\pm 7.94)\ \text{FPSA} + 8.62(\pm 0.83),$$
$$n = 24,\ r = 0.94,\ s = 0.35,\ F = 83.44, \quad (8.27)$$

$$pK_{1a} = 5.74(\pm 0.65)\ V_{\text{dif}(1a)} - 2.14(\pm 0.44)\ f\text{-}P\pi\text{-}\pi(\text{C}-\text{O}) + 8.61(\pm 035),$$
$$n = 34,\ r = 0.86,\ s = 0.41,\ F = 43.24, \quad (8.28)$$

$$pK_{1b} = 0.017(\pm 0.0018)\ V_{\text{in}(1b)} - 0.0052(\pm 0.00062)\ \text{WPSA-1} + 5.59(\pm 042),$$
$$n = 30,\ r = 0.89,\ s = 0.43,\ F = 52.26, \quad (8.29)$$

$$pK_{1d} = 7.08(\pm 0.68)\ V_{\text{in}(1d)/V_{\text{mol}}} + 52.18(\pm 14.08)\ f\text{-}P\sigma\text{-}\pi(\text{C}-\text{X}),$$
$$n = 31,\ r = 0.90,\ s = 0.48,\ F = 59.84, \quad (8.30)$$

where ABIC2 is the average bonding information content (order 2), which encodes the branching ratio and constitutional diversity of a molecule; FPSA and WPSA-1 are, respectively, the surface weighted charged partial surface area and the fractional charged partial surface area computed on the whole ligands; f-Pπ–π(C–O) and f-Pσ–π(C–X) are, respectively, the π- and σ–π bond order of a C–O or C–N pair of atoms of the fragments, which modulate the availability of the lone electron pairs of the heteroatoms for intermolecular interactions.

Some improvement in the statistics of these bilinear regression models was achieved with respect to previous linear equations. Moreover, the possibility to generate in a fast way a large pool of QSAR models able to describe a biological property assumes a relevant importance in particular with respect to the prediction problem. However, the results of both the validation techniques and the predictive power for binding affinity and selectivity of the considered compounds were quite similar for simple and bilinear regressions [68,97].

As for the modeling of *in vitro* and *in vivo* potencies and selectivities, some interesting models were obtained on a subset of representative ligands. In fact, the QSAR analysis of functional activities for this set of compounds, although numerically limited, yields successful correlations [97]. Among these linear correlations, the following equation constitutes a significant example:

$$-\log(DBU/UP) = -9.4(\pm 1.07)\ V_{dif(1a)} + 2.12(\pm 0.36),$$
$$n = 8,\ r = 0.96,\ s = 0.17,\ F = 76.94, \quad (8.31)$$

where $-\log(DBU/UP)$ is the *in vivo* selectivity studied in a dog model, UP represents the dose active in inhibiting by 50% of the urethral contractions induced by noradrenaline and DBU is the dose active in lowering diastolic blood pressure by 25%.

The best statistical parameters were obtained by correlating the *in vivo* selectivity with the V_{dif} descriptor defined with respect to the α_{1a}-AR supermolecule. It is worth noting that the α_{1a} is the adrenergic receptor subtype of functional relevance for the urethra tissue (dog model) [8]. Thus, ligands showing high potency and selectivity for the lower urinary tract are those, which better fit the volume of the supermolecule that represents the binding site of the α_{1a}-AR subtype.

In summary, from the extensive QSAR modeling, it was inferred that MO indexes describing the small variation of the molecular electronic structure, which contributes to the recognition step (electrostatic interaction between the protonated amine function and a primary nucleophilic site of the receptor), are more suitable for rationalizing the binding affinity of subsets of congeneric ligands or to be used in connection with selectivity data. On the contrary, the strict requirements for shape complementarity between the ligand and the receptor are encoded by ad hoc-defined size and shape descriptors.

A very similar QSAR approach for the modeling and prediction of selectivity of α_1-AR antagonists has recently been carried out by Eric *el al*. It confirms the usefulness of the supermolecules approach and of the ad hoc shape descriptors in the rationalization of α_1-ARs affinity and α_1-AR subtype selectivity [98].

Finally, very recently Pallavicini *et al.*, in continuation of a previous study on *ortho*-monosubstituted compounds, designed and synthesized a series of 2-[(2-phenoxyethyl) aminomethyl]-1,4-benzodioxanes *ortho*-disubstituted at the phenoxy moiety [99]. The disubstituted analogues were tested for their binding affinities at the three α_1-AR subtypes and for the 5-HT$_{1A}$-R. The affinity values of the new compounds were compared with those of the enantiomers of the 2,6-dimethoxyphenoxy analogue, the well-known α_1-AR antagonist WB-4101 (Scheme 8.1), and of the *ortho*-monosubstituted derivatives. The results suggested some distinctive aspects in the interaction of the phenoxy moiety of monosubstituted and disubstituted compounds with the α_{1a}-AR and the 5-HT$_{1A}$ receptors. A classical (Hansch) QSAR analysis was applied to the whole set of the (S)-enantiomers of the *ortho*-, mono- and disubstituted WB4101 analogues (i.e. 26 compounds), finding a very good correlation for the α_{1a}-AR affinity. In this respect, a significant parabolic relationship was also found between α_{1a}-AR affinity and vdW volume of the two *ortho*-substituents. In contrast, the same size/shape descriptor exhibited low or insignificant relationship with the 5-HT_{1A}-R binding affinity.

8.4.2
Ligand-Based Subtype Pharmacophores

One of the early examples of pharmacophore modeling of the α_{1a}- and α_{1b}-AR subtypes was carried out by means of the Apex-3D software [100]. The common features of the pharmacophore models of the two α_1-AR subtypes were a protonated amine, an aromatic ring and a polar region. The main difference between the two pharmacophores was the distance between the aromatic ring and the amine function. This distance was shorter for the α_{1a} subtype. Successive advances in the pharmacophore models consisted in extending the set of ligands to α_{1d}-antagonists and replacing the use of Apex-3D with the Catalyst program. In an attempt to generate general (cross-chemotype) α_1-AR pharmacophores, applicable to all the structural classes of ligands, the complete training set of 38 antagonists was selected. Protonated forms of the antagonists at the common amine function were built to predefine at least this one common feature to all molecules, as this is postulated to be an essential feature for α_1-AR ligand binding [101]. Conformational searching was performed and the chemical features used in the pharmacophoric generation step included the following elements: positive ion, hydrogen bond acceptor and donor, aromatic ring and hydrophobic (aliphatic) groups. However, the resulting hypotheses, which included the positive ion feature, had several problems. The correlations between predicted and experimental affinities were fairly low for the compounds of the training sets. Of special concern in this respect was the fact that high errors (>fivefold) were observed for binding affinity estimations of compounds selective for the subtype under consideration. The Catalyst program was also employed in a recent study by MacDougall and Griffith [102] to build ligand-based pharmacophores from three training sets of α_{1a}, α_{1b} and α_{1d}, selective antagonists available in the literature. Four-feature pharmacophores were developed for the α_{1a} and α_{1b} subtype-selective antagonists, and a five-feature pharmacophore was developed for the α_{1d} subtype-selective antagonists. The four features constituting the α_{1a} pharmacophore were (a) one positive ion, (b) one aromatic ring, (c) one hydrogen bond acceptor and (d) a hydrophobic (aliphatic) group. Such a pharmacophore is representative of antagonists with at least 100-fold selectivity over the α_{1b}-AR and 40-fold selectivity over the α_{1d}. The training set for this pharmacophore included the two classes of antagonists as described by Klabunde and Evers [96], as well as structurally different compounds, which fitted into neither classification [102]. The α_{1b} pharmacophore was generated from antagonists showing selectivity over the other two subtypes, which were predominantly prazosin analogues. This pharmacophore did not include a positive ion feature, as it may have been expected, but comprised the following features: (a) two H-bonding acceptors, (b) one aromatic ring and (c) one hydrophobic group [102].

Finally, the five-feature α_{1d} pharmacophore was generated from those α_{1d} antagonists, which were 100-fold selective over α_{1a} and 40-fold selective over α_{1b} receptors. The pharmacophore consisted of the following features: (a) one positive ion, (b) one aromatic ring, (c) one H-bonding acceptor and (d) two hydrophobic groups [102].

Collectively, it was inferred that one reason for subtype selectivity is the size of the antagonist. The average molecular weights of the training sets (α_{1a}, 485; α_{1b}, 443; α_{1d}, 400) appeared suggestive of higher capacity of the α_{1a}-AR binding site compared to the other two subtypes. These inferences, given the antitarget propensity of α_{1a}-AR, are in line with those from our ligand-based and receptor-based α_1-AR modeling studies [41,68,95]. Validation of the α_{1a} and α_{1d} pharmacophore was carried out by using a test set of 15 α_1-AR antagonists differing in structure and affinity. In contrast, validation of the α_{1b}-AR pharmacophore could not be done since all available α_{1b}-selective antagonists were included in the training set. Collectively, the best performance was achieved by the α_{1d} pharmacophore, despite the fact that the training set contained essentially BMY-7378 (i.e. compound **20** in Scheme 8.1), whereas the test set comprised structurally different compounds [102]. The most striking difference between the pharmacophore hypotheses obtained by MacDougall *et al.* and the ones extracted from the three α_1-AR supermolecules (Figure 8.2) is the lack of the second aromatic feature in the former. It is worth noting that, different from the strategy employed by MacDougall *et al.*, subtype selectivity was not taken into account for the choice of the compounds constituting the supermolecules [68]. In this respect, being selectivity/specificity a derived molecular property defined as binding affinity ratio, it is more appropriate to use high affinity ligands as reference molecules for target binding site modeling instead of selective ligands.

In a very recent paper, Vistoli *et al.* reported a study aimed at correlating pharmacological properties with molecular parameters derived from the physicochemical property space of bioactive molecules [103]. A data set of 36 ligands of the α_{1a}-, α_{1b}- and α_{1d}-adrenoceptors as published by Bremner *et al.* [101] was used. One thousand conformers were generated for each ligand by Monte Carlo conformational analysis, and four 3D-dependent physicochemical properties were computed for each conformer of each ligand, namely virtual lipophilicity (clog P), dipole moment, polar surface area and solvent-accessible surface area. Thus, a space of four physicochemical properties was obtained for each ligand. These spaces were assessed by two descriptors, namely, their range and their sensitivity (i.e. the variation amplitude of a given physicochemical property for a given variation in molecular geometric properties). Neither the range nor the sensitivity of any of the four physicochemical properties correlated with receptor affinities. In contrast, range and sensitivity showed promising correlations with ΔpK_{a-b} (i.e. the α_{1a}/α_{1b} selectivity) for the complete data set. The correlations were lower for ΔpK_{a-d} (i.e. the α_{1a}/α_{1d} selectivity), whereas there was no correlation at all with ΔpK_{b-d}. Another Catalyst-based pharmacophore for the α_{1a}-AR has recently been proposed by Li *et al.* [104]. Almost similar to the study by MacDougall *et al.* [102], the best scoring pharmacophore hypothesis consisted of four important chemical features: positively charged, hydrogen bonding donor, aromatic and hydrophobic.

8.4.3
Receptor-Based Subtype Pharmacophore and Ligand–Target/Antitarget Interaction-Based QSAR

A challenging task in computational modeling of GPCRs is to define intermolecular interaction descriptors and computational protocols, which could be used for virtual functional screening [105]. Pioneers in this respect, we have used computational experiments aimed at defining intermolecular interaction descriptors on the ligand–receptor complexes, which could correlate linearly with biological data such as binding affinities or efficacies (see [21,55–60,95,106] for review). In this respect, an intermolecular interaction descriptor, somewhat related to the binding energy and, hence, called BE, was used to rationalize and predict the binding affinities and selectivities of different GPCR ligands (reviewed in [21]). This descriptor, which ill-defines all the entropic effects, is computed on the ligand–receptor energy-minimized complexes according to the following formula: $BE = IE + E_R + E_L$, where IE is the ligand–receptor interaction energy and E_R and E_L are the distortion energies of the receptor and the ligand, respectively, calculated as differences between the energies of the bound and the free optimized molecular forms. The BE index was employed to build QSAR models for a series of 16 selective and nonselective α_1-AR antagonists (i.e. compounds **1, 4, 6, 9, 11, 18–20, 24, 29, 31, 32, 34** and **37, 39** and **40** in Scheme 8.1), which were docked into the average minimized structures of the three α_1-AR subtypes. For each ligand–receptor minimized complex, the computed BE were correlated with the experimental binding affinities measured on the three cloned subtypes, leading to the following equations:

$$pK_{1a} = 5.45(\pm 0.37) - 0.08(\pm 0.01)BE_{1a}, \quad n = 16, r = 0.90, s = 0.52, \tag{8.32}$$

$$pK_{1b} = 4.20(\pm 0.32) - 0.12(\pm 0.01)BE_{1b}, \quad n = 16, r = 0.94, s = 0.37, \tag{8.33}$$

$$pK_{1d} = 4.43(\pm 0.36) - 0.09(\pm 0.01)BE_{1d}, \quad n = 16, r = 0.92, s = 0.39. \tag{8.34}$$

Consistency was found between the affinity/selectivity predictions by the intermolecular interaction descriptor BE and those by a ligand-based approach, that is, the supermolecule approach [41,68,71]. Consistent with the inferences from ligand-based QSAR modeling, it was indeed found that, while the protonated nitrogen atom is an essential pharmacophoric element for the long-range electrostatic recognition and productive interaction with D3.32 of the α_1-AR binding site, its contribution to the interaction energy is relevant but almost constant [95]. In contrast, short-range forces modulate both ligand affinity and selectivity. Thus, the modulation of the binding affinity by a wide noncongeneric series of α_1-AR ligands can be described and explained by the variation of the ligand size–shape features that are related to the short-range acting forces. In this respect, the intermolecular interaction descriptor, BE, is rather independent of the reliability of the ligand–receptor complexes.

Finally, in a recent study, more than 75 compounds structurally related to BMY-7378 (i.e. compound **20** in Scheme 8.1) were designed and synthesized. Structural variations of each part of the reference molecule were introduced, obtaining highly selective ligands for the α_{1d}-AR. In this respect, α_{1d}-AR selectivity was carried by the 2′,5′-dichloro substitution on the phenylpiperazine [107]. The integration of an extensive QSAR analysis with docking simulations using the rhodopsin-based models of the three α_1-AR subtypes and of the 5-HT$_{1A}$-R provided significant insights into the characterization of the receptor binding sites as well as into the molecular determinants of ligand selectivity at the target α_{1d}-AR with respect to the antitargets α_{1a}-AR, α_{1b}-AR and 5-HT$_{1A}$-R. The results of multiple copies simultaneous search (MCSS) on the substituted phenylpiperazines together with those of manual docking of selected compounds into the putative binding sites of the α_{1a}-AR, α_{1b}-AR, α_{1d}-AR and 5-HT$_{1A}$-R suggested that the phenylpiperazine moiety would dock into a site formed by amino acids in helices 3, 4, 5, 6 and in the extracellular loop 2 (i.e. overlapping, at least in part, with the norepinephrine binding site (Figure 8.1)), whereas the spirocyclic ring of the ligand docks into a site formed by amino acids of helices 1, 2, 3 and 7. This docking mode is consistent with the QSAR data previously discussed. The results of docking simulations also suggest that the second and third extracellular loops may act as selectivity filters for the substituted phenylpiperazines. Furthermore, the key features of ligand selectivity toward the α_{1d}-AR reside in a combination between conformational preferences of the substituted phenylpiperazine and structural features of its binding site.

8.5
Antitarget Modeling of Biogenic Amine-Binding GPCRs: Common Features and Subtle Differences

The subfamily of biogenic amine binding GPCRs has provided excellent drug targets for the treatment of numerous diseases. However, their central role in cell signaling also raises the problem that many new drug candidates manifest side affinities toward different receptor sites of the same (sub)family. Hence, drug candidates have the potential to interfere with the physiological signaling processes, thus, causing undesired side effects. This situation is due to general and peculiar features of this GPCR subfamily. In fact, (a) all GPCRs share a common structural architecture and a common molecular strategy of signal transduction; (b) GPCRs are allosterically modulated proteins that exist as complex conformational ensembles, and hence they are very dynamic objects that show a peculiar 'molecular plasticity' and binding site adaptability; (c) emerging evidence suggests that GPCRs exist as homo- and heterodimers/oligomers, thus, protein–protein interactions are key regulatory features; (d) biogenic amine binding GPCRs are characterized by high sequence similarity and they share, in the natural agonist binding site, the crucial aspartate (D3.32) residue; (e) the protonated amine group is a crucial common pharmacophoric feature of the ligands of this GPCR subfamily, together with the presence of aromatic/heteroaromatic rings and/or hydrophobic groups that modulate the 3D pharmacophoric features and selectivity. Collectively, points a–e depict the main interrelated features of this GPCR subfamily, which constitutes a group of very important and interesting target and antitarget receptors. The difficult tasks of pharmacophore/QSAR modeling for selective ligand design purposes and 'cross-chemotype' pharmacophoric modeling (ligand- and receptor-based) for *in silico* antitarget screening, are, in fact, the two faces of the same coin. In this respect, the examples of selective modeling shown in Sections 8.3 and 8.4 can also be seen as an attempt of ligand optimization versus a single target against the other receptors that assume the role of antitargets. In this way, the effort of model building of selective 3D pharmacophores (by considering noncongeneric ligands belonging to different and heterogeneous chemical classes) and their QSAR validations become a formidable tool for both selective ligand design and *in silico* antitarget affinity screening. In this context, the α_{1a}-AR constitutes an emblematic example. In fact, the different approaches discussed above consistently generated and validated 3D pharmacophores on a wide number of different chemical series of ligands, confirming the high adaptability of the α_{1a}-AR binding site and, hence, its strong antitarget propensity. Reliable *in silico* tools and computational models have recently been proposed to identify compounds with strong antitarget affinity and to guide the chemical (drug) optimization (described in Chapter 6) [96,108,109].

In conclusion, the 3D pharmacophore and QSAR models presented can be easily used for *in silico* antitarget screening of compound databases to identify ligands with side affinity and potential α1-mediated side effects.

8.6
Conclusions

8.6.1
From Molecules to Pharmacophores to Descriptors to Models

The pharmacophore concept, like that of molecular/chemical structure, assumes different meanings, and consequently different definitions, according to the accuracy and resolution level of its information content.

However, the official definition proposed by an IUPAC working group is 'A pharmacophore is the ensemble of steric and electronic features that are necessary to ensure the optimal supramolecular interactions with specific biological target structure and to trigger (or to block) its biological response' [110]. If 'the ensemble of steric and electronic features' is referred to a molecular level of description of the objects that influence the biological response (as the 'optimal supramolecular interactions' suggest), the 'ensemble of steric and electronic features', that is the pharmacophore, constitutes a virtual (not physically observable) structural concept.

A great variety of methods and techniques are now available for pharmacophore model generation. These methods are based on the appropriate choice of the ligand set (ligand-based, when much information is available on different ligand classes), and/or on one or more target/antitarget proteins (structure-based, particularly useful when none or poor information is available on the ligands). The aim is to (a) describe and predict the molecular recognition propensities of different ligand classes for different target/antitarget protein family; (b) translate these propensities into (sub)molecular descriptors/predictors; and (c) generate quantitative statistical models to guide drug design and discovery by appropriate scoring functions, QSAR analysis and database virtual screening. The different aspects of this complex matter are treated, in this chapter, with different emphasis. However, the results presented here and obtained by different approaches, coherently integrate both the pharmacophoric elements responsible for the α_1-AR subtype selectivity and the antitarget modeling of some amine GPCRs that interfere with the α_1-ARs, thus inducing unwanted side effects.

8.7
Perspectives

8.7.1
Pharmacophore Combination Approach: From Lock and Key to Passe–Partout Model

It is now widely accepted that a simultaneous modulation of multiple protein targets can provide a better therapeutic effect and side effect profile than the employment of a selective ligand. Rational design in which pharmacophoric features from selective ligands are combined in a single structure has produced multiple ligands that cover a wide variety of targets. A key challenge in the design of multiple ligands is a programmed and controlled combination and integration of different pharmacophores to maintain a balanced activity at each target of interest and, at the same time, achieve a wider selectivity. A simple approach is to consider a primary well-validated target and conjugate secondary activities (antitarget activities) to enhance efficacy.

In this scenario, the importance of both target and antitarget pharmacophore modeling as an essential prerequisite is evident for a rational combination/mixing of pharmacophores to obtain multiple ligands designed for modulating multiple targets. In this respect, the next challenge is to create new concepts and methods for the modeling and development of nonmixed, but really highly integrated (conceptually a single entity, i.e. a selective Passe–Partout model) multiple pharmacophores, devoted to the design and discovery of multitarget drugs.

Finally, our view is that selective Passe–Partout pharmacophoric generation, rather than single target/antitarget selectivity evaluation, needs to capture the different peculiar and determinant dynamic aspects of multiligand/multitarget ensemble of multicomplex interactions and translate these dynamic features into multipharmacophoric elements, useful for multiligand design and discovery.

Abbreviations

GPCRs	G-protein-coupled receptors
SAR	structure–activity relationships

QSAR	quantitative structure–activity relationship
α_1-ARs	α_1-adrenergic receptors
5-HT	5-hydroxy-tryptamine (serotonine)
5-HT$_{1A}$-R	5-HT$_{1A}$-receptor
α_2-AR	α_2-adrenergic receptor
D$_2$-R	dopamine receptor
3D	three dimensional
$\Delta G°$	Gibbs standard free energy
QSPR	quantitative structure–property relationship
V_{mol}	van der Waals (vdW) molecular volume
V_{in}, V_{out}	intersection and the outer vdW volume of the ligand considered with respect to the volume of the reference supermolecule, respectively
V_{sup}	vdW molecular volume of the reference supermolecule
V_{dif}	normalized difference vdW volume computed according to the following formula: $V_{dif} = (V_{in} - V_{out})/V_{sup}$
n	number of compounds
r	correlation coefficient
r_{cv}	cross-validated r
s	standard deviation
F	Fisher ratio
Vf(H)	hydrogen atom valency of the protonated nitrogen atom
SL(N)	nucleophilic superdelocalizability of the protonated nitrogen atom
HACA-1 = $q_A S_A$	q_A is the charge on the hydrogen bonding acceptor atom and S_A is its surface area
TMSA	total molecular surface area
E_{LUMO}	lowest unoccupied molecular orbital
8-OH-DPAT	8-hydroxy-2-(di-n-propylammino)-tetralin
MO	molecular orbital
μh	total hybridization component of the molecular dipole
f-PNSA-1	fractional partial negative surface area
q_{tot}(H)	net atomic charge of the proton
E_{st}(N)	minimum atomic state energy for the protonated nitrogen atom
PNSA-1	partial negative surface area
CoMFA	comparative molecular field analysis
Rotbonds	number of rotatable bonds
CHI	topological family of descriptors
ABIC2	average bonding information content (order 2)
FPSA	surface weighted charged partial surface area
WPSA-1	fractional charged partial surface area
f-Pπ-π(C−O) and f-Pσ-π(C−X)	the π−π and σ−π bond order of a C−O or C−N pair of atoms, respectively
UP	dose active in inhibiting by 50% of the urethral contractions induced by noradrenaline

DBU   dose active in lowering diastolic blood pressure by 25%
clog P   virtual lipophilicity
BE   binding energy
IE   ligand–receptor interaction energy
E_R and E_L distortion energies of the receptor and of the ligand, respectively

References

1 Brady, A.E. and Limbird, L.E. (2002) G protein-coupled receptor interacting proteins: emerging roles in localization and signal transduction. *Cellular Signalling*, **14**, 297–309.

2 Pierce, K.L., Premont, R.T. and Lefkowitz, R.J. (2002) Seven-transmembrane receptors. *Nature Reviews Molecular Cell Biology*, **3**, 639–650.

3 Zhang, M. and Wang, W. (2003) Organization of signaling complexes by PDZ-domain scaffold proteins. *Accounts of Chemical Research*, **36**, 530–538.

4 Palczewski, K., Kumasaka, T., Hori, T., Behnke, C.A., Motoshima, H., Fox, B.A., Le Trong, I., Teller, D.C., Okada, T., Stenkamp, R.E., Yamamoto, M., Miyano, M. (2000) Crystal structure of rhodopsin: a G protein-coupled receptor. *Science*, **289**, 739–745.

5 Okada, T. (2004) X-ray crystallographic studies for ligand–protein interaction changes in rhodopsin. *Biochemical Society Transactions*, **32**, 738–741.

6 Kenakin, T. (2001) Inverse, protean, and ligand-selective agonism: matters of receptor conformation. *FASEB Journal*, **15**, 598–611.

7 Hancock, A.A. (1996) α1-Adrenoceptor adrenoceptor subtypes: a synopsis of their pharmacology and molecular biology. *Drug Development Research*, **39**, 54–107.

8 Leonardi, A., Testa, R., Motta, G., De Benedetti, P.G., Heible, J.P. and Giardinà, D. (1996) α1-Adrenoceptors subtype- and organ-selectivity of different agents, in *Perspectives in Receptor Research* (eds D. Giardinà, A. Piergentili and M. Pigini), Elsevier Science B.V, Amsterdam, pp. 135–152.

9 Rosini, M., Bolognesi, M.L., Giardina, D., Minarini, A., Tumiatti, V. and Melchiorre, C. (2007) Recent advances in alpha1-adrenoreceptor antagonists as pharmacological tools and therapeutic agents. *Current Topics in Medicinal Chemistry*, **7**, 147–162.

10 Michel, M.C., Kenny, B. and Schwinn, D.A. (1995) Classification of alpha 1-adrenoceptor subtypes. *Naunyn-Schmiedeberg's Archives of Pharmacology*, **352**, 1–10.

11 Bockaert, J. and Pin, J.P. (1999) Molecular tinkering of G protein-coupled receptors: an evolutionary success. *The EMBO Journal*, **18**, 1723–1729.

12 Lefkowitz, R.J. (2000) The superfamily of heptahelical receptors. *Nature Cell Biology*, **2**, E133–E136.

13 Gether, U. (2000) Uncovering molecular mechanisms involved in activation of G protein-coupled receptors. *Endocrine Reviews*, **21**, 90–113.

14 Kristiansen, K. (2004) Molecular mechanisms of ligand binding, signaling, and regulation within the superfamily of G-protein-coupled receptors: molecular modeling and mutagenesis approaches to receptor structure and function. *Pharmacology & Therapeutics*, **103**, 21–80.

15 Tyndall, J.D., Pfeiffer, B., Abbenante, G. and Fairlie, D.P. (2005) Over one hundred peptide-activated G protein-coupled receptors recognize ligands with

turn structure. *Chemical Reviews*, **105**, 793–826.
16. Ballesteros, J.A. and Weinstein, H. (1995) Integrated methods for the construction of three-dimensional models and computational probing of structure-function relations in G protein-coupled receptors. *Methods in Neuroscience*, **25**, 366–428.
17. Cavalli, A., Fanelli, F., Taddei, C., De Benedetti, P.G. and Cotecchia, S. (1996) Amino acids of the α1B-adrenergic receptor involved in agonist binding: further differences in docking catecholamines to receptor subtypes. *FEBS Letters*, **399**, 9–13.
18. Hwa, J. and Perez, D.M. (1996) The unique nature of the serine interactions for alpha 1-adrenergic receptor agonist binding and activation. *The Journal of Biological Chemistry*, **271**, 6322–6327.
19. Graham, R.M., Perez, D.M., Hwa, J. and Piascik, M.T. (1996) Alpha 1-adrenergic receptor subtypes. Molecular structure, function, and signaling. *Circulation Research*, **78**, 737–749.
20. Rossier, O., Abuin, L., Fanelli, F., Leonardi, A. and Cotecchia, S. (1999) Inverse agonism and neutral antagonism at alpha(1a)- and alpha(1b)-adrenergic receptor subtypes. *Molecular Pharmacology*, **56**, 858–866.
21. Fanelli, F. and De Benedetti, P.G. (2005) Computational modeling approaches to structure–function analysis of G protein-coupled receptors. *Chemical Reviews*, **105**, 3297–3351.
22. Albert, A. (1985) *Selective Toxicity: The Physico-Chemical Basis of Therapy*, 7th edn, Chapman and Hall, London.
23. van de Waterbeemd, H. and Testa, B. (1987) Advances in Drug Research, Vol. **16**, Academic Press, London, pp. 85–225.
24. De Benedetti, P.G., Menziani, M.C., Cocchi, M. and Fanelli, F. (1995) Prototropic molecular forms and theoretical descriptors in QSAR analysis. *Journal of Molecular Structure (Theochem)*, **333**, 1–17.
25. Bouvier, M. (2001) Oligomerization of G-protein-coupled transmitter receptors. *Nature Reviews Neuroscience*, **2**, 274–286.
26. Milligan, G. (2001) Oligomerisation of G-protein-coupled receptors. *Journal of Cell Science*, **114**, 1265–1271.
27. Rios, C.D., Jordan, B.A., Gomes, I. and Devi, L.A. (2001) G-protein-coupled receptor dimerization: modulation of receptor function. *Pharmacology & Therapeutics*, **92**, 71–87.
28. George, S.R., O'Dowd, B.F. and Lee, S.P. (2002) G-protein-coupled receptor oligomerization and its potential for drug discovery. *Nature Reviews Drug discovery*, **1**, 808–820.
29. Agnati, L.F., Ferre, S., Lluis, C., Franco, R. and Fuxe, K. (2003) Molecular mechanisms and therapeutical implications of intramembrane receptor/receptor interactions among heptahelical receptors with examples from the striatopallidal GABA neurons. *Pharmacological Reviews*, **55**, 509–550.
30. Franco, R., Canals, M., Marcellino, D., Ferre, S., Agnati, L., Mallol, J., Casado, V., Ciruela, F., Fuxe, K., Lluis, C., Canela, E.I. (2003) Regulation of heptaspanning-membrane-receptor function by dimerization and clustering. *Trends in Biochemical Sciences*, **28**, 238–243.
31. Park, P.S., Filipek, S., Wells, J.W. and Palczewski, K. (2004) Oligomerization of G protein-coupled receptors: past, present, and future. *Biochemistry*, **43**, 15643–15656.
32. Terrillon, S. and Bouvier, M. (2004) Roles of G-protein-coupled receptor dimerization. *EMBO Reports*, **5**, 30–34.
33. Bulenger, S., Marullo, S. and Bouvier, M. (2005) Emerging role of homo- and heterodimerization in G-protein-coupled receptor biosynthesis and maturation. *Trends in Pharmacological Sciences*, **26**, 131–137.
34. Maggio, R., Novi, F., Scarselli, M. and Corsini, G.U. (2005) The impact of G-protein-coupled receptor hetero-oligomerization on function and

pharmacology. *FEBS Journal*, **272**, 2939–2946.

35 Stanasila, L., Perez, J.B., Vogel, H. and Cotecchia, S. (2003) Oligomerization of the alpha 1a- and alpha 1b-adrenergic receptor subtypes. Potential implications in receptor internalization. *The Journal of Biological Chemistry*, **278**, 40239–40251.

36 Jordan, B.A. and Devi, L.A. (1999) G-protein-coupled receptor heterodimerization modulates receptor function. *Nature*, **399**, 697–700.

37 Freire, E. (2000) Can allosteric regulation be predicted from structure? *Proceedings of the National Academy of Sciences of the United States of America*, **97**, 11680–11682.

38 Onaran, H.O., Scheer, A., Cotecchia, S. and Costa, T. (2000) A look into receptor efficacy. From the signalling network of the cell to the intramolecular motion of the receptor, in *Handbook of Experimental Pharmacology* (eds T. Kenakin and J. Angus), Springer, Heidelberg, Vol. 148, pp. 217–280.

39 Kenakin, T. (2002) Efficacy at G-protein-coupled receptors. *Nature Reviews Drug discovery*, **1**, 103–110.

40 Johnson, M.A. and Maggiora, G.M. (1990) *Concepts and Applications of Molecular Similarity*, John Wiley & Sons, Inc., New York.

41 De Benedetti, P.G., Fanelli, F., Menziani, M.C. and Cocchi, M. (2000) The ad hoc supermolecule approach to receptor ligand design. *Journal of Molecular Structure (Theochem)*, **503**, 1–16.

42 Dean, P.M. (1987) *Molecular Foundation of Drug–Receptor Interactions*, Cambridge University Press, Cambridge.

43 Behr, J.-P. (1994) *The Lock-and-Key Principle: The State of the Art-100 Years*, John Wiley & Sons, Inc., New York.

44 Leach, A.R. (1996) *Molecular Modelling: Principles and Applications*, Longman, Edimburg.

45 Kubinyi, H. (1993) *Hansch Analysis and Related Approaches*, Weinheim, New York.

46 Kubinyi, H. (1993) *3D QSAR in Drug Design*, ESCOM, Leiden.

47 Hammett, L.P. (1937) The Effect of structure upon the reactions of organic compounds. Benzene derivatives. *Journal of the American Chemical Society*, **59**, 96–103.

48 Baroni, M., Costantino, G., Cruciani, G., Riganelli, D., Valigi, R. and Clementi, S. (1993) Generating optimal linear PLS estimations (GOLPE): an advanced chemometric tool for handling 3D-QSAR problems. *Quantitative Structure–Activity Relationships*, **12**, 9–20.

49 Cocchi, M., Menziani, M.C., De Benedetti, P.G. and Cruciani, G. (1992) Theoretical versus empirical molecular descriptors in monosubstituted benzenes. A chemometric study. *Chemometrics and Intelligent Laboratory System*, **14**, 209–224.

50 Cocchi, M., Menziani, M.C., Fanelli, F. and De Benedetti, P.G. (1995) Theoretical quantitative structure–activity relationship analysis of congeneric and non congeneric α1-adrenoceptor antagonists: a chemometric study. *Journal of Molecular Structure (Theochem)*, **331**, 79–93.

51 Cramer, R.D., Patterson, D.E. and Bunce, J.D. (1988) Comparative molecular field analysis (CoMFA). 1. Effect of shape on binding of steroids to carrier proteins. *Journal of the American Chemical Society*, **110**, 5959–5967.

52 Weinstein, H., Osman, R., Topiol, S. and Green, J.P. (1980) *Chemical Applications of Atomic and Electrostatic Potential*, Plenum, New York.

53 Menziani, M.C., De Benedetti, P.G., Gago, F. and Richards, W.G. (1989) The binding of benzenesulfonamides to carbonic anhydrase enzyme. A molecular mechanics study and quantitative structure–activity

relationships. *Journal of Medicinal Chemistry*, **32**, 951–956.

54 Menziani, M.C. and De Benedetti, P. (1991) Direct and indirect theoretical QSAR modelling in sulfonamide carbonic anhydrase inhibitors, in *QSAR: Rational Approaches on the Design of Bioactive Compounds* (eds C. Silipo and A. Vittoria), Elsevier, Amsterdam, pp. 331.

55 De Benedetti, P.G., Menziani, M.C., Fanelli, F. and Cocchi, M. (1993) The heuristic-direct approach to quantitative structure–activity relationship analysis. *Journal of Molecular Structure (Theochem)*, **285**, 147–153.

56 Fanelli, F., Menziani, M.C., Carotti, A. and De Benedetti, P.G. (1994) Theoretical quantitative structure–activity relationship analysis on three dimensional models of ligand-m1 muscarinic receptor complexes. *Bioorganic & Medicinal Chemistry*, **2**, 195–211.

57 Fanelli, F., Menziani, M.C., Cocchi, M., Leonardi, A. and De Benedetti, P.G. (1994) The heuristic-direct approach to theoretical quantitative structure activity relationship analysis of α1-adrenoceptor ligands. *Journal of Molecular Structure (Theochem)*, **314**, 265–276.

58 Fanelli, F., Menziani, M.C. and De Benedetti, P.G. (1995) Molecular dynamics simulations of m3-muscarinic receptor activation and QSAR analysis. *Bioorganic & Medicinal Chemistry*, **3**, 1465–1477.

59 Menziani, M.C., Cocchi, M., Fanelli, F. and De Benedetti, P.G. (1995) Quantitative structure–affinity/selectivity relationship analysis on three-dimensional models of the complexes between the ETA and ETB receptors and C-terminal endothelin hexapeptide antagonists. *Journal of Molecular Structure (Theochem)*, **333**, 243–248.

60 Cappelli, A., Anzini, M., Vomero, S., Menziani, M.C., De Benedetti, P.G., Sbacchi, M., Clarke, G.D. and Mennuni, L. (1996) Synthesis, biological evaluation, and quantitative receptor docking simulations of 2-[(acylamino) ethyl]-1,4-benzodiazepines as novel tifluadom-like ligands with high affinity and selectivity for kappa-opioid receptors. *Journal of Medicinal Chemistry*, **39**, 860–872.

61 Menziani, M.C., Fanelli, F., Cocchi, M. and De Benedetti, P.G. (1996) The heuristic-direct approach to quantitative structure–activity relationship analysis of ligand-G protein coupled receptor complexes, in *Membrane Protein Models: Experiment, Theory and Speculation* (ed. J. Findlay), Bios Science Publications, pp. 113–131.

62 Karelson, M., Lobanov, V.S. and Katritzky, A.R. (1996) Quantum-chemical descriptors in QSAR/QSPR studies. *Chemical Reviews*, **96**, 1027–1044.

63 Rastelli, G., Fanelli, F., Menziani, M.C., Cocchi, M. and De Benedetti, P.G. (1991) Conformational analysis, molecular modeling and quantitative structure–activity relationships studies of 2,4-diamino-6,7-dimethoxy-2-substituted quinazoline α1-adrenergic antagonists. *Journal of Molecular Structure (Theochem)*, **251**, 307–318.

64 Venturelli, P., Menziani, M.C., Cocchi, M., Fanelli, F. and De Benedetti, P.G. (1992) Molecular modeling and quantitative structure–activity relationship analysis using theoretical descriptors of 1,4-benzodioxan (WB-4101) related compounds α1-adrenergic antagonists. *Journal of Molecular Structure (Theochem)*, **276**, 327–340.

65 De Benedetti, P.G., Cocchi, M., Menziani, M.C. and Fanelli, F. (1993) Theoretical quantitative structure–activity analysis and pharmacophore modelling of selective non congeneric α1a-adrenergic antagonists. *Journal of Molecular Structure (Theochem)*, **280**, 283–290.

66 De Benedetti, P.G., Cocchi, M., Menziani, M.C. and Fanelli, F. (1994) Theoretical quantitative size and shape activity and selectivity analyses of 5-HT1A serotonin and α-adrenergic receptor ligands. *Journal of Molecular Structure (Theochem)*, **305**, 101–110.

67 Cocchi, M., Fanelli, F., Menziani, M.C. and De Benedetti, P.G. (1997) Conformational analysis and theoretical quantitative size and shape–affinity relationships of N4-protonated N1-arylpiperazine 5-HT1A serotoninergic ligands. *Journal of Molecular Structure (Theochem)*, **397**, 129–145.

68 Montorsi, M., Menziani, M.C., Cocchi, M., Fanelli, F. and De Benedetti, P.G. (1998) Computer modeling of size and shape descriptors of α1-adrenergic receptor antagonists and quantitative structure–affinity/selectivity relationships. *Methods*, **14**, 239–254.

69 Cocchi, M. and De Benedetti, P.G. (1998) Use of the supermolecule approach to derive molecular similarity descriptors for QSAR analysis. *Journal of Molecular Modelling*, **4**, 113–131.

70 Leonardi, A., Motta, G., Boi, C., Testa, R., Poggesi, E., De Benedetti, P.G. and Menziani, M.C. (1999) Synthesis, pharmacological evaluation, and structure–activity relationship and quantitative structure–activity relationship studies on novel derivatives of 2,4-diamino-6,7-dimethoxyquinazoline alpha1-adrenoceptor antagonists. *Journal of Medicinal Chemistry*, **42**, 427–437.

71 Barlocco, D., Cignarella, G., Piaz, V.D., Giovannoni, M.P., De Benedetti, P.G., Fanelli, F., Montesano, F., Poggesi, E., Leonardi, A. (2001) Phenylpiperazinylalkylamino substituted pyridazinones as potent alpha(1) adrenoceptor antagonists. *Journal of Medicinal Chemistry*, **44**, 2403–2410.

72 De Mairinis, R.M., Wise, M., Hieble, J.P. and Ruffolo, R.R.J. (1987) Structure–activity relationships for alpha-1 adrenergic receptor agonists and antagonists, in *The alpha1-Adrenergic Receptor* (ed. R.R.J. Ruffolo), Humana Press, Clifton, pp. 211–265.

73 Alabaster, V.A., Campbell, S.F., Danilewicz, J.C., Greengrass, C.W. and Plews, R.M. (1987) 2,4-Diamino-6,7-dimethoxyquinazolines. 2.2-(4-Carbamoylpiperidino) derivatives as alpha 1-adrenoceptor antagonists and antihypertensive agents. *Journal of Medicinal Chemistry*, **30**, 999–1003.

74 Campbell, S.F., Davey, M.J., Hardstone, J.D., Lewis, B.N. and Palmer, M.J. (1987) 2,4-Diamino-6,7-dimethoxyquinazolines. 1.2-[4-(1,4-Benzodioxan-2-ylcarbonyl)piperazin-1-yl] derivatives as alpha 1-adrenoceptor antagonists and antihypertensive agents. *Journal of Medicinal Chemistry*, **30**, 49–57.

75 Campbell, S.F. and Plews, R.M. (1987) 2,4-Diamino-6,7-dimethoxyquinazolines. 3.2-(4-Heterocyclylpiperazin-1-yl) derivatives as alpha 1-adrenoceptor antagonists and antihypertensive agents. *Journal of Medicinal Chemistry*, **30**, 1794–1798.

76 Campbell, S.F., Danilewicz, J.C., Greengrass, C.W. and Plews, R.M. (1988) 2,4-Diamino-6,7-dimethoxyquinazolines. 4.2-[4-(Substituted oxyethoxy)piperidino] derivatives as alpha 1-adrenoceptor antagonists and antihypertensive agents. *Journal of Medicinal Chemistry*, **31**, 516–520.

77 Giardina, D., Brasili, L., Gregori, M., Massi, M., Picchio, M.T., Quaglia, W. and Melchiorre, C. (1989) Structure–activity relationships in prazosin-related compounds. Effect of replacing a piperazine ring with an alkanediamine moiety on alpha 1-adrenoreceptor blocking activity. *Journal of Medicinal Chemistry*, **32**, 50–55.

78 Giardina, D., Crucianelli, M., Romanelli, R., Leonardi, A., Poggesi, E. and Melchiorre, C. (1996) Synthesis and biological profile of the enantiomers of [4-(4-amino-6,7-dimethoxyquinazolin-2-

yl)-*cis*-octahydroquinoxalin-1-yl]furan-2-ylmethanone (cyclazosin), a potent competitive alpha 1B-adrenoceptor antagonist. *Journal of Medicinal Chemistry*, **39**, 4602–4607.

79 Giardina, D., Gulini, U., Massi, M., Piloni, M.G., Pompei, P., Rafaiani, G. and Melchiorre, C. (1993) Structure–activity relationships in prazosin-related compounds. 2. Role of the piperazine ring on alpha-blocking activity. *Journal of Medicinal Chemistry*, **36**, 690–698.

80 Campbell, S.F., Hardstone, J.D. and Palmer, M.J. (1988) 2,4-Diamino-6,7-dimethoxyquinoline derivatives as alpha 1-adrenoceptor antagonists and antihypertensive agents. *Journal of Medicinal Chemistry*, **31**, 1031–1035.

81 Bordner, J., Campbell, S.F., Palmer, M.J. and Tute, M.S. (1988) 1,3-Diamino-6,7-dimethoxyisoquinoline derivatives as potential alpha 1-adrenoceptor antagonists. *Journal of Medicinal Chemistry*, **31**, 1036–1039.

82 De Benedetti, P.G., Menziani, M.C., Rastelli, G. and Cocchi, M. (1991) Molecular orbital study of the basicity of prasozin analogues in relation to their α-adrenoceptor binding affinity. *Journal of Molecular Structure (Theochem)*, **233**, 343–351.

83 Kinsella, G.K., Watson, G.W. and Rozas, I. (2006) Theoretical proton affinities of alpha1 adrenoceptor ligands. *Bioorganic & Medicinal Chemistry*, **14**, 1580–1587.

84 Melchiorre, C., Brasili, L., Giardina, D., Pigini, M. and Strappaghetti, G. (1984) 2-[[[2-(2,6-Dimethoxyphenoxy)ethyl] amino]-methyl]-1,4-benzoxathian: a new antagonist with high potency and selectivity toward alpha 1-adrenoreceptors. *Journal of Medicinal Chemistry*, **27**, 1535–1536.

85 Pigini, M., Brasili, L., Giannella, M., Giardina, D., Gulini, U., Quaglia, W. and Melchiorre, C. (1988) Structure–activity relationships in 1,4-benzodioxan-related compounds. Investigation on the role of the dehydrodioxane ring on alpha 1-adrenoreceptor blocking activity. *Journal of Medicinal Chemistry*, **31**, 2300–2304.

86 Quaglia, W., Pigini, M., Giannella, M. and Melchiorre, C. (1990) 3-Phenyl analogues of 2-[[[2-(2,6-dimethoxyphenoxy)ethyl]amino]-methyl]-1,4-benzodioxan (WB 4101) as highly selective alpha 1-adrenoceptor antagonists. *Journal of Medicinal Chemistry*, **33**, 2946–2948.

87 Dijkstra, G.D.H. (1993) Conformational analysis of 1-arylpiperazines and 4-arylpiperidines. *Recueil des Travaux Chimiques des Pays-Bas*, **112**, 151–160.

88 Kuipers, W., van Wijngaarden, I., Kruse, C.G., ter Horst-van Amstel M., Tulp, M.T. and IJzerman, AP. (1995) N4-Unsubstituted N1-arylpiperazines as high-affinity 5-HT1A receptor ligands. *Journal of Medicinal Chemistry*, **38**, 1942–1954.

89 Ho, B.Y., Karschin, A., Branchek, T., Davidson, N. and Lester, H.A. (1992) The role of conserved aspartate and serine residues in ligand binding and in function of the 5-HT1A receptor: a site-directed mutation study. *FEBS Letters*, **312**, 259–262.

90 Menziani, M.C., De Benedetti, P.G. and Karelson, M. (1998) Theoretical descriptors in quantitative structure–affinity and selectivity relationship study of potent N4-substituted arylpiperazine 5-HT1A receptor antagonists. *Bioorganic & Medicinal Chemistry*, **6**, 535–550.

91 Lopez-Rodriguez, M.L., Morcillo, M.J., Fernandez, E., Rosado, M.L., Pardo, L. and Schaper, K. (2001) Synthesis and structure–activity relationships of a new model of arylpiperazines. Study of the 5-HT(1a)/alpha(1)-adrenergic receptor affinity by classical Hansch analysis, artificial neural networks, and computational simulation of ligand recognition. *Journal of Medicinal Chemistry*, **44**, 198–207.

92 Lopez-Rodriguez, M.L., Rosado, M.L., Benhamu, B., Morcillo, M.J., Fernandez, E. and Schaper, K.J. (1997) Synthesis and

structure–activity relationships of a new model of arylpiperazines. 2. Three-dimensional quantitative structure–activity relationships of hydantoin-phenylpiperazine derivatives with affinity for 5-HT1A and alpha 1 receptors. A comparison of CoMFA models. *Journal of Medicinal Chemistry*, **40**, 1648–1656.

93 Barbaro, R., Betti, L., Botta, M., Corelli, F., Giannaccini, G., Maccari, L., Manetti, F., Strappaghetti, G., Corsano, S. (2001) Synthesis, biological evaluation, and pharmacophore generation of new pyridazinone derivatives with affinity toward alpha(1)- and alpha(2)-adrenoceptors. *Journal of Medicinal Chemistry*, **44**, 2118–2132.

94 Maccari, L., Magnani, M., Strappaghetti, G., Corelli, F., Botta, M. and Manetti, F. (2006) A genetic-function-approximation-based QSAR model for the affinity of arylpiperazines toward alpha1 adrenoceptors. *Journal of Chemical Information and Modeling*, **46**, 1466–1478.

95 De Benedetti, P.G., Fanelli, F., Menziani, M.C., Cocchi, M., Testa, R. and Leonardi, A. (1997) Alpha 1-adrenoceptor subtype selectivity: molecular modelling and theoretical quantitative structure–affinity relationships. *Bioorganic & Medicinal Chemistry*, **5**, 809–816.

96 Klabunde, T. and Evers, A. (2005) GPCR antitarget modeling: pharmacophore models for biogenic amine binding GPCRs to avoid GPCR-mediated side effects. *Chembiochem: A European Journal of Chemical Biology*, **6**, 876–889.

97 Menziani, M.C., Montorsi, M., De Benedetti, P.G. and Karelson, M. (1999) Relevance of theoretical molecular descriptors in quantitative structure–activity relationship analysis of alpha1-adrenergic receptor antagonists. *Bioorganic & Medicinal Chemistry*, **7**, 2437–2451.

98 Eric, S., Solmajer, T., Zupan, J., Novic, M., Oblak, M. and Agbaba, D. (2004) Quantitative structure–activity relationships of alpha1 adrenergic antagonists. *Journal of Molecular Modelling*, **10**, 139–150.

99 Pallavicini, M., Fumagalli, L., Gobbi, M., Bolchi, C., Colleoni, S., Moroni, B., Pedretti, A., Rusconi, C., Vistoli, G., Valoti, E. (2006) QSAR study for a novel series of *ortho* disubstituted phenoxy analogues of alpha1-adrenoceptor antagonist WB4101. *European Journal of Medicinal Chemistry*, **41**, 1025–1040.

100 Bremner, J.B., Coban, B. and Griffith, R. (1996) Pharmacophore development for antagonists at alpha 1 adrenergic receptor subtypes. *Journal of Computer-Aided Molecular Design*, **10**, 545–557.

101 Bremner, J.B., Coban, B., Griffith, R., Groenewoud, K.M. and Yates, B.F. (2000) Ligand design for alpha1 adrenoceptor subtype selective antagonists. *Bioorganic & Medicinal Chemistry*, **8**, 201–214.

102 MacDougall, I.J. and Griffith, R. (2006) Selective pharmacophore design for alpha1-adrenoceptor subtypes. *Journal of Molecular Graphics & Modelling*, **25**, 146–157.

103 Vistoli, G., Pedretti, A., Villa, L. and Testa, B. (2005) Range and sensitivity as descriptors of molecular property spaces in dynamic QSAR analyses. *Journal of Medicinal Chemistry*, **48**, 4947–4952.

104 Li, M.Y., Tsai, K.C. and Xia, L. (2005) Pharmacophore identification of alpha (1A)-adrenoceptor antagonists. *Bioorganic & Medicinal Chemistry Letters*, **15**, 657–664.

105 Jimonet, P. and Jager, R. (2004) Strategies for designing GPCR-focused libraries and screening sets. *Current Opinion in Drug Discovery & Development*, **7**, 325–333.

106 Reaper, C.M., Fanelli, F., Buckingham, S.D., Millar, N.S. and Sattelle, D.B. (1998) Antagonist profile and molecular dynamic simulation of a *Drosophila melanogaster* muscarinic acetylcholine receptor. *Receptors Channels*, **5**, 331–345.

107 Leonardi, A., Barlocco, D., Montesano, F., Cignarella, G., Motta, G., Testa, R., Poggesi, E., Seeber, M. *et al.*(2004) Synthesis, screening, and molecular modeling of new potent and selective antagonists at the alpha 1d adrenergic receptor. *Journal of Medicinal Chemistry*, **47**, 1900–1918.

108 Evers, A., Hessler, G., Matter, H. and Klabunde, T. (2005) Virtual screening of biogenic amine-binding G-protein coupled receptors: comparative evaluation of protein- and ligand-based virtual screening protocols. *Journal of Medicinal Chemistry*, **48**, 5448–5465.

109 Evers, A. and Klabunde, T. (2005) Structure-based drug discovery using GPCR homology modeling: successful virtual screening for antagonists of the alpha1A adrenergic receptor. *Journal of Medicinal Chemistry*, **48**, 1088–1097.

110 Wermuth, C.G., Ganellin, C.R., Lindberg, P. and Mitcher, L.A. (1998) Glossary of terms used in medicinal chemistry (UPAC Recommendations 1997). *Annual Reports in Medicinal Chemistry*, **33**, 385–395.

III
Antitargets: Cytochrome P450s and Transporters

Antitargets. Edited by R. J. Vaz and T. Klabunde
Copyright © 2008 WILEY-VCH Verlag GmbH & Co. KGaA, Weinheim
ISBN: 978-3-527-31821-6

9
Cytochrome P450s: Drug–Drug Interactions
Dan Rock, Jan Wahlstrom, Larry Wienkers

9.1
Introduction

Drug metabolism represents an integral contributor to the many physiological processes that govern the pharmacokinetic/dispositional fate of most therapeutic agents. In rudimentary terms, drug metabolism can be described as the biological transformation of lipophilic, nonpolar molecules (also known as drugs) to more hydrophilic, polar molecules (also known as metabolites), which are in turn readily eliminated from the body. On the basis of the chemical nature of biotransformation, drug metabolism reactions have been grouped into two broad categories: phase I (oxidative) reactions, which typically involve the addition or removal of a functional group to increase a compound's polarity or phase II (conjugative) reactions that involve the addition of a hydrophilic moiety to the lipophilic molecule to make it more water soluble. Irrespective of the elimination pathways, the majority of therapeutic agents prescribed today are cleared via phase I oxidative metabolism. As a consequence, understanding the role of phase I enzymes, in particular cytochrome P450 enzymes, associated with the metabolism of new chemical entities (NCEs) is mandated by global regulatory agencies as part of new drug applications.

Oxidation metabolism of xenobiotics (detailed in Figure 9.1) has been described for over 50 years [1–3]. The enzymatic basis of these reactions was first linked to a unique carbon monoxide binding pigment found in rat liver microsomes first reported in 1958 [4]. Later, Sato and Omura [5] identified additional properties of the system in 1962 and developed the name P450 (first used with the hyphen, 'P-450') for 'pigment 450' because of the A_{max} at 450 nm when carbon monoxide was bound to the enzyme in the reduced state. In a relatively short period, experimental evidence for the presence of more than one form of microsomal cytochrome P450 began to accumulate [6]. The first successful purification and reconstitution of a CYP system by Lu and Coon [7] was followed by extensive efforts by a number of laboratories to purify CYP enzymes and catalog oxidative reactions associated with the enzymes from experimental animals and humans [8,9].

Antitargets. Edited by R. J. Vaz and T. Klabunde
Copyright © 2008 WILEY-VCH Verlag GmbH & Co. KGaA, Weinheim
ISBN: 978-3-527-31821-6

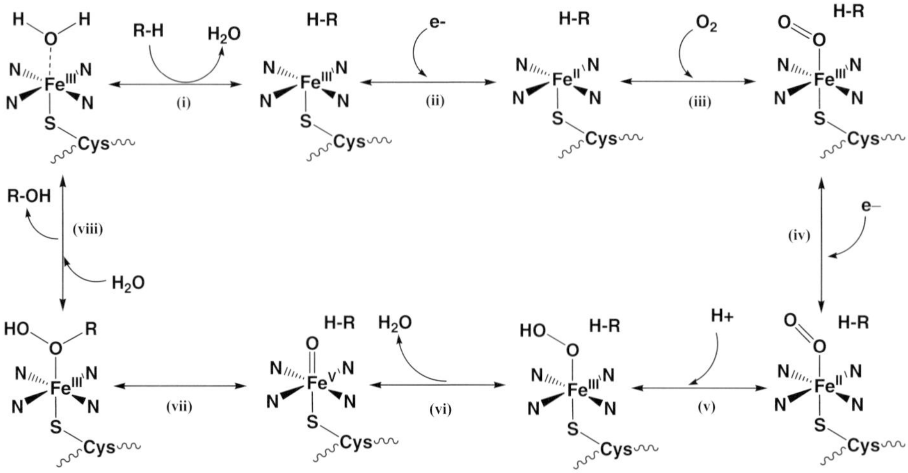

Figure 9.1 CYP catalytic cycle. The sequential two-electron reduction of CYP and the various transient intermediates were first described in the late 1960s [206]. The sequence of events that make up the CYP catalytic cycle is shown. The simplified CYP cycle begins with heme iron in the ferric state. In step (i), the substrate (R–H) binds to the enzyme, somewhere near the distal region of the heme group and disrupts the water lattice within the enzymes active site [207]. The loss of water elicits a change in the heme iron spin state (from low spin to high spin) [208]. Step (ii) involves the transfers of an electron from NADPH via the accessory flavoprotein NADPH-CYP reductase, with the electron flow going from the reductase prosthetic group FAD to FMN to the CYP enzyme [206,209]. The ferrous CYP can now bind oxygen (iii) [210,211]. The resulting iron–oxygen complex is unstable and can generate ferric iron and superoxide anion, or a second electron may enter the system in step (iv). This second electron may come from NADPH-CYP reductase or, in some cases, from cytochrome b_5 [212,213]. A proton is added in step (v) and the O—O bond is cleaved in step (vi) [214], generating H_2O (with the addition of a proton) and an entity shown as FeO^{3+} [215]. In step (vii), this electron-deficient complex either abstracts a hydrogen atom or an electron from the substrate [216,217]. A subsequent collapse of the intermediate or intermediate pair generates the metabolite (R—OH), and the metabolite subsequently dissociates from the enzyme in step (viii) [218].

As the number and diversity of the CYP enzymes increased, a systematic classification based on protein sequence was developed [10]. This system allowed CYP enzymes to be grouped into families, subfamilies and single enzymes. In this system, the protein sequences within a given gene family are at least 40% identical (e.g. CYP2) and the sequences within a given subfamily are more than 55% identical (e.g. CYP2B and CYP2C), with the last number designating the particular CYP enzyme (e.g. CYP2B6) [11]. In humans, at least 53 different CYP genes have been categorized into 18 CYP families [12].

With respect to drug-metabolizing enzymes, the majority of the CYPs responsible for phase I metabolism are concentrated in liver. The CYPs considered here are all found in the endoplasmic reticulum (isolated as 'microsomes'). Of the 18 human CYP families known, the bulk of xenobiotic biotransformation processes are carried out by enzymes from the CYP1, CYP2 and CYP3 families. In humans, realistically,

Table 9.1 Selected listing of important web sites regarding cytochrome CYP metabolism/substrates.

www.themedicalletter.com
www.drug-interactions.com
www.druginteractioninfo.org
www.imm.ki.se/CYPalleles/
www.medicine.iupui.edu/flockhart
www.atforum.com
www.fda.gov/cder/drug/drugReactions
www.fda.gov/cder/drug/drugInteractions

the most important CYPs from the point of view of drug metabolism are CYP1A2, CYP2C9, CYP2C19, CYP2D6 and CYP3A4/5 [13]. A comprehensive listing of all CYP substrates is beyond the scope of this book; however, Table 9.1 details a selected listing of web sites that serve as an up-to-date compendium of CYP-mediated drug reactions.

9.1.1
CYP1A Subfamily

Human CYP1A accounts for ~10–15% of the total CYPs expressed in the liver [14,15]. Two members of the 1A family, CYP1A1 and CYP1A2, catalyze the activation of procarcinogenic chemicals, namely polycyclic aromatic hydrocarbons (PAHs) and heterocyclic aromatic amines (HAAs), as well as metabolism of drugs such as caffeine [16], bropirimine [17], tacrine [18], clozapine [19], mexiletine [20], ropinirole [21] and theophylline [22]. In light of the of the unique role of CYP1A enzymes in drug metabolism and the metabolic activation of carcinogens, variations in CYP1A activities in human tissues have been proposed as key indicators of an individuals susceptibility to chemical carcinogenesis [23,24].

A primary feature associated with the variability in CYP1A metabolism of xenobiotics is the enzymes' inducibility by a variety of chemicals [25,26]. As stated earlier, the human CYP1A subfamily consists of two members, CYP1A1 and CYP1A2. CYP1A2 is one of the major CYPs in human liver and CYP1A1 is mainly expressed in human lung, placenta and lymphocytes in a much less abundance [27]. Although both CYP1A1 and CYP1A2 are inducible, CYP1A1 is generally more sensitive to inducers than CYP1A2. The induction of CYP1A enzymes by PAHs was observed by Conney *et al.* [28] 50 years ago. Initial evidence of the involvement of the aryl hydrocarbon receptor (AhR) in the induction of CYP1A enzymes was established through the identification of a hepatic cytosolic protein that exhibits stereospecific and high-affinity binding to 2,3,7,8-tetrachlorodibenzo-*p*-dioxin (TCDD), an environmental toxin [29]. This discovery led to the identification of a gene that encodes the AhR. Subsequently, the role of AhR in CYP1A induction was confirmed by the studies in inbred strains of mice. TCDD is about 10-fold more potent in inducing CYP1A1 in AhR-more-responsive strains (C57B/6) than in AhR-less-responsive strains (DBA/2) [30].

As CYP1A2 is involved in the metabolism of many drugs, the plasma concentrations of these drugs depend on the activity of this enzyme in the liver. Given the large interindividual and intraindividual variability associated with CYP1A2 activity in humans, some of the substances metabolized by CYP1A2 that have a narrow therapeutic range (e.g. theophylline and clozapine) require individualized dose regimens [31]. In view of the apparent importance of CYP1A induction in human drug metabolism and perhaps cancer risk, it is important to examine the major determinants responsible for human variability in CYP1A metabolism.

9.1.2
CYP2C Subfamily

The CYP2C subfamily consists of three enzymes (CYP2C8, CYP2C9 and CYP2C19) and is the second most abundant CYP subfamily expressed in human liver, representing about 20% of the total CYP enzymes [15]. CYP2C8 is expressed at relatively low levels and does not play a significant role in drug metabolism except for the oncology drug, taxol [32], the antiepileptic agent carbamazepine [33] and amodiaquine [34].

CYP2C9 is the major CYP2C enzyme expressed in human liver [35] and plays a major role in the metabolism of numerous therapeutics including many weak acidic drugs such as the nonsteroidal anti-inflammatory compounds (e.g. flurbiprofen) [36] and antidiabetics (e.g. tolbutamide) [37]. In addition, CYP2C9 also metabolizes (S)-warfarin [38], lornoxicam [39] and phenytoin [37].

Further, CYP2C9 has been shown to be genetically polymorphic with different alleles that produce protein variants [40]. The most common variants, *CYP2C9*2* and *CYP2C9*3*, represent the predominant alleles with clinical consequences [41]. In this case, the enzymes encoded by the CYP2C9 allelic variants exhibit altered apparent affinity constants (K_m) or intrinsic clearance (V_{max}/K_m) for a variety of substrates [42]. The clinical importance of CYP2C9 genetic variability is best demonstrated with the individual differences in the metabolism of common oral anticoagulant warfarin, a narrow therapeutic index drug that possesses concentration-dependent adverse effects (e.g. the risk of bleeding) [43]. For many patients, warfarin therapy is often fraught with difficulties in initial dosing predictions, as well as maintenance dosing regimens. The basis for many events may be attributed to the kinetics of oxidation for *CYP2C9*2* allele, which results in an attenuation in rates of oxidation of S-warfarin [40]. Interestingly, though *CYP2C9*2* effects appear to be more substrate specific, the *CYP2C9*3* variants demonstrate reduced catalytic activity across the majority of CYP2C9 substrates, with lowered maximum catalytic rates or lower affinity for substrates overall [44]. For either of these variants, the maintenance dose of warfarin in persons heterozygous for *CYP2C9*2/*3* is lower than the dose in patients who are homozygous for the normal enzyme [45].

Alternatively, the last CYP2C enzyme, CYP2C19, is a polymorphically expressed CYP enzyme, where the population is divided into extensive metabolizer (EM) and poor metabolizer (PM) phenotypes, the latter expressing less active or completely inactive enzyme. This polymorphism has a large ethnic component, with the prevalence of PMs reported to be 2–5% in Caucasians [46], 4–8% in Africans [47] and 11–23% in Asians [48]. In addition to metabolizing (S)-mephenytoin, CYP2C19 is also involved in the metabolism of a variety of drugs, most notably omeprazole [49], (R)-warfarin [50] and proguanil [51].

9.1.3
CYP2D6

In the liver, CYP2D6 represents between 1% and 5% of the total CYP enzymes [52]. Although CYP2D6 represents a relatively small portion of the total hepatic CYP enzymes, this enzyme is responsible for metabolizing a significant number of medications [53]. For example, CYP2D6 is responsible for the metabolism of certain opioids [54], antidepressants [55], neuroleptics [56] and cardiac medications [57]. Because these medications are often used in combination therapy and may have narrow therapeutic indices, CYP2D6 is often associated with clinically significant pharmacokinetic interactions [58,59]. A common feature of drugs metabolized by CYP2D6 is the presence of at least one basic nitrogen atom at a distance of 5–7 Å from the site of oxidation [60].

CYP2D6 was the first example of genetic polymorphism identified within the superfamily of CYP enzymes [61,62]. Four phenotypic subpopulations that exist define the rate of drug metabolism by CYP2D6: persons with a poor, an intermediate, an extensive or an ultrarapid ability to metabolize [63]. To date, at least 46 variants of CYP2D6 have been identified; many result in inactive enzyme, whereas some reduce the catalytic activity of the enzyme. Five to 10% of Caucasians have a poor ability to metabolize (homozygous for null variants), as do 1–2% of Southeast Asians. Moreover, CYP2D6 is susceptible to gene duplication, which may range from 3 to 13 copies, leading to exaggerated or ultrarapid CYP2D6 activity [64].

CYP2D6 polymorphisms are clinically important mainly because of the greater likelihood of adverse reactions among persons with poor metabolism due to the high plasma concentration of the affected drug and the lack of efficacy among persons with ultrarapid metabolism, due to the consequently low plasma concentration of the affected drug. For example, CYP2D6 is responsible for the conversion of codeine to morphine, an active metabolite; adequate experimental or clinical pain relief is more difficult to achieve in persons with poor metabolism [65]. In contrast, several case reports have noted morphine-like adverse effects among persons with the phenotype of ultrarapid metabolism [66]. In the light of the numerous drugs on the market that are metabolized by CYP2D6 and the polymorphic nature of the expression of this protein, many pharmaceutical companies seek to avoid CYP2D6-mediated metabolic pathways associated with NCEs in early drug discovery.

9.1.4
CYP3A Subfamily

CYP3A is usually considered the most important of all drug-metabolizing enzymes as it accounts for nearly 40% of the total CYP enzymes expressed in the liver [15] and has the ability to metabolize a multitude of structurally diverse molecules from almost every therapeutic class of drug [67]. It is likely that CYP3A is involved in the clearance of more than half the therapeutic agents that undergo CYP-mediated drug oxidation [13]. The CYP3A subfamily consists of three members: CYP3A4, CYP3A5 and CYP3A7 [12]. Within this subfamily, CYP3A4 is the most abundant form found in human liver and is expressed in several tissues, with the expression in the liver and small intestine of primary interest because of their importance in first-pass clearance of orally administered drugs [68]. CYP3A5 is a highly related polymorphic form of CYP3A, with 25–40% of individuals expressing appreciable levels of this enzyme in the liver [69]. The members of the CYP3A subfamily have overlapping substrate specificities and participate in about half of the CYP drug oxidation reactions [70]. Substrates of CYP3A4 vary in size from small molecules such as acetaminophen to extremely large compounds such as cyclosporin A (molecular weights of 151 and 1201 g/mol, respectively) [71]. In addition, CYP3A4 substrates span most therapeutic classes, including CNS (e.g. midazolam, reboxetine) [72,73], antimicrobal (e.g. clindamycin) [74], cardiovascular (e.g. diltiazem) [75] and antiviral (e.g. delavirdine) [76]. Although constitutive variability in humans is about fivefold, CYP3A activity is susceptible to modulation through several factors (e.g. inhibition/induction), leading to large effects on interindividual variation both in terms of bioavailability and drug–drug interactions (DDIs) [59].

Drug interactions may reduce CYP3A activity through inhibition or may increase metabolic activity through induction. Such interactions can expand the range of variability to about 400-fold [70]. Obviously, variability in drug levels of this magnitude represents a major therapeutic problem in dosage optimization. For example, dosage of the immunosuppressive drug cyclosporin generally must be reduced by greater than 70% to prevent unacceptably high drug levels and subsequent cyclosporin toxicity in patients concomitantly receiving the antifungal agent ketoconazole [70]. Beyond cyclosporin, there are many other known potent inhibitors of CYP3A that, even when administered at customary doses, are likely to increase the plasma concentrations of drugs metabolized by CYP3A enzymes (in particular HIV therapy) [77]. However, today, many adverse effects are predictable and can be avoided through dosage adjustments.

In contrast to drugs that inhibit CYP3A activity, treatment with drugs such as the rifamycins and some anticonvulsants may result in a marked reduction in the plasma concentrations of certain drugs administered concurrently through the upregulation (e.g. induction) of CYP3A proteins [78,79]. The mechanism by which CYP3A4 is upregulated involves intracellular binding of the inducer to the pregnane X receptor (PXR) or the steroid X receptor [80]. This receptor forms a heterodimer with the retinoid X receptor (RXR), which subsequently functions as a transcription factor by interacting with cognate response elements located in the 5′ regulatory region of the CYP3A4 gene [81]. Under conditions that promote CYP3A4 induction, drug dosages that were previously effective become ineffective. For example, patients taking oral contraceptives may become pregnant if they ingest inducing agents [82].

An interesting aspect to CYP3A4-mediated drug oxidations is the susceptibility of the enzyme to exhibit a variety of atypical kinetic profiles including positive cooperativity [83] and substrate inhibition [84]. Several factors may contribute to the atypical kinetics associated with CYP3A4 reactions. Molecular modeling [85] and mechanistic studies [86,87] suggest that the complex effects observed with selected CYP3A4 substrates may be attributable to the binding of multiple substrates within the active site of the enzyme [88]. Consistent with the hypothesis of multiple CYP3A4 binding sites, recent *in vitro* interaction profiles with various CYP3A4 substrates have been shown to be highly dependent on the substrate as well as on the coenzyme present in the microsomal matrix [89,90].

Given the restricted number of CYP enzymes responsible for the metabolism of a large number of drugs, it is not surprising that many drugs possess overlapping CYP enzyme specificities. As a consequence, for many drug therapies in which it is common for a patient to receive two or more drugs simultaneously, a situation exists that may manifest itself in a CYP-based drug–drug interaction [91]. The clinical significance of a pharmacokinetically based drug interaction is primarily dependent on how well the change in drug concentration is tolerated by the system. For compounds with a narrow therapeutic index, an increase in drug exposure, through enzyme inhibition, may precipitate untoward side effects. The magnitude of change in the concentration of a drug is dependent on the number of elimination pathways associated with clearance [92].

For all intent and purpose, inhibition of CYP enzymes can be classified into two categories: reversible (e.g. competitive and noncompetitive) inhibition and mechanism-based (e.g. quasi-irreversible and irreversible) inhibition [93]. The remainder of this chapter will be divided to reflect these two categories of inhibition and will focus solely on CYP-based drug interactions.

9.2
Reversible Inhibition

The DDI potential of high-throughput screening hits, lead optimization chemotypes and clinical candidates are continually assessed as compounds move from discovery through development and finally into the clinic. Each stage of advancement requires different rigors of assessment: single-point screening of hits, IC_{50} data for compound selection and K_i estimation for kinetics characterization (Figure 9.2)

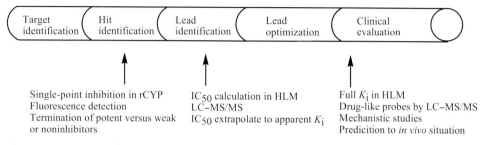

Figure 9.2 Assessment of DDI potential at varying stages of drug development [93].

[93]. While the level of characterization of DDI potential increases as the compound proceeds toward clinical dosing, so does the amount of information known about the compound. Because of the lack of molecular characterization, early discovery DDI screening presents a unique challenge in terms of balancing decision making and experimental design.

Many factors may confound the assessment of the DDI potential of early discovery compounds [93]. Limited or no solubility data exist to understand the likelihood that the compound will precipitate out of an *in vitro* incubation. The compounds have generally not been analyzed from a spectroscopic perspective; their characteristics may interfere with a fluorogenic DDI assay. Metabolism data are typically not available. The binding of a compound to plasma proteins or microsomal incubation constituents is not well understood, which may lead to underprediction of its inhibitory potential. The compounds are typically delivered in DMSO, which may cause solvent-related inhibition of the enzymatic assay. Also, since little is known about *in vivo* concentrations or projected dose, framing the consequences of an early DDI *in vitro* experiment may be difficult. With these factors in mind, general experimental paradigms have been developed to help minimize their potential impact.

Generally, four design features have the greatest impact on the outcome of an *in vitro* DDI experiment: bioanalytical technique (LC–MS/MS versus plate reader), reagent selection (probe substrate and CYP reagent), reaction conditions (buffer conditions, organic solvent and total protein content) and experiment type and data evaluation (single-point screen, IC_{50} or K_i). The factors affecting the selection of these features and potential confounding issues will be the focus of the following section.

9.2.1
Bioanalytical Techniques

In vitro techniques for studying DDI potential are based on the metabolism of known marker substrates. Two assay types are typically used to study DDIs: the turnover of drug-like probes monitored by LC–MS/MS methods or the use of spectrophotometer (plate reader) based methods. As each technique has unique advantages and shortcomings, assay use has not been standardized across the industry. Although techniques based on the turnover of radiolabeled substrates have also been developed [94–97], these methods are used infrequently and will not be discussed further.

Gentest (now BD Biosciences) was the first to develop spectrophotometric assays to study CYP inhibition [98]. These assays are based on the turnover of mildly fluorescent substrate probes to moderately fluorescent metabolites, where metabolite formation is monitored by an increase in fluorescence using a plate reader [99,100]. Problems with these methods include background interference due to low signal-to-noise ratio, chemotype-specific interference and fluorescence quenching. Aurora Biosciences (now Vertex) has designed probes that exhibit larger fluorescence increases to help alleviate interference issues [100,101]. Luminogenic assays have also been developed to reduce the impact of interference. The CYP-Glo technology couples CYP-based enzymatic activity with firefly luciferase luminescence to study CYP inhibition [102]. This technique combines the speed of plate reader assays with selectivity similar to LC–MS/MS methods, though specialized plate readers are required to monitor the chemiluminescence signal [103]. Inhibition readouts are often similar between plate reader and LC–MS/MS methods, though chemotype-specific differences do exist [104–106]. Although the spectrophotometric methods are high throughput, the substrate probes are not CYP specific and require that separate incubations be run with individual recombinant CYPs (rCYPs). Other plate reader methods that have recently been developed are based on the coupling of the generation of reactive oxygen species from *in vitro* incubations to oxidation of a probe to generate a fluorescent metabolite. These methods must be viewed with caution because of the complexities of uncoupling mechanisms associated with CYP activity and the unclear relationship of reactive oxygen species formation with CYP inhibition [107–110].

LC/MS techniques are based on the direct quantitation of known metabolites and are therefore not sensitive to spectrophotometric interference or efficiency of a coupling reaction as are the plate reader assays. However, LC/MS assays are limited by the need for separation of the analyte from the incubation constituents. Several techniques have been used to improve the throughput of LC/MS-based methods. Multiple incubations have been combined in a single injection to monitor more than one analyte simultaneously [111]. Cocktail assays are an extension of this methodology. For cocktail assays, the activities of multiple substrates are assessed in the same incubation so that more than one analyte can undergo quantitation in a single LC/MS run [112,113]. Although slower than plate reader based assays, multiple incubations using individual rCYPs are avoided using the cocktail approach because the substrate probes are selective.

Other techniques to improve throughput are instrumentation based and may involve multiple HPLC systems. The simplest method involves the automated use of solid phase extraction cartridges for sample cleanup followed by direct injection into the mass spectrometer [114]. Coupling of multiple HPLC systems to one mass spectrometer allows one column to equilibrate and separate while another column to flow into the mass spectrometer. Multiple HPLC systems may be configured such that the mass spectrometer is only exposed to each serial HPLC eluent as the analyte of interest is eluted [115,116]. Although multiple HPLC-based methods may increase throughput, they also typically decrease sensitivity and may confound data workup and interpretation.

Both plate reader and mass spectrometry based methods are commonly used for screening. The selection of a technique depends on instrument availability, throughput needs and the stage of compound advancement. For the characterization of compounds in drug development and clinical candidates, assays carried out using drug-like probes and analyzed by LC–MS/MS methods are considered the gold standard [117].

9.2.2
Reagent Selection and Reaction Conditions

Substrate probe selection is the key design decision for the DDI experiment and several factors impact the selection of the substrate probe. From a kinetics perspective, the probe should be selective for a single CYP, ideally form a single metabolite and exhibit a Michaelis–Menten kinetic profile. The probe should exhibit reasonable aqueous solubility so that organic concentrations can be minimized. The probe should have a high turnover number so that the incubation time and protein concentrations can be minimized, thus reducing the risk of inhibitor depletion and nonspecific binding. The probe metabolite should be amenable to LC–MS/MS (or other rapid analysis techniques) to avoid the need for analytical separation. Ideally, the probe and metabolite are commercially available and reasonably priced. Because of the nature of the CYP active site, multiple substrates may need to be used to characterize the inhibition potential of a compound fully, particularly for CYP3A4 and CYP2C9 [90,118]. Typical substrate probes, their metabolic pathway and commonly used control inhibitors are shown in Table 9.2 [117]. Ideal control inhibitors are CYP selective and reversible and exhibit minimal CYP-based turnover and binding to microsomal constituents in an *in vitro* incubation matrix.

The development of biological tools to support DDI studies has paralleled the development of bioanalytical techniques. To better understand *in vitro–in vivo* (IVIV) correlations, the effects of differences in enzyme preparations and incubation conditions must be understood. Differences between enzyme preparations include nonspecific binding, the ratio of accessory proteins (cytochrome b_5 and reductase) to CYPs and genetic variability; differences in incubation conditions include buffer strength, the presence of inorganic cations and solvent effects. Understanding how biology influences enzymatic activity is crucial to accurate and consistent prediction of the inhibition potential.

CYP reagents typically fall into three categories: recombinant preparations, human liver microsomes (HLMs) and hepatocytes. Selection depends on the ease of use, substrate probe selectivity, the need for an intact cellular system and availability of material. The two most commonly used reagents are HLMs and rCYPs. HLMs maintain levels of accessory proteins similar to the *in vivo* system and are easy to use and may be derived from pooled donors to minimize interdonor genetic differences. rCYPs typically have accessory protein levels that differ from the *in vivo* situation but are easy to use and exhibit consistent kinetic behavior if purchased from the same commercial vendor. Recombinant preparations may also be used to study differences in the enzymatic activity due to genetic variability, as these studies may suffer from the lack of available donor material [119]. Hepatocytes are more difficult to prepare, exhibit interbatch differences and may contain constituents to which many compounds exhibit high protein binding, because of which they are not often used for DDI determinations [120–122].

The key factor in understanding *in vitro* DDI data is to determine how the apparent kinetic measurements match the unbound concentration of an inhibitor in the

Table 9.2 Summary of enzyme kinetic parameters and inhibitor potencies for 11 human CYP activities in pooled human liver microsomes [117].

Enzyme	Assay	K_m (μM)	V_{max} (pmol/mg/min)	Inhibitor	IC_{50} (μM)
CYP1A2	Phenactein O-deethylase	47.0	688	Furafylline	1.76
CYP2A6	Coumarin 7-hydroxylase	0.841	940	Tranylcypromine	0.449
CYP2B6	Bupriopion hydroxylase	81.7	413	PPP	7.74
CYP2C8	Amodiaquine N-deethylase	1.89	1480	Quercitin	3.06
CYP2C9	Diclofenac 4'-hydroxylase	4.04	1670	Sulfaphenazole	0.272
CYP2C19	(S)-Mephenytoin 4'-hydroxylase	57.2	58.3	(+)-N-3-Benzylnirvanol	0.414
CYP2D6	Dextromethorphan O-demethylase	4.64	202	Quinidine	0.0579
CYP2E1	Chloroxazone 6-hydroxylase	73.9	2360	Tranylcypromine	8.94
CYP3A	Felodipine dehydrogenase	2.81	1630	Ketoconazole	0.0163
CYP3A	Midazolam 1'-hydroxylase	2.27	1220	Ketoconazole	0.0187
CYP3A	Testoserone 6β-hydroxylase	45.4	5260	Ketoconazole	0.0261

Table 9.3 HLM protein concentrations and ketoconazole K_i determination [203].

CYP3A4 Substrate	Protein Concentration (mg/ml)	In Vitro K_i (μM)
Midazolam	0.035	0.015
Sildenafil	0.1	0.02
Triazolam	0.125	0.006
Terfenadine	0.2	0.024
Dihydroqinghaosu	0.2	0.28
Quinidine	0.5	0.15
Vinblastine	0.5	0.17
β-Arteether	0.5	0.33
Brotizolam	1.0	0.5
Buprenorphin	1.0	0.6
Cortisol	3.0	0.9

incubation matrix. On the basis of the free drug hypothesis, the amount of free drug available at an enzyme active site is responsible for enzymatic activity. Nonspecific binding to a matrix is dependent on both the physicochemical characteristics of the compound [123] and the concentration of phospholipid in the incubation [124]. CYPs are membrane bound proteins; approximately 60% of the membrane matrix is phospholipid [125]. Compounds that are cationic (amines) or are highly lipophilic are particularly susceptible to nonspecific binding. The effect of varying protein concentrations on the observed inhibition potential for ketoconazole is shown in Table 9.3. To minimize nonspecific binding, incubation conditions should use the minimal amount of protein required, typically 0.2–0.03 mg/ml [117]. The unbound fraction of the inhibitor in the incubation should also be measured [126].

Two key accessory proteins, CYP reductase and cytochrome b_5, mediate electron transfer and modulate CYP activity through induced protein–protein interactions. CYP reductase is essential for electron transfer between CYPs and NADPH [127,128]. The active physiological pairing is 1:1, although varying ratios are found in microsomal preparations. Differences in CYP reductase:CYP ratios have been identified as the key reason for the differences between different rCYP preparations and between recombinant and microsomal preparations [129,130]. Low CYP reductase:CYP ratios have been implicated in reduced activity in lymphocytes when compared to HLMs [129]. Cytochrome b_5 may increase electron transfer rates for some isoform–substrate combinations and induce CYP-based allosteric changes due to protein–protein interactions [131]. Varying ratios of b_5:CYP have been reported between microsomal and rCYP preparations [129,130]. A graphical representation of the variability in enzyme kinetic parameters for recombinant CYP2C9 due in part to varying levels of accessory proteins is shown in Figure 9.3. For these reasons, kinetic characterization for each DDI probe should be carried out if the rCYP vendor is changed or if rCYP preparations and microsomal preparations are interchanged. For drug development studies, pooled human liver microsomes are the preferred reagent.

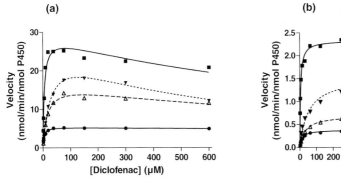

Enzyme preparation	Ratio of P450 reductase/CYP2C9	Ratio of cytochrome b_5/CYP2C9
Baculosomes®	1.3	None
Supersomes™	0.4	1.3
RECO®	3.7	0.16
Reconstituted	2.0	1.0

Figure 9.3 The effects of varying levels of accessory proteins on CYP2C9 kinetics using diclofenac (a) or (S)-warfarin (b) as substrate probe [219].

Changing the ionic strength of a buffer may influence CYP architecture, thereby causing differences in binding to accessory proteins or CYP-based allosteric changes that may alter substrate kinetics or inhibitor binding in an CYP-dependent manner [132]. Typically, 100 mM potassium phosphate buffer, pH 7.4, is used as an incubation medium for microsomal or rCYP preparations [117]. The addition of inorganic cations, such as magnesium, may also alter the kinetic profile of a substrate probe in a similar manner depending on the enzyme preparation and the concentration of cation [133]. Typically, 3 mM magnesium chloride is added to microsomal incubations. Organic solvents have mixed effects on enzymatic activity [134,135]. DMSO has particularly deleterious effect on CYP activity (Table 9.4) [136], while the addition of 2–4% acetonitrile or acetone may stimulate CYP activity in an isoform-dependent manner [137,138]. In general, minimizing the amount of organic solvent (<1% organic content) and maintaining a consistent set of experimental conditions is critical for obtaining reproducible data.

Table 9.4 The effects (% inhibition) of organic solvents on rCYP activities in human lymphoblast cell microsomes [136].

Enzyme	Solvent	Solvent		
		0.3%	1%	3%
CYP1A2	Methanol	0	9	12
	Ethanol	2	3	30
	DMSO	(−1)	0	12
	Acetonitrile	2	1	36
CYP2B6	Methanol	3	19	42
	Ethanol	5	80	97
	DMSO	52	27	50
	Acetonitrile	13	42	84
CYP2C8	Methanol	(−4)	(−8)	(−3)
	Ethanol	4	(−2)	16
	DMSO	2	7	10
	Acetonitrile	1	(−4)	19
CYP2C9	Methanol	1	12	13
	Ethanol	6	6	7
	DMSO	10	12	18
	Acetonitrile	1	(−12)	(−14)
CYP2C19	Methanol	2	9	37
	Ethanol	53	72	91
	DMSO	30	62	88
	Acetonitrile	2	10	43
CYP2D6	Methanol	7	26	57
	Ethanol	24	59	83
	DMSO	38	67	87
	Acetonitrile	3	19	48
CYP3A4	Methanol	0	4	25
	Ethanol	0	33	40
	DMSO	25	43	67
	Acetonitrile	2	11	37

The architecture of various CYPs may accommodate entities of different shapes [139]. Several CYPs, particularly CYP3A4 [140,141] and CYP2C9 [142], may exhibit atypical (non-Michaelis–Menten) kinetics such as heterotropic activation, homotropic activation, substrate inhibition and partial inhibition, all in a substrate–effector-dependent manner [143]. Several hypotheses have been proposed to account for the observation of atypical kinetics, including simultaneous occupancy of the CYP active site by two substrates (or one substrate and one effector simultaneously) [144] and allosteric changes in CYP architecture due to binding of an effector [145,146]. Along

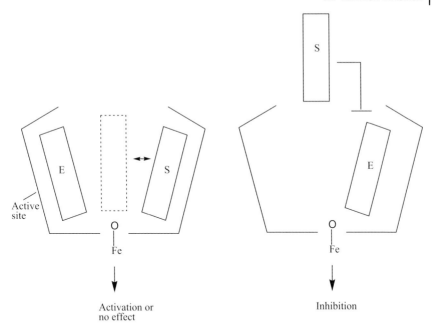

Figure 9.4 Possible reversible outcomes on addition of an effector [93].

similar lines, a CYP active site may contain more than one binding region for typical substrates (Figure 9.4) [90,93]. This has been postulated for both CYP3A4 [90] and CYP2C9 [118]. Because of these phenomena, multiple probe molecules may need to be used to characterize the inhibitory behavor of an NCE.

CYPs may be subject to stimulation on the basis of homotropic (self-activation) or heterotropic (activated by an effector) cooperativity. This kinetic behavior may be due to structural and/or electronic changes with the enzyme. The kinetic manifestation of activation is an increase in the V_{max} estimation or a decrease in the K_m estimation. The regioselectivity (site of metabolism) of metabolite formation may switch; it is possible to activate one metabolic pathway while inhibiting another (Figure 9.5) [84]. The most common explanation of homotropic and heterotropic behaviors is that two compounds occupy the CYP active site simultaneously, though other proposals include multiple CYP conformations or multiple binding sites accompanied by multiple CYP conformations [145]. It should be noted that some fluorescent substrate probes, because of their small size and planar nature, are particularly susceptible to heteroactivation [104]. This behavior accounts for experimental results where greater substrate turnover is observed with effector present than in the control. Although activation kinetics may confound extrapolation to the *in vivo* situation, the observed effects of activation *in vivo* have been minimal so far [147].

Partial inhibition occurs when an inhibitor partially inhibits the turnover of a substrate, but is not able to fully inhibit substrate turnover even at saturating conditions [143]. The most common explanation for partial inhibition is simultaneous occupancy of the CYP active site by both the effector and substrate probe in

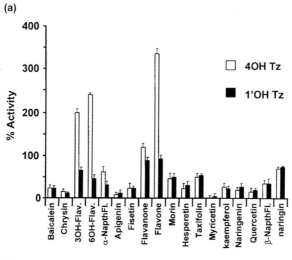
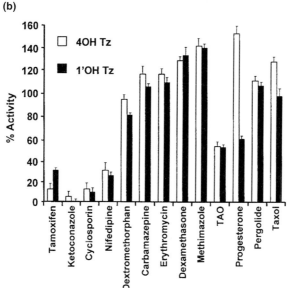

Figure 9.5 Percentage inhibition of 1- and 4'-hydroxylation of triazolam in the presence of varying inhibitors [220].

differing binding regions within the CYP active site [144]. The multiple binding site hypothesis is further strengthened by the fact that a compound may exhibit differences in K_m for two different metabolites (such as triazolam, which has K_m values of 42 and 241 μM, respectively, for the alpha and 4-hydroxy metabolites) [84]. In partial inhibition, both the substrate and the inhibitor may have access to the reactive oxygen species with a substrate–enzyme–inhibitor complex that is active.

9.2.3
Experimental Design

Mechanistic interpretation of inhibition data depends on the level of characterization. The simplest experiment is a single-point screen (percent inhibition) (Box 9.1). Single-point screen experiments are typically carried out at a single inhibitor ($\sim$1–10 µM) and substrate concentration ($\sim K_m$) and are used to determine if a CYP inhibition liability may be present. Little mechanistic interpretation may be derived from these experiments. The next level of inhibition characterization is the IC_{50} (Box 9.1). The IC_{50} experiments are typically carried out at a single concentration of substrate (usually at the K_m of the substrate), but multiple inhibitor concentrations are included (8–10 inhibitor concentrations that span a 10-fold range surrounding the anticipated IC_{50}). When carried out in this manner, the IC_{50} experiment does not provide for more mechanistic interpretation than the single-point screen but allows for more accurate determination of inhibition potential. Estimates of K_i from the IC_{50} determination may be made using the Cheng Prusoff equation (Box 9.1) [148].

The inhibitory constant (K_i) experiment provides reasonable information on both the mechanism and the potency of inhibition. For typical K_i experiments, four substrate (0.25 K_m, 0.5 K_m, 1 K_m, 2 K_m) and five inhibitor concentrations (spanning a 10-fold range surrounding the anticipated K_i, including a zero inhibitor control) are used. Data workup of the K_i experiment allows further determination of inhibition mechanism. Both Dixon ($1/v$ versus [I]; Figure 9.6) [149] and Lineweaver–Burk

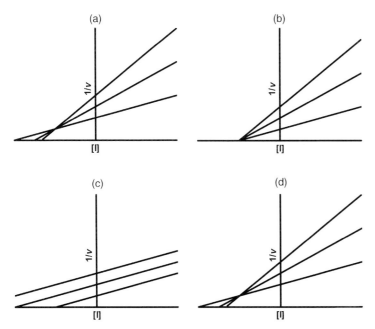

Figure 9.6 Dixon plot representations of competitive (a), noncompetitive (b), uncompetitive (c) and mixed (d) inhibition.

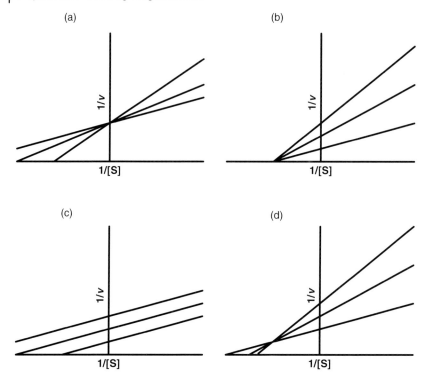

Figure 9.7 Lineweaver–Burke plot representations of competitive (a), noncompetitive (b), uncompetitive (c) and mixed (d) inhibition.

($1/v$ versus $1/[S]$; Figure 9.7) [150] visualizations may aid in distinguishing between the common types of reversible inhibition: competitive, noncompetitive, uncompetitive and mixed [151]. In competitive inhibition, the substrate and inhibitor compete directly for the active site of the enzyme. The V_{max} of the reaction remains unchanged, but the apparent K_m increases. For noncompetitive inhibition, binding of the inhibitor does not change the affinity of the substrate for the enzyme and binding of the substrate does not change the affinity of binding of the inhibitor. The apparent V_{max} decreases for noncompetitive inhibition, while the K_m remains unchanged. For uncompetitive inhibition, the inhibitor binds ES but not free enzyme. Both the apparent V_{max} and K_m decrease in the presence of an uncompetitive inhibitor. Finally, mixed inhibition is similar to noncompetitive inhibition but occurs when binding of the inhibitor influences the affinity of substrate binding and binding of the substrate influences affinity for the inhibitor to the same degree. While V_{max} decreases, the relative K_m may increase or decrease depending on the relative affinities of the substrate and inhibitor. The appropriate equations for estimating K_i using nonlinear regression are listed in Box 9.2 [151].

Box 9.1
Inhibition equations
Single-point screen:

$$\text{percent inhibition} = \frac{\%\text{activity(inhibitor)}}{\%\text{activity(control)}}.$$

K_i estimation from IC$_{50}$ data [148]:

$$IC_{50} = K_i 1 + \frac{[S]}{K_m},$$

when incubation conditions are run at or near the K_m, this simplifies to IC$_{50}$ = 2 K_i.
Equations for nonlinear fit of K_i:

$$v = \frac{V_{max}[S]}{K_m 1 + \frac{[I]}{K_i} + [S]} \quad \text{competitive,}$$

$$v = \frac{V_{max}[S]}{K_m 1 + \frac{[I]}{K_i} + [S]1 + \frac{[I]}{K_i}} \quad \text{noncompetitive,}$$

$$v = \frac{V_{max}[S]}{K_m + [S]1 + \frac{[I]}{K_i}} \quad \text{uncompetitive,}$$

$$v = \frac{V_{max}[S]}{K_m 1 + \frac{[I]}{K_i} + [S]1 + \frac{[I]}{K'_i}} \quad \text{linear mixed,}$$

$$v = \frac{V_{max}[S]}{K_m 1 + \frac{[I]}{K_i}/1 + \frac{[I]}{\alpha K_i} + [S]}$$

Individual variability due to genetic polymorphism may also play a role in DDI predictions. Genetic predisposition may lead to altered levels of protein expression or expression of protein that has similar, reduced or null activity when compared to the wild type. Expression of a CYP variant often leads to reduced clearance and a resulting increase in area under the drug concentration–time curve (AUC) for the victim drug. The three issues of greatest concern when dealing with turnover and inhibition of a polymorphic enzyme are the therapeutic index of the drug, the fraction of the drug that undergoes metabolism by a particular CYP and the frequency of expression of the polymorphism (Table 9.5) [155]. Drugs that are cleared by multiple enzymatic pathways or have large therapeutic indexes are less susceptible to AUC-based adverse event, as other mechanisms may provide a compensatory route for drug clearance [154,156]. Incidence of polymorphic expression is also important, as the lower the level of polymorphic expression, the fewer the individuals who will be affected by individual variability, as discussed in Section 9.1.

Box 9.2
In vitro **DDI predictions:**

The default properties of CYP catalysis can be defined by Michaelis–Menten kinetics:

$$v = \frac{V_{max}[S]}{K_m + [S]}, \qquad (9.1)$$

where v equals the observed rate of metabolite formation, V_{max} represents the maximum rate of metabolite formation, K_m is the substrate concentration at which the rate of metabolite formation is half-maximal and [S] is the substrate concentration.

Competitive reversible inhibition results from inhibitor competing with substrate at the site of catalysis. In this instance, the rate is expressed by the following equation:

$$v = \frac{V_{max}[S]}{K_m\left(1 + \frac{[I]}{K_i}\right) + [S]}, \qquad (9.2)$$

Table 9.5 Examples of human CYP polymorphisms for CYP2C9, CYP2C19 and CYP2D6 et al. (1998) and Rodrigues et al. (2002). [155,204,205].

CYP	Variant alleles	Mutation	Consequence	References
CYP2C9	CYP2C9*2	Arg^{144} → Cys^{144}	Reduced activity	[204]
	CYP2C9*3	Ile^{359} → Leu^{359}	Reduced activity	[204]
	CYP2C9*4	Ile^{359} → Thr^{359}	Reduced activity	[205]
	CYP2C9*5	Asp^{360} → Glu^{360}	Reduced activity	[205]
CYP2C19	CYP2C19*2	Splicing defect	Inactive enzyme	[204]
	CYP2C19*3	Premature stop codon	Inactive enzyme	[204]
	CYP2C19*4	GTG initiation codon	Inactive enzyme	Ibeanu *et al.* (1998)
	CYP2C19*5	Arg^{433} → Trp^{433}	Inactive enzyme	Ibeanu *et al.* (1998)
	CYP2C19*6	Arg^{132} → Gln^{132}	Inactive enzyme	[155]
	CYP2C19*7	Splicing defect	Inactive enzyme	[155]
	CYP2C19*8	Trp^{120} → Arg^{120}	Reduced activity	[155]
CYP2D6	CYP2D6*2xn	Gene duplication	Increased activity	[204]
	CYP2D6*4	Splicing defect	Inactive enzyme	[204]
	CYP2D6*5	Gene deletion	No enzyme	[204]
	CYP2D6*10	Pro^{34} → Ser^{34}	Unstable enzyme	[204]
		Ser^{486} → Thr^{486}	Unstable enzyme	[204]
	CYP2D6*17	Thr^{107} → Ile^{107}	Reduced activity	[204]
		Arg^{296} → Cys^{296}	Reduced activity	[204]
		Ser^{486} → The^{486}	Reduced activity	[204]

Box 9.2 (cont.)

where [I] is the inhibitor concentration and K_i is the dissociation constant of the EI complex.

With first-order reaction kinetics, the rate of metabolism is proportional to the substrate; the following relation can be expressed:

$$\text{Cl}_{int} = \frac{v}{[S]} = \frac{V_{max}}{K_m}. \tag{9.3}$$

For competitive inhibition, the intrinsic clearance of a substrate by an inhibited enzyme can be described with the following equation:

$$\text{Cl}_{int} = \frac{V_{max}}{K_m\left(1 + \frac{[I]}{K_i}\right)}. \tag{9.4}$$

In this case, [I] represents the inhibitor concentration at the enzyme-active site and K_i is the inhibition constant for a single enzyme:

$$\frac{\text{Cl}_{int}}{\text{Cl}_{int}(i)} = 1 + \frac{[I]}{K_i}, \tag{9.5}$$

where $\text{Cl}_{int}(i)$ is the intrinsic clearance in the presence of inhibitor.

The methodology to extrapolate an *in vitro* estimate to an *in vivo* prediction is detailed in Box 9.1 and highlights the relationship of $[I]/K_i$. Increasing either the concentration of inhibitor at the site of enzymatic activity or the potency of inhibition increases the risk of a DDI. Although experimental details of how to obtain a consistent *in vitro* K_i have been worked out for the most part, the question of which concentration to use for [I] still remains. Several concentrations have been used and compared to *in vivo* DDI data, including average systemic plasma concentration after multiple dosage, maximum systemic plasma concentration after multiple dosage and maximum hepatic input concentration [152]. From a conservative standpoint, the use of the maximum hepatic input concentration for calculations minimized false negatives, though some false positives were still obtained using this concentration. Considering the contribution of parallel pathways of elimination and the impact of fraction metabolized by the inhibited CYP significantly improved the false positive rate [153,154].

9.3
Irreversible Inhibition

Interpretation of inhibitory mechanisms (competitive versus irreversible) impacts the approaches used to estimate clinical DDI potential [153,157]. Normal variance in kinetic data may prohibit the distinct differentiation of different kinetic models based on curve fitting and thus become kinetically indistinguishable. Therefore, multiple

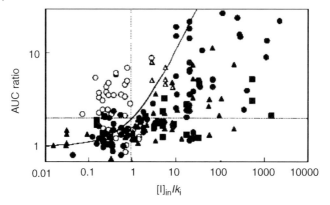

Figure 9.8 Drug–drug interaction predictions involving reversible (closed symbols) or mechanism-based inhibition (open symbols). CYP3A4 (○), CYP2D6 (▲) and CYP2C9 (■) [152].

kinetic experiments may be required to investigate the 'true' mechanism of inhibition. For example, Sesardic et al. [158] identified furafylline as a selective inhibitor of CYP1A2. Under the investigating conditions, furafylline appeared to inhibit phenacetin O-deethylation via a noncompetitive mechanism. Further investigation revealed that furafylline operates as a mechanism-based inhibitor (MBI) of CYP1A2 [159] and can result in a 17-fold increase in caffeine half-life in vivo [160]. The example of furafylline highlights the difficulty in defining the mechanism of CYP inhibition and the necessity to use numerous approaches to identify the inhibition mechanism. To clearly distinguish between reversible or irreversible (MBI) inhibition, the underlying experiments must take advantage of the unique traits that represent each type of CYP inhibition.

For reversible inhibitors, the relationship between $[I]/K_i$ on the in vivo AUC is illustrated in Figure 9.8 [152]. In this figure, the solid line represents the exponential fit of $[I]/K_i$ to AUC changes based on Equation 9.5 from Box 9.1. In general, if the $[I]/K_i$ ratio is below 0.1, there is a low risk of significant in vivo DDI (change in AUC > 2), a medium risk of interaction if the ratio is between 0.1 and 1.0 and a high risk if the $[I]/K_i$ is more than 1.0. However, the kinetic expression is predicated on the reversible nature of inhibition, and the expression is no longer valid when the inhibitor is irreversible. The open circles in Figure 9.8 represent irreversible inhibitors and illustrate the poor fit generated by the competitive inhibitor equation. The net result is underprediction of the change in AUC for compounds that behave as MBIs [152]. Instead, for irreversible inhibition, the more appropriate indicator of DDI risk may be the k_{inact}/K_I ratio [153]. Although this ratio has not been validated as an estimator of potency like the $[I]/K_i$ ratio, it appears in many cases that a k_{inact}/K_I ratio below 0.001 has a lower potential of manifesting a clinically meaningful change in AUC (<2).

The accurate in vivo extrapolation of irreversible (MBI) inhibition based on in vitro DDI data requires two key elements: the ability to differentiate reversible from irreversible inhibition (MBI) and thereafter to correctly express the impact of the

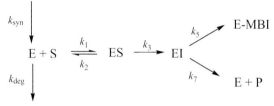

Figure 9.9 Kinetic scheme for MBI. k_{syn} represents the zero-order rate of CYP synthesis; subsequently, k_{deg} is the first-order rate constant for the natural degradation of CYP *in vivo*. k_1/k_2 represents the binding constant, k_d, for the system. The ES is the enzyme–substrate complex. EI constitutes the enzyme–intermediate complex, which has two fates: enzyme inactivation, k_5, or product release that may be reactive, k_7. The ratio of k_5/k_7 represents the partition ratio.

irreversible mechanisms *in vivo*. In practice, MBI is the result of a chemical transformation of an inhibitor to an intermediate that irreversibly binds to and inactivates the CYP prior to exiting the active site [161]. The impact of this mechanism is distinct from competitive inhibition. Foremost, when a competitive inhibitor is eliminated, the potential for a drug interaction is gone, whereas for MBI the potential persists well after the compound has been eliminated [162]. This relates to the second difference, that is the MBI perturbs the steady-state levels of the CYP it inactivates and levels recover only with the natural synthesis rate within the organism [163]. Numerous mechanistic models have been developed to describe this relationship [164–166]. Figure 9.9 represents a kinetic diagram depicting the dominant features associated with irreversible inhibition. This representation is the basis for the kinetic arguments used to extrapolate the data for *in vitro* MBI (Box 9.3) [166].

Box 9.3
***In vitro* DDI predictions for irreversible (MBI) inhibitors:**

The enzyme inactivation rate can be defined as:

$$k_{obs} = \frac{k_{inact}[I]}{K_I + [I]}, \tag{9.6}$$

where k_{obs} is the observed rate of enzyme inactivation, k_{inact} represents the maximum rate of enzyme inactivation, K_I is the inhibitor concentration when $k_{obs} = k_{inact}/2$ and [I] is the inhibitor concentration.

The inactivation must be placed in context with the steady-state enzyme synthesis rate, which can be simply described:

$$E_{ss} \propto \frac{k_{syn}}{k_{deg}}, \tag{9.7}$$

where k_{syn} represents the zero rate constant for enzyme synthesis and k_{deg} is the first-order rate constant for enzyme degradation. The MBI rate can then be combined with the natural turnover of the enzyme:

$$E'_{ss} \propto \frac{k_{syn}}{k_{deg} + k_{obs}}. \tag{9.8}$$

Substituting Equation 9.6 into Equation 9.7 yields the following:

$$\frac{E'_{ss}}{E_{ss}} = \frac{k_{deg}}{k_{deg} + ((k_{inact}[I])/(K_I + [I]))}. \tag{9.9}$$

Since Cl_{int} is proportional to E_{ss} and because Cl_{int} is inversely proportional to AUC, the following relationship is valid:

$$\frac{AUCpo}{AUCpo'} = \frac{Cl'_{int}}{Cl_{int}} = \frac{[E]'_{ss}}{[E]_{ss}} = \frac{k_{deg}}{k_{deg} + \frac{k_{inact}[I]}{K_I + [I]}}. \tag{9.10}$$

This use of Equation 9.10 to relate AUCpo to k_{deg} is valid under the assumption that only hepatic elimination occurs, well-stirred model conditions and complete absorption occurs in the GI tract.

A thorough definition of what constitutes an MBI has been provided (Silverman, 1988). In total, seven criteria have been deemed necessary for a compound to be classified as an MBI [161]. However, defining all seven criteria can be cumbersome, and given that the bulk of DDI predictions arise from the K_I and k_{inact} rate constants, some MBI criteria such as stoichiometry and reversibility are commonly overlooked. In some instances, this confounds the correct classification of a compound as an MBI [167]. A poignant example is the question of reversibility. For CYPs, one of the more common MBI mechanisms occurs through the formation of an MIC (Franklin, 1977). One method for forming this complex is through the oxidation of a primary amine to a nitroso group, which subsequently coordinates to the heme iron. Alternatively, methylene dioxy compounds can be oxidized to carbenes, which continue to produce an iron complex [168,169]. For MI complexes, the stability has been shown to be variable, so it becomes important to define the stability and extent of reversibility over time [170,171]. Specifically, if reversibility occurs on the time-scale similar to the half-life of the inhibitor, from a prediction point of view, the inhibitor is then no different than a reversible inhibitor.

9.3.1
Identification of Mechanism-Based Inhibitors

The first clue regarding molecules' ability to undergo bioactivation, the precursor to MBI, is often determined from its chemical structure. For example, certain substructures are prone to forming reactive intermediates capable of alkylating protein nucleophiles including CYP, as in the case of MBI. A comprehensive look at different chemical structures prone to CYP bioactivation has been reviewed recently [172,173].

However, the temptation to link glutathione (GSH) adducts to MBI must be avoided since the identification of metabolite adducts with GSH does *not* confirm MBI, albeit it may be an indicator of other adverse drug reactions (Walgren et al., 2005). By definition, exogenous nucleophiles such as GSH should offer no protection from MBI [161], and in principle, even when GSH adducts are formed from the same chemical entity responsible for MBI, the exogenous adducts themselves may not be related to the enzyme-inactivating species [174].

An interesting example comes from the thiazolinedione, troglitazone, which was withdrawn from the market after 3 years of use in 2000 as a result of an increased incidence of hepatoxicity [175]. One hypothesis for the observed hepatotoxicity is related to CYP-mediated bioactivation of troglitazone [176]. Subsequently, the inhibitory mechanism of troglitazone was determined to be competitive in nature [177,178]. In an alternate study, the inhibition of troglitazone was observed to be time dependent [176]. In these experiments, the addition of GSH produced several GSH–troglitazone adducts and completely protected the enzyme from inactivation, thus excluding troglitazone as an MBI. Despite these initial findings, troglitazone was further investigated as a potential MBI on the basis of its metabolite data with GSH. The results found troglitazone to be an MBI with a K_I of 5 µM and a k_{inact} of 0.0355 min^{-1} [179]. Other examples, albeit less pronounced, have shown the protective effects of GSH against time-dependent inhibition (TDI) including clorgyline against CYP1A2 and raloxifene with CYP3A4 [180,181]. These findings may validate the common use of formation of GSH adducts to initiate investigations into MBI.

Alternatives to GSH adducts that elicit MBI investigations, pharmacokinetic studies that manifest unexpected increase in AUC (e.g. accumulation inconsistent with half-life and dose) on repeat dosing could also prompt investigations into MBI [169]. When a compound is suspected to operate as an MBI, a reasonable first experiment to perform is an IC_{50} shift assay [182]. Theoretically, this assay format provides a straightforward method to identify if preincubating the inhibitor alters the inhibitory profile of a particular CYP isoform. A left shift upon preincubation indicates that the inhibitor has been modified, presumably through metabolism, to an entity that has increased inhibitor potency (Figure 9.10) [183]. However, under these experimental conditions, the left shift does *not* distinguish the potential contribution of a reversible inhibitor (metabolite) that formed during the preincubation step (Figure 9.11). An additional caveat to this experiment is the diagnostic left shift that is predicated on the ability of the incubation system (inhibitor concentration, time and enzyme source) to metabolize the compound. Therefore, an important control in an IC_{50} shift experiment is to measure the percentage turnover of the inhibitor during the preincubation step. This determines if the preincubation conditions generated sufficient turnover to investigate the potential for MBI. For example, if the compound in question has less than 5% turnover, it may be necessary to increase the preincubation time and change protein concentration, tissue fraction (S-9, microsomes or hepatocytes) or any combination thereof. Although seemingly simple, the IC_{50} shift experiment should not be trivialized, as the proper control experiments validate the experimental results. Fortunately, if executed

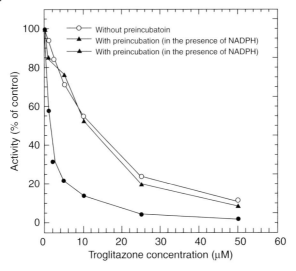

Figure 9.10 Preincubation effect on enzyme activity [183].

properly, information from the IC_{50} shift experiments such as inhibitor and protein concentrations will be useful for future studies required to complete the characterization of the inhibitor.

9.3.2
Characterization of Mechanism-Based Inhibition

After a chemical scaffold tests positive in the IC_{50} shift assay, it becomes necessary to provide a quantitative measure of the potency and rate of the suspected MBI. A TDI

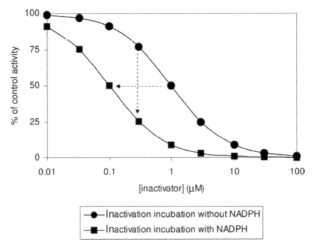

Figure 9.11 Time-dependent inhibition of CYP3A4 by troglitazone [176].

assay should be performed to estimate the maximum rate of enzyme inactivation (k_{inact}) and the inhibitor concentration to have maximal inactivation (K_I). The assay differs from the IC_{50} shift assay in several ways. First, the preincubation step uses a primary incubation to activate the inhibitor in question, from which a dilution is made into a secondary incubation (10–100-fold) that contains the selective marker substrate for the particular CYP isoform under investigation. The second difference is that multiple preincubation times of the inhibitor are included. With these differences in mind, several other variables must also be considered: (1) the competitive inhibition at the secondary stage of the incubation is minimized by sufficiently diluting the primary to secondary incubations, (2) appropriate number of inhibitor concentrations are used, (3) protein concentration is optimized to minimize nonspecific binding yet enough to produce sufficient turnover in the primary and secondary incubations and (4) the secondary incubations are carried out under linear conditions with saturating amounts (V_{max} conditions) of the appropriate substrate probes. Careful attention to these design features helps in reducing bias in the assay format and provides the necessary end points to quantify the effect of time and metabolism of the inhibitor in question [184].

Visual inspection of the results serves as a diagnostic to ensure that the criteria for a successful experiment were satisfied. First, each inhibitor concentration and its time course should be plotted as the percent remaining relative to control on the ordinate axis versus the preincubation time on the abscissa. These data can be presented in a semi-log plot or, if the changes are small, with a linear plot to facilitate the visualization of any aberrant data points (Figure 9.12a and b). In addition, this plot provides a semiquantitative view to the extent of competitive inhibition observed in the secondary incubations. For example, the highest concentration inhibitor sample in Figure 9.12b illustrates the impact of competitive inhibition. This is diagnosed by the first time point (no preincubation) at which the remaining activity should be close to 100%, but in this example it is only 60% of the control. These data could confound the accurate determination of the initial inactivation rates. Next, the data should be transformed to a linear plot (Kitz–Wislon) by plotting $1/k_{obs}$ on the ordinate versus 1/inhibitor on the abscissa [185] (Figure 9.12c). The intercept of the ordinate axis yields $1/k_{inact}$ and the slope of the line estimates the k_{inact}/K_I. The range in data should be such that a single point does not dominate the linear regression. Secondly, any deviation from linearity may indicate limitations of inhibitor solubility or a mixed mechanism of inhibition. When possible, the K_I and k_{inact} values should be derived from nonlinear regression (Box 9.3 and Figure 9.12d). At the end of a TDI experiment, there are at least four graphs that should be used to interrogate the data fitness from the TDI experiment.

Typically, microsomal fractions are used for TDI analysis, but in many cases TDI is performed in rCYPs. The ability to use rCYPs for TDI relies on the accurate ability to extrapolate the rCYP data [186]. The second requirement for successful extrapolation of rCYP data is that the clearance pathways of the inhibitor in question are understood. For example, raloxifene has been shown to be a strong MBI of CYP3A4 in microsomes and rCYPs, but in hepatocytes the most predominant pathway is glucuronadation, which would effectively mask any CYP3A4 inactivation observed in

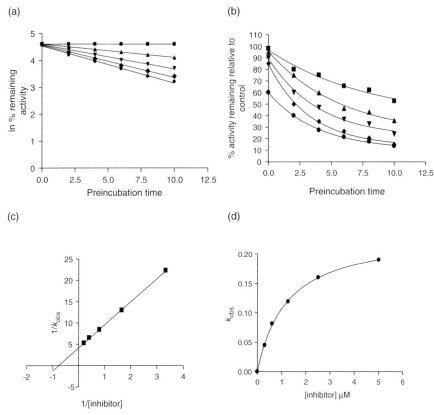

Figure 9.12 Plots for interpreting TDI. (a) Natural log of percent activity remaining versus preincubation time; (b) percent activity relative to control versus preincubation time (provides clarity regarding amount of competitive inhibition at chosen inhibitor concentrations; (c) Kitz–Wilson plot provides visual inspection of data to ensure that appropriate inhibitor concentrations were chosen and provides estimates of K_i and k_{inact}; and (d) nonlinear regression plot used to determine K_i and k_{inact}.

the other systems [187]. Furthermore, rCYPs alone cannot provide the level of detail that is necessary to context the potential risk of MBI without the knowledge of the relevant metabolic pathways. As Figure 9.13 illustrates, the fraction of the drug metabolized by the enzyme is paramount to the extent to which the drug interaction is manifested and therefore these data become crucial to an accurate assessment of the *in vivo* impact of an MBI [157]. The use of hepatocytes has also been examined for TDI [188]. Arguably, hepatocytes present the most complete *in vitro* system to study *in vivo* metabolism. Caution should be extended with the use of hepatocytes since this system does contain the cellular machinery necessary for CYP induction. Therefore, generating inhibition constants with hepatocytes may be significantly more challenging if the inhibitor is also an inducer.

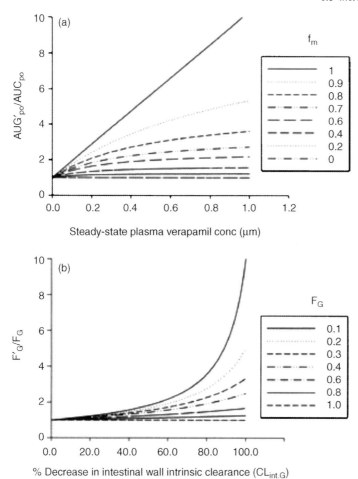

Figure 9.13 Impact of fraction metabolized on change in AUC [157].

9.3.3
Additional Mechanistic Tools for Investigating Mechanism-Based Inhibition

In addition to the TDI experiment, the partition ratio measures the TDI efficiency. Specifically, the partition ratio is the number of inactivation kinetic events (k_{inact}) versus the number of substrate turnover events per unit enzyme (k_{cat}) [161]. Thus, the most potent partition ratio is zero. The most common experimental setup for determining the partition ratio is the titration method that increases the inhibitor concentration relative to a known amount of enzyme. After the incubations, a secondary incubation containing a probe substrate similar to the TDI experiment is used to define the remaining activity. For accurate determination of the partition ratio from the titration method, it is assumed that the inhibitor is 100% metabolized;

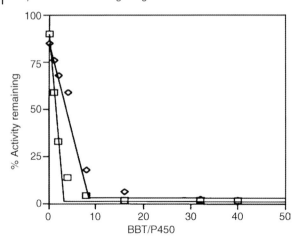

Figure 9.14 Example of titration method for determining partition ratio [189].

therefore, either a sufficient dilution of the primary incubation or dialysis prior to measuring the remaining CYP activity can be used. The extrapolation of the zero activity to the X-axis defines the partition ratio for the inhibitor (Figure 9.14) [189]. As the partition ratio is not used for *in vitro* extrapolations, these experiments are not run routinely. Furthermore, the k_{inact}/K_I has been shown to correlate extremely well to the partition ratio and thus also provides information regarding the efficiency of inactivation [190]. More recently, the apparent partition ratio (APR) was used as a diagnostic for investigating TDI of different compounds [179]. This experimental format may prove to supplement to the IC_{50} shift assay, as it is easily automated and able to identify several compounds previously overlooked as TDIs (Figure 9.10).

An additional step in differentiating TDI from MBI is to define the extent of reversibility associated with the inhibition. The classic method for monitoring reversibility is to incubate with the suspected MBI and recover the incubated protein for dialysis. Once dialysis is complete and the reversible inhibitors have been removed, the remaining enzymatic activity can be assessed with the percent loss attributed to irreversible enzyme binding. If the experiment is run with radiolabeled inhibitor, the reversibility experiment can also serve to measure the stoichiometry of inactivation, assuming that proper controls were run. In some cases, the reaction is a simple 1 : 1 stoichiometry, whereas other cases have reported stoichiometry as high as 5 : 1 [191]. On a more qualitative basis, gel electrophoresis provides an indication of irreversibility [192]. For CYP alkylation, the radiolabel should coincide with a strong phosphor-imaged band around ∼55–57 000 Da as indicated in Figure 9.15 [189]. The use of rCYP incubation provides an unmatched positive control for this experiment. Furthermore, specific CYP inhibitors can be added to the MBI to examine the reduction in signal on the gel, indicating the ability of competitive inhibition to block MBI adduct formation [193].

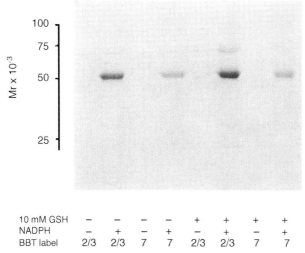

Figure 9.15 Incorporation of covalent radiolabel into CYP [189].

Mass spectrometry of intact CYPs has recently been incorporated into the characterization of inactivation and has the advantage of determining irreversibility and stoichiometry in the same experiment [194,195]. Figure 9.16 illustrates the mass spectrometry data from CYP3A4 covalently labeled by L-754,394 [194]. The deconvoluted masses shown in Figure 9.16 are indicative of parent mass 56 276 Da and the adducted protein mass 56 961 Da, indicating a 1 : 1 stoichiometry between L-754,394 and CYP3A4. Furthermore, the mass change from the deconvoluted spectra show that the L-754,394 adduct has two oxygen atoms incorporated into the adduct, which provides mechanistic information about the adduct. Although this example successfully illustrates the mass difference of the inactivating species, it is unlikely that the current mass spectrometry technology can differentiate the adduction of a parent molecule from that of its dehydrated intermediate (-2 Da). Another limitation to the use of mass spectrometry for adduct characterization is that rCYP is typically required. Furthermore, the instrumentation and techniques used for CYP mass spectrometry can be highly specialized [194]. However, advances in hybrid mass spectrometry equipment and technology may soon facilitate the use of other matrices (HLM) to assess covalent binding in the near future.

Even with the knowledge of the reactive moieties that are suspected to trigger MBI, there are numerous potential pathways for the chemistry to lead to protein inactivation [174,196]. Differentiating these mechanisms can facilitate the generation of alternate and safer chemical scaffolds. The UV–Vis spectrophotometer has been a key instrument in activity and functional characterization for CYPs for the past 40 years, as indicated by the derivation of its name, 'pigment 450' being the signature UV band present when reduced in the presence of CO [5,197]. This technique has been

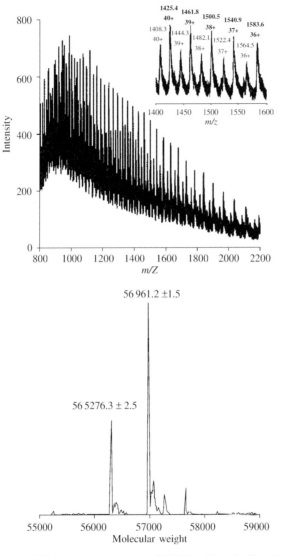

Figure 9.16 Intact mass spectrum of CYP3A4 adducted with L-754,394 [194].

extended to differentiate apoadducted protein from heme-based MBI (Figure 9.17) [198]. For instance, when CO-difference spectra are monitored before and after incubation, the loss in CO binding from the start of incubation is suspected to coincide with the amount of alkylated heme. However, as in the example in Figure 9.17, 1-[(2-ethyl-4-methyl-1H-imidazol-5-yl)methyl]-4-[4-(trifluoromethyl)-2-pyridinyl]piperazine showed no sign of heme alkylation, despite showing approximately 35% loss of CO binding capacity. Data have been accumulated to suggest

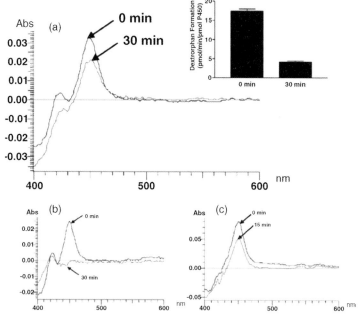

Figure 9.17 Measurements of CO spectra before and after incubation with MBI. (a) CO measurement indicating partial loss of CO binding despite being an apoprotein adduct; (b) positive heme adduct, ABT, showing heme alkylation and subsequent disruption of CO binding; and (c) MBI, midazolam thought to result from apoprotein adduct also showing marginal loss of CO binding [198].

that the disappearance of CO-difference spectra does not directly implicate heme alkylation. For example, tienilic acid was shown to form an apoprotein adduct exclusively, yet CO spectral-binding studies showed approximately 50% loss in CO binding capacity [196]. Although this discrepancy has not been sufficiently explained, it may be possible that the apoprotein adducts destabilize the folded state of the protein, resulting in protein unfolding to the point where CO no longer binds. Therefore, the reduced CO spectra may not adequately distinguish between apoprotein adducts compared to heme alkylation. Alternatively, quantitation of heme and its adducts may be possible with LC/UV and mass spectrometry (Figure 9.18) [199]. In this instance, UV detection at 400 nm was used to determine free heme (Figure 9.18a) from CYP4B1 covalently modified heme (Figure 9.18b). 1-ABT forms a covalent adduct with the heme of CYP and has been measured directly using mass spectrometry [200]. Combined with NMR data, the exact details of the heme adduct and its structure provided the basis for the mechanism for 1-ABT heme alkylation. The advantage of this technique is that information regarding inactivation based on heme adduct formation can be determined directly. Several other heme adducts have been identified using this strategy [201,202].

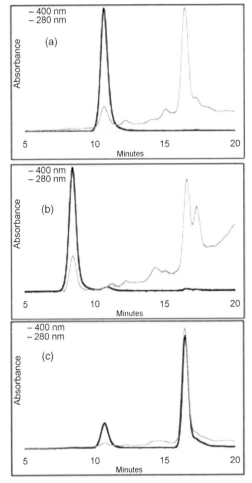

Figure 9.18 Characterization of heme with UV detection at 400 nm to illustrate different heme retention times (a) native heme, (b) monohydroxyheme and (c) CYP4B1 covalently bound heme. Modified from Zheng et al. [199].

9.4
Conclusion

In the light of the success with polypharmacy in treating many disease states, the potential for serious drug–drug interactions will remain a serious health threat. However, through the careful use of *in vitro* experiments and thoughtful clinical studies, the level of understanding of a new chemical entity with respect to drug inhibition liabilities is vastly increased. With regard to *in vitro* studies, key advances have been made in understanding the kinetic mechanisms of inhibition (reversible and irreversible), developing new technologies, improving analytical capabilities, and refining *in vitro–in vivo* extrapolation techniques. It is anticipated that this information will significantly improve the safety and therapeutic outcome for patients receiving polypharmacy.

Abbreviations

LC/MS	liquid chromatography and mass spectrometry
CYP or P450	cytochrome P450
NCE	new chemical entity
HLM	human liver microsome
AhR	aryl hydrocarbon receptor
PK	pharmacokinetics
rCYP	recombinant CYP
GSH	glutathione
NADPH	reduced nicotinamide adenine dinucleotide phosphate
AUC	area under the curve
CO	carbon monoxide
TDI	time-dependent inhibition
MBI	mechanism-based inhibition
DDI	drug–drug interaction
HTS	high-throughput screening
APR	apparent partition ratio.

References

1 Brodie, B.B., Gillette, J.R. and La Du, B.N. (1958) Enzymatic metabolism of drugs and other foreign compounds. *Annual Review of Biochemistry*, **27** (3), 427–454.

2 Axelrod, J. (1955) The enzymatic conversion of codeine to morphine. *Journal of Pharmacology and Experimental Therapeutics*, **115** (3), 259–267.

3 Boyland, E. and Levi, A.A. (1935) Metabolism of polycyclic compounds: production of dihydroxy-dihydroanthracene from anthracene. *Biochemical Journal*, **29** (12), 2679–2683.

4 Klingenberg, M. (1958) Pigments of rat liver microsomes. *Archives of Biochemistry and Biophysics*, **75** (2), 376–386.

5 Omura, T. and Sato, R. (1962) A new cytochrome in liver microsomes. *Journal of Biological Chemistry*, **237**, 1375–1376.

6 Sladek, N.E. and Mannering, G.J. (1969) Induction of drug metabolism. I. Differences in the mechanisms by which polycyclic hydrocarbons and phenobarbital produce their inductive effects on microsomal N-demethylating systems. *Molecular Pharmacology*, **5** (2), 174–185.

7 Lu, A.Y. and Coon, M.J. (1968) Role of hemoprotein P-450 in fatty acid omega-hydroxylation in a soluble enzyme system from liver microsomes. *Journal of Biological Chemistry*, **243** (6), 1331–1332.

8 Mannering, G.J., Renton, K.W., el Azhary, R. and Deloria, L.B. (1980)

Effects of interferon-inducing agents on hepatic cytochrome P-450 drug metabolizing systems. *Annals of the New York Academy of Sciences*, **350**, 314–331.

9 Guengerich, F.P. (1989) Characterization of human microsomal cytochrome P-450 enzymes. *Annual Review of Pharmacology and Toxicology*, **29**, 241–264.

10 Nebert, D.W., Adesnik, M., Coon, M.J., Estabrook, R.W., Gonzalez, F.J., Guengerich, F.P., Gunsalus, I.C., Johnson, E.F., Kemper, B., Levin, W. et al.(1987) The P450 gene superfamily: recommended nomenclature. *DNA*, **6** (1), 1–11.

11 Nelson, D.R., Koymans, L., Kamataki, T., Stegeman, J.J., Feyereisen, R., Waxman, D.J., Waterman, M.R., Gotoh, O., Coon, M.J., Estabrook, R.W., Gunsalus, I.C. and Nebert, D.W. (1996) P450 superfamily: update on new sequences, gene mapping, accession numbers and nomenclature. *Pharmacogenetics*, **6** (1), 1–42.

12 Nelson, D.R. (1999) Cytochrome P450 and the individuality of species. *Archives of Biochemistry and Biophysics*, **369** (1), 1–10.

13 Williams, J.A., Hyland, R., Jones, B.C., Smith, D.A., Hurst, S., Goosen, T.C., Peterkin, V., Koup, J.R. and Ball, S.E. (2004) Drug–drug interactions for UDP-glucuronosyltransferase substrates: a pharmacokinetic explanation for typically observed low exposure (AUCi/AUC) ratios. *Drug Metabolism and Disposition*, **32** (11), 1201–1208.

14 Hakkola, J., Pasanen, M., Purkunen, R., Saarikoski, S., Pelkonen, O., Maenpaa, J., Rane, A. and Raunio, H. (1994) Expression of xenobiotic-metabolizing cytochrome P450 forms in human adult and fetal liver. *Biochemical Pharmacology*, **48** (1), 59–64.

15 Shimada, T., Yamazaki, H., Mimura, M., Inui, Y. and Guengerich, F.P. (1994) Interindividual variations in human liver cytochrome P-450 enzymes involved in the oxidation of drugs, carcinogens and toxic chemicals: studies with liver microsomes of 30 Japanese and 30 Caucasians. *Journal of Pharmacology and Experimental Therapeutics*, **270** (1), 414–423.

16 Bloomer, J.C., Clarke, S.E. and Chenery, R.J. (1995) Determination of P4501A2 activity in human liver microsomes using [3-14C-methyl]caffeine. *Xenobiotica*, **25** (9), 917–927.

17 Wynalda, M.A., Hauer, M.J. and Wienkers, L.C. (1998) Human biotransformation of bropirimine. Characterization of the major bropirimine oxidative metabolites formed *in vitro*. *Drug Metabolism and Disposition*, **26** (10), 1048–1051.

18 Spaldin, V., Madden, S., Adams, D.A., Edwards, R.J., Davies, D.S. and Park, B. K. (1995) Determination of human hepatic cytochrome P4501A2 activity *in vitro* use of tacrine as an isoenzyme-specific probe. *Drug Metabolism and Disposition*, **23** (9), 929–934.

19 Bertilsson, L., Carrillo, J.A., Dahl, M.L., Llerena, A., Alm, C., Bondesson, U., Lindstrom, L., Rodriguez de la Rubia, I., Ramos, S. and Benitez, J. (1994) Clozapine disposition covaries with CYP1A2 activity determined by a caffeine test. *British Journal of Clinical Pharmacology*, **38** (5), 471–473.

20 Nakajima, M., Kobayashi, K., Shimada, N., Tokudome, S., Yamamoto, T. and Kuroiwa, Y. (1998) Involvement of CYP1A2 in mexiletine metabolism. *British Journal of Clinical Pharmacology*, **46** (1), 55–62.

21 Kaye, C.M. and Nicholls, B. (2000) Clinical pharmacokinetics of ropinirole. *Clinical Pharmacokinetics*, **39** (4), 243–254.

22 Ha, H.R., Chen, J., Freiburghaus, A.U. and Follath, F. (1995) Metabolism of theophylline by cDNA-expressed human cytochromes P-450. *British Journal of Clinical Pharmacology*, **39** (3), 321–326.

23 Kiyohara, C., Nakanishi, Y., Inutsuka, S., Takayama, K., Hara, N., Motohiro, A., Tanaka, K., Kono, S. and Hirohata, T. (1998) The relationship between CYP1A1 aryl hydrocarbon hydroxylase activity and lung cancer in a Japanese population. *Pharmacogenetics*, **8** (4), 315–323.

24 Cauchi, S., Stucker, I., Solas, C., Laurent-Puig, P., Cenee, S., Hemon, D., Jacquet, M., Kremers, P., Beaune, P. and Massaad-Massade, L. (2001) Polymorphisms of human aryl hydrocarbon receptor (AhR) gene in a French population: relationship with CYP1A1 inducibility and lung cancer. *Carcinogenesis*, **22** (11), 1819–1824.

25 Tantcheva-Poor, I., Zaigler, M., Rietbrock, S. and Fuhr, U. (1999) Estimation of cytochrome P-450 CYP1A2 activity in 863 healthy Caucasians using a saliva-based caffeine test. *Pharmacogenetics*, **9** (2), 131–144.

26 Kall, M.A. and Clausen, J. (1995) Dietary effect on mixed function P450 1A2 activity assayed by estimation of caffeine metabolism in man. *Human and Experimental Toxicology*, **14** (10), 801–807.

27 Whitlock, J.P., Jr (1999) Induction of cytochrome P4501 A1. *Annual Review of Pharmacology and Toxicology*, **39**, 103–125.

28 Conney, A.H., Miller, E.C. and Miller, J.A. (1957) Substrate-induced synthesis and other properties of benzpyrene hydroxylase in rat liver. *Journal of Biological Chemistry*, **228** (2), 753–766.

29 Poland, A., Clover, E., Kende, A.S., DeCamp, M. and Giandomenico, C.M. (1976) 3,4,3′,4′-Tetrachloro azoxybenzene and azobenzene: potent inducers of aryl hydrocarbon hydroxylase. *Science*, **194** (4265), 627–630.

30 Miller, A.G., Israel, D. and Whitlock, J.P., Jr (1983) Biochemical and genetic analysis of variant mouse hepatoma cells defective in the induction of benzo(a) pyrene-metabolizing enzyme activity. *Journal of Biological Chemistry*, **258** (6), 3523–3527.

31 Balogh, A., Harder, S., Vollandt, R. and Staib, A.H. (1992) Intra-individual variability of caffeine elimination in healthy subjects. *International Journal of Clinical Pharmacology, Therapy, and Toxicology*, **30** (10), 383–387.

32 Harris, J.W., Rahman, A., Kim, B.R., Guengerich, F.P. and Collins, J.M. (1994) Metabolism of taxol by human hepatic microsomes and liver slices: participation of cytochrome P450 3A4 and an unknown P450 enzyme. *Cancer Research*, **54** (15), 4026–4035.

33 Kerr, B.M., Thummel, K.E., Wurden, C.J., Klein, S.M., Kroetz, D.L., Gonzalez, F.J. and Levy, R.H. (1994) Human liver carbamazepine metabolism. Role of CYP3A4 and CYP2C8 in 10,11-epoxide formation. *Biochemical Pharmacology*, **47** (11), 1969–1979.

34 Li, X.Q., Bjorkman, A., Andersson, T.B., Ridderstrom, M. and Masimirembwa, C.M. (2002) Amodiaquine clearance and its metabolism to *N*-desethylamodiaquine is mediated by CYP2 C8: a new high affinity and turnover enzyme-specific probe substrate. *Journal of Pharmacology and Experimental Therapeutics*, **300** (2), 399–407.

35 Goldstein, J.A., Faletto, M.B., Romkes-Sparks, M., Sullivan, T., Kitareewan, S., Raucy, J.L., Lasker, J.M. and Ghanayem, B.I. (1994) Evidence that CYP2C19 is the major (*S*)-mephenytoin 4′-hydroxylase in humans. *Biochemistry*, **33** (7), 1743–1752.

36 Tracy, T.S., Marra, C., Wrighton, S.A., Gonzalez, F.J. and Korzekwa, K.R. (1996) Studies of flurbiprofen 4′-hydroxylation. Additional evidence suggesting the sole involvement of cytochrome P450 2 C9. *Biochemical Pharmacology*, **52** (8), 1305–1309.

37 Hall, S.D., Hamman, M.A., Rettie, A.E., Wienkers, L.C., Trager, W.F., Vandenbranden, M. and Wrighton, S.A. (1994) Relationships between the levels

of cytochrome P4502C9 and its prototypic catalytic activities in human liver microsomes. *Drug Metabolism and Disposition*, **22** (6), 975–978.
38 Rettie, A.E., Korzekwa, K.R., Kunze, K.L., Lawrence, R.F., Eddy, A.C., Aoyama, T., Gelboin, H.V., Gonzalez, F.J. and Trager, W.F. (1992) Hydroxylation of warfarin by human cDNA-expressed cytochrome P-450: a role for P-4502C9 in the etiology of (S)-warfarin–drug interactions. *Chemical Research in Toxicology*, **5** (1), 54–59.
39 Bonnabry, P., Leemann, T. and Dayer, P. (1996) Role of human liver microsomal CYP2C9 in the biotransformation of lornoxicam. *European Journal of Clinical Pharmacology*, **49** (4), 305–308.
40 Rettie, A.E., Wienkers, L.C., Gonzalez, F.J., Trager, W.F. and Korzekwa, K.R. (1994) Impaired (S)-warfarin metabolism catalysed by the R144C allelic variant of CYP2 C9. *Pharmacogenetics*, **4** (1), 39–42.
41 Lee, C.R., Goldstein, J.A. and Pieper, J.A. (2002) Cytochrome P450 2C9 polymorphisms: a comprehensive review of the *in-vitro* and human data. *Pharmacogenetics*, **12** (3), 251–263.
42 Iida, I., Miyata, A., Arai, M., Hirota, M., Akimoto, M., Higuchi, S., Kobayashi, K. and Chiba, K. (2004) Catalytic roles of CYP2C9 and its variants (CYP2C9*2 and CYP2C9*3) in lornoxicam 5'-hydroxylation. *Drug Metabolism and Disposition*, **32** (1), 7–9.
43 Heimark, L.D., Wienkers, L., Kunze, K., Gibaldi, M., Eddy, A.C., Trager, W.F., O'Reilly, R.A. and Goulart, D.A. (1992) The mechanism of the interaction between amiodarone and warfarin in humans. *Clinical Pharmacology and Therapeutics*, **51** (4), 398–407.
44 Steward, D.J., Haining, R.L., Henne, K.R., Davis, G., Rushmore, T.H., Trager, W.F. and Rettie, A.E. (1997) Genetic association between sensitivity to warfarin and expression of CYP2C9*3. *Pharmacogenetics*, **7** (5), 361–367.

45 Becquemont, L., Verstuyft, C. and Jaillon, P. (2006) Pharmacogenetics and interindividual variability in drug response: cytochrome P-450 2C9 and coumarin anticoagulants. *Bulletin de l'Academie Nationale de Medecine*, **190** (1), 37–49. discussion 50-3.
46 Wedlund, P.J., Aslanian, W.S., McAllister, C.B., Wilkinson, G.R. and Branch, R.A. (1984) Mephenytoin hydroxylation deficiency in Caucasians: frequency of a new oxidative drug metabolism polymorphism. *Clinical Pharmacology and Therapeutics*, **36** (6), 773–780.
47 Goldstein, J.A., Ishizaki, T., Chiba, K., de Morais, S.M., Bell, D., Krahn, P.M. and Evans, D.A. (1997) Frequencies of the defective CYP2C19 alleles responsible for the mephenytoin poor metabolizer phenotype in various Oriental, Caucasian, Saudi Arabian and American black populations. *Pharmacogenetics*, **7** (1), 59–64.
48 Wedlund, P.J. (2000) The CYP2C19 enzyme polymorphism. *Pharmacology*, **61** (3), 174–183.
49 Abelo, A., Andersson, T.B., Antonsson, M., Naudot, A.K., Skanberg, I. and Weidolf, L. (2000) Stereoselective metabolism of omeprazole by human cytochrome P450 enzymes. *Drug Metabolism and Disposition*, **28** (8), 966–972.
50 Wienkers, L.C., Wurden, C.J., Storch, E., Kunze, K.L., Rettie, A.E. and Trager, W.F. (1996) Formation of (R)-8-hydroxywarfarin in human liver microsomes. A new metabolic marker for the (S)-mephenytoin hydroxylase, P4502 C19. *Drug Metabolism and Disposition*, **24** (5), 610–614.
51 Helsby, N.A., Ward, S.A., Howells, R.E. and Breckenridge, A.M. (1990) *in vitro* metabolism of the biguanide antimalarials in human liver microsomes: evidence for a role of the mephenytoin hydroxylase (P450 MP)

enzyme. *British Journal of Clinical Pharmacology*, **30** (2), 287–291.
52 Evans, W.E. and Relling, M.V. (2004) Moving towards individualized medicine with pharmacogenomics. *Nature*, **429** (6990), 464–468.
53 Zanger, U.M., Raimundo, S. and Eichelbaum, M. (2004) Cytochrome P450 2 D6: overview and update on pharmacology, genetics, biochemistry. *Naunyn-Schmiedeberg's Archives Of Pharmacology*, **369** (1), 23–37.
54 Wilcox, R.A. and Owen, H. (2000) Variable cytochrome P450 2D6 expression and metabolism of codeine and other opioid prodrugs: implications for the Australian anaesthetist. *Anaesthesia and Intensive Care*, **28** (6), 611–619.
55 Brachtendorf, L., Jetter, A., Beckurts, K.T., Holscher, A.H. and Fuhr, U. (2002) Cytochrome P450 enzymes contributing to demethylation of maprotiline in man. *Pharmacology and Toxicology*, **90** (3), 144–149.
56 Shin, J.G., Soukhova, N. and Flockhart, D.A. (1999) Effect of antipsychotic drugs on human liver cytochrome P-450 (CYP) isoforms *in vitro*: preferential inhibition of CYP2 D6. *Drug Metabolism and Disposition*, **27** (9), 1078–1084.
57 Botsch, S., Gautier, J.C., Beaune, P., Eichelbaum, M. and Kroemer, H.K. (1993) Identification and characterization of the cytochrome P450 enzymes involved in *N*-dealkylation of propafenone: molecular base for interaction potential and variable disposition of active metabolites. *Molecular Pharmacology*, **43** (1), 120–126.
58 Naranjo, C.A., Sproule, B.A. and Knoke, D.M. (1999) Metabolic interactions of central nervous system medications and selective serotonin reuptake inhibitors. *International Clinical Psychopharmacology*, **14** (Suppl. 2), S35–S47.
59 Wilkinson, G.R. (2005) Drug metabolism and variability among patients in drug response. *New England Journal of Medicine*, **352** (21), 2211–2221.
60 de Groot, M.J., Bijloo, G.J., Martens, B.J., van Acker, F.A. and Vermeulen, N.P. (1997) A refined substrate model for human cytochrome P450 2 D6. *Chemical Research in Toxicology*, **10** (1), 41–48.
61 Eichelbaum, M., Spannbrucker, N. and Dengler, H.J. (1979) Influence of the defective metabolism of sparteine on its pharmacokinetics. *European Journal of Clinical Pharmacology*, **16** (3), 189–194.
62 Kimura, S., Umeno, M., Skoda, R.C., Meyer, U.A. and Gonzalez, F.J. (1989) The human debrisoquine 4-hydroxylase (CYP2D) locus: sequence and identification of the polymorphic CYP2D6 gene, a related gene, and a pseudogene. *American Journal of Human Genetics*, **45** (6), 889–904.
63 Ingelman-Sundberg, M. (2005) Genetic polymorphisms of cytochrome P450 2D6 (CYP2D6): clinical consequences, evolutionary aspects and functional diversity. *Pharmacogenomics Journal*, **5** (1), 6–13.
64 Xie, H.G., Kim, R.B., Wood, A.J. and Stein, C.M. (2001) Molecular basis of ethnic differences in drug disposition and response. *Annual Review of Pharmacology and Toxicology*, **41**, 815–850.
65 Sindrup, S.H. and Brosen, K. (1995) The pharmacogenetics of codeine hypoalgesia. *Pharmacogenetics*, **5** (6), 335–346.
66 Gasche, Y., Daali, Y., Fathi, M., Chiappe, A., Cottini, S., Dayer, P. and Desmeules, J. (2004) Codeine intoxication associated with ultrarapid CYP2D6 metabolism. *New England Journal of Medicine*, **351** (27), 2827–2831.
67 Wacher, V.J., Wu, C.Y. and Benet, L.Z. (1995) Overlapping substrate specificities and tissue distribution of cytochrome P450 3A and P-glycoprotein: implications for drug delivery and

activity in cancer chemotherapy. *Molecular Carcinogenesis*, **13** (3), 129–134.

68 Hall, S.D., Thummel, K.E., Watkins, P.B., Lown, K.S., Benet, L.Z., Paine, M.F., Mayo, R.R., Turgeon, D.K., Bailey, D.G., Fontana, R.J. and Wrighton, S.A. (1999) Molecular and physical mechanisms of first-pass extraction. *Drug Metabolism and Disposition*, **27** (2), 161–166.

69 Gibson, G.G., Plant, N.J., Swales, K.E., Ayrton, A. and El-Sankary, W. (2002) Receptor-dependent transcriptional activation of cytochrome P4503A genes: induction mechanisms, species differences and interindividual variation in man. *Xenobiotica*, **32** (3), 165–206.

70 Thummel, K.E. and Wilkinson, G.R. (1998) in vitro and in vivo drug interactions involving human CYP3A. *Annual Review of Pharmacology and Toxicology*, **38**, 389–430.

71 Guengerich, F.P., Parikh, A., Turesky, R.J. and Josephy, P.D. (1999) Interindividual differences in the metabolism of environmental toxicants: cytochrome P450 1A2 as a prototype. *Mutation Research*, **428** (1–2), 115–124.

72 Thummel, K.E., Shen, D.D., Podoll, T.D., Kunze, K.L., Trager, W.F., Bacchi, C.E., Marsh, C.L., McVicar, J.P., Barr, D.M., Perkins, J.D., Carithers, R.L. (1994) Use of midazolam as a human cytochrome P450 3A probe. II. Characterization of inter- and intraindividual hepatic CYP3A variability after liver transplantation. *Journal of Pharmacology and Experimental Therapeutics*, **271** (1), 557–566.

73 Wienkers, L.C., Allievi, C., Hauer, M.J. and Wynalda, M.A. (1999) Cytochrome P-450-mediated metabolism of the individual enantiomers of the antidepressant agent reboxetine in human liver microsomes. *Drug Metabolism and Disposition*, **27** (11), 1334–1340.

74 Wynalda, M.A., Hutzler, J.M., Koets, M.D., Podoll, T. and Wienkers, L.C. (2003) in vitro metabolism of clindamycin in human liver and intestinal microsomes. *Drug Metabolism and Disposition*, **31** (7), 878–887.

75 Sutton, D., Butler, A.M., Nadin, L. and Murray, M. (1997) Role of CYP3A4 in human hepatic diltiazem N-demethylation: inhibition of CYP3A4 activity by oxidized diltiazem metabolites. *Journal of Pharmacology and Experimental Therapeutics*, **282** (1), 294–300.

76 Voorman, R.L., Payne, N.A., Wienkers, L.C., Hauer, M.J. and Sanders, P.E. (2001) Interaction of delavirdine with human liver microsomal cytochrome P450: inhibition of CYP2C9, CYP2C19, and CYP2D6. *Drug Metabolism and Disposition*, **29** (1), 41–47.

77 Flexner, C. (2000) Dual protease inhibitor therapy in HIV-infected patients: pharmacologic rationale and clinical benefits. *Annual Review of Pharmacology and Toxicology*, **40**, 649–674.

78 Kolars, J.C., Schmiedlin-Ren, P., Schuetz, J.D., Fang, C. and Watkins, P.B. (1992) Identification of rifampin-inducible P450IIIA4 (CYP3A4) in human small bowel enterocytes. *Journal of Clinical Investigation*, **90** (5), 1871–1878.

79 Lin, J.H. and Lu, A.Y. (2001) Interindividual variability in inhibition and induction of cytochrome P450 enzymes. *Annual Review of Pharmacology and Toxicology*, **41**, 535–567.

80 Jones, S.A., Moore, L.B., Shenk, J.L., Wisely, G.B., Hamilton, G.A., McKee, D.D., Tomkinson, N.C., LeCluyse, E.L., Lambert, M.H., Willson, T.M., Kliewer, S.A. and Moore, J.T. (2000) The pregnane X receptor: a promiscuous xenobiotic receptor that has diverged during evolution. *Molecular Endocrinology*, **14** (1), 27–39.

81 Willson, T.M. and Kliewer, S.A. (2002) PXR, CAR and drug metabolism. *Nature Reviews. Drug Discovery*, **1** (4), 259–266.

82 Henderson, L., Yue, Q.Y., Bergquist, C., Gerden, B. and Arlett, P. (2002) St John's

wort (*Hypericum perforatum*): drug interactions and clinical outcomes. *British Journal of Clinical Pharmacology*, **54** (4), 349–356.

83 Ueng, Y.F., Kuwabara, T., Chun, Y.J. and Guengerich, F.P. (1997) Cooperativity in oxidations catalyzed by cytochrome P450 3 A4. *Biochemistry*, **36** (2), 370–381.

84 Schrag, M.L. and Wienkers, L.C. (2001) Triazolam substrate inhibition: evidence of competition for heme-bound reactive oxygen within the CYP3A4 active site. *Drug Metabolism and Disposition*, **29** (1), 70–75.

85 Harlow, G.R. and Halpert, J.R. (1998) Analysis of human cytochrome P450 3A4 cooperativity: construction and characterization of a site-directed mutant that displays hyperbolic steroid hydroxylation kinetics. *Proceedings of the National Academy of Sciences of the United States of America*, **95** (12), 6636–6641.

86 Schrag, M.L. and Wienkers, L.C. (2001) Covalent alteration of the CYP3A4 active site: evidence for multiple substrate binding domains. *Archives of Biochemistry and Biophysics*, **391** (1), 49–55.

87 Dabrowski, M.J., Schrag, M.L., Wienkers, L.C. and Atkins, W.M. (2002) Pyrene–pyrene complexes at the active site of cytochrome P450 3 A4: evidence for a multiple substrate binding site. *Journal of the American Chemical Society*, **124** (40), 11866–11867.

88 Shou, M., Grogan, J., Mancewicz, J.A., Krausz, K.W., Gonzalez, F.J., Gelboin, H. V. and Korzekwa, K.R. (1994) Activation of CYP3 A4: evidence for the simultaneous binding of two substrates in a cytochrome P450 active site. *Biochemistry*, **33** (21), 6450–6455.

89 Jushchyshyn, M.I., Hutzler, J.M., Schrag, M.L. and Wienkers, L.C. (2005) Catalytic turnover of pyrene by CYP3 A4: evidence that cytochrome b5 directly induces positive cooperativity. *Archives of Biochemistry and Biophysics*, **438** (1), 21–28.

90 Kenworthy, K.E., Bloomer, J.C., Clarke, S.E. and Houston, J.B. (1999) CYP3A4 drug interactions: correlation of 10 *in vitro* probe substrates. *British Journal of Clinical Pharmacology*, **48** (5), 716–727.

91 Bertz, R.J. and Granneman, G.R. (1997) Use of *in vitro* and *in vivo* data to estimate the likelihood of metabolic pharmacokinetic interactions. *Clinical Pharmacokinetics*, **32** (3), 210–258.

92 Rowland, M., Benet, L.Z. and Graham, G.G. (1973) Clearance concepts in pharmacokinetics. *Journal of Pharmacokinetics and Biopharmaceutics*, **1** (2), 123–136.

93 Wienkers, L.C. and Heath, T.G. (2005) Predicting *in vivo* drug interactions from *in vitro* drug discovery data. *Nature Reviews. Drug Discovery*, **4** (10), 825–833.

94 Di Marco, A., Marcucci, I., Chaudhary, A., Taliani, M. and Laufer, R. (2005) Development and validation of a high-throughput radiometric cyp2c9 inhibition assay using tritiated diclofenac. *Drug Metabolism and Disposition*, **33** (3), 359–364.

95 Di Marco, A., Marcucci, I., Verdirame, M., Perez, J., Sanchez, M., Pelaez, F., Chaudhary, A. and Laufer, R. (2005) Development and validation of a high-throughput radiometric CYP3A4/5 inhibition assay using tritiated testosterone. *Drug Metabolism and Disposition*, **33** (3), 349–358.

96 Delaporte, E., Slaughter, D.E., Egan, M. A., Gatto, G.J., Santos, A., Shelley, J., Price, E., Howells, L., Dean, D.C. and Rodrigues, A.D. (2001) The potential for CYP2D6 inhibition screening using a novel scintillation proximity assay-based approach. *Journal of Biomolecular Screening*, **6** (4), 225–231.

97 Moody, G.C., Griffin, S.J., Mather, A.N., McGinnity, D.F. and Riley, R.J. (1999) Fully automated analysis of activities catalysed by the major human liver cytochrome P450 (CYP) enzymes: assessment of human CYP inhibition potential. *Xenobiotica*, **29** (1), 53–75.

98 Crespi, C.L., Miller, V.P. and Penman, B.W. (1997) Microtiter plate assays for inhibition of human, drug-metabolizing cytochromes P450. *Analytical Biochemistry*, **248** (1), 188–190.

99 Stresser, D.M., Blanchard, A.P., Turner, S.D., Erve, J.C., Dandeneau, A.A., Miller, V.P. and Crespi, C.L. (2000) Substrate-dependent modulation of CYP3A4 catalytic activity: analysis of 27 test compounds with four fluorometric substrates. *Drug Metabolism and Disposition*, **28** (12), 1440–1448.

100 Friden, M., Vanaja, K. and Nandi, V.N. (2006) Drug–drug interactions of anti-infective drugs: utility of fluorescence CYP inhibition assays in drug discovery. *Drug Metabolism and Drug Interactions*, **21** (3–4), 163–185.

101 Trubetskoy, O.V., Gibson, J.R. and Marks, B.D. (2005) Highly miniaturized formats for *in vitro* drug metabolism assays using vivid fluorescent substrates and recombinant human cytochrome P450 enzymes. *Journal of Biomolecular Screening*, **10** (1), 56–66.

102 Cali, J.J., Ma, D., Sobol, M., Simpson, D.J., Frackman, S., Good, T.D., Daily, W.J. and Liu, D. (2006) Luminogenic cytochrome P450 assays. *Expert Opinion on Drug Metabolism and Toxicology*, **2** (4), 629–645.

103 Didenko, V.V. and Hornsby, P.J. (1996) A quantitative luminescence assay for nonradioactive nucleic acid probes. *Journal of the Histochemistry Society*, **44** (6), 657–660.

104 Hummel, M.A., Tracy, T.S., Hutzler, J.M., Wahlstrom, J.L., Zhou, Y. and Rock, D.A. (2006) Influence of fluorescent probe size and cytochrome b5 on drug–drug interactions in CYP2 C9. *Journal of Biomolecular Screening*, **11** (3), 303–309.

105 Turpeinen, M., Korhonen, L.E., Tolonen, A., Uusitalo, J., Juvonen, R., Raunio, H. and Pelkonen, O. (2006) Cytochrome P450 (CYP) inhibition screening: comparison of three tests. *European Journal of Pharmaceutical Sciences*. **29** (2), 130–138.

106 Cohen, L.H., Remley, M.J., Raunig, D. and Vaz, A.D. (2003) *in vitro* drug interactions of cytochrome p450: an evaluation of fluorogenic to conventional substrates. *Drug Metabolism and Disposition*, **31** (8), 1005–1015.

107 Ahmed, S.S., Napoli, K.L. and Strobel, H.W. (1995) Oxygen radical formation during cytochrome P450-catalyzed cyclosporine metabolism in rat and human liver microsomes at varying hydrogen ion concentrations. *Molecular and Cellular Biochemistry*, **151** (2), 131–140.

108 Atkins, W.M. and Sligar, S.G. (1988) Deuterium isotope effects in norcamphor metabolism by cytochrome P-450cam: kinetic evidence for the two-electron reduction of a high-valent iron-oxo intermediate. *Biochemistry*, **27** (5), 1610–1616.

109 Denisov, I.G., Grinkova, Y.V., Baas, B.J. and Sligar, S.G. (2006) The ferrous-dioxygen intermediate in human cytochrome P450 3 A4. Substrate dependence of formation and decay kinetics. *Journal of Biological Chemistry*, **281** (33), 23313–23318.

110 Hutzler, J.M., Wienkers, L.C., Wahlstrom, J.L., Carlson, T.J. and Tracy, T.S. (2003) Activation of cytochrome P450 2C9-mediated metabolism: mechanistic evidence in support of kinetic observations. *Archives of Biochemistry and Biophysics*, **410** (1), 16–24.

111 Turpeinen, M., Uusitalo, J., Jalonen, J. and Pelkonen, O. (2005) Multiple P450 substrates in a single run: rapid and comprehensive *in vitro* interaction assay. *European Journal of Pharmaceutical Sciences*, **24** (1), 123–132.

112 Weaver, R., Graham, K.S., Beattie, I.G. and Riley, R.J. (2003) Cytochrome P450 inhibition using recombinant proteins and mass spectrometry/multiple reaction monitoring technology in a

cassette incubation. *Drug Metabolism and Disposition*, **31** (7), 955–966.

113 Dierks, E.A., Stams, K.R., Lim, H.K., Cornelius, G., Zhang, H. and Ball, S.E. (2001) A method for the simultaneous evaluation of the activities of seven major human drug-metabolizing cytochrome P450s using an *in vitro* cocktail of probe substrates and fast gradient liquid chromatography tandem mass spectrometry. *Drug Metabolism and Disposition*, **29** (1), 23–29.

114 Zimmer, D., Pickard, V., Czembor, W. and Muller, C. (1999) Comparison of turbulent-flow chromatography with automated solid-phase extraction in 96-well plates and liquid–liquid extraction used as plasma sample preparation techniques for liquid chromatography-tandem mass spectrometry. *Journal of Chromatography A*, **854** (1–2), 23–35.

115 Lindqvist, A., Hilke, S. and Skoglund, E. (2004) Generic three-column parallel LC–MS/MS system for high-throughput *in vitro* screens. *Journal of Chromatography A*, **1058** (1–2), 121–126.

116 Briem, S., Pettersson, B. and Skoglund, E. (2005) Description and validation of a four-channel staggered LC–MS/MS system for high-throughput *in vitro* screens. *Analytical Chemistry*, **77** (6), 1905–1910.

117 Walsky, R.L. and Obach, R.S. (2004) Validated assays for human cytochrome P450 activities. *Drug Metabolism and Disposition*, **32** (6), 647–660.

118 Kumar, V., Wahlstrom, J.L., Rock, D.A., Warren, C.J., Gorman, L.A. and Tracy, T.S. (2006) CYP2C9 inhibition: impact of probe selection and pharmacogenetics on *in vitro* inhibition profiles. *Drug Metabolism and Disposition*. **34** (12), 1966–1975.

119 Crespi, C.L. and Miller, V.P. (1997) The R144C change in the CYP2C9*2 allele alters interaction of the cytochrome P450 with NADPH: cytochrome P450 oxidoreductase. *Pharmacogenetics*, **7** (3), 203–210.

120 Li, A.P. (1999) Overview: hepatocytes and cryopreservation – a personal historical perspective. *Chemico-Biological Interactions*, **121** (1), 1–5.

121 Ponsoda, X., Donato, M.T., Perez-Cataldo, G., Gomez-Lechon, M.J. and Castell, J.V. (2004) Drug metabolism by cultured human hepatocytes: how far are we from the *in vivo* reality? *Alternatives to Laboratory Animals*, **32** (2), 101–110.

122 McGinnity, D.F., Tucker, J., Trigg, S. and Riley, R.J. (2005) Prediction of CYP2C9-mediated drug–drug interactions: a comparison using data from recombinant enzymes and human hepatocytes. *Drug Metabolism and Disposition*, **33** (11), 1700–1707.

123 Austin, R.P., Barton, P., Mohmed, S. and Riley, R.J. (2005) The binding of drugs to hepatocytes and its relationship to physicochemical properties. *Drug Metabolism and Disposition*, **33** (3), 419–425.

124 Obach, R.S. (1996) The importance of nonspecific binding in *in vitro* matrices, its impact on enzyme kinetic studies of drug metabolism reactions, and implications for *in vitro–in vivo* correlations. *Drug Metabolism and Disposition*, **24** (10), 1047–1049.

125 Margolis, J.M. and Obach, R.S. (2003) Impact of nonspecific binding to microsomes and phospholipid on the inhibition of cytochrome P4502 D6: implications for relating *in vitro* inhibition data to *in vivo* drug interactions. *Drug Metabolism and Disposition*, **31** (5), 606–611.

126 Tran, T.H., Von Moltke, L.L., Venkatakrishnan, K., Granda, B.W., Gibbs, M.A., Obach, R.S., Harmatz, J.S. and Greenblatt, D.J. (2002) Microsomal protein concentration modifies the apparent inhibitory potency of CYP3A inhibitors. *Drug Metabolism and Disposition*, **30** (12), 1441–1445.

127 Grunau, A., Paine, M.J., Ladbury, J.E. and Gutierrez, A. (2006) Global effects of the energetics of coenzyme binding:

NADPH controls the protein interaction properties of human cytochrome P450 reductase. *Biochemistry*, **45** (5), 1421–1434.

128 Davydov, D.R., Kariakin, A.A., Petushkova, N.A. and Peterson, J.A. (2000) Association of cytochromes P450 with their reductases: opposite sign of the electrostatic interactions in P450BM-3 as compared with the microsomal 2B4 system. *Biochemistry*, **39** (21), 6489–6497.

129 Venkatakrishnan, K., von Moltke, L.L., Court, M.H., Harmatz, J.S., Crespi, C.L. and Greenblatt, D.J. (2000) Comparison between cytochrome P450 (CYP) content and relative activity approaches to scaling from cDNA-expressed CYPs to human liver microsomes: ratios of accessory proteins as sources of discrepancies between the approaches. *Drug Metabolism and Disposition*, **28** (12), 1493–1504.

130 Kumar, V., Rock, D.A., Warren, C.J., Tracy, T.S. and Wahlstrom, J.L. (2006) Enzyme source effects on CYP2C9 kinetics and inhibition. *Drug Metabolism and Disposition*. **34** (11), 1903–1908.

131 Guengerich, F.P. (2005) Reduction of cytochrome b5 by NADPH-cytochrome P450 reductase. *Archives of Biochemistry and Biophysics*, **440** (2), 204–211.

132 Gemzik, B., Halvorson, M.R. and Parkinson, A. (1990) Pronounced and differential effects of ionic strength and pH on testosterone oxidation by membrane-bound and purified forms of rat liver microsomal cytochrome P-450. *Journal of Steroid Biochemistry*, **35** (3–4), 429–440.

133 Schrag, M.L. and Wienkers, L.C. (2000) Topological alteration of the CYP3A4 active site by the divalent cation Mg(2+). *Drug Metabolism and Disposition*, **28** (10), 1198–1201.

134 Chauret, N., Gauthier, A. and Nicoll-Griffith, D.A. (1998) Effect of common organic solvents on *in vitro* cytochrome P450-mediated metabolic activities in human liver microsomes. *Drug Metabolism and Disposition*, **26** (1), 1–4.

135 Hickman, D., Wang, J.P., Wang, Y. and Unadkat, J.D. (1998) Evaluation of the selectivity of *in vitro* probes and suitability of organic solvents for the measurement of human cytochrome P450 monooxygenase activities. *Drug Metabolism and Disposition*, **26** (3), 207–215.

136 Busby, W.F., Jr, Ackermann, J.M. and Crespi, C.L. (1999) Effect of methanol, ethanol, dimethyl sulfoxide, and acetonitrile on *in vitro* activities of cDNA-expressed human cytochromes P-450. *Drug Metabolism and Disposition*, **27** (2), 246–249.

137 Tang, C., Shou, M. and Rodrigues, A.D. (2000) Substrate-dependent effect of acetonitrile on human liver microsomal cytochrome P450 2C9 (CYP2C9) activity. *Drug Metabolism and Disposition*, **28** (5), 567–572.

138 Palamanda, J., Feng, W.W., Lin, C.C. and Nomeir, A.A. (2000) Stimulation of tolbutamide hydroxylation by acetone and acetonitrile in human liver microsomes and in a cytochrome P-450 2C9-reconstituted system. *Drug Metabolism and Disposition*, **28** (1), 38–43.

139 Hollenberg, P.F. (2002) Characteristics and common properties of inhibitors, inducers, and activators of CYP enzymes. *Drug Metabolism Reviews*, **34** (1–2), 17–35.

140 Atkins, W.M., Wang, R.W. and Lu, A.Y. (2001) Allosteric behavior in cytochrome p450-dependent *in vitro* drug–drug interactions: a prospective based on conformational dynamics. *Chemical Research in Toxicology*, **14** (4), 338–347.

141 Wang, R.W., Newton, D.J., Liu, N., Atkins, W.M. and Lu, A.Y. (2000) Human cytochrome P-450 3 A4: *in vitro* drug–drug interaction patterns are substrate-dependent. *Drug Metabolism and Disposition*, **28** (3), 360–366.

142 Hutzler, J.M., Hauer, M.J. and Tracy, T.S. (2001) Dapsone activation of CYP2C9-mediated metabolism:

evidence for activation of multiple substrates and a two-site model. *Drug Metabolism and Disposition*, **29** (7), 1029–1034.

143 Tracy, T.S. (2003) Atypical enzyme kinetics: their effect on *in vitro–in vivo* pharmacokinetic predictions and drug interactions. *Current Drug Metabolism*, **4** (5), 341–346.

144 Korzekwa, K.R., Krishnamachary, N., Shou, M., Ogai, A., Parise, R.A., Rettie, A.E., Gonzalez, F.J. and Tracy, T.S. (1998) Evaluation of atypical cytochrome P450 kinetics with two-substrate models: evidence that multiple substrates can simultaneously bind to cytochrome P450 active sites. *Biochemistry*, **37** (12), 4137–4147.

145 Shou, M., Lin, Y., Lu, P., Tang, C., Mei, Q., Cui, D., Tang, W., Ngui, J.S., Lin, C.C., Singh, R., Wong, B.K., Yergey, J.A., Lin, J.H., Pearson, P.G., Baillie, T.A., Rodrigues, A.D. and Rushmore, T.H. (2001) Enzyme kinetics of cytochrome P450-mediated reactions. *Current Drug Metabolism*, **2** (1), 17–36.

146 Davydov, D.R., Halpert, J.R., Renaud, J.P. and Hui Bon Hoa, G. (2003) Conformational heterogeneity of cytochrome P450 3A4 revealed by high pressure spectroscopy. *Biochemical and Biophysical Research Communications*, **312** (1), 121–130.

147 Hutzler, J.M., Frye, R.F., Korzekwa, K.R., Branch, R.A., Huang, S.M. and Tracy, T.S. (2001) Minimal *in vivo* activation of CYP2C9-mediated flurbiprofen metabolism by dapsone. *European Journal of Pharmaceutical Sciences*, **14** (1), 47–52.

148 Cheng, Y. and Prusoff, W.H. (1973) Relationship between the inhibition constant (K1) and the concentration of inhibitor which causes 50 per cent inhibition (I50) of an enzymatic reaction. *Biochemical Pharmacology*, **22** (23), 3099–3108.

149 Dixon, M. (1952) The determination of enzyme inhibition constants. *Biochemical Journal*, **55**, 170–171.

150 Lineweaver, H. and Burk, D. (1934) The determination of enzyme dissociation constants. *Journal of the American Chemical Society*, **56**, 658–666.

151 Segel, I.H. (1993) *Enzyme Kinetics: Behavior and Analysis of Rapid Equilibrium and Steady-State Enzyme Systems*, Wiley-Interscience New York.

152 Ito, K., Brown, H.S. and Houston, J.B. (2004) Database analyses for the prediction of *in vivo* drug–drug interactions from *in vitro* data. *British Journal of Clinical Pharmacology*, **57** (4), 473–486.

153 Brown, H.S., Ito, K., Galetin, A. and Houston, J.B. (2005) Prediction of *in vivo* drug–drug interactions from *in vitro* data: impact of incorporating parallel pathways of drug elimination and inhibitor absorption rate constant. *British Journal of Clinical Pharmacology*, **60** (5), 508–518.

154 Ito, K., Hallifax, D., Obach, R.S. and Houston, J.B. (2005) Impact of parallel pathways of drug elimination and multiple cytochrome P450 involvement on drug–drug interactions: CYP2D6 paradigm. *Drug Metabolism and Disposition*, **33** (6), 837–844.

155 Rodrigues, A.D. and Rushmore, T.H. (2002) Cytochrome P450 pharmacogenetics in drug development: *in vitro* studies and clinical consequences. *Current Drug Metabolism*, **3** (3), 289–309.

156 Rowland, M. and Matin, S.B. (1973) Kinetics of drug–drug interactions. *Journal of Pharmacokinetics and Biopharmaceutics*, **1**, 553–567.

157 Wang, Y.H., Jones, D.R. and Hall, S.D. (2004) Prediction of cytochrome P450 3A inhibition by verapamil enantiomers and their metabolites. *Drug Metabolism and Disposition*, **32** (2), 259–266.

158 Sesardic, D., Boobis, A.R., Murray, B.P., Murray, S., Segura, J., de la Torre, R. and Davies, D.S. (1990) Furafylline is a potent and selective inhibitor of cytochrome P450IA2 in man. *British*

Journal of Clinical Pharmacology, **29** (6), 651–663.

159 Kunze, K.L. and Trager, W.F. (1993) Isoform-selective mechanism-based inhibition of human cytochrome P450 1A2 by furafylline. *Chemical Research in Toxicology*, **6** (5), 649–656.

160 Tarrus, E., Cami, J., Roberts, D.J., Spickett, R.G., Celdran, E. and Segura, J. (1987) Accumulation of caffeine in healthy volunteers treated with furafylline. *British Journal of Clinical Pharmacology*, **23** (1), 9–18.

161 Silverman, R.B. (1995) Mechanism-based enzyme inactivators. *Methods in Enzymology*, **249**, 240–283.

162 Waley, S.G. (1985) Kinetics of suicide substrates. Practical procedures for determining parameters. *Biochemical Journal*, **227** (3), 843–849.

163 Correia, M.A. (1991) Cytochrome P450 turnover. *Methods in Enzymology*, **206**, 315–325.

164 Waley, S.G. (1980) Kinetics of suicide substrates. *Biochemical Journal*, **185** (3), 771–773.

165 Funaki, T., Takanohashi, Y., Fukazawa, H. and Kuruma, I. (1991) Estimation of kinetic parameters in the inactivation of an enzyme by a suicide substrate. *Biochimica et Biophysica Acta*, **1078** (1), 43–46.

166 Mayhew, B.S., Jones, D.R. and Hall, S.D. (2000) An *in vitro* model for predicting *in vivo* inhibition of cytochrome P450 3A4 by metabolic intermediate complex formation. *Drug Metabolism and Disposition*, **28** (9), 1031–1037.

167 Masubuchi, Y. (2006) Metabolic and non-metabolic factors determining troglitazone hepatotoxicity: a review. *Drug Metabolism and Pharmacokinetics*, **21** (5), 347–356.

168 Hutzler, J.M., Melton, R.J., Rumsey, J.M., Schnute, M.E., Locuson, C.W. and Wienkers, L.C. (2006) Inhibition of cytochrome P450 3A4 by a pyrimidineimidazole: evidence for complex heme interactions. *Chemical Research in Toxicology*, **19** (12), 1650–1659.

169 Ferrero, J.L., Thomas, S.B., Marsh, K.C., Rodrigues, A.D., Uchic, J.T. and Buko, A.M. (2002) Implication of P450–metabolite complex formation in the nonlinear pharmacokinetics and metabolic fate of $(\pm)$-$(1'R*,3R*)$-3-phenyl-1-[$(1',2',3',4'$-tetrahydro-5',6'-methylene-dioxy-1'-naphthalenyl) methyl] pyrrolidine methanesulfonate (ABT-200) in dogs. *Drug Metabolism and Disposition*, **30** (10), 1094–1101.

170 Schalk, M., Cabello-Hurtado, F., Pierrel, M.A., Atanossova, R., Saindrenan, P. and Werck-Reichhart, D. (1998) Piperonylic acid, a selective, mechanism-based inactivator of the *trans*-cinnamate 4-hydroxylase: a new tool to control the flux of metabolites in the phenylpropanoid pathway. *Plant Physiology*, **118** (1), 209–218.

171 Ma, B., Prueksaritanont, T. and Lin, J.H. (2000) Drug interactions with calcium channel blockers: possible involvement of metabolite–intermediate complexation with CYP3A. *Drug Metabolism and Disposition*, **28** (2), 125–130.

172 Kalgutkar, A.S., Gardner, I., Obach, R.S., Shaffer, C.L., Callegari, E., Henne, K.R., Mutlib, A.E., Dalvie, D.K., Lee, J.S., Nakai, Y., O'Donnell, J.P., Boer, J. and Harriman, S.P. (2005) A comprehensive listing of bioactivation pathways of organic functional groups. *Current Drug Metabolism*, **6** (3), 161–225.

173 Zhou, S., Chan, E., Duan, W., Huang, M. and Chen, Y.Z. (2005) Drug bioactivation, covalent binding to target proteins and toxicity relevance. *Drug Metabolism Reviews*, **37** (1), 41–213.

174 Kent, U.M., Lin, H.L., Mills, D.E., Regal, K.A. and Hollenberg, P.F. (2006) Identification of 17-alpha-ethynylestradiol-modified active site peptides and glutathione conjugates

formed during metabolism and inactivation of P450s 2B1 and 2 B6. *Chemical Research in Toxicology*, **19** (2), 279–287.

175 Graham, D.J., Green, L., Senior, J.R. and Nourjah, P. (2003) Troglitazone-induced liver failure: a case study. *The American Journal of Medicine*, **114** (4), 299–306.

176 Kassahun, K., Pearson, P.G., Tang, W., McIntosh, I., Leung, K., Elmore, C., Dean, D., Wang, R., Doss, G. and Baillie, T.A. (2001) Studies on the metabolism of troglitazone to reactive intermediates *in vitro* and *in vivo*. Evidence for novel biotransformation pathways involving quinone methide formation and thiazolidinedione ring scission. *Chemical Research in Toxicology*, **14** (1), 62–70.

177 Yamazaki, H., Suzuki, M., Tane, K., Shimada, N., Nakajima, M. and Yokoi, T. (2000) *in vitro* inhibitory effects of troglitazone and its metabolites on drug oxidation activities of human cytochrome P450 enzymes: comparison with pioglitazone and rosiglitazone. *Xenobiotica*, **30** (1), 61–70.

178 Sahi, J., Black, C.B., Hamilton, G.A., Zheng, X., Jolley, S., Rose, K.A., Gilbert, D., LeCluyse, E.L. and Sinz, M.W. (2003) Comparative effects of thiazolidinediones on *in vitro* P450 enzyme induction and inhibition. *Drug Metabolism and Disposition*, **31** (4), 439–446.

179 Lim, H.K., Duczak, N., Jr, Brougham, L., Elliot, M., Patel, K. and Chan, K. (2005) Automated screening with confirmation of mechanism-based inactivation of CYP3A4, CYP2C9, CYP2C19, CYP2D6, and CYP1A2 in pooled human liver microsomes. *Drug Metabolism and Disposition*, **33** (8), 1211–1219.

180 Chen, Q., Ngui, J.S., Doss, G.A., Wang, R.W., Cai, X., DiNinno, F.P., Blizzard, T.A., Hammond, M.L., Stearns, R.A., Evans, D.C., Baillie, T.A. and Tang, W. (2002) Cytochrome P450 3A4-mediated bioactivation of raloxifene: irreversible enzyme inhibition and thiol adduct formation. *Chemical Research in Toxicology*, **15** (7), 907–914.

181 Polasek, T.M., Elliot, D.J., Somogyi, A.A., Gillam, E.M., Lewis, B.C. and Miners, J.O. (2006) An evaluation of potential mechanism-based inactivation of human drug metabolizing cytochromes P450 by monoamine oxidase inhibitors, including isoniazid. *British Journal of Clinical Pharmacology*, **61** (5), 570–584.

182 Atkinson, A., Kenny, J.R. and Grime, K. (2005) Automated assessment of time-dependent inhibition of human cytochrome P450 enzymes using liquid chromatography-tandem mass spectrometry analysis. *Drug Metabolism and Disposition*, **33** (11), 1637–1647.

183 Obach, R.S., Walsky, R.L. and Venkatakrishnan, K. (2007) Mechanism-based inactivation of human cytochrome p450 enzymes and the prediction of drug–drug interactions. *Drug Metabolism and Disposition*, **35** (2), 246–255.

184 Yang, J., Jamei, M., Yeo, K.R., Tucker, G.T. and Rostami-Hodjegan, A. (2005) Kinetic values for mechanism-based enzyme inhibition: assessing the bias introduced by the conventional experimental protocol. *European Journal of Pharmaceutical Sciences*, **26** (3–4), 334–340.

185 Kitz, R. and Wilson, I.B. (1962) Esters of methanesulfonic acid as irreversible inhibitors of acetylcholinesterase. *Journal of Biological Chemistry*, **237**, 3245–3249.

186 Proctor, N.J., Tucker, G.T. and Rostami-Hodjegan, A. (2004) Predicting drug clearance from recombinantly expressed CYPs: intersystem extrapolation factors. *Xenobiotica*, **34** (2), 151–178.

187 Kemp, D.C., Fan, P.W. and Stevens, J.C. (2002) Characterization of raloxifene glucuronidation *in vitro*: contribution of intestinal metabolism to presystemic

clearance. *Drug Metabolism and Disposition*, **30** (6), 694–700.

188 McGinnity, D.F., Berry, A.J., Kenny, J.R., Grime, K. and Riley, R.J. (2006) Evaluation of time-dependent cytochrome P450 inhibition using cultured human hepatocytes. *Drug Metabolism and Disposition*, **34** (8), 1291–1300.

189 Kent, U.M., Bend, J.R., Chamberlin, B.A., Gage, D.A. and Hollenberg, P.F. (1997) Mechanism-based inactivation of cytochrome P450 2B1 by N-benzyl-1-aminobenzotriazole. *Chemical Research in Toxicology*, **10** (5), 600–608.

190 Ernest, C.S., II, Hall, S.D. and Jones, D.R. (2005) Mechanism-based inactivation of CYP3A by HIV protease inhibitors. *Journal of Pharmacology and Experimental Therapeutics*, **312** (2), 583–591.

191 Jushchyshyn, M.I., Kent, U.M. and Hollenberg, P.F. (2003) The mechanism-based inactivation of human cytochrome P450 2B6 by phencyclidine. *Drug Metabolism and Disposition*, **31** (1), 46–52.

192 Guengerich, F.P. (1990) Mechanism-based inactivation of human liver microsomal cytochrome P-450 IIIA4 by gestodene. *Chemical Research in Toxicology*, **3** (4), 363–371.

193 Voorman, R.L., Maio, S.M., Payne, N.A., Zhao, Z., Koeplinger, K.A. and Wang, X. (1998) Microsomal metabolism of delavirdine: evidence for mechanism-based inactivation of human cytochrome P450 3A. *Journal of Pharmacology and Experimental Therapeutics*, **287** (1), 381–388.

194 Bateman, K.P., Baker, J., Wilke, M., Lee, J., Leriche, T., Seto, C., Day, S., Chauret, N., Ouellet, M. and Nicoll-Griffith, D.A. (2004) Detection of covalent adducts to cytochrome P450 3A4 using liquid chromatography mass spectrometry. *Chemical Research in Toxicology*, **17** (10), 1356–1361.

195 Regal, K.A., Schrag, M.L., Kent, U.M., Wienkers, L.C. and Hollenberg, P.F. (2000) Mechanism-based inactivation of cytochrome P450 2B1 by 7-ethynylcoumarin: verification of apo-P450 adduction by electrospray ion trap mass spectrometry. *Chemical Research in Toxicology*, **13** (4), 262–270.

196 Koenigs, L.L., Peter, R.M., Hunter, A.P., Haining, R.L., Rettie, A.E., Friedberg, T., Pritchard, M.P., Shou, M., Rushmore, T.H. and Trager, W.F. (1999) Electrospray ionization mass spectrometric analysis of intact cytochrome P450: identification of tienilic acid adducts to P450 2C9. *Biochemistry*, **38** (8), 2312–2319.

197 Omura, T. and Sato, R. (1964) The carbon monoxide-binding pigment of liver microsomes. I. Evidence for its hemoprotein nature. *Journal of Biological Chemistry*, **239**, 2370–2378.

198 Hutzler, J.M., Steenwyk, R.C., Smith, E.B., Walker, G.S. and Wienkers, L.C. (2004) Mechanism-based inactivation of cytochrome P450 2D6 by 1-[(2-ethyl-4-methyl-1H-imidazol-5-yl)methyl]-4-[4-(trifluoromethyl)-2-pyridinyl]piperazine: kinetic characterization and evidence for apoprotein adduction. *Chemical Research in Toxicology*, **17** (2), 174–184.

199 Zheng, Y.M., Baer, B.R., Kneller, M.B., Henne, K.R., Kunze, K.L. and Rettie, A.E. (2003) Covalent heme binding to CYP4B1 via Glu310 and a carbocation porphyrin intermediate. *Biochemistry*, **42** (15), 4601–4606.

200 Ortiz de Montellano, P.R. and Mathews, J.M. (1981) Autocatalytic alkylation of the cytochrome P-450 prosthetic haem group by 1-aminobenzotriazole. Isolation of an NN-bridged benzyne-protoporphyrin IX adduct. *Biochemical Journal*, **195** (3), 761–764.

201 He, K., Falick, A.M., Chen, B., Nilsson, F. and Correia, M.A. (1996) Identification of the heme adduct and an active site peptide modified during mechanism-based inactivation of rat liver cytochrome P450 2B1 by secobarbital. *Chemical Research in Toxicology*, **9** (3), 614–622.

202 Lin, H.L. and Hollenberg, P.F. (2007) The inactivation of cytochrome P450 3A5

by 17{alpha}-ethynylestradiol is cytochrome b5 dependent: metabolic activation of the ethynyl moiety leads to the formation of glutathione conjugates, a heme adduct and covalent binding to the apoprotein. *Journal of Pharmacology and Experimental Therapeutics*.**321** (1), 276–287.

203 Wienkers, L.C. (2002) Factors confounding the successful extrapolation of *in vitro* CYP3A inhibition information to the *in vivo* condition. *European Journal of Pharmaceutical Sciences*, **15** (3), 239–242.

204 Ingelman-Sundberg, M. (2001) Implications of polymorphic cytochrome p450-dependent drug metabolism for drug development. *Drug Metabolism and Disposition*, **29** (4 Pt 2), 570–573.

205 Dickmann, L.J., Rettie, A.E., Kneller, M.B., Kim, R.B., Wood, A.J., Stein, C.M., Wilkinson, G.R. and Schwarz, U.I. (2001) Identification and functional characterization of a new CYP2C9 variant (CYP2C9*5) expressed among African Americans. *Molecular Pharmacology*, **60** (2), 382–387.

206 Vermilion, J.L. and Coon, M.J. (1978) Purified liver microsomal NADPH-cytochrome P-450 reductase. Spectral characterization of oxidation–reduction states. *Journal of Biological Chemistry*, **253** (8), 2694–2704.

207 Sligar, S.G. (1976) Coupling of spin, substrate, and redox equilibria in cytochrome P450. *Biochemistry*, **15** (24), 5399–5406.

208 Poulos, T.L. (1996) Ligands and electrons and haem proteins. *Nature Structural Biology*, **3** (5), 401–403.

209 Bhattacharyya, A.K., Lipka, J.J., Waskell, L. and Tollin, G. (1991) Laser flash photolysis studies of the reduction kinetics of NADPH: cytochrome P-450 reductase. *Biochemistry*, **30** (3), 759–765.

210 Oprian, D.D., Gorsky, L.D. and Coon, M.J. (1983) Properties of the oxygenated form of liver microsomal cytochrome P-450. *Journal of Biological Chemistry*, **258** (14), 8684–8691.

211 Estabrook, R.W., Hildebrandt, A.G., Baron, J., Netter, K.J. and Leibman, K. (1971) A new spectral intermediate associated with cytochrome P-450 function in liver microsomes. *Biochemical and Biophysical Research Communications*, **42** (1), 132–139.

212 Pompon, D. and Coon, M.J. (1984) On the mechanism of action of cytochrome P-450. Oxidation and reduction of the ferrous dioxygen complex of liver microsomal cytochrome P-450 by cytochrome b5. *Journal of Biological Chemistry*, **259** (24), 15377–15385.

213 Hildebrandt, A. and Estabrook, R.W. (1971) Evidence for the participation of cytochrome b 5 in hepatic microsomal mixed-function oxidation reactions. *Archives of Biochemistry and Biophysics*, **143** (1), 66–79.

214 Karuzina, I.I. and Archakov, A.I. (1994) Hydrogen peroxide-mediated inactivation of microsomal cytochrome P450 during monooxygenase reactions. *Free Radical Biology and Medicine*, **17** (6), 557–567.

215 Shaik, S., Kumar, D., de Visser, S.P., Altun, A. and Thiel, W. (2005) Theoretical perspective on the structure and mechanism of cytochrome P450 enzymes. *Chemical Reviews*, **105** (6), 2279–2328.

216 Groves, J.T. and McClusky, G.A. (1978) Aliphatic hydroxylation by highly purified liver microsomal cytochrome P-450. Evidence for a carbon radical intermediate. *Biochemical and Biophysical Research Communications*, **81** (1), 154–160.

217 Augusto, O., Beilan, H.S. and Ortiz de Montellano, P.R. (1982) The catalytic mechanism of cytochrome P-450. Spin-trapping evidence for one-electron substrate oxidation. *Journal of Biological Chemistry*, **257** (19), 11288–11295.

218 Guengerich, F.P. (1990) Enzymatic oxidation of xenobiotic chemicals. *Critical Reviews in Biochemistry and Molecular Biology,* **25** (2), 97–153.

219 Kumar, V., Rock, D.A., Warren, C.J., Tracy, T.S. and Wahlstrom, J.L. (2006) Enzyme source effects on CYP2C9 kinetics and inhibition. *Drug Metabolism and Disposition,* **34** (11), 1903–1908.

220 Schrag, M.L. and Wienkers, L.C. (2001) Triazolam substrate inhibition: evidence of competition for heme-bound reactive oxygen within the CYP3A4 active site. *Advances in Experimental Medicine and Biology,* **500**, 347–350.

10
Site of Metabolism Predictions: Facts and Experiences
Ismael Zamora

10.1
Introduction

Drug metabolism has been recognized as one of the key factors in the discovery of new chemical entities. A lead compound needs to not only interact with the target enzyme/receptor but also remain over a certain threshold concentration at the site of action for a defined period to produce the desired therapeutic effect. Drug metabolism together with absorption, distribution and excretion are among the factors that influence the final time–concentration relationship of drugs and therefore the potential efficacy of the compound [1].

There are several aspects of drug metabolism that need to be considered for the optimization of a lead compound to a drug candidate, for example each enzyme system responsible for a metabolic reaction, the turnover rate of the transformation (V_{max}), the affinity of the compound to the enzyme system (K_M), the structure of the compound formed in the metabolic reaction and the inherent properties of the metabolite formed such as its chemical reactivity. There are several enzyme systems responsible for xenobiotic transformation. Among them, the cytochrome P450 (CYP) superfamily of enzymes plays a fundamental role in phase I oxidative metabolism, most of the transformations being due to a subset of six human CYPs: CYP1A2, CYP2C9, CYP2C19, CYP2D6, CYP2E1 and CYP3A4 [2]. This enzyme family generally produces a chemical oxidation by a mechanism that involves the interaction of the compound with an iron-containing porphyrin system called heme. The mechanism for this interaction still remains unclear, though a hypothetical one has been proposed as shown in Scheme 10.1 [3].

The reaction hypothetically occurs in several steps:
(1) compound binding in the enzyme cavity;
(2) molecular oxygen diffusion to bind to the iron atom;
(3) one electron transfer to produce a water molecule and a hydroxy–ferryl species;
(4) further electron transfer and oxidation of the compound;
(5) exit of the compound from the enzyme cavity.

Antitargets. Edited by R. J. Vaz and T. Klabunde
Copyright © 2008 WILEY-VCH Verlag GmbH & Co. KGaA, Weinheim
ISBN: 978-3-527-31821-6

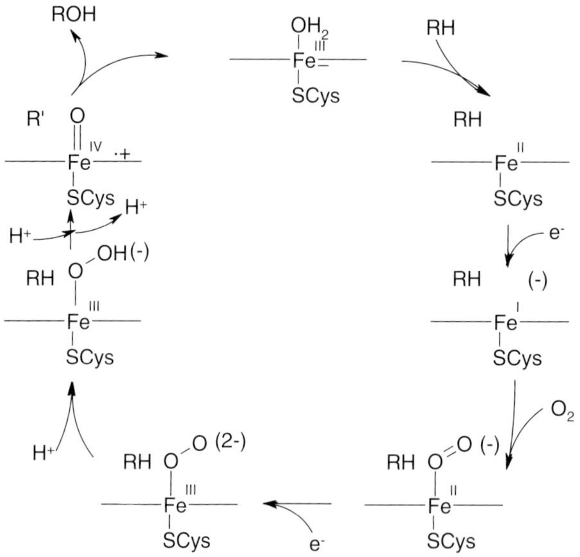

Scheme 10.1 Proposed mechanism of CYP-mediated oxidative process.

Several kinetic parameters can be measured on different experimental systems to account for the interaction of a compound with CYPs. For example when studying the metabolic stability of a compound, it could be measured in a recombinant CYP system, in human liver microsomes, in hepatocytes and so on. Each system increases in biological complexity. Although in the recombinant CYP system only the cytochrome under consideration is studied, in the case of the human liver microsomes, there is a pool of enzyme present that includes several CYPs, and finally in the hepatocyte cell system, metabolizing enzymes play an important role in the metabolic compound stability. In addition, transport systems are also present that could involve recirculation or other transport phenomena. The more complex the experimental system, the more difficult it is to extract information on the protein/ligand interaction, albeit it is closer to the *in vivo* real situation and therefore to the mechanism that is actually working in the body.

10.2
Factors That Influence the Site of Metabolism Prediction by Cytochrome P450s

There are at least two factors that could influence the turnover rate, the site of metabolism (hot spot) and the affinity of a compound toward these enzymes: the protein/ligand (substrate or inhibitor) interaction and the chemical reactivity of the compound towards oxidation. Because of the interaction of the protein with the potential ligand, certain atoms of the compound could be exposed to the heme group, and depending on the chemical nature of these moieties the oxidative reaction will take place at different rates, for example celecoxib is metabolized by CYP2C9 at the benzylic position as the cytochrome can orient this group toward the heme and also because the methyl group is reactive toward oxidation. When this chemical moiety is modified to a trifluoromethyl group, a more stable moiety toward chemical oxidation, it is observed that the compound is no longer a substrate for CYP2C9. Nevertheless, the celecoxib derivative keeps its interactions with the protein and therefore becomes an inhibitor of the enzyme. In this case, the CYP450 orients a chemical group toward the heme that is not able to be oxidized under these conditions, and therefore the compound is stable [4].

10.2.1
Chemical Reactivity

The most common reactions mediated by the CYPs involve aliphatic and aromatic hydroxylation, epoxidation, heteroatom oxygenation or dealkylation, oxidative group transfer, cleavage of esters and dehydrogenation [5]. These reactions have different chemical mechanisms and therefore a general description of the chemical reactivity is difficult to achieve; instead a specific description would be needed for each reaction type. Moreover, different oxidant species may also be involved in each chemical reaction [6]. In the case of aliphatic hydroxylation, the mechanism proposed (rebound mechanism) requires an initial hydrogen atom abstraction and radical formation. Therefore, the analysis of the relative stability of the different potential radicals formed during the process could help in ranking the different atoms in the molecule, indicating the atom with the lowest energy for radical formation as the probable site for oxidation [7].

The theory predicting the stability of the radical formed assumes that the rate-limiting step is the extraction of the hydrogen atom to form a radical, and this hypothesis is in principle valid for aliphatic hydroxylation, but it might not be the case for other reactions. Several examples in the literature show that this method is useful for the prediction of the site of metabolism for compounds undergoing metabolism by CYP3A4 [8]. Nevertheless, there is a lack of a theory that could explain all the different metabolism reactions and mechanisms that may or may not involve radical formation.

Another approach is to consider the bond orders to the different possible sites of metabolism. The bond order is defined as the degree of bonding between two atoms relative to that of a normal single bond, that is the bond provided by one localized electron pair, which has been normalized for different bonds between standard atoms for reactions involving oxygen. The hydrogen atoms have been ranked according to their normalized reactivity, and a penalty term is taken to be proportional to the solvent accessible surface of each hydrogen atom [9].

10.2.2
Protein/Compound Interaction

In general, the CYP enzymes have low substrate specificity compared to other protein systems in the human body. General rules have been described in the literature for a compound to be a substrate of the different CYPs, for example substrates for CYP2D6 are generally positively charged compounds at physiological pH with a tertiary nitrogen at 5–7 Å from a lipophilic moiety, or even more general descriptions such as lipophilic compounds with a $\log D > 3$ are generally metabolized by CYP3A4 [10]. Nevertheless, the different enzymes have distinct substrate specificity, which means that the protein in fact has an effect on the orientation of the compound in the CYP cavity and therefore on the site and turnover rate of the oxidative reaction. Nowadays, several mammalian and human CYP450 crystal structures are available in the Brookhaven Protein Data Bank (PDB). For mammalian CYPs, two different isoforms have been crystallized, CYP2C5 (1dt6 [11], 1n6b [12] and 1nr6 [13]) and CYP2B4 (1po5 [14], 1suo [15] and 2bdm [16]), and for human CYPs six different isoforms have been crystallized, CYP1A2 (1hi4 [17]), CYP2A6 (1z10 [18]), CYP2C9 (1r9o [19], 1og5 and 1og2 [20]), CYP2C8 (1pq2 [21]), CYP2D6 (2f9q [22]) and CYP3A4 (1w0e, 1w0f and 1w0g [23], 1tqn [24], 2j0d, and 2j0c [25]). This structural information could help to elucidate the structural elements relevant for CYP/ligand interaction. Nevertheless, some of these crystal structures show that these enzymes have a high degree of flexibility and can accommodate multiple copies of substrates or inhibitors in quite different structural conformations or binding modes. Moreover, the enzymes can present different conformations leading to productive binding modes that produce a metabolite, or alternatively they could adopt nonproductive binding modes that would not yield any chemical transformation of the compound. Despite the number of crystal structures available to date, they probably do not exhibit exhaustive information because, for example, conformational freedom may be limited by crystal packing, substrate characteristics and crystallization conditions. Therefore, homology models might still be of importance.

10.3
Methods to Predict the Site of Metabolism

Several methods have been published to predict the site of metabolism (hot spots). All use one or several of the different factors described above. It is important to distinguish between methods that consider one specific human CYP isoform and those that consider all the metabolic transformations that a compound may undergo in different biological systems. In the isoform-specific case, one could distinguish between the methods that consider only the protein structure, such as docking into a crystal structure or homology model, and those that consider only the chemical reactivity (radical stability/bond order). Certain methods such as MetaSite, described in Chapter 12 of this book, consider both the chemical reactivity and the protein structure at the same time. In the case of unspecific prediction of biotransformation pathways, several methods have been published that perform statistical analysis of the presence or absence of fragments in databases of compounds with known metabolic pathways. On the basis of the fragment frequency, these methods assign a probability to the atom to be metabolized. Here, we will describe briefly the different methods used.

10.3.1
Knowledge-Based Methods

Knowledge-based methods are those based on the application of certain rules to describe the metabolism. These rules could be defined as chemical reactions relating structure and biotransformations to predict the metabolic fate of a query chemical structure, as in the Meteor approach [26], or alternatively they could be obtained by fragment analysis of a metabolic database as performed in the SPORCalc (Substrate Product Occurrence Ratio Calculator) system [27].

The Meteor system contains a dictionary of approximately 200 general reaction types. The reasoning engine contains two types of rules [28]:

(a) The absolute ones that evaluate the probability of a metabolic transformation on five levels: probable, plausible, equivocal, doubtful and improbable.
(b) The relative ones that assign priorities to potentially competing reactions (e.g. primary alcohols are oxidized in preference to secondary alcohols).

Another approach to knowledge-based metabolic predictions is SPORCalc [29]. This is a general procedure for interrogating reaction databases that could be applied to biochemical transformation as produced in the metabolism of xenobiotics. The method aims to discriminate between the most likely substructures in diverse substrates and products, comparing normal distributions of substrate and products in a given enzyme reaction class compared to an entire database of substrates, intermediates and products. From this fragmental analysis the methodology generates individual occurrence ratios for a given enzyme reaction that determine the probability of a certain reaction to occur at a given atom in a query molecule.

10.3.2
From Protein Structure to Chemical Reactivity

10.3.2.1 Docking
Docking, based on different algorithms, is a method to predict the site of metabolism, totally dependent on the structure of the protein and the orientation of the compounds in the enzyme cavity. Zhou et al. [30] used the docking software GLUE distributed in the GRID package [31] to predict the site of metabolism for a set of known CYP3A4 substrates. In the GLUE procedure, the active site is mapped using several probes to analyze the hydrophobic, hydrogen bond donor/acceptor and electrostatic interactions. From the GRID minima points, several pharmacophores are defined and used as templates to compare with the ligand. Moreover, the procedure identifies the polar and hydrophobic heavy atoms of the ligand and computes several pharmacophores from the atomic positions. Then, the pharmacophores in the ligand are compared to the ones defined in the interaction fields aligning the ligand in the enzyme cavity, and an energy computation follows. If there are any conflicting contacts between the ligand and

the protein, an induced fit process is started to accommodate the substrate in the protein cavity.

A similar procedure has been used in the case of CYP2D6 analysis [32]. In this particular case, the molecules under analysis were studied to understand whether they should be in a protonated state or not considering the pK_a of the compound, because the electrostatic interactions between the basic nitrogen in the ligands used and the negatively charged Glu216 on top of the active site in CYP2D6 are thought to be important for the orientation of this set of compounds.

Other docking algorithms such as Glide [33], Dock [34] and GOLD [35] have also been used to predict the site of metabolism, and a recent study compared the performance of multiple algorithms in the field of cytochrome P450s [36].

10.3.2.2 MetaSite

The MetaSite methodology [37,38] considers structural complementarity between the active site of the enzyme and the ligand as well as reactivity of the ligand to predict the site of metabolism. Two sets of descriptors are calculated to compare the enzyme and ligand. For the enzyme, these are based on flexible GRID molecular interaction fields (GRID-MIFs). To generate the descriptor set of the ligand, each atom of the molecule is classified as a GRID probe and the distances between them are calculated. The resulting fingerprint of the ligand is compared with the description of the enzyme active site and the most optimal orientations are obtained. On the basis of this, all atoms in the molecule are ranked according to their relative accessibility to the heme. In addition to this, a reactivity factor based on fragment recognition is added. In summary, the most likely site of metabolism is described by a probability index that is the product of the similarity between the ligand and protein (i.e. accessibility of one atom to the heme) and the reactivity. This method is thoroughly discussed in Chapter 12 of this book.

The protein factor in MetaSite is based on homology models of the CYP450s, but it offers the possibility for the user to introduce any CYP450 structure. Several analyses have been done comparing the results obtained using the crystal structures and the homology models of the cytochromes.

10.3.2.3 Quantum Chemistry

These methods assume that the site of metabolism is largely dependent on the electronic environment of the hydrogen atoms. Moreover, in the methods that rely on hydrogen abstraction, it is assumed that the rate-limiting step is an abstraction of the proton. Singh *et al.* [8] computed the energy for a given hydrogen atom as the difference between the heat of formation of the native substrate and that of its radical. The heat of formation is computed by the authors using the AM1 method. As this computation for each hydrogen atom as a potential site of metabolism in a molecule would be time consuming, the authors proposed the use of a QSAR type of approach for the estimation of the heat of formation. Alternatively, when the positively charged radical is formed, the site of metabolism will be where the spin hole is located in the positively charged radical [9]. The spin distribution is calculated at the B3LYP/3-21G level of theory and the magnitude of the Fermi contact term is taken as a measure for spin density on each hydrogen atom. The Fermi contact value is used as the basis for the ranking in combination with the steric constraint described above. The single-point molecular orbital calculations are done at the B3LYP/3-21G level of theory in the Gaussian03 package [39].

Another way to compute the reactivity of each hydrogen atom assumes that aliphatic hydroxylation occurs on the weakest bonded hydrogen in the molecule [9]. The bond strength is calculated from the sum of atom pair bond orders in a molecular orbital calculation. The atom pair bond orders are calculated from the wave function at the B3LYP/3-21G level of theory in this work. The atom pair bond orders are normalized with respect to a standard value calculated from model systems and the bond strength is calculated to be proportional to the deviation from the average value. A steric hindrance restriction on the reactivity is also applied. Thus, if a hydrogen atom is deeply buried, a penalty to the reactivity is applied. The penalty term is taken to be proportional to a function that is dependent on the solvent accessible surface of each hydrogen atom. Thus, the normalized and scaled reactivity is used to rank all the hydrogens in the molecule with respect to the likelihood of being the site of metabolism.

10.4
The Influence of the Protein Structure on the Site of Metabolism

10.4.1
Comparative Analysis Between the Different CYP Crystal Structures

There are several structures of mammalian (CYP2C5, CYP2B4) and human cytochromes (CYP2C9, CYP2D6 and CYP3A4) that have been crystallized with different ligands (Scheme 10.2). A comparative analysis of these crystal structures can provide an idea about the flexibility of these types of structures and perhaps partially explain the broad substrate specificity of these enzymes. Moreover, for example in CYP2D6, where there is only one protein structure deposited in the Protein Data Bank, there are homology models that can also be compared to the crystal structures.

Cytochrome P450s are membrane-bound proteins. To crystallize these enzymes, the N-terminal amino acids that are inserted in the membrane are cleaved from the protein sequence. These enzymes contain a heme group in the catalytic site, responsible for the oxidation of the substrate. The iron atom in this group binds with a sulfur atom of a cysteine amino acid. These enzymes generally have one helix-rich and one β-sheet regions. The heme group is typically sandwiched between helix L and helix I. The protein/ligand interaction usually involves the secondary elements: B–C and the F–G loops, helix F and helix G (Figure 10.1).

10.4.1.1 CYP1A2
The CYP1A2 crystal structure (2hi4) has the enzyme bound to 2-phenyl-4H-benzo[h]chromen-4-one (11), a flat CYP1A2 inhibitor: (a) this compound binds with the naphthyl group in a hydrophobic pocket composed of the F226, F260, F125, F256,

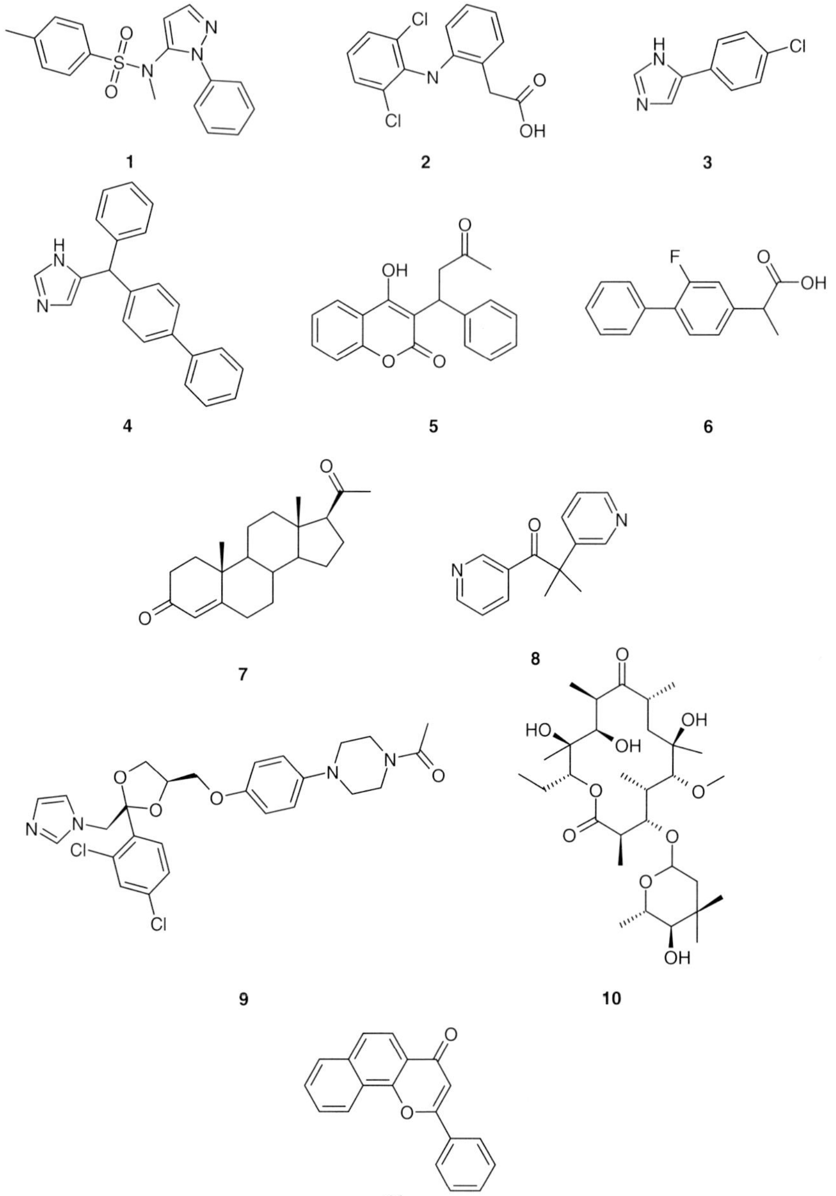

Scheme 10.2 Compounds that have been crystallized with different CYPs.

G316 amino acids; (b) the carbonyl group binds to a water molecule and a T223; (c) the phenyl ring interacts with the A317, L497, I386, T321 and T498. The presence of several threonine residues allows modulation of the hydrophobic nature of the cavity, depending on the side chain orientation towards the compound (methyl or hydroxyl

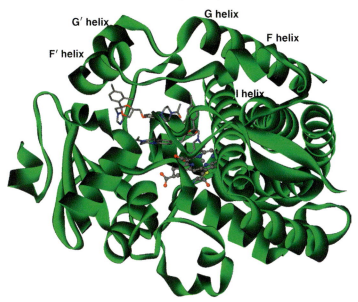

Figure 10.1 A ribbon diagram showing the I, F, G, F' and G' helices in the CYP3A4 structure stored with 2 copies of the Retoconazole molecule.

group). Typical CYP1A2 substrates are flat and hydrophobic, like kinase inhibitors [40].

10.4.1.2 CYP2C5

The mammalian CYP2C5 was the first CYP crystallized, and there are three crystal structures of the mammalian CYP2C5 reported in the Protein Data Bank: the protein without any ligand (1dt6), cocrystalized with a derivative of sulfafenazole (4-methyl-N-methyl-N-(2-phenyl-2H-pyrazol-3-yl) benzenesulfonamide (**1**), DMZ) (1n6b) and with diclofenac (**2**) (1nr6). As discussed by Wester *et al.* [13], the overall conformation of the proteins are very similar, and the root-mean-square deviation (rmsd) values for the α-carbon and the backbone are around 1 Å (Table 10.1). Nevertheless, there are several conformational changes upon the binding of the substrates; the protein seems to adopt a conformation that can accommodate the substrate with the best

Table 10.1 Structural comparison of α-carbons and backbone atoms of CYP2C5 from different crystal structures (rmsd values are given in Å).

α-Carbon	1dt6	1n6b	1nr6	Backbone	1dt6	1n6b	1nr6
1dt6	0	1.26	1.02	1dt6	0	1.28	1.05
1n6b		0	1.1	1n6b		0	1.11
1nr6			0	1nr6			0

interaction. In the case of the DMZ ligand, the B–C loop adopts a conformation to optimize the hydrophobic interaction with the ligand moving the B' helix to the G helix, closing the possible access channel for the ligand as proposed for the protein without the ligand. Moreover, a second solvent channel described in 1dt6, seems also to be closed upon ligand binding by the movement of helix F toward helix I, which is also reflected in the F–G loop. In the case of diclofenac, a smaller and negatively charged ligand, the authors described that the B–C and the F–G loops adopt a different conformation compared to 1dt6 and 1n6b to accommodate the compound and the extensive water network.

10.4.1.3 CYP2B4

CYP2B4 has not been reported as a major enzyme in the metabolism of xenobiotics, but it has been frequently used as a model for these membrane proteins. There are three crystal structures reported in the Protein Data Bank: without any ligand (1po5), with a selective inhibitor (4-(4-chloro-phenyl) imidazole (3) (CPI, 1suo) and with a broad spectrum antifungal agent (bifonazole (4), 2dbm). The degree of plasticity of CYP2B4 is exceptional, as can be observed from the rmsd values for the α-carbon of the protein backbones (Table 10.2). The structure of the protein without any ligand crystallizes as a dimer, where a histidine of one monomer interacts with the iron atom of the heme from the other monomer, producing a structure where the secondary elements are conserved, but the orientation between these elements are very different compared to the other known mammalian CYP structures. The structure reveals an open cleft that extends from the protein surface to the heme. The binding of the small selective inhibitor CPI directly to the iron atom yields a type II UV spectrum, indicating a direct coordination of the ligand to the iron atom in the heme group. In contrast to the structure without any ligand, the complex adopted a closed form similar to the CYP2C structures. The secondary elements are ordered in such a way that they close the cleft to optimize the interaction of the protein with the small ligand. As seen in the 2dbm structure, bifonazole binds as a type II inhibitor, and the secondary elements are maintained to accommodate this compound in the proximity of the iron atom.

10.4.1.4 CYP2C9

CYP2C9 was the first human cytochrome P450 that was crystallized on the basis of the CYP2C5 experience. Nevertheless, the construct in this initial crystal structure

Table 10.2 Structural comparison of α-carbons and backbone atoms of CYP2B4 from different crystal structures aligned by heme group (rmsd values are given in Å).

α-Carbon	1po5	1suo	2dbm	Backbone	1po5	1suo	2dbm
1po5	0	3.86	4.14	1po5	0	3.86	4.12
1suo		0	4.71	1suo		0	4.68
2dbm			0	2dbm			0

10.4 The Influence of the Protein Structure on the Site of Metabolism

Table 10.3 Structural comparison of α-carbons and backbone atoms of CYP2C9 from different crystal structures aligned by heme group (rmsd values are given in Å).

α-Carbon	1og2	1og5	1r9o	Backbone	1og2	1og5	1r9o
1og2	0	0.45	1.33	1og2	0	0.48	1.35
1og5		0	1.15	1og5		0	1.17
1r9o			0	1r9o			0

had some amino acid mutations to increase solubility, which could affect the binding of different substrates. There are also three CYP2C9 crystal structures reported in the Protein Data Bank: without any ligand (1og2), with warfarin (**5**) (1og5) and with flurbiprofen (**6**) (1r9o). As in the CYP2C5 and CYP2B4 cases, the secondary elements of the three crystal structures are very similar (Table 10.3). Nevertheless, the 1r9o structure showed differences with 1og5 in the B–C and the F–G loops. The changes in the F–G region could be because the 1og5 construct introduced amino acid substitutions in this area. As for CYP2C5 (1dt6), these mutations were introduced to improve the solubility of the protein and facilitate the crystallization process. Moreover, a significant difference between the 1r9o and the 1og5 structures in the B–C region is the orientation of arginine R108. In the latter case, the side chain points away from the protein cavity, while in the former case it points into the protein cavity and forms a salt bridge interaction with the carboxylic acid of flurbiprofen (**6**). This effect could be because the protein could optimize the interaction in this region to accommodate negatively charged ligands.

10.4.1.5 CYP2D6

In the case of CYP2D6, there is only one crystal structure reported in the Protein Data Bank without a ligand (2f9q). In this crystal structure, the orientation of I106 and F483 makes a narrow hydrophobic pocket, forming a channel from the heme to the surface. The orientation of F120 into the active site also contributes to the narrow shape of the pocket. The narrow nature of the pocket observed in this structure could be because there is no ligand cocrystalized inside the enzyme cavity; therefore, the hydrophobic amino acids collapse, closing the pocket. Moreover, the E216 and D301 amino acids that are thought to have a major impact on the orientation of the substrate due to electrostatic interactions, mainly with positively charged nitrogen atoms, do not seem to point into the enzyme cavity; only the calculation of MIF with the N1+ probe shows that the position of the E216 contributes to a small electrostatic region within the cavity. The secondary elements of the CYP2D6 are conserved as compared to the other CYP isoenzymes.

10.4.1.6 CYP3A4

There are six crystal structures reported in the Protein Data Bank: two without ligand (1tqn and 1w0e), one cocrystallized with progesterone (**7**) (1w0f), one with mytrapone (**8**) (1w0g), one with ketoconazole (**9**) (2j0c) and one with erythromycin (**10**) (2j0d).

Table 10.4 Structural comparison of α-carbons and backbone atoms of CYP3A4 from different crystal structures aligned by heme group (rmsd values are given in Å).

α-Carbon	1tqn	1w0e	1w0f	1w0g	2j0c	2j0d
1tqn	0	0.6	0.72	0.83	1.56	1.07
1w0e		0	0.65	0.91	1.61	0.97
1w0f			0	0.93	1.78	1.01
1w0g				0	1.58	1.38
2j0c					0	1.8
2j0d						0

The two structures without cocrystallized ligand or inhibitor (1tqn and 1w0e) are similar (Table 10.4). The greatest difference between these structures is the orientation of arginine R212, which points into the active site in 1tqn. Moreover, there are also smaller differences in the side chain orientations of lysine K173 and arginine R255. The position of R212 in 1tqn results in a split hydrophobic field over the heme group whereas in the other structures it is more compact. In the structure with progesterone (**7**) (1w0f), the substrate is located far away from the heme active site. The protein structure is very similar to the apo structure (1w0e). Moreover, the binding of a small ligand like mytrapone (**8**) does not change the overall protein structure (1w0g). In the structure with erythromycin (**10**) present in the binding site (2j0d), an analysis of the hydrophobic interaction reveals a larger pocket with a wide access channel from the surface. In this structure, R212 is pointing out of the active site, similar to the 1w0e structure. Moreover, the erythromycin (**10**) structure differs from all other CYP3A4 apo or complex structures: when the F–G loop is moved toward the surface increasing the pocket volume, the residues 214–217 are disordered and not resolved. An estimation of the solvent accessible volume suggests an increase in the volume of the active site pocket to ~2000 Å^3 compared to 950 Å^3 in the apo structure. In the complex structure with ketoconazole (**9**), a potent inhibitor known to interact directly with the heme iron (type II binder), the active site volume is around 1600 Å^3. The active site can accommodate two inhibitor molecules. One inhibitor makes a direct interaction with the iron atom in the heme and the second one interacts with the helix F'. Both structures (2j0c and 2j0d) have a different arrangement of the secondary elements and show that the protein can accommodate compounds in different ways. In the ketoconazole structure (2j0c), the F–G loop is down, whereas in the complex with erythromycin (2j0d), the intervening helices are moved up.

The analysis of the cytochrome crystal structures reveals extensive differences among the mammalian and the human CYPs that may reflect the structural flexibility of these enzymes and therefore the broad substrate specificity observed. In general, the secondary elements and the overall structure of the CYP are conserved. Nevertheless, it seems that in most enzymes, the most flexible regions are located between the F–G helixes and the B–C loop. For example in the CYP2B4 apoprotein (1po5) the movement of these elements opens a cleft to the heme active site that would allow large ligands to access the heme group without the constrictions of an access channel. Moreover, in most of the structures cocrystallized with large and small ligands (CYP2B4 and CYP3A4), two forms of the enzymes can be recognized, an open one that can accommodate large compounds and a closed one that usually optimized the interactions of the amino acids with the compounds bound. Nevertheless, it is worth noting that the open forms could also differ depending on the ligand bound, as it happens in the ketoconazole (**9**) or erythromycin (**10**) cases with CYP3A4.

10.4.2
The Effect of the Structure in the Different Methods to Predict the Site of Metabolism

Zhou et al. [30] described for the first time the effect of a crystal structure in the prediction of the site of metabolism by using the MetaSite and docking methods. Obviously, the docking method depending only on the protein structure yields more mixed results with respect to a correct prediction of the site of metabolism. Interestingly enough, the prediction rate for the first three ranked atoms from the docking method using the GLUE program and the 1w0e crystal structure had the worst results (47%), whereas predictions based on a homology model developed by De Rienzo et al. [41] were better (57%). The MetaSite methodology that considers both the protein and the reactivity factors is less dependent on the protein structure. The same authors describe that both the homology model and the prediction based on 1w0e gave a similar prediction rate: $\sim$70% when the reactivity correction was applied and $\sim$40% when no reactivity correction is used.

Moreover, Afzelius et al. [9] described the prediction rate for the site of metabolism, comparing different computational methods for a diverse set of compounds based on reactivity (bond order computations biased for surface accessible area), knowledge-based approaches such as SPORCalc, the combined method MetaSite, a docking method based on the Glide software, a prediction made by a scientist with more than 20 years of experience on metabolite identification and a random unique hydrogen atom for CYP2C9 and CYP3A4. Moreover, the authors describe the effect of the protein structure on the prediction rate depending on several crystal structures for CYP3A4 (2j0c and 2j0d).

A study has been performed to compare the MetaSite prediction rates based on different CYP crystal structures and the homology models for CYP2C9 (Figure 10.2a), CYP2D6 (Figure 10.2b) and CYP3A4 (Figure 10.2c). The protein structures were transformed into MetaSite vectors using the application to import protein structures in version 2.7.5. The three top ranked solutions were collected for each case using a set of substrates collected by Molecular Discovery Ltd. Moreover, the combination columns are obtained if the right site of metabolism was in any of the first three ranked positions in any of the proteins.

The introduction of the reactivity correction improves the rate of correct prediction in all cases. Nevertheless, the impact seems to be similar on CYP3A4 (average difference between the reactive and nonreactive is 30%) and CYP2C9 (29%) and

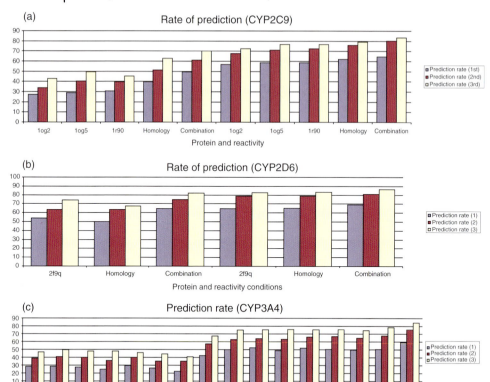

Figure 10.2 Prediction rates for (a) CYP2C9, (b) CYP2D6 and (c) CYP3A4.

smaller for CYP2D6 (15%) and even smaller for the combination case. Not all compounds that are well predicted by one protein structure are well predicted by the other protein structures for the same enzyme. In Table 10.5, the percentage of compounds in common between the different proteins is shown.

10.5
Conclusions

The are several clearance and toxicological aspects that have to be considered in the drug discovery process such as metabolic stability, enzyme selectivity, CYP inhibition and type of inhibition. Among these factors, the prediction of the site of metabolism has become one of the most successful parameters for prediction. The knowledge of the site of metabolism enhances the opportunity to chemically modify the molecule to improve the metabolic stability. There are several approaches based on database mining, chemical reactivity, protein interaction or both that have been developed for the prediction of this property, with different degree of success and applicability.

Table 10.5 Percentage of compounds well predicted among the first three ranked position with and without reactivity correction for the pair-wise comparison.

	CYP2C9			
With reactivity	1og2	1og5	1r90	Homology
1og2	100.00	97.18	95.77	88
1og5	100.00	100.00	98.59	90.67
1r90	98.55	98.59	100.00	90.67
Homology	95.65	95.77	95.77	100.00
Without reactivity	1og2	1og5	1r90	Homology
1og2	100.00	71.67	76.36	59.21
1og5	82.69	100.00	83.64	67.11
1r90	80.77	76.67	100.00	60.53
Homology	86.54	85.00	83.64	100.00

	CYP2D6	
With reactivity	2f9q	Homology
2f9q	100.00	96.33
Homology	95.45	100.00
Without reactivity	2f9q	Homology
2f9q	100.00	80.61
Homology	88.76	100.00

	CYP3A4						
Without reactivity	1tqn	1w0e	1w0f	1w0g	2j0c	2j0d	Homology
1tqn	100.00	85.71	87.37	84.21	85.33	91.78	81.48
1w0e	90.32	100.00	92.63	87.37	89.33	94.52	77.78
1w0f	89.25	89.80	100.00	84.21	88.00	91.78	76.54
1w0g	86.02	84.69	84.21	100.00	85.33	91.78	72.84
2j0c	75.27	79.59	77.89	73.68	100.00	75.34	77.78
2j0d	80.65	79.59	80.00	81.05	89.33	100.00	69.14
Homology	70.97	64.29	65.26	62.11	76.00	76.71	100.00
With Reactivity	1tqn	1w0e	1w0f	1w0g	2j0c	2j0d	Homology
1tqn	100.00	95.27	98.65	95.27	94.59	95.95	93.92
1w0e	94.63	100.00	95.30	93.96	97.32	93.96	93.96
1w0f	97.33	94.67	100.00	94.67	94.00	96.00	94.67
1w0g	94.63	93.96	95.30	100.00	94.63	95.97	94.63
2j0c	94.48	97.24	95.17	95.17	100.00	95.17	95.17
2j0d	97.87	97.87	98.58	98.58	97.16	100.00	100.00
Homology	89.68	90.32	91.61	90.97	90.97	90.32	100.00

Recently, several CYP crystal structures have been deposited in the PDB and could be used to understand the different metabolic properties. From the analysis of the different protein structures, it could be concluded that in general CYPs maintain the secondary elements across the different subfamilies, but they have a quite high flexibility, accommodating the protein structures to the ligand bound in the enzyme cavity.

The evaluation of the impact of the different crystal structures on the prediction of the site of metabolism by the MetaSite methodology without reactivity correction has been considered. In this case one can conclude that overall the rate of good prediction (number of experimentally determined sites of metabolism among the first three solutions predicted by MetaSite) is similar for the different structures. Nevertheless, there is only a moderate overlap between the compounds that are well predicted by each crystal structure, and a combination of all structures usually yield higher prediction rates. When using the reactivity correction, the difference between the prediction rates for each crystal structure is reduced, suggesting a diminishing rate of the crystal structure used for the analysis.

Abbreviations

CPI	(4-(4-chloro-phenyl) imidazole
CYP450	cytochrome P450
log D	octanol/water partition coefficient at given pH
MIF	molecular interaction field
PDB	Protein Data Bank
QSAR	quantitative structure activity relationship
SPORCalc	substrate product occurrence ratio calculator

Acknowledgments

We thank our collaborator, Dr Roy Vaz, for providing the crystal information needed to perform some of the computations shown in this chapter of the book. The authors would also like to thank Dr Lovisa Afzelius for correcting the original manuscript.

References

1 van de Waterbeemd, H. and Gifford, E. (2003) ADMET *in silico* modelling: towards prediction paradise? *Nature Reviews. Drug Discovery*, **2** (3), 192–204.

2 Evans, W.E. and Relling, M.V. (1999) Pharmacogenomics: translating functional genomics into rational therapeutics. *Science*, **286**, 487–491.

3 Davydov, R., Makris, T.M., Kofman, V., Werst, D.E., Sligar, S.G. and Hoffman, B. M. (2001) Hydroxylation of camphor by reduced oxy-cytochrome P450cam: mechanistic implications of EPR and ENDOR studies of catalytic intermediates in native and mutant enzymes. *Journal of the American Chemical Society*, **123**, 1403–1415.

4 Ablström, M., Ridderström, M., Luthman, K. and Zamora, I. (2007) CYP2C9 structure-metabolism relationships: optimizing the metabolic stability of cox-2. *Journal of Medicinal Chemistry*. **50** (18), 4444–4452.

5 Parkinson, A. (2001) The Biotransformation of Xenobiotics, in *Casarett & Doull's Toxicology: The Basic Science of Poisons*, (ed. C.D. Klaasen), 5th edn., McGraw-Hill.

6 Ogliaro, F., de Visser, S.P., Cohen, S., Sharma, P.K. and Shaik, S. (2002) Searching for the second oxidant in the catalytic cycle of cytochrome P450: a theoretical investigation of the iron (III)-hydroperoxo species and its epoxidation pathways. *Journal of the American Chemical Society*, **124**, 2806–2816.

7 He, X. and Ortiz de Montellano, P.R. (2004) Radical rebound mechanism in cytochrome P-450 catalyzed hydroxylation of multifaceted radical clocks α-and β-thujone. *The Journal of Biological Chemistry*, **279**, 39479–39484.

8 Singh, S.B., Shen, L.Q., Walker, M.J. and Sheridan, R.P. (2003) A model for predicting likely sites of CYP3A4-mediated metabolism on drug-like molecules. *Journal of Medicinal Chemistry*, **46**, 1330–1336.

9 Afzelius, L., Arnby, C.H., Broo, A., Carlsson, L., Isaksson, C., Jurva, U., Kjellander, B., Kolmodin, K., Nilsson, K., Raubacher, F. and Weidolf, L. (2007) State-of-the-art tools for computational site of metabolism predictions: comparative analysis, mechanistic insights, and future applications. *Drug Metabolism Reviews*, **39** (1), 61–86.

10 Lewis, D.F. (2000) On the recognition of mammalian microsomal cytochrome P450 substrates and their characteristics: towards the prediction of human p450 substrate specificity and metabolism. *Biochemical Pharmacology*, **60** (3), 293–306.

11 Williams, P.A., Cosme, J., Sridhar, V., Johnson, E.F. and McRee, D.E. (2000) Mammalian microsomal cytochrome P450 monooxygenase: structural adaptations for membrane binding and functional diversity. *Molecules and Cells*, **5**, 121–131.

12 Wester, M.R., Johnson, E.F., Marques-Soares, C., Dansette, P.M., Mansuy, D. and Stout, C.D. (2003) Structure of a substrate complex of mammalian cytochrome P450 2C5 at 2.3 Å resolution: evidence for multiple substrate binding modes. *Biochemistry*, **42**, 6370–6379.

13 Wester, M.R., Johnson, E.F., Marques-Soares, C., Dijols, S., Dansette, P.M., Mansuy, D. and Stout, C.D. (2003) Structure of mammalian cytochrome P450 2C5 complexed with diclofenac at 2.1 Å resolution: evidence for an induced fit model of substrate binding. *Biochemistry*, **42**, 9335–9345.

14 Scott, E.E., He, Y.A., Wester, M.R., White, M.A., Chin, C.C., Halpert, J.R., Johnson, E. F. and Stout, C.D. (2003) An open conformation of mammalian cytochrome P450 2B4 at 1.6-Å resolution. *Proceedings of the National Academy of Sciences of the United States of America*, **100**, 13196–13201.

15 Scott, E.E., White, M.A., He, Y.A., Johnson, E.F., Stout, C.D. and Halpert, J. R. (2004) Structure of mammalian cytochrome P450 2B4 complexed with 4-(4-chlorophenyl)imidazole at 1.9-A resolution: insight into the range of P450 conformations and the coordination of redox partner binding. *The Journal of Biological Chemistry*, **279**, 27294–27301.

16 Zhao, Y., White, M.A., Muralidhara, B.K., Sun, L., Halpert, J.R. and Stout, C.D. (2006) Structure of microsomal cytochrome P450 2B4 complexed with the antifungal drug bifonazole: insight into P450 conformational plasticity and membrane interaction. *The Journal of Biological Chemistry*, **281**, 5973–5981.

17 Sansen, S., Yano, Y.K., Reynald, R.L., Schoch, G.A., Stout, C.D. and Johnson, E. F. (2007) Adaptations for the oxidation of polycyclic aromatic hydrocarbons

exhibited by the structure of human P450 1A2. *The Journal of Biological Chemistry*, **289** (19), 14348–14355.

18 Yano, J.K., Hsu, M.H., Griffin, K.J., Stout, C.D. and Johnson, E.F. (2005) Structures of human microsomal cytochrome P450 2A6 complexed with coumarin and methoxsalen. *Nature Structural & Molecular Biology*, **12**, 822–823.

19 Wester, M.R., Yano, J.K., Schoch, G.A., Yang, C., Griffin, K.J., Stout, C.D. and Johnson, E.F. (2004) The structure of human cytochrome P450 2C9 complexed with flurbiprofen at 2.0-Å resolution. *The Journal of Biological Chemistry*, **279**, 35630–35637.

20 Williams, P.A., Cosme, J., Ward, A., Angove, H.C., Matak Vinkovic, D. and Jhoti, H. (2003) Crystal structure of human cytochrome P450 2C9 with bound warfarin. *Nature*, **424**, 464–468.

21 Schoch, G.A., Yano, J.K., Wester, M.R., Griffin, K.J., Stout, C.D. and Johnson, E.F. (2004) Structure of human microsomal cytochrome P450 2 C8. Evidence for a peripheral fatty acid binding site. *The Journal of Biological Chemistry*, **279**, 9497–9503.

22 Rowland, P., Blaney, F.E., Smyth, MG M. G., Jones, J.J., Leydon, V.R., Oxbrow, A.K., Lewis, C.J., Tennant, M.G., Modi, S., Eggleston, D.S., Chenery, R.J. and Bridges, A.M. (2006) Crystal structure of human cytochrome P450 2 D6. *The Journal of Biological Chemistry*, **281**, 7614–7622.

23 Williams, P.A., Cosme, J., Vinkovic, D.M., Ward, A., Angove, H.C., Day, P.J., Vonrhein, C., Tickle, J.J. and Jhoti, H. (2004) Crystal structures of human cytochrome P450 3A4 bound to metyrapone and progesterone. *Science*, **305**, 683–686.

24 Yano, J.K., Wester, M.R., Schoch, G.A., Griffin, K.J., Stout, C.D. and Johnson, E.F. (2004) The structure of human microsomal cytochrome P450 3A4 determined by X-ray crystallography to 2.05-Å resolution. *The Journal of Biological Chemistry*, **279**, 38091–38094.

25 Ekroos, M. and Sjogren, T. (2006) Structural basis for ligand promiscuity in cytochrome P450 3 A4. *Proceedings of the National Academy of Sciences of the United States of America*, **103**, 13682–13687.

26 Judson, P.N. and Vessey, J.D. (2003) *Journal of Chemical Information and Computer Sciences*, **43**, 1356.

27 Smith, J. Prediction of putative reaction centres in Xenobiotics from metabolite using substructural fingerprinting. *Transactions of the Biochemical Society*. http://www.chemie.uni-erlangen.de/clark/smith/SPORCalc.html.

28 Testa, B., Balmat, A.L., Long, A. and Judson, P. (2005) Predicting drug metabolism-an evaluation of the expert system METEOR. *Chemical and Biodiversity*, **2**, 872–885.

29 Boyer, S. and Zamora, I. (2002) New methods in predictive metabolism. *Journal of Computer-Aided Molecular Design*, **16** (5/6), 403–413.

30 Zhou, D., Afzelius, L., Grimm, S.W., Andersson, T.B., Zauhar, R.J. and Zamora, I. (2006) Comparison of methods for the prediction of the metabolic sites for CYP3A4-mediated metabolic reactions. *Drug Metabolism and Disposition: The Biological Fate of Chemicals*, **34**, 976–983.

31 Goodford, P.J. (1985) A computational procedure for determining energetically favorable binding sites on biologically important macromolecules. *Journal of Medicinal Chemistry*, **28**, 849–857.

32 Kjellander, B., Masimirembwa, C.M. and Zamora, I. (2007) Exploration of enzyme–ligand interactions in CYP2D6 & 3A4 homology models and crystal structures using a novel computational approach. *Journal of Chemical Information and Modeling*, **47** (3), 1234–1247.

33 Friesner, R.A.B., Murphy, R.B., Halgren, T. A., Klicic, J.J., Mainz, D.T., Repasky, M.P., Knoll, E.H., Shelly, M., Perry, J.K., Shaw, D. E., Francis, P. and Shenkin, P.S. (2004) Glide: a new approach for rapid, accurate docking and scoring. 1. Method and

assessment of docking accuracy. *Journal of Medicinal Chemistry*, **47**, 1739–1749.
34 Ewing, T.J., Makino, S., Skillman, A.G. and Kuntz, I.D. (2001) DOCK 4.0: search strategies for automated molecular docking of flexible molecule databases. *Journal of Computer-Aided Molecular Design*, **15** (5), 411–428.
35 Verdonk, M.L., Cole, J.C., Hartshorn, M.J., Murray, C.W. and Taylor, R.D. (2003) Improved protein–ligand docking using GOLD. *Proteins*, **52** (4), 609–623.
36 de Graaf, C. and Vermeulen, N.P. (2005) Cytochrome p450 *in silico*: an integrative modeling approach. *Journal of Medicinal Chemistry*, **48** (8), 2725–2755.
37 Zamora, I., Afzelius, L. and Cruciani, G. (2003) Predicting drug metabolism: a site of metabolism prediction tool applied to the cytochrome P450 2 C9. *Journal of Medicinal Chemistry*, **46**, 2313–2324.
38 Cruciani, G., Carosati, E., De Boeck, B., Ethirajulu, K., Mackie, C., Howe, T. and Vianello, R. (2005) MetaSite: understanding metabolism in human cytochromes from the perspective of the chemist. *Journal of Medicinal Chemistry*, **48**, 6970–6979.
39 Frisch, M.J. (2003) *Gaussian 03 Revision B.05*, Gaussian, Inc, Pittsburgh, PA.
40 Uckun, F.M., Thoen, J., Chen, H., Sudbeck, E., Mao, C., Malaviya, R., Liu, X.-P. and Chen, C.-L. (2002) CYP1A-mediated metabolism of the janus kinase-3 inhibitor 4-(4′-hydroxyphenyl)-amino-6,7-dimethoxyquinazoline: structural basis for inactivation by regioselective *o*-demethylation. *Drug Metabolism and Disposition: The Biological Fate of Chemicals*, **30** (1), 74–85.
41 De Rienzo, F., Fanelli, F., Menziani, M.C. and De Benedetti, P.G. (2000) Theoretical investigation of substrate specificity for cytochromes P450 IA2, P450 IID6 and P450 II IA4. *Journal of Computer-Aided Molecular Design*, **14**, 93–116.

11
Irreversible Cytochrome P450 Inhibition: Common Substructures and Implications for Drug Development
Sonia M. Poli

11.1
Introduction

Cytochrome P450s (CYPs) are widely distributed enzymes responsible for the metabolism of endogenous and exogenous compounds [1]. A small subset of the CYPs is involved mainly in the metabolism of xenobiotics. The rate of metabolism by human CYPs is one of the major determinants of the *in vivo* clearance of drugs and therefore inhibition of these enzymes can change the pharmacokinetic parameters of a drug and cause clinically significant drug–drug interactions that can lead to severe consequences for patients [2]. It has also been reported that irreversible CYP inhibition can result in nonlinear pharmacokinetic behavior of certain drugs that act as mechanism-based inhibitors (MBIs) of their own metabolism [3].

Irreversible inhibition of CYPs is particularly worrisome as its consequences cannot be predicted easily or quantified from *in vitro* data: the *in vivo* effect of an irreversible inhibitor is usually greater than that predicted based on affinity alone. Moreover, irreversible inhibition is generally the consequence of the production of reactive metabolites (electrophiles), which can also bind covalently to endogenous proteins and, in rare cases, trigger serious autoimmune reactions [4].

To develop safer drugs, it is important to identify irreversible CYP inhibitors at an early stage in the drug discovery process. Certain substructures, reviewed in this chapter, are frequently responsible for irreversible CYP inhibition and either their use in medicinal chemistry programs should be limited or their potential for irreversible binding to CYPs should be carefully monitored.

Antitargets. Edited by R. J. Vaz and T. Klabunde
Copyright © 2008 WILEY-VCH Verlag GmbH & Co. KGaA, Weinheim
ISBN: 978-3-527-31821-6

11.2
Overview

11.2.1
Characteristics of Irreversible CYP Inhibition

Irreversible CYP inhibition can arise from different chemical mechanisms. However, a common initial step is the metabolic activation of a substrate into a reactive metabolite that is trapped within the active site of the CYP to form a tightly bound complex causing a long-lasting inactivation of enzyme activity. Enzymatic activity can be restored only through the new synthesis of the enzyme. For this reason, irreversible CYP inhibition is often referred to as 'mechanism-based inhibition', 'metabolite-based inhibition' or 'suicide inhibition'.

Mechanism-based CYP inhibition is time dependent, as it is a consequence of metabolic activation. For the same reason, it also depends on substrate concentration. It also, usually, depends on the presence of NADPH in the incubation system (or an NADPH regeneration system), as most of the CYP-catalyzed reactions require NADPH to occur [5].

These three simple observations (time, substrate and NADPH dependence) can be used to detect potential MBIs early in the drug discovery process, as simple experiments can be designed from established screening protocols used to measure CYP inhibition [6].

When reactive metabolites are formed by metabolic activation, some of them can escape from the active site and bind to external protein residues or be trapped by reduced glutathione (GSH) or other nucleophiles. The remaining molecules that are not released from the active site will cause the suicide inhibition [7]. The ratio of the number of reactive molecules remaining in the active site and those escaping is a measure of the reactivity of the intermediates formed. The addition of scavengers or GSH to the incubation mixture does not affect and cannot prevent the CYP mechanism-based inhibition. However, GSH can reduce the extent of the nonspecific covalent binding to proteins by those reactive molecules that escape from the active site. In contrast, addition of substrates or inhibitors that compete for the same catalytic center usually results in reduction of the extent of inhibition.

The interaction within the active site can be either in the form of covalent binding or in the form of quasi-irreversible (tight but slowly reversible) binding, and it can involve the protein residues, the porphyrin moiety or the catalytic center (heme iron) [8]. CYP inactivation follows a stoichiometry of one substrate molecule per enzyme molecule inactivated. To measure the stoichiometry of the inactivation, it is necessary to trap all molecules that are not specifically bound to the active site, by using an appropriate scavenger, normally GSH.

Owing to the irreversible nature of the inactivation, the alkylated CYP protein can be isolated, dialyzed or purified and digested with proteases to identify the peptide, or in the best case, the amino acid where the binding occurs. This approach has been performed on several MBIs, and it is worth noting that the results show that most of the MBIs bind either to the porphyrin ring or to the I helix, which is a highly conserved region in all CYPs, rich in threonine and serine residues, involved in the catalytic cycle and activation of molecular oxygen [9,10]. The specificity of an MBI for a CYP is determined mainly by the affinity of the substrate molecule and its orientation in the active site, rather than by the specific protein residues present in close proximity of the reactive moiety [11].

11.2.2
Quantitative Prediction of Drug–Drug Interaction Caused by Irreversible CYP Inhibitors

Because irreversible inhibition usually destroys CYP activity, a new pool of enzyme must be synthesized to reestablish the basal native activity; this process takes from 1 to 3 days. Thus, even after the drug has been cleared from the body, normal CYP activity does not return until the physiological CYP concentration has been reestablished.

For this reason, the *in vivo* effects of an MBI cannot be predicted solely on the basis of its affinity K_i (dissociation constant of the enzyme–substrate complex) for the CYP, and they are usually more pronounced for a given K_i value than that observed with a reversible inhibitor [12].

The quantitative *in vitro* to *in vivo* prediction using the $[I]/K_i$ approach [12,13] usually leads to underprediction of the potency of an MBI. The maximum inactivation rate constant (k_{inact}), which is a measurement of the enzyme degradation through mechanism-based inhibition, must be compared to the rate of *de novo* synthesis of the CYP. Some progress in estimating the quantitative inhibition has been made recently by Galetin *et al.* [13], who have shown the importance of the accuracy and precision of the CYP degradation rate estimation for those substrates that are metabolized by CYP3A4 by more than 50%.

11.2.3
Irreversible CYP Inhibition and Autoimmune Diseases

A rare but serious event that can result from irreversible CYP inhibition is the development of a hypersensitivity reaction. The bioactivation of a drug and the formation of a covalent adduct between the activated substrate and the enzyme can lead to hapten formation and eventually to an idiosyncratic autoimmune response (usually in the form of autoimmune hepatitis) [14]. The hapten formation is the first key step toward the autoimmune response. The CYP macromolecule is made immunogenic ('foreign') by the covalent binding of the electrophilic metabolites, and the immune reaction follows with the production of autoantibodies against the target molecule (not necessarily alkylated).

There are several examples of autoimmune hepatitis caused by mechanism-based inhibitors in the field of CYPs: hepatitis induced by halothane (CYP2E1) [15,16], by tienilic acid (CYP2C9) [17] and by dihydralazine (CYP1A2) [18,19].

In the case of tienilic acid induced hepatitis, Beaune *et al.* [20] have shown that autoantibodies against CYP2C9 are present in the serum of patients suffering from the disease. Additional *in vitro* evidence has shown the potential for tienilic acid to alkylate CYP2C9 [21].

An autoimmune response is rarely detected during preclinical drug development because of the lack of good predictive animal models for the human immune system. Owing to their low incidence and idiosyncratic nature, autoimmune diseases quite often appear as a serious issue only after product launch [2,22].

From the examples cited above, it is clear that the formation of reactive metabolites and alkylation of CYP are key steps in the initiation of the cascade of events, which leads to an autoimmune response. However, in the absence of a better understanding of the mechanisms involved in the triggering of the idiosyncratic reactions, the interpretation of the significance of the reactive species', formation in the development of an autoimmune response is not straightforward. In fact, there are many marketed drugs that covalently bind to macromolecules without causing an immune response. The alkylation of a macromolecule leading to a foreign protein is not enough to trigger the autoimmune response; there are, in fact, additional important factors in the development of the immune response, such as cellular stress induced by the reactive metabolites by glutathione depletion [23–25].

11.3
Structural Features Often Responsible for Mechanism-Based CYP Inhibition

Most reactive metabolites produced by CYP metabolic activation are electrophilic in nature, which means that they can react easily with the nucleophiles present in the protein side chains. Several functional groups are recurrent structural features in MBIs. These groups have been reviewed by Fontana *et al.* [26] and can be summarized as follows: terminal (ω or $\omega - 1$) acetylenes, olefins, furans and thiophenes, epoxides, dichloro- and trichloroethylenes, secondary amines, benzodioxoles (methylenedioxyphenyl, MDP), conjugated structures, hydrazines, isothiocyanates, thioamides, dithiocarbamates and, in general, Michael acceptors (Scheme 11.1).

The presence of one of these substructures in the molecule is only an alert. The affinity of the molecule for the CYP and its orientation within the active site are important factors in determining if a reactive metabolite can be formed. However, when considering potential MBIs of CYP3A4, one should remember that the CYP3A4 active site is very large and that it can accommodate small molecules in several orientations [27]. Therefore, when one of the structural features mentioned above is present in a CYP3A4 inhibitor, there is a high probability that it reaches the heme iron for subsequent oxidation and formation of reactive metabolites. That may be one reason why the literature contains many examples of CYP3A4 MBIs, while, for instance, there are very few examples of MBIs produced by CYP2D6, a much more specific enzyme.

Because of the reactive nature of the intermediates trapped during metabolic activation, it has been difficult to identify the mechanism of the reactions. Some of the most common intermediates have been studied in detail, for example, acetylenes, where the mechanism is well established and accepted [28]. For other reactions, the pathways are less clear; thus, only the most widely accepted hypotheses are presented here.

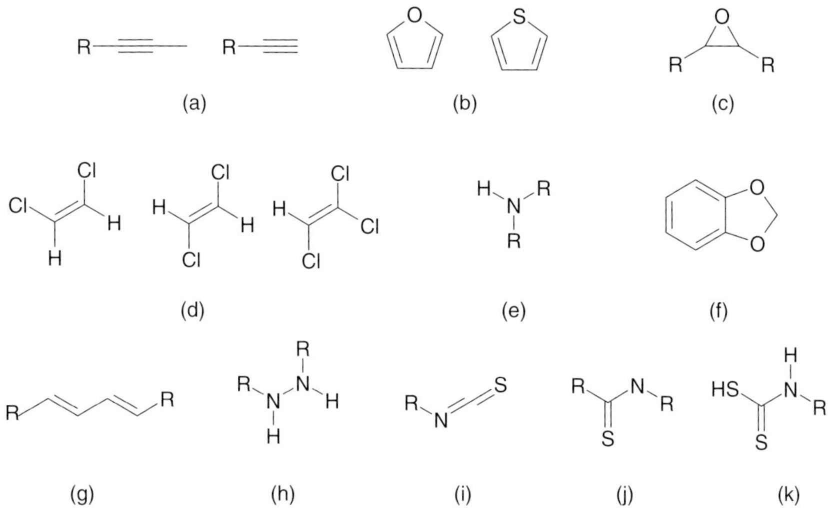

Scheme 11.1 The most common substructures causing mechanism-based inactivation of cytochrome P450s: (a) acetylenes; (b) furans and thiophenes; (c) epoxides; (d) dichloro- and trichloroethylenes; (e) secondary amines; (f) benzodioxoles (methylenedioxyphenyl); (g) conjugated systems; (h) hydrazines; (i) isothiocyanates; (j) thioamides; (k) dithiocarbamate.

11.3.1
Terminal Acetylenes (ω or $\omega - 1$)

Numerous mechanism-based CYP inhibitors belong to this group. Because of the high reactivity of the intermediates, the exact mechanism of inactivation is unknown, but there is ample experimental evidence to indicate that the inactivation starts with the triple bond oxidation giving rise to an oxirene that rearranges to a highly reactive ketene that, in turn, is attacked by nucleophilic protein side chains or by the heme nitrogen [28–30]. Several classes of acetylenic compounds have been identified as MBIs of specific CYPs: planar, aromatic aryl acetylenes have been reported to be MBIs for CYP1A, CYP1B and CYP2B (4-propynyl biphenyl, 2-ethynyl naphthalene, 5-phenyl pentyne, 2-propynyl phenantrene, 2-ethynyl pyrene and 1-propynyl pyrene [30]), fatty acid acetylenes for CYP4A [31,32] and steroid structures for CYP3A (17-alpha-ethynylestradiol, gestodene and mifepristone [33–35]; see Scheme 11.2).

11.3.2
Alkenes

Alkenes may be oxidized to epoxides that are reactive metabolites because of ring strain [36] and can undergo nucleophilic attack. Epoxides are not always highly reactive species. In fact, some of them are relatively unreactive; for example, the arene oxides that derive from oxidation of phenyl rings. Most drugs containing a phenyl

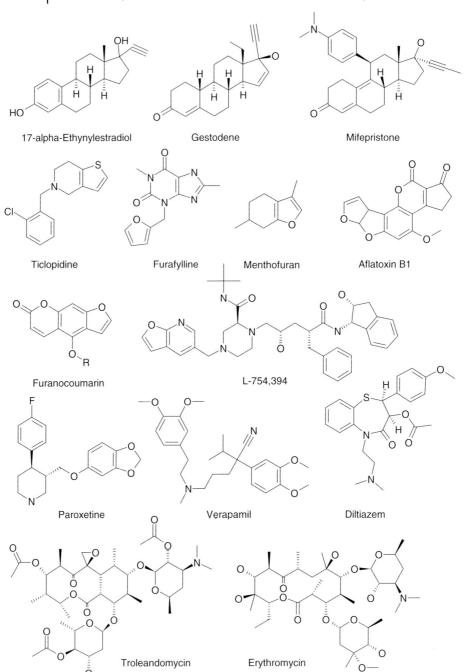

Scheme 11.2 Mechanism-based inhibitors on human CYP isoforms.

ring are oxidized to phenols through the arene oxide intermediate, without any consequence. However, for instance, some polycyclic aromatic hydrocarbons (PAHs) strongly inhibit their own and other PAH metabolism catalyzed by cytochrome P450 1A1, 1A2 and 1B1 in a mechanism-based manner [37,38].

The alkene class also covers furans and thiophenes, discussed in more detail in the next section, and the dichloro- and trichloroethylenes that are also oxidized to reactive epoxides.

11.3.3
Furans and Thiophenes

Furans and thiophenes are believed to be oxidized to epoxides, following the general reaction described for alkenes. However, at least in the case of tienilic acid, experimental evidence shows that the reactive intermediate is an S-oxide [39]. Another example of CYP inactivation by a thiophene derivative is the covalent binding of ticlopidine to CYP2C19 [40].

In the case of furans, the epoxidation can lead to hydroxylated furans or to ring opening and final formation of a butene-1,4-dial, which is a Michael acceptor [41].

The furan moiety is found in many drug molecules and natural products. Examples of enzyme inactivation include the CYP1A2 inactivation by furafylline [36] and the CYP2A6 inactivation by menthofuran (see Scheme 11.2) [42].

The potent carcinogen aflatoxin B1 is activated by generation of an epoxide on a dihydrofuran moiety [43]. The potent MBIs contained in grapefruit juice and in the herbal medication St John's Wort are furanocoumarin compounds [44–46]. A similar moiety is also incorporated in the HIV protease inhibitor L-754,394, a potent CYP3A4 MBI [47].

11.3.4
Secondary and Tertiary Amines

Secondary and tertiary amines are very common substructures in drugs and natural compounds but their efficiency and specificity as MBIs are highly variable (see Scheme 11.2).

The stepwise oxidation of alkylamine, which leads to N-dealkylation, generates nitrones that form tightly bound complexes with the heme iron [48]. These heme iron complexes give rise to characteristic changes in the UV–Vis spectrum of the CYP.

Many drugs such as verapamil, diltiazem, troleandomycin and erythromycin contain alkylamine groups and are able to form tightly bound complexes, in particular with CYP3A4 [49–53].

11.3.5
Benzodioxoles (MDP)

Benzodioxole's methylene group is activated by the two oxygen atoms and therefore is more reactive than an alkyl methylene group.

Compounds containing a benzodioxole moiety are oxidized by CYPs to generate a carbene that forms a tightly bound complex with the heme iron, easily measurable by UV–Vis spectroscopy. There are numerous examples in the literature: paroxetine, MDP alkylamines, MDP benzothiazines, MDP benzothiazolines and MDP piperazines (see Scheme 11.2) [54–60].

11.4
Conclusions

Irreversible CYP inhibition, especially when considered as a cause of chemically reactive metabolites, raises a number of issues that should be considered early in the drug discovery process. Careful evaluation of the structural features of the lead series as well as screening for MBIs early in the drug development process should guide series selection and optimization with the goal of avoiding the introduction of potentially risky moieties into new chemical entities.

Because of the disproportionate drug–drug interaction and the uncertainties and potential toxicological and immunological consequences of reactive metabolites, mechanism-based inhibitors of CYPs should be developed only after a careful risk/benefit assessment. The decision to bring forward an MBI into preclinical and clinical development must be evaluated together with a number of other important factors including the therapeutic use, the lack of existing medications for the target indication, the target population, the dose and the dosing regimen.

Abbreviations

CYP	cytochrome P450
GSH	reduced glutathione
MBI	mechanism-based inhibitor
MDP	methylenedioxyphenyl

Acknowledgments

The author wishes to thank Dr Mark Epping-Jordan for his assistance and advice in reviewing the manuscript.

References

1 Isin, E.M. and Guengerich, F.P. (2007) *Biochimica et Biophysica Acta*, **1770** (3), 314–329.

2 Dresser, G.K., Spence, J.D. and Biley, D.G. (2000) *Clinical Pharmacokinetics*, **38** (1), 41–57.

3 Lin, J.H. et al. (2000) *Drug Metabolism and Disposition*, **28** (4), 460–466.
4 Knowles, S.R., Uetrecht, J. and Shear, N.H. (2000) *Lancet*, **356** (9241), 1587–1591.
5 Silverman, R.B. (1995) *Methods in Enzymology*, **249**, 240–283.
6 Naritomi, Y., Teramura, Y., Terashita, S. and Kagayama, A. (2004) *Drug Metabolism and Pharmacokinetics*, **19** (1), 55–61.
7 Kent, U.M., Jushchyshyn, M.I. and Hollenberg, P.F. (2001) *Current Drug Metabolism*, **2** (3), 215–243.
8 Chan, W.K., Sui, Z. and Ortiz de Montellano P.R. (1993) *Chemical Research in Toxicology*, **6** (1), 38–45.
9 Sridar, C., Harleton, E. and Hollenberg, P.F. (2005) *Biochemical and Biophysical Research Communications*, **338** (1), 386–393.
10 Blobaum, A.L., Lu, Y., Kent, U.M., Wang, S. and Hollenberg, P.F. (2004) *Biochemistry*, **43** (38), 11942–11952.
11 Strobel, S.M., Szklarz, G.D., He, Y., Foroozesh, M., Alworth, W.L., Roberts, E.S., Hollenberg, P.F. and Halpert, J.R. (1999) *Journal of Pharmacology and Experimental Therapeutics*, **290** (1), 445–451.
12 Venkatakrishnan, K. and Obach, R.S. (2007) *Current Drug Metabolism*, **8** (5), 449–462.
13 Galetin, A., Burt, H., Gibbons, L. and Houston, J.B. (2006) *Drug Metabolism and Disposition*, **34** (1), 166–175.
14 Li, A.P. (2002) *Chemico-Biological Interactions*, **142** (1–2), 7–23.
15 Bourdi, M., Chen, W., Peter, R.M., Martin, J.L., Buters, J.T., Nelson, S.D. and Pohl, L.R. (1996) *Chemical Research in Toxicology*, **9** (7), 1159–1166.
16 Vergani, D., Mieli-Vergani, G., Alberti, A., Neuberger, J., Eddleston, A.L., Davis, M. and Williams, R. (1980) *The New England Journal of Medicine*, **303** (2), 66–71.
17 Lecoeur, S., Bonierbale, E., Challine, D., Gautier, J.C., Valadon, P., Dansette, P.M., Catinot, R., Ballet, F., Mansuy, D. and Beaune, P.H. (1994) *Chemical Research in Toxicology*, **7** (3), 434–442.
18 Bourdi, M. et al. (1990) *Journal of Clinical Investigation*, **85** (6), 1967–1973.
19 Masubuchi, Y. and Horie, T. (1999) *Chemical Research in Toxicology*, **12**, 1028–1032.
20 Beaune, P., Dansette, P.M., Mansuy, D., Kiffel, L., Finck, M., Amar, C., Leroux, J.P. and Homberg, J.C. (1987) *Proceedings of the National Academy of Sciences of the United States of America*, **84** (2), 551–555.
21 Bonierbale, E., Valadon, P., Pons, C., Desfosses, B., Dansette, P.M. and Mansuy, D. (1999) *Chemical Research in Toxicology*, **12** (3), 286–296.
22 Zhou, S., Yung Chan, S., Cher Goh, B., Chan, E., Duan, W., Huang, M. and McLeod, H.L. (2005) *Clinical Pharmacokinetics*, **44** (3), 279–304.
23 Matzinger, P. (1994) *Annual Review of Immunology*, **12**, 991–1045.
24 Matzinger, P. (1998) *Seminars in Immunology*, **10** (5), 399–415.
25 Gallucci, S. and Matzinger, P. (2001) *Current Opinion in Immunology*, **13** (5), 114–119.
26 Fontana, E., Dansette, P.M. and Poli, S.M. (2005) *Current Drug Metabolism*, **6** (5), 413–454.
27 Shou, M., Dai, R., Cui, D., Korzekwa, K.R., Baillie, T.A. and Rushmore, T.H. (2001) *Journal of Biological Chemistry*, **276** (3), 2256–2262.
28 Foroozesh, M., Primrose, G., Guo, Z., Bell, L.C., Alworth, W.L. and Guengerich, F.P. (1997) *Chemical Research in Toxicology*, **10** (1), 91–102.
29 Beebe, L.E., Roberts, E.S., Fornwald, L.W., Hollenberg, P.F. and Alworth, W.L. (1996) *Biochemical Pharmacology*, **52** (10), 1507–1513.
30 Shimada, T., Yamazaki, H., Foroozesh, M., Hopkins, N.E., Alworth, W.L. and Guengerich, F.P. (1998) *Chemical Research in Toxicology*, **11** (9), 1048–1056.
31 Xu, F. et al. (2002) *American Journal of Physiology – Regulatory Integrative and Comparative Physiology*, **283** (3), R710–R720.

32 Helvig, C., Alayrac, C., Mioskowski, C., Koop, D., Poullain, D., Durst, F. and Salaün, J.P. (1997) *Journal of Biological Chemistry*, **272** (1), 414–421.

33 Lin, H.-L., Kent, U.M. and Hollenberg, P.F. (2002) *Journal of Pharmacology and Experimental Therapeutics*, **301** (1), 160–167.

34 Guengerich, F.P. (1990) *Chemical Research in Toxicology*, **3** (4), 363–371.

35 He, K., Woolf, T.F. and Hollenberg, P.F. (1999) *Journal of Pharmacology and Experimental Therapeutics*, **288** (2), 791–797.

36 Kunze, K.L. and Trager, W.F. (1993) *Chemical Research in Toxicology*, **6** (5), 649–656.

37 Guengerich, F.P. and MacDonald, J.S. (2007) *Chemical Research in Toxicology*, **20** (3), 344–369.

38 Shimada, T., Murayama, N., Okada, K., Funae, Y., Yamazaki, H. and Guengerich, F.P. (2007) *Chemical Research in Toxicology*, **20** (3), 489–496.

39 Valadon, P., Dansette, P.M., Girault, J.P., Amar, C. and Mansuy, D. (1996) *Chemical Research in Toxicology*, **9** (8), 1403–1413.

40 Liu, Z.C. and Uetrecht, J.P. (2000) *Drug Metabolism and Disposition*, **28** (7) 726–730.

41 Chen, L.J., Hecht, S.S., Peterson, L.A. (1995) *Chemical Research in Toxicology*, **8** (7), 903–906.

42 Khojasteb-Bakht, S.C., Koenigs, L.L., Peter, R.M., Trager, W.F. and Nelson, S.D. (1998) *Drug Metabolism and Disposition*, **26** (7), 701–704.

43 Guengerich, F.P. *et al.* (1998) *Mutation Research*, **402** (1-2), 121–128.

44 Guo, L.-Q. *et al.* (2001) *Japanese Journal of Pharmacology*, **85** (4), 399–408.

45 Malhotra, S. *et al.* (2001) *Clinical Pharmacology and Therapeutics*, **69** (1), 14–23.

46 Guo, L.-Q. *et al.* (2000) *Drug Metabolism and Disposition*, **28** (7), 766–771.

47 Lightning, L.K., Jones, J.P., Friedberg, T., Pritchard, M.P., Shou, M., Rushmore, T.H. and Trager, W.F. (2000) *Biochemistry*, **39** (15), 4276–4287.

48 Mayhew, B.S., Jones, D.R. and Hall, S.D. (2000) *Drug Metabolism and Disposition*, **28** (9), 1031–1037.

49 Yeo, K.R. and Yeo, W.W. (2001) *British Journal of Clinical Pharmacology*, **51** (5), 461–470.

50 Ma, B., Prueksaritanont, T. and Lin, J.H. (2000) *Drug Metabolism and Disposition*, **28** (2), 125–130.

51 Jones, D.R., Gorski, J.C., Hamman, M.A., Mayhew, B.S., Rider, S. and Hall, S.D. (1999) *Journal of Pharmacology and Experimental Therapeutics*, **290** (3), 1116–1125.

52 Yamazaki, H. and Shimada, T. (1998) *Drug Metabolism and Disposition*, **26** (11), 1053–1057.

53 Kanamitsu, S., Ito, K., Green, C.E., Tyson, C.A., Shimada, N. and Sugiyama, Y. (2000) *Pharmaceutical Research*, **17** (4), 419–426.

54 Usia, T., Watabe, T., Kadota, S. and Tezuka, Y. (2005) *Life Sciences*, **76** (20), 2381–2391.

55 Mathews, J.M., Etheridge, A.S. and Black, S.R. (2002) *Drug Metabolism and Disposition*, **30** (11), 1153–1157.

56 Diamond, S. and Christ, D.D. (1999) *ISSX Proceedings*, **15**, 101.

57 Iwata, H., Tezuka, Y., Kadota, S., Hiratsuka, A. and Watabe, T. (2004) *Drug Metabolism and Disposition*, **32** (12), 1351–1358.

58 Heydari, A., Rowland Yeo, K., Lennard, M.S., Elli, S.W., Tucker, G.T. and Rostami-Hodjegan, A. (2004) *Drug Metabolism and Disposition*, **32** (11), 1213–1217.

59 Nakajima, M., Suzuki, M., Yamaji, R., Takashina, H., Shimada, N., Yamazaki, H. and Yokoi, T. (1999) *Xenobiotica*, **29** (12), 1191–1202.

60 Bertelsen, K.M., Venkatakrishnan, K., von Moltke Lisa, L., Obach, R.S. and Greenblatt, D.J. (2003) *Drug Metabolism and Disposition*, **31** (3), 289–293.

12
MetaSite: Understanding CYP Antitarget Modeling for Early Toxicity Detection

Yasmin Aristei, Gabriele Cruciani, Sergio Clementi, Emanuele Carosati, Riccardo Vianello, Paolo Benedetti

12.1
Introduction

The membrane-attached cytochrome P450 enzymes are involved in a significant fraction of events associated with drug metabolism. Most of the cytochrome P450 (CYP) catalyzed reactions lead to the detoxification of xenobiotics, by forming hydrophilic metabolites that can be readily excreted from the body.

Identification of metabolic reactions at an early phase can significantly affect the drug discovery process, because bioavailability, activity, toxicity, distribution and final elimination all depend on metabolic biotransformations [1]. Once obtained, this information can help researchers judge whether or not a potential candidate should be eliminated from the pipeline or modified to reduce the affinity for CYP antitarget enzymes.

However, CYPs can also produce reactive metabolites acting as irreversible inhibitors, one of the most common causes of toxicity and drug–drug interactions [2]. The action of the products of these enzymes is a common cause of adverse drug reactions and failures in drug development. Adverse drug reactions ranked fourth as the most common cause of death in the United States, costing the healthcare industry an estimated $136 billion per annum [3]. Thus, there has been a recent emphasis placed on predicting CYP–ligand interactions, to have chemical structures with no pronounced affinity for CYP antitargets.

Human UDP-glucuronosyl transferases (UGTs) are dominant phase II conjugative metabolism enzymes that process a large number of endogenous compounds and xenobiotics, transforming them into biologically inactive, water-soluble and rapidly excreted compounds. They can also be considered as antitarget enzymes. UGTs act by regulating the level of many potent endobiotic signaling molecules and drugs from all major therapeutic classes. Influencing UGT–drug recognition may allow for smaller doses of compounds to be administered to patients, resulting in higher efficacy and fewer side effects [4].

Antitargets. Edited by R. J. Vaz and T. Klabunde
Copyright © 2008 WILEY-VCH Verlag GmbH & Co. KGaA, Weinheim
ISBN: 978-3-527-31821-6

In crop protection as well, understanding plant metabolism is of paramount importance to increase selectivity and to address resistance of chemical compounds. Moreover, dissipation of a compound in the aquatic ecosystem is very similar to the excretion phenomena of the bodies. An extensive amount of evidence has been accumulated to support the involvement of CYPs in the metabolism and detoxification of herbicides, fungicides and insecticides. The understanding of their biotransformations at the molecular level may be extremely helpful for herbicide- or insecticide-synergistic development.

In the above-mentioned examples, the prediction of CYP-mediated compound interactions is a starting point in any metabolic pathway prediction or enzyme inactivation. This chapter presents an evolution of a standard method [1], widely used in pharmaceutical research in the early-ADMET (absorption, distribution, metabolism, excretion and toxicity) field, which provides information on the biotransformations produced by CYP-mediated substrate interactions. The methodology can be applied automatically to all the cytochromes whose 3D structure can be modeled or is known, including plants as well as phase II enzymes. It can be used by chemists to detect molecular positions that should be protected to avoid metabolic degradation, or to check the suitability of a new scaffold or prodrug. The fully automated procedure is also a valuable new tool in early-ADMET where metabolite- or mechanism based inhibition (MBI) must be evaluated as early as possible.

12.2
The CYPs as Antitarget Enzymes

The CYP family is composed of a large group of monooxygenases that mediate the metabolism of xenobiotics and endogenous compounds. If a drug is to be orally active, it should be both chemically and metabolically stable. Metabolism normally only takes place at a specific position of a molecular skeleton and, unfortunately, metabolic regularities are exceptions. Experienced chemists also find it very difficult to predict where metabolism occurs in a molecule [1].

It is obvious that when a molecule is metabolically labile, it will not be sufficiently available 'at the site of action'. The problem may be addressed when the labile site(s) in a molecule is(are) known, because metabolic transformation may be reduced, if not blocked, by appropriate modification of the chemical structure. The same applies in the field of crop protection science. If an insecticide is metabolically labile and transformed by plant CYPs, it will not be available for protection against the targeted insects.

For these reasons, researchers have recently focused on developing faster robotic systems and more sensitive analytical metabolite identification tools [5–9]. However, such techniques are usually resource demanding, consuming a considerable amount of compound, and cannot be used before compound synthesis. Also, because of the increasing number of potential candidates, experimental metabolite identification remains a huge challenge.

Some *in silico* procedures are able to report the most likely position(s) of biotransformations in a molecular skeleton, thus suggesting the path forward to prevention [1]. By using this information, fast metabolism can be modulated and

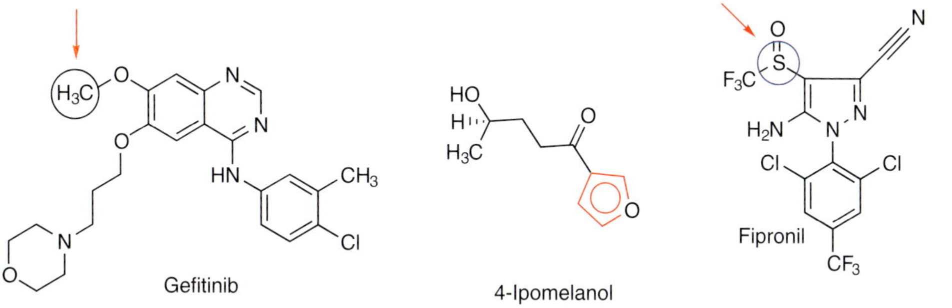

Figure 12.1 The *in silico* prediction of the site of metabolism for Gefitinib and Fipronil, and the *in silico* prediction of MBI for 4-ipomelanol, showing the problematic molecular moiety.

the site of metabolism oriented toward acceptable molecular site(s) by synthesizing appropriate molecular analogues.

Sometimes CYPs can also produce reactive metabolite species that, instead of undergoing the normal detoxification pathway, can act as irreversible CYP inhibitors, thus causing toxicity. Such reactive metabolites that cause CYP inactivation are called MBI and are described in Chapter 9. Mechanism-based enzyme inhibition is associated with irreversible or quasi-irreversible loss of enzyme function, requiring synthesis of new enzymes before activity is restored. The consequences of MBI could be auto-inhibition of the clearance of the inactivator itself or prolonged inhibition of the clearance of other drugs that are cleared by the same isozyme. There may also be serious immunotoxicological consequences if a reactive intermediate is covalently bound to the enzyme. Therefore, screening of new compounds for MBI is now a standard practice within the pharmaceutical industry.

However, simple knowledge of the inhibition mechanism or inhibition level is not helpful in designing a safer compound. Researchers need to understand why a specific compound leads to MBI at the molecular level to design a safer compound. This can be achieved only by understanding which molecular group is exposed to the CYP–heme, to make ligand–heme interactions more difficult (Figure 12.1).

12.3
The UGTs as Antitarget Enzymes

The UGT family is composed of a large group of transmembrane enzymes localized predominantly in the endoplasmic reticulum and nuclear membranes. UGTs are not only most abundantly expressed in the liver but also present in a variety of other tissues. Similar to CYPs, UGTs are promiscuous enzymes that share a similar catalytic mechanism. These enzymes are involved in the metabolism of a wide range of drugs and xenobiotics, where UGTs catalyze the covalent addition of glucuronic acid sugar moieties to a host molecule. The predominant functional group used for conjugation is a hydroxyl group, but other similar groups can also be used for conjugation [10]. This reaction frequently results in the bio-inactivation of a compound, followed by rapid excretion. Table 12.1 shows some classes of compounds

Table 12.1 Examples of different classes of compounds together with the identification of the site of glucuronidation.

Type	On Molecule	2D Structure with Glucuronidation Site	References
O-Glucuronide			
Phenol	Acetaminophen		[11]
Alcohol	Chloramphenicol		[12]
Carboxyl	Fenoprofen		[13]
N-Glucuronide			
Amine	Desipramine		[14]
Carbamate	Meprobamate		[15]
Sulfonamide	Sulfadimethoxine		[16]

Table 12.1 (Continued)

Type	On Molecule	2D Structure with Glucuronidation Site	References
S-Glucuronide Sulfhydryl	Methimazole		[17]
Carbodithioic acid	Disulfiram		[18]
C-Glucuronide	Phenylbutazone		[19]

forming glucuronides with different functional groups used for conjugation. Glucuronidation is the most common mammalian conjugation pathway and is one of the most important phase II reaction transformation.

As reported in Table 12.1, conjugation reactions take place primarily with hydroxyl, carboxyl, amino, heterocyclic nitrogen and thiol groups. If these groups are not present in the compound, they will be introduced or unmasked with high likelihood by CYPs or other phase I enzymes. There is some evidence that UGTs are closely associated with CYPs, so as a drug is oxidized by the phase I cytochrome, the metabolites are efficiently conjugated. For this reason, the prediction of phase II conjugation reactions is very difficult, because it depends not only on the molecular structure but also on the possible biotransformation that occurs in phase I metabolism.

A variety of important drugs such as the anticancer drug epirubicin are inactivated via UGT-mediated glucuronidation [20]. A UGT inhibitor coadministered with these drugs may allow for smaller doses of such drugs to be administered to patients, resulting in higher efficacy and fewer side effects. Unfortunately, these inhibitors are likely to possess serious side effects because a number of critical biological processes use nucleotide sugars [4].

In silico methods able to highlight the most likely position(s) for conjugation may facilitate the development of new molecular scaffold with decreased UGT sensitivity, thus making the compounds more 'resistant' to UGT conjugation. This may be another route for the development of more effective compounds, made possible by additional structural information on UGT enzymes available (Figure 12.2).

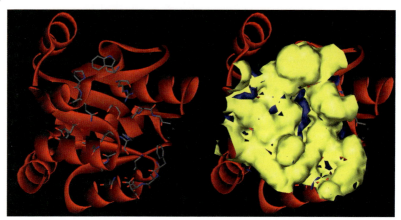

Figure 12.2 The X-ray structure of human UGT2B7 (left) showing the UDPGA-binding site (left), and their molecular interaction fields (right) obtained using GRID force field [21], showing the large cavity and the hydrophilic regions (in blue).

12.4
The MetaSite Technology

MetaSite is a software program that could have a significant effect on drug discovery. It has been developed to predict which part of a potential drug compound will be metabolized by one of the major human CYP enzymes. If knowledge of a compound's degradation site could be predicted in advance, medicinal chemists could protect that site chemically and thus prolong the compound's half-life in the body. Chemists could also use the information to prescreen new drug candidates for suitability, design prodrugs (inactive compounds that become drugs when metabolized) or help assess a drug's ADMET properties [22].

The standard MetaSite methodology involves the calculation of two sets of descriptors, one for the enzyme (i.e. the particular CYP or the UGT enzyme) and the other for the potential substrate, representing the chemical fingerprint of the enzyme and the substrate, respectively. The set of descriptors used to characterize the enzyme is based on GRID flexible molecular interaction fields (GRID-MIFs) [23]. Flexible molecular interaction fields, reported in Figure 12.3, are, in fact, independent of the initial side chain position of the CYP 3D structure and better suited to simulate the adaptation of the enzyme to the substrate structure. To derive the GRID-MIF interaction, the 3D structures of the CYP enzymes are required. These 3D structures can be either modeled or obtained from X-ray crystallography.

The majority of enzyme substrates contain flexible moieties. As the 3D structure of the substrate to be analyzed (the conformation) has a reasonable impact on the outcome of the method, a precise protocol is used to build it. Each substrate is subjected to a conformational search followed by energy minimization by means of a software developed at Molecular Discovery Ltd. The population of conformers is

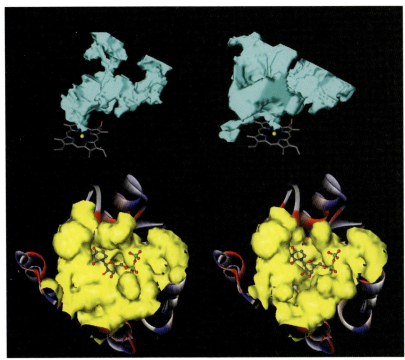

Figure 12.3 Rigid and flexible molecular interaction field maps with the hydrogen probe in the active-site cavity for CYP2C9 and UGT2B7 enzymes. It is worth noting that, with flexible side chains, the overall cavity volume changes considerably. This demonstrates the important role played by MIF-flexibility calculations in enzyme–substrate recognition.

generated under the constraint of obtaining 3D structures induced by the *interaction fields* and shape of the enzyme-active site.

Starting from a conformation of the substrate molecule, GRID-MIFs are computed around the substrate molecules. Then, all the substrate atoms are classified into GRID probe categories depending on their hydrophobic, hydrogen bond donor or acceptor capabilities. Their distances in space are then binned and transformed into clustered distances. One set of descriptors is computed for each category of atom type: hydrophobic, hydrogen bond acceptor and hydrogen bond donor, yielding a fingerprint for each atom in the molecule. The distances between the different atom positions classified using the previous criteria are then transformed into binned distances. The distances between the different atoms are calculated and a value of one or zero is assigned to each bin distance, indicating the presence or the absence of such a distance in the substrate.

The MIFs were subsequently transformed and simplified according to this scheme: in a first step, the regions close to the binding site but not accessible to the substrates were removed from the analysis. Then, the selected interaction points were used to calculate a new set of descriptors using the GRIND technology developed by Pastor *et al.* [24]. For each enzyme–probe interaction, this approach transforms the interaction energies at a certain spatial position (MIF descriptors) into a number of histograms that capture the 3D pharmacophoric interaction patterns of the flexible protein (correlograms). The GRIND technology requires an anchor point inside the enzyme cavity to produce correlograms of interaction energies. For CYP enzymes, the anchor point is represented by the iron of the heme group. For UGT enzymes, the anchor point is represented by the phosphorous atom of the uridine-5-diphosphate-α-D-glucuronic acid (UDPGA).

Once the protein interaction pattern is translated from Cartesian coordinates into distances from the reactive center of the enzyme and the structure of the ligand has been described with similar fingerprints, both sets of descriptors can be compared [25]. The hydrophobic complementarity, the complementarity of charges and H-bonds for the protein and the substrates are all computed using Carbó similarity indices [26]. The prediction of the site of metabolism (either in CYP or in UGT) is based on the hypothesis that the distance between the reactive center on the protein (iron atom in the heme group or the phosphorous atom in UDP) and the interaction points in the protein cavity (GRID-MIF) should correlate to the distance between the reactive center of the molecule (i.e. positions of hydrogen atoms and heteroatoms) and the position of the different atom types in the molecule [27].

Finally, the different atoms in each substrate are assigned a similarity score. Because of the computational mechanism, the score is proportional to the exposure of such substrate atoms toward the reactive heme or UDPGA (in CYP or in UGT, respectively) and represents the accessibility component.

The accessibility component, called E_i, represents the recognition between the specific protein and the ligand when the ligand is positioned in the protein and exposes the atom i to the enzyme anchor point. It depends on the 3D structure, conformation and chirality of the ligand and the 3D structure and side chain flexibility of the enzyme. Thus, the E_i score is proportional to the exposure of the ligand atom i to the anchor point of a specific enzyme.

Although UGTs catalyze only glucuronic acid conjugation, CYPs catalyze a variety of oxidative reactions. Oxidative biotransformations include aromatic and side chain hydroxylation, *N*-, *O*-, *S*-dealkylation, *N*-oxidation, sulfoxidation, *N*-hydroxylation, deamination, dehalogenation and desulfation. The majority of these reactions require the formation of radical species; this is usually the rate-determining step for the reactivity process [28]. Hence, reactivity contributions are computed for CYPs, but a different computation is performed with the UGT enzyme (as described in Section 12.4.2).

When R_i is the reactivity of the atom i in the appropriate reaction mechanism, it represents the activation energy required to produce the reactive intermediate. It depends on the ligand 3D structure and on the mechanism of reaction. Therefore, R_i is a score proportional to the reactivity of the ligand atom i in a specific reaction mechanism.

There are minimal experimental data available for the R_i component of drug-like compounds. However, the quantification of the R_i component can be approximated using *different* methods or fragmental approaches. Fragments can be classified according to their liability: stable, nonreactive, medium, moderate and strongly reactive. After recognition, the reactivity component R_i can be assigned.

Summing up, the site of metabolism can be described by a probability function P_{SM} that is correlated to, and can be considered, the free energy of the overall process:

$$P_{SM\ i} = E_i R_i,$$

where P_i is the probability of the atom i to be the site of metabolism due to the CYP–heme group; E_i is the exposition of the atom i to the heme group; and R_i is the reactivity of the atom i in the actual mechanism of reaction, as stated above.

For the same ligand, the P_{SM} function assumes different values for different atoms according to the E_i and R_i components.

12.4.1
Mechanism-Based Inhibitors

X-ray crystal structures (see Figure 12.4) show that compounds that act as MBIs should initially bind in the active site in close proximity of the heme group, where the initial metabolism reaction occurs [29]. The heme group is considered crucial in activating, orienting as well as reacting with the substrate.

In comparison with the metabolism algorithm of the site of prediction, MetaSite uses a slightly different approach to evaluate potential MBI. First, the fingerprint similarity between the CYP active site and the tested ligand molecule is computed (as for the site of metabolism prediction). If the maximum similarity between CYP and ligand fingerprints occurs when certain ligand chemical group faces the heme, the ligand can be considered as inhibitor or substrate depending on whether its chemical

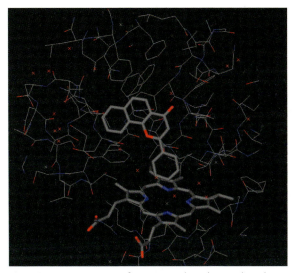

Figure 12.4 X-ray structure for a potential mechanism-based inhibitor cocrystallized in CYP1A2 [29].

group is labile or stable toward oxidation. When the chemical group facing the heme is stable toward oxidation, the reaction would not occur and the ligand is a potential inhibitor. However, when the chemical group facing the heme could lead to a reactive intermediate (in general, an electrophilic species), then the ligand could covalently react with the heme or be activated, leading to an MBI. It is important to note that this approach can be considered 'structure based', as it depends on the CYP 3D structure and not on the training set.

As the method reported is not based on QSAR and thus does not use training information, it cannot be used to predict quantitative inhibition levels (IC_{50}, percentage inhibition or similar information). The method can only be used to find 'potential' mechanism-based inhibitors (binary scale). Despite this limitation, an important advantage over other techniques is that the method suggests the site of the molecule responsible for the inhibition mechanism. With this information researchers may change the molecular structure to maintain activity at the expense of inhibitory effects.

Figure 12.5 indicates the results of this technique when applied to three MBIs and three non-MBI compounds. The three MBIs are clearly recognized and the electrophilic intermediate leading to MBI clearly reported. The three non-MBI compounds are also well predicted (as non-MBIs) despite the presence on the molecular scaffold of well-known potential electrophilic intermediates. Similar to the prediction of the site of metabolism, the example demonstrates that the recognition component E_i is more important than the reactivity component R_i to determine if a compound may lead to MBI.

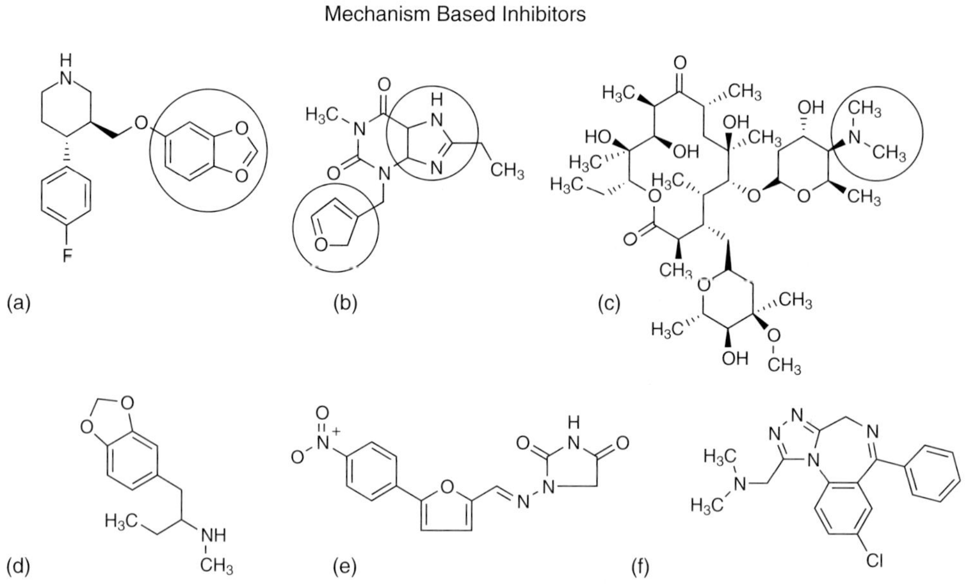

Figure 12.5 Three MBIs ((a) paroxetine, CYP2D6; (b) furafylline, CYP1A2; (c) erythromycin, CYP3A4) [30] and three substrates ((d) MBDB, CYP2D6; (e) dantrolene, CYP3A4; (f) adinazolam, CYP3A4) of different CYP isoforms.

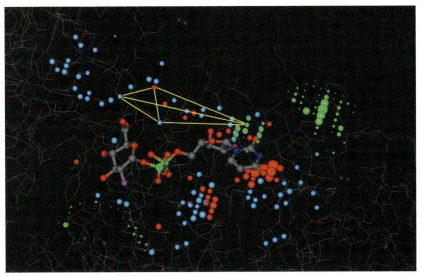

Figure 12.6 MIFs in proximity to the UDPGA-binding site.

12.4.2
Phase II Metabolism by UGTs

Figure 12.6 shows the molecular interaction fields obtained from UGT2B7 studied in proximity of the UDPGA-binding site. The UDPGA-binding site is highly conserved in human UGTs.

Graphical inspection of Figure 12.6 shows that, when the movements of the flexible side chains are considered, the H-bond acceptor blue regions are clustered in particular 3D regions of the enzyme. From the locations obtained by using flexible GRID-MIF, a pharmacophoric pattern for the UGT enzymes can be created by using the FLAP (fingerprints for ligands and proteins) method [31]. The procedure is then repeated using the maps derived by the hydrophobic probe, H-bond donor, H-bond acceptor and charged probes. The global pharmacophoric pattern produced is then compared with the pharmacophore of a substrate, left conformationally free to fit the enzyme pharmacophoric positions. A local similarity between the two matching pharmacophores is then computed. If the procedure is repeated for all the potential glucuronidation sites, each local similarity score may be then used to rank the substrate position for glucuronidation (Figure 12.7).

12.4.3
The Flowchart of the Overall Method

The MetaSite procedure is fully automated and does not require any user assistance. All the work can be submitted and handled in a batch computation. The flexible MIFs for CYP structures and those for the UGT structures obtained from the GRID package are precomputed and stored. Once the structures of the

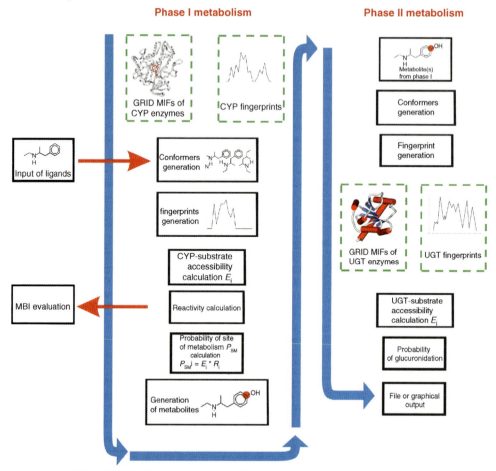

Figure 12.7 The flowchart of the overall method.

xenobiotic compounds are provided, the conformers are generated in the enzyme cavities, and the fingerprints are calculated (for every conformer). For CYP site of metabolism prediction, the accessibility component E_i is computed by comparing the fingerprints of the enzymes with those of the ligand conformers.

The obtained accessibility E_i is also used for MBI calculation. The reactivity component for exposed groups is then computed, considering that R_i values may be different according to the reaction mechanism under consideration. Therefore, different values of the reactivity component are assigned to an exposed group when the group is oxidized or when a reactive metabolite is produced (for inhibition). After the computation of these two components, the global probability function can now be computed, allowing the ranking of different metabolites. The metabolite structures are then generated, ready for subsequent analyses. If the compound is not an MBI, it may be transformed into a potential MBI after the reaction with the CYP enzyme. Therefore, search for potential inhibitors may be started again from all the 'predicted' metabolites (only those metabolite with high probability to be produced in CYP metabolism prediction).

The predicted metabolites are also the starting point for the phase II metabolic prediction, to find where glucuronidation could occur. All the probable metabolites obtained from CYP metabolism reactions are submitted to a possible phase II reaction catalyzed by UGTs, using the UGT structure(s) as a template. The accessibility component is computed in the UGT cavity to prioritize glucuronic acid transfer. The final metabolite structures are then reported in graphical output or saved to a file.

12.5 Conclusions

A methodology is described to predict the site of metabolism and the potential MBI by CYPs for compounds as well as subsequent possible phase II metabolism by UGTs. On average, for about 85% of the cases, the method predicted the right site of metabolism within the first two atoms in the ranking list and for more than 80% of the MBI inhibitors. The same methodology can also be applied to predict phase II, UGT-mediated metabolism.

The method is based on flexible MIFs generated by the GRID force field on the CYPs and UGT homology modeled structures that were treated and filtered to extract the most relevant information.

The methodology is very fast and completely automated. To predict the site of metabolism for drug-like substrates, the method requires few seconds per molecule. It is important to point out that the method uses neither any training set nor any statistical model or supervised technique, and it has proven to be predictive for extensively diverse validation sets preformed in different pharmaceutical companies.

The methodology can be used either to suggest modifications in compounds to avoid a certain metabolic profile or to check the suitability of a prodrug. Also, this procedure can be used to filter potential interactions of virtual compounds for early toxicity monitoring. Because of the increasing number of potential candidates, identification of the experimental metabolite remains a huge challenge. It has also been clearly demonstrated that the use of this new *in silico* approach to predict a hypothetical metabolite structure can speed up the process of metabolite identification by focusing experimental work on specific target structures, thus improving the method of metabolite structure confirmation and elucidation.

Although the current software package was prepared to work with the five most important human cytochromes, this procedure can be adopted to work with any CYP structure and applied to humans, bacteria, fish and plant CYPs. There are more than 120 P450 families, and more than 1000 P450 enzymes. All these structures can in theory be imported, processed and used for the prediction of the site of metabolism. The procedure is totally automated and does not require user assistance. It only requires the availability of the 3D structure of the enzyme and input of the compound(s) to be evaluated.

12.6
Software Package

The overall procedure is called MetaSite (site of metabolism prediction) [32]. The MIF calculations, pharmacophoric recognition, descriptor handling, similarity computation, reactivity computation, inhibition and subselectivity are calculated automatically once the ligand compound(s) is(are) provided. The complete calculation is performed in few seconds in IRIX Linux or Windows environments.

Abbreviations

ADMET	absorption, distribution, metabolism, excretion, toxicity
CYP	cytochrome P450
MBI	mechanism/metabolite-based inhibition/inhibitors
MIF	molecular interaction field
UDPGA	uridine-5-diphosphate-α-D-glucuronic acid
UGT	UDP-glucuronosyl transferases

References

1 Cruciani, G., Carosati, E., De Boeck, B., Ethirajulu, K., Mackie, C., Howe, T. and Vianello, R. (2005) MetaSite: understanding metabolism in human cytochromes from the perspective of the chemist. *Journal of Medicinal Chemistry*, 48, 6970–6979.

2 Fontana, E., Dansette, P.M. and Poli, S.M. (2005) Cytochrome P450 enzymes mechanism based inhibitors: common sub-structures and reactivity. *Current Drug Metabolism*, 6, 413–454.

3 Johnson, J.A. and Bootman, J.L. (1995) Drug-related morbidity and mortality. A cost of illness model. *Archives of Internal Medicine*, 155 (18), 1949–1956.

4 Miley, M.J., Zielinska, A.K., Keenan, J.E., Bratton, S.M., Pandya, A.R. and Redinbol, M.R. (2007) Crystal structure of the cofactor-binding domain of the human phase II drug-metabolism enzyme UDP-glucuronosyltransferase 2B7. *Journal of Molecular Biology*, 369 (2), 498–511.

5 Thomas, S.R. and Gerhard, U. (2004) Open-access high-resolution mass spectrometry in early drug discovery. *Journal of Mass Spectrometry*, 39, 942–948.

6 Kantharaj, E., Tuytelaars, A., Proost, P., Ongel, Z., Assouw, H.P. and Gilissen, R.A. (2003) Simultaneous measurement of drug metabolic stability and identification of metabolities using ion-trap mass spectrometry. *Rapid Communications in Mass Spectrometry*, 17, 2661–2668.

7 Kostiainen, R., Kotiano, T., Kuuranne, T. and Auriola, S. (2003) Liquid chromatography-atmospheric pressure ionization/mass spectrometry (LC-API/MS) in drug metabolism studies. *Journal of Mass Spectrometry*, 38, 357–372.

8 Corcoran, O. and Spraul, M. (2003) LC–NMR–MS in drug discovery. *Drug Discovery Today*, 8, 624–631.

9 Nassar, A.E.F. and Talaat, R.E. (2004) Strategies for dealing with metabolite elucidation in drug discovery and

development. *Drug Discovery Today*, **9**, 317–327.
10. Silverman, R.B. (2004) *The Organic Chemistry of Drug Design and Drug Action*, 2nd edn (ed. R.B. Silverman) Elsevier Academic Press, Burlington, MA.
11. Cummings, A.J., King, M.L. and Martin, B.K. (1967) Pharmacokinetics of paracetamol (acetaminophen) after intravenous and oral administration. *British Journal of Pharmacology and Chemotherapy*, **29**, 150.
12. Nakagawa, T., Masada, M. and Uno, T.J. (1975) Gas chromatographic determination and gas chromatographic-mass spectrophotometric analysis of chloramphenicol, thiamphenicol and their metabolites. *Journal of Chromatography*, **111**, 355.
13. Rubin, A., Warrick, P., Wolen, R.L., Chernish, S.M., Ridolfo, A.S. and Gruber, C.M. Jr. (1972) Physiological disposition of fenprofen in man. III. Metabolism and protein binding of phenprofen. *Journal of Pharmacology and Experimental Therapeutics*, **183**, 449.
14. Bickel, M.H., Minder, R. and diFrancesco, C. (1973) Formation of *N*-glucuronide of desmethylimipramine in the dog. *Experientia*, **29**, 960.
15. Tsukamoto, H., Yoshimura, H. and Tatsumi, K. (1963) Metabolism of drugs. XXXV. Metabolic fate of meprobamate. (3). A new metabolic pathway of carbamate group – the formation of meprobamate *N*-glucuronide in animal body. *Chemical & Pharmaceutical Bulletin*, **11**, 421.
16. Adamson, R.H., Bridges, J.W., Kibby, M.R., Walker, S.R. and Williams, R. (1970) The fate of sulphadimethoxine in primates compared with other species. *The Biochemical Journal*, **118**, 41.
17. Sitar, D.S. and Thornhill, D.P. (1973) Methimazole absorption, metabolism and excretion in the albino rat. *Journal of Pharmacology and Experimental Therapeutics*, **184**, 432.
18. Dutton, G.J. and Illing, H.P.A. (1972) Mechanism of biosynthesis of thio-β-D-glucuronides and thio-β-D-glucosides. *The Biochemical Journal*, **129**, 539.
19. Dieterle, W., Faigle, J.W., Frueh, F., Mory, H., Theobald, W., Alt, K.O. and Richter, W.J. (1976) Metabolism of phenylbutazone in man. *Arzneimittel Forschung*, **26**, 572.
20. Innocenti, F., Iyer, L., Ramirez, J., Green, M.D. and Ratain, M.J. (2001) Epirubicin glucuronidation is catalyzed by human UDP-glucuronosyltransferase 2B7. *Drug Metabolism and Disposition*, **29**, 686–692.
21. Carosati, E., Sciabola, S. and Cruciani, G. (2004) Hydrogen bonding interactions of covalently bonded fluorine atoms: from crystallographic data to a new angular function in the GRID force field. *Journal of Medicinal Chemistry*, **47**, 5114–5125.
22. http://pubs.acs.org/cen/news/83/i42/8342notw8.html.
23. Goodford P.J.(1998) Atom movements during drug–receptor interactions. in *Rational Molecular Design in Drug Research*, (eds Liljefors, T., Jorgensen, F.S. and Krogsgaard-Larsen, P.) Alfred Benzon Symposium 42, Munkgaard, Copenhagen pp. 215–230.
24. Pastor, M., Cruciani, G., McLay, I., Pickett, S. and Clementi, S. (2000) GRid-INdependent descriptors (GRIND): a novel class of alignment-independent three-dimensional molecular descriptors. *Journal of Medicinal Chemistry*, **43**, 3233–3243.
25. Cruciani, G., Aristei, Y., Vianello, R. and Baroni, M. (2005) GRID-derived molecular interaction fields for predicting the site of metabolism in human cytochromes. in *Molecular Interaction Fields* (ed. G. Cruciani) Wiley-VCH Verlag GmbH, Weinheim, pp. 273–290.
26. Amat, L. and Carbó-Dorca, R. (1999) Fitted electronic density functions from H to Rn for use in quantum similarity measures: *cis*-diamminedichloroplatinum(II) complex as an application example.

27 Zamora, I., Afzelius, L. and Cruciani, G. (2003) Predicting drug metabolism: a site of metabolism prediction tool applied to the cytochrome P450 2C9. *Journal of Medicinal Chemistry*, **46**, 2313–2324.

28 Guengerich, F.P. (2003) Cytochrome P450 oxidations in the generation of reactive electrophiles: epoxidation and related reactions. *Archives of Biochemistry and Biophysics*, **409**, 59–71.

29 Sansen, S., Yano, J.K., Reynald, R.L., Schoch, G.S., Griffin, K.J., Stout, C.D. and Johnson, E.F. (2007) Adaptation for the oxidation of polycyclic aromatic hydrocarbons exhibited by the structure of human P450 1A2. *J. Biol. Chem.*, **282**, 14348–14355.

30 Obach, R.S., Walsky, R.L. and Venkatakrishnan, K. (2007) Mechanism-based inactivation of human cytochrome p450 enzymes and the prediction of drug–drug interactions. *Drug Metabolism and Disposition*, **35**, 246–255.

31 Baroni, M., Cruciani, G., Sciabola, S., Perruccio, F. and Mason, J.S. (2007) A common reference framework for analyzing/comparing proteins and ligands: FLAP theory and application. *Journal of Chemical Information and Modeling*, **47**, 279–294.

32 MetaSite is available at www.moldiscovery.com.

13
Orphan Nuclear Receptor PXR-Mediated Gene Regulation in Drug Metabolism and Endobiotic Homeostasis

Jie Zhou, Yonggong Zhai, Wen Xie

13.1
Cloning and Initial Characterization of PXR

In 1998, several groups published the cloning and initial characterization of the orphan nuclear receptor (NR) pregnane X receptor (PXR). Kliewer and colleagues identified mouse PXR on the basis of its sequence homology with other NRs. It was named PXR based on its activation by various natural and synthetic pregnanes [1]. The cloning of human PXR (hPXR) was first reported by Blumberg and colleagues and was initially called 'steroid and xenobiotic receptor (SXR) by this group (led by Dr Ronald M. Evans at the Salk Institute) based on its activation by steroids and xenobiotics [2]. The hPXR was also cloned by Bertilsson and colleagues and was referred to as the pregnane-activated receptor (PAR) [3].

PXR shares a common NR modular structure with a conserved N-terminal DNA-binding domain (DBD) and a C-terminal ligand-binding domain (LBD). The DBD contains two zinc-finger motifs, which mediate interaction with specific DNA sequences known as hormone response elements (HREs). The LBD, in addition to determining ligand binding specificity, contains a ligand-inducible transactivation function (AF2) and a motif that directs binding to the common heterodimerization partner retinoid X receptor (RXR). Although DBDs of the mammalian PXRs are highly conserved, sharing more than 95% amino acid identity, the LBDs of PXRs are much more divergent across species than those of other NRs. For example, the human and rat PXRs share only 76% amino acid identity in their LBD, whereas most human and rodent NR orthologs share more than 90% amino acid identity [4]. The LBD sequence divergence of PXR between species is believed to be responsible for the species-specific response to ligands (see below for more discussion).

PXR is highly expressed in the liver and small intestine, where most drug-metabolizing enzymes are also highly expressed and induced. In rodents, lower levels of PXR mRNA have also been detected in the kidney, stomach, lung, uterus, ovary and placenta. PXR can be activated by xenobiotics (including many clinical drugs) and endobiotics (such as bile acids and some hormones, including

Antitargets. Edited by R. J. Vaz and T. Klabunde
Copyright © 2008 WILEY-VCH Verlag GmbH & Co. KGaA, Weinheim
ISBN: 978-3-527-31821-6

glucocorticoids, antiglucocorticoids and estrogens), which underlie the significance of PXR in both xenobiotic and endobiotic responses. PXR orthologs also exhibit interesting and striking species-dependent ligand specificity. For example, the antiglucocorticoid pregnenolone-16α-carbonitrile (PCN) is an effective activator of mouse and rat PXRs but has little effect on human PXR. In contrast, the antibiotic rifampicin (RIF) can activate the human and rabbit PXRs, but not the mouse and rat PXRs. The PXR ligand species specificity and its biological significance will be elaborated later on.

Like many other orphan nuclear receptors, the activation of PXR by ligand binding leads to the formation of PXR/RXR heterodimer and transcriptional activation of target genes by binding to specific PXR DNA response elements (PXREs) (Figure 13.1). The classical PXRE is DR-3, which is composed of two copies of direct repeat of the consensus AG(G/T)TCA NR binding motif and separated by three nucleotides. Other PXREs, including DR-4, DR-5, ER-6 (everted repeat spaced by six nucleotides), ER-8 (everted repeat separated by eight nucleotides) and IR-0 (inverted repeat without a spacing nucleotide), have also been shown to be bound by the PXR/RXR heterodimers [5–7].

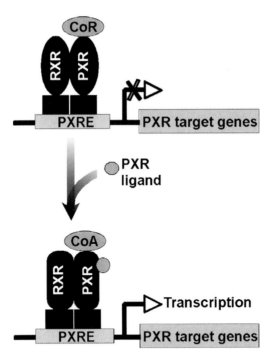

Figure 13.1 Current model of PXR-mediated gene regulation. Function of PXR requires the heterodimerization with the RXR. Binding of ligands to PXR results in the disassociation of corepressors (CoRs) and recruitment of coactivator (CoA) and subsequent activation of target gene transcription. The regulation is mediated by PXRE present in the promoter regions of the target genes.

13.2
PXR and Its Regulation of Drug-Metabolizing Enzymes and Transporters

Physiological homeostasis requires the detoxification and removal of endogenous hormones and xenobiotic compounds with biological activity. Much of the detoxification and elimination are carried out by the phase I cytochrome P450 (CYP) enzymes, phase II conjugation enzymes and 'phase III' drug transporters. The products of phase I metabolism are generally more polar and readily excreted than the parent compounds. The phase I enzyme-mediated functionalization also often enables xenobiotics to become better substrates for the phase II conjugating enzymes. The phase II metabolism involves conjugation of hydrophilic moieties, such as sugar, glutathione and sulfate, to further increase polarity and water solubility of xenobiotics and therefore promoting their excretion and elimination. In addition to phase I and phase II enzymes, equally important is a group of transporter proteins expressed in various tissues, such as the liver, intestine, brain and kidney, which modulate the absorption, distribution and excretion of many drugs.

13.2.1
PXR in Phase I CYP Enzyme Regulation

The oxidative CYP enzymes catalyze the metabolic conversion of xenobiotics to polar derivatives that are more readily eliminated. Among the 78 mouse and 65 human cytochrome P450 enzymes itemized within the current build of the Unigene sequence databank, a significant number of these enzymes are known to be induced by various environmental pollutants, pharmaceuticals or steroid metabolites. A more current list of CYPs is available at http://drnelson.utmem.edu/CytochromeP450.html. Among CYP enzymes, the CYP3A isozymes are of particular medical significance as they are involved in the metabolism of more than 50% of clinical drugs as well as neutraceuticals and herbal medicines [8]. In 1998, PXR was isolated as a candidate xenobiotic receptor (xenosensor) postulated to regulate CYP3A gene expression [1–3,9]. PXR regulates the expression of both human and rodent CYP3A genes by directly binding to the xenobiotic response elements (XREs) ER-6 and DR-3 localized in the promoter regions of the human and rodent CYP3A genes, respectively. Both 'loss-of-function' gene knockout and 'gain-of-function' transgenic mouse studies have provided genetic and pharmacological evidence to support the role of PXR in CYP3A-mediated xenobiotic responses. Targeted disruption of the mouse PXR locus abolished the CYP3A xenobiotic response to prototypic inducers such as PCN and dexamethasone [10,11]. In contrast, hepatic expression of an activated form of hPXR in transgenic mice resulted in sustained induction of CYP3A enzymes and enhanced protection against xenobiotic toxicants such as zoxazolamine and tribromoethanol [10].

In addition to CYP3A isozymes, PXR has also been shown to regulate other CYP enzymes, including CYP2B6 [12–14], CYP2B9 [15], CYP2C8 [16] and CYP2C9 [15].

13.2.2
PXR in Phase II Enzyme Regulation

The phase II conjugating enzymes include broad-specificity transferases such as UDP-glucuronosyltransferase (UGT), glutathione S-transferase (GST) and sulfotransferases (SULTs) [17–20]. Phase II conjugation has a dual role in drug metabolism. First, conjugation reactions not only terminate the ability of electrophiles to react with DNA and proteins, but also prevent nucleophiles from interacting with receptor proteins. Second, conjugation increases the water solubility of the compounds, which, in turn, promotes renal and biliary excretion [21]. Therefore, the phase II reactions play a critical role in detoxifying both exogenous and endogenous chemicals.

13.2.2.1 PXR in UGT Regulation

Glucuronidation, a major metabolic pathway for many endo- and xenobiotics, is catalyzed by enzymes belonging to the family of membrane-bound UGTs. Using UDP-glucuronic acid as a sugar donor, UGTs catalyze the transfer of glucuronic acid to a variety of substrates and thus convert small lipophilic molecules to water-soluble glucuronides. UGT-mediated glucuronidation functions as the principal means of eliminating steroid, heme metabolites, environmental toxins and drugs from the body [20,22–24]. Among UGTs, UGT1A1 is one of the most characterized UGT isoform. UGT1A1 is the crucial enzyme responsible for the detoxification of bilirubin. Deficiency in the expression and/or activity of UGT1A1 can lead to severe accumulation of unconjugated bilirubin, medically termed hyperbilirubinemia, a hallmark of the Crigler–Najjar (CN) syndrome [20].

Our own study showed that UGT1A1 is under the positive control of PXR. UGT1A1 mRNA and protein expression were upregulated in transgenic mice that express the activated hPXR (VP-hPXR) and in RIF-treated 'humanized' hPXR transgenic mice [25]. The microsomal glucuronidation activity toward β-estradiol (a UGT1A1 substrate), thyroid hormones, corticosterone and xenobiotics (such as 4-nitrophenol and 4-OH-PhIP) was also increased in the VP-hPXR transgenic mice. Chen and colleagues showed that PCN, a rodent-specific PXR agonist, could induce UGT expression and increase UGT enzymatic activity in wild-type mice and this induction was abolished in PXR-null mice [26]. The identification of a DR-3-like PXR-responsive element (*GGTTCATAAAGGGTA*) in the human UGT1A1 promoter further established UGT1A1 as a direct target gene of PXR. In addition to UGT1A1, the expression of UGT1A9, but not UGT1A2 and 2B5, was also increased by PXR activation, suggesting that the UGT induction by PXR is isoform specific [26].

In addition to PXR, the expression of UGT1A1 and several other UGT isoforms have also been reported to be regulated by several other nuclear receptors, including constitutive androstane receptor (CAR) [25,27,28] and peroxisome proliferator activated receptor α (PPARα) [29,30].

13.2.2.2 PXR in SULT Regulation

The cytosolic SULTs are another important family of phase II enzymes that catalyze the conjugation of nucleophilic compounds. SULTs catalyze the transfer of a sulfonyl group from a sulfate donor 3'-phosphoadenosine 5'-phosphosulfate (PAPS) to hydroxyl or amino groups of acceptor molecules, forming sulfate or sulfamate conjugates. SULTs are specific to sulfonate small lipophilic molecules such as steroids, bioamines and therapeutic drugs. It was thought that, though SULTs and UGTs have a similar substrate spectrum, SULTs have higher affinities but lower turnover rates than UGTs [31]. Thus, UGT activities are likely to play a more significant role when substrates are abundant because of the higher turnover rate of UGTs, whereas SULTs may play a leading role in cases where concentrations of the substrate are lower. The SULT-mediated sulfonation is known to play a significant role in the homeostasis of sex hormones that are present in low concentrations in plasma [32].

The expression of rodent hepatic Sult2a9, also called Sult2a1 or dehydroepiandrosterone sulfotransferase (DHEA SULT or STD), is subjected to transcriptional regulation by PXR and an IR-0 element located in the 5'-flanking region of rodent Sult2a gene is required for its activation by PXR [33]. The expression of Sult2a9/2a1/STD has also been shown to be regulated by farnesoid X receptor (FXR) [34], CAR [35] and liver X receptor (LXR) [36]. Interestingly, all four receptors share the same IR-0 response element to regulate the expression of this SULT isoform. The hierarchy and relative contribution of individual nuclear receptors in Sult2a regulation remain to be determined.

13.2.2.3 PXR in GST Regulation

GSTs are soluble homo- and heterodimeric enzymes that use reduced glutathione in conjugation and reduction reactions. GSTs play an important role in the conjugation of electrified chemicals. Since chemical carcinogens are often highly nucleophilic, they also represent potential substrates for GST. For this reason, GSTs are believed to play an important role to protect cells from genotoxic compounds. The cytosolic GST isozymes of rodents and humans can be grouped into several classes, such as Alpha, Mu, Pi, Theta-Omega and Zeta, based on their amino acid sequences, immunological properties and substrate specificities [37,38]. The Alpha, Mu and Pi classes are the most abundantly expressed GSTs.

GST regulation by PXR was first hinted by several studies on general profiling of gene expression [39,40]. It was also reported that GSTA2 expression was induced by PXR and an IR-6 response element was responsible for this transcriptional regulation [37]. A more systemic and comparative analysis of GST regulation by PXR was reported in transgenic mice that bear the expression of activated PXR in the liver and intestine [41]. In this study, it was shown that the expressions of GST Alpha, Pi and Mu classes are all under the control of PXR. Interestingly, PXR-mediated GST regulation exhibited clear isoform, tissue and gender specificities: (1) GST Mu was the only isoform that was upregulated by PXR in both liver and intestine in both sexes; (2) GST Alpha was induced in the small intestine, but not in the liver; and (3) PXR had an opposite effect on hepatic GST Pi expression, inducing this class in females, but suppressing it in males.

Paradoxically, although the overall GST expression and activity were increased, activation of PXR sensitized the response to the oxidative xenotoxicant paraquat *in vivo* and in cultured cancer cells. Moreover, heightened paraquat sensitivity in transgenic mice was female specific. Whether the PXR-mediated, gender-specific GST regulation accounts for the intact paraquat sensitivity in transgenic males remains to be determined. Nevertheless, the regulation of GSTs by PXR suggests that this regulatory pathway may be relevant to carcinogenesis by sensitizing normal and cancerous tissues to oxidative cellular damage [41].

13.2.3
PXR in Drug Transporter Regulation

In addition to regulating phase I CYP enzymes and phase II conjugating enzymes, PXR also participates in the regulation of drug transporters, which are responsible for uptake and efflux of endogenous and exogenous chemicals, including many clinically prescribed drugs; thus, the expression and activity of transporters affect drug efflux and clearance [42]. Organic anion-transporting polypeptides (Oatps) are a family of major uptake drug transporters in the liver. Oatps are localized to the basolateral membrane of hepatocytes and transport compounds, including organic anions, organic cations and neutral compounds, from blood into hepatocytes. Oatps family includes 9 human, 13 rat and 15 mouse members [43]. PXR ligand PCN administration induced Oatp2 mRNA expression in the livers of wild-type mice, but not PXR-null mice [44]. A DR-3 type PXR response element has been identified in the rat Oatp2 promoter [45].

Multidrug resistance-associated proteins (Mrps) are efflux transporters for structurally diverse amphipathic chemicals and organic anions. The liver-enriched Mrp2 is localized to the canalicular membrane of hepatocytes and is responsible for the hepatobiliary excretion of amphipathic anions [46,47]. Natural mutations in MRP2 (multidrug resistance-associated protein 2) cause Dubin–Johnson syndrome/hyperbilirubinemia II, a disorder characterized by impaired transfer of anionic conjugates into the bile. Induction of Mrp2 by PXR ligand PCN and dexamethasone was observed in primary hepatocytes isolated from the livers of wild-type mice, but not from PXR-null mice, implicating a role of PXR in the regulation of mouse Oatp2 [47]. An ER-8 type of NR response element was identified in the promoter of the rat Mrp2 gene that confers the induction of Mrp2 by PXR [47]. Mrp3 was also reported to be induced by PXR [48]. In the intestine, PXR has been shown to stimulate the expression of multidrug resistance protein 1 (Mdr1), which encodes an ATP-dependent efflux pump that transports a wide variety of xenobiotics, including many widely used prescription drugs. In humans, activation of PXR in the intestine may decrease intestinal drug absorption by increasing the expression of *Mdr1*. A DR-4-type PXR response element was identified in *Mdr1* gene promoter [49].

13.3
Crosstalk Between PXR and Other Nuclear Receptors

Emerging evidence points to the existence of receptor crosstalk between PXR and other nuclear receptors. An early example is the PXR–CAR crosstalk in regulating

CYP3A and CYP2B genes. PXR and CAR were initially shown to regulate CYP3A and CYP2B genes, respectively. Interestingly, it was demonstrated that PXR and CAR can cross-regulate each other's target genes by binding the response elements reciprocally. Thus, PXR can bind to the previously shown CAR-responsive phenobarbital response element (PBRE) to regulate *CYP2B* gene expression, whereas CAR can also activate CYP3A through binding to the PXR response elements [50].

In addition to regulating phase I enzymes, PXR also cooperates with other nuclear receptors to regulate the expression of phase II enzymes and drug transporters. PXR and CAR have been shown to induce the expression of UGT1A1, OATP2, GSTA1 and GSTA2, all of which are important for bilirubin detoxification. The sharing of bilirubin detoxifying target genes may have accounted for a similar resistance to experimental hyperbilirubinemia in mice with PXR or CAR genetically or pharmacologically activated [25,27,28]. In another example of receptor crosstalk, MRP2, a transporter mediating the efflux of several conjugated compounds across the apical membrane of the hepatocytes into the bile canaliculi, is regulated by FXR, PXR and CAR [47]. An ER-8 type of NR response element was identified in the promoter of the rat *Mrp2* gene. This element is capable of binding to and conferring PXR, CAR and FXR responsiveness on a heterologous thymidine kinase promoter [47]. This is reminiscent of the coregulation of the rodent Sult2a9/2a1 by FXR, PXR, CAR and LXR, in which an IR-0 was the shared response element [33–36].

Another interesting example of functional interplay between xenobiotic receptors was revealed in the characterization of PXR-null mice in bilirubin clearance [51]. After knowing that activation of PXR is sufficient to detoxify bilirubin, it was surprising to find that the PXR-null mice also showed increased bilirubin clearance. We proposed that, when both PXR and CAR are present, ligand-free PXR suppresses the constitutive activity of CAR, maintaining a basal capacity of bilirubin detoxification. The increased bilirubin clearance in PXR-null mice was probably the result of CAR derepression as a consequence of the PXR loss. This notion was supported by the observation that the pattern of enzyme and transporter regulation in the PXR-null mice was remarkably similar to that of transgenic mice expressing activated CAR [51].

In a more recent example, we showed that PXR, LXR and PPARγ cooperate to regulate the free fatty acid transporter CD36 and to promote fatty acid mediated hepatic lipogenesis. In this case, three distinct NR response elements were found for each of the three receptors and they are clustered in 500-bp sequences in the mouse *Cd36* gene promoter (J. Zhou and W. Xie, unpublished results). Consistent with the coregulation of *Cd36*, activation of both PXR and LXR had an additive effect to induce hepatic triglyceride content (J. Zhou and W. Xie, unpublished results).

13.4
Implications of PXR-Mediated Gene Regulation for Drug Metabolism and Pathophysiology

The implications of PXR-mediated gene regulation for drug metabolism and drug interaction have been recognized since the initial cloning of this xenobiotic receptor.

Consistent with the notion that xenobiotic enzymes and transporters are also involved in the biotransformation and homeostasis of many endogenous chemicals, accumulating evidence has also pointed to a role of PXR in normal physiology and diseases.

13.4.1
PXR in Drug Metabolism and Drug–Drug Interactions

Our body encounters numerous xenobiotic chemicals, including prescription drugs, over-the-counter and herbal medicines. Many of them, especially when accumulated in excess, may exert toxic effects through various mechanisms. The process of drug metabolism is known to largely depend upon a concerted action of phase I and II enzymes, as well as drug transporters. Expressed mainly in the liver and intestine, these enzymes and transporters are capable of recognizing an amazing diversity of xenobiotics to promote their clearance.

13.4.1.1 PXR in Drug Metabolism

PXR can be activated by a variety of xenobiotics and activation of PXR leads to the regulation of phase I and phase II enzymes and drug transporters [5–7]. The identification of RIF as a potent hPXR agonist has provided an explanation why this drug is a good inducer of drug-metabolizing enzymes and thus prone to drug–drug interaction, a phenomenon that has long been recognized in the clinic. Another example of PXR-activating clinical drug is paclitaxel (Taxol), one of the most commonly used antineoplastic agents. Paclitaxel is subject to metabolic inactivation by hepatic CYP3A4 and CYP2C8. In addition to being inactivated by hepatic P450 enzymes, paclitaxel is excreted from the intestine via P-glycoprotein (p-gp or ABCB1), a broad-specificity efflux pump protein encoded by the gene *MDR1* (also known as *ABCB1*). Synold and colleagues showed that paclitaxel-activated SXR/hPXR can induce hepatic expression of *CYP3A4* and *CYP2C8*, as well as *MDR1* expression in intestinal tumor cells. This hPXR-mediated drug clearance pathway may enhance metabolism and efflux of paclitaxel, which may lead to increased intestinal drug excretion and resistance [52]. In contrast, docetaxel (Taxotere), a closely related antineoplastic agent, did not activate SXR/hPXR and thus displayed superior pharmacokinetic properties [52].

Other than RIF and Taxol, many other commonly used clinical drugs have also been shown to activate PXR. These include peptide-mimetic HIV protease inhibitors [53], the cholesterol-lowering lovastatin and the anti-inflammatory dexamethasone [54]. A more comprehensive analysis of the effect of commonly used clinical drugs on PXR activation has recently been published by Sinz and colleagues [55].

13.4.1.2 PXR in Drug–Drug Interactions

The regulation of drug-metabolizing enzymes by PXR is involved in clinical drug–drug interactions, in which one drug accelerates the metabolism of a second medicine and may change or cause adverse results. Because CYP enzymes can recognize a large spectrum of pharmaceutical substrates, a CYP gene-inducing drug is potentially capable of affecting the metabolism and clearance of any coconsumed drugs. As mentioned earlier, the identification of RIF as a potent hPXR agonist has provided an explanation why this antibiotic drug is prone to drug–drug interactions. In another example, St John's wort (SJW), a popular herbal remedy for depression, has been reported to trigger severe adverse interactions with several clinical drugs such as oral contraceptives, the HIV protease inhibitor indinavir and the immunosuppressant cyclosporin. Such drug–drug interactions are likely to result from the activation of PXR and consequent induction of CYP3A by SJW and the subsequent increased metabolism and/or decreased bioavailability of cometabolized drugs. In the case of birth control pills, the use of SJW enhances drug clearance, increasing contraceptive failure and thus the birth of 'miracle babies'. The identification of SJW as a PXR agonist offered a plausible explanation for the propensity of SJW to cause drug–drug interactions [56].

Several traditional Chinese medicines (TCMs), essential components of alternative medicines, have also been implicated in drug–drug interactions. One clinical concern about such herbal products is their effect on the metabolism of coadministered drugs. We showed that two TCM herbs Wu Wei Zi (*Schisandra chinensis* Baill) and Gan Cao (*Glycyrrhiza uralensis* Fisch) can activate PXR and induce the expression of several drug-metabolizing enzymes and transporters, including CYP3A, 2C9 and MRP2 in reporter gene assays and in primary hepatocyte cultures [57]. The anticoagulant warfarin is known to be metabolized by CYP2C9 in humans [58]. As expected, administration of Wu Wei Zi and Gan Cao extracts in rats resulted in an increased metabolism of coadministered warfarin, reinforcing concerns involving the safe use of herbal medicines and other neutraceuticals to avoid PXR-mediated drug–drug interactions [57].

After knowing the potential of PXR-activating agents in causing drug–drug interactions, it is also important to emphasize that PXR activation alone may not be sufficient to predict the propensity of drug–drug interactions. Sinz and colleagues recently published the evaluation of 170 xenobiotics in an hPXR transactivation assay and compared these results to known clinical drug–drug interactions. Of the 170 xenobiotics tested, 54% of them demonstrated some level of hPXR transactivation. However, by taking into consideration cell culture conditions (solubility, cytotoxicity and appropriate drug concentration in media), as well as *in vivo* pharmacokinetics (therapeutic plasma concentration or C_{max}, distribution, route of administration, dosing regimen, liver exposure and potential to inhibit CYP3A4), the risk potential of CYP3A4 enzyme induction for most compounds reduced dramatically. By employing this overall interpretation strategy, the final percentage of compounds predicted to induce CYP3A4 significantly reduced to 5%, all of which are known to cause drug–drug interactions in the clinic [55].

13.4.1.3 Enantiospecificity of PXR-Activating Drugs and its Implication in Drug Development

We have recently reported the synthesis and identification of novel PXR agonists from a library of peptide isosteres [59]. One of the most interesting findings in this study was the identification of compound **S20**, a C-cyclopropylalkylamide, as a unique PXR agonist with both enantiomer- and species-specific selectivity. **S20** has three chiral carbons and was resolved into two enantiomers. The individual **S20** enantiomers exhibited striking mouse/human-specific PXR activation, in which the enantiomer (+)-**S20** preferentially activated hPXR, while enantiomer (−)-**S20** was a better activator for mPXR. To our knowledge, **S20** represents the first compound whose enantiomers have opposite species preference in activating a xenobiotic receptor.

The enantiospecificity of PXR agonism has several potential implications in pharmaceutical development. First of all, for pharmaceutical agents whose therapeutic effects do not rely on PXR, the choice of PXR-neutral but therapeutically effective enantiomers may help to avoid unwanted drug–drug interactions. However, for drugs with PXR as therapeutic target, an hPXR-specific enantiomer will be necessary to achieve intended therapeutic effects in humans. Thus, the identification of species- and enantiomer-specific ligands for PXR can be both desirable and valuable. This work takes a first step in this direction by demonstrating exquisite species-dependent enantiomer-based selectivity.

13.4.2
Endobiotic Function of PXR

Even though PXR was initially identified as a 'xenobiotic receptor', emerging evidence has pointed to an equally important role of PXR as an 'endobiotic receptor' that responds to a wide array of endogenous chemicals (endobiotics). Moreover, the activation of PXR by endogenous or xenobiotic ligands has implications in several important physiological and pathological conditions. For this reason, there has been a lot of discussion on whether PXR can be explored as a therapeutic target [7,60].

13.4.2.1 PXR in Bile Acid Detoxification and Cholestasis

One family of endogenous PXR ligands identified shortly after the cloning of PXR is bile acids, which are catabolic end products of cholesterol metabolism. They are physiologically important in the formation of bile and solubilizing biliary lipids and promoting their absorption. However, excessive bile acids are potentially toxic. For example, the secondary bile acid lithocholic acid (LCA) has been shown to cause cholestasis in experimental animals and has long been suspected of doing the same in humans. As an average human releases 600 ml of bile a day, the potential for disrupting bile flow (cholestasis) and the resultant accumulation of toxic by-products is significant. Therefore, excess bile acid should be efficiently eliminated to avoid the toxic effect.

PXR has been demonstrated to act as an LCA sensor and plays an essential role in detoxification of cholestatic bile acids [11,61]. Studies in different animal models showed that activation of PXR protected against severe liver damage induced by LCA. Pretreatment of wild-type mice, but not the PXR-null mice, with PCN reduced the toxic effects of LCA. Moreover, genetic activation of PXR by expressing the activated

PXR in the liver of transgenic mice was sufficient to confer resistance to LCA hepatotoxicity. The cholestatic preventive effect of PXR was initially thought to be due to the activation of CYP3A, an important CYP enzyme responsible for bile acid hydroxylation [11,61]. Subsequent identification of SULT2A, a bile acid detoxifying hydroxysteroid sulfotransferase, as a PXR target gene suggested that additional PXR target genes may also have contributed to the phenotype [33]. Several follow-up studies, including those using mice with individual or combined loss of PXR and CAR, have suggested that PXR-responsive bile acid transporter regulation may also play a role in preventing cholestasis [62–64]. As activation of PXR was sufficient to prevent cholestasis, it has been suggested that PXR agonists may prove useful in the treatment of human cholestatic liver disease, a notion that has been supported by several clinical observations. Both RIF and SJW have been empirically used to treat cholestatic liver diseases [5]. The relief from cholestasis-associated pruritus and amelioration of cholestasis by RIF was associated with increased 6α-hydroxylation of bile acids, which in turn facilitates glucuronidation by the UGTs at the 6α-hydroxy position. Both RIF and SJW are potent agonists of hPXR and both CYP3A and UGT are PXR target genes, suggesting the anticholestatic effects of RIF and SJW are mediated by PXR.

13.4.2.2 PXR in Bilirubin Detoxification and Clearance

Bilirubin is the catabolic by-product of heme proteins such as β-globin and CYP enzymes. Accumulation of bilirubin in the blood is potentially hepato- and neurotoxic. For example, an insufficiency in expression of UGT1A1, a key enzyme for the conjugation of bilirubin, in the Crigler–Najjar syndrome and Gilbert's disease results in severe hyperbilirubinemia. Deficiency of MRP2, a drug transporter responsible for the hepatic excretion of conjugated bilirubin, leads to Dubin–Johnson syndrome, characterized by the accumulation of glucuronidated bilirubin. PXR has been shown to induce the expression of multiple key components in the clearance pathway, including UGT1A1, OATP2, GSTA1 and GSTA2 and MRP2. OATP2 facilitates bilirubin uptake from blood into hepatocytes [42]. GSTA1 and GSTA2 reduce bilirubin back efflux from hepatocytes into blood. MRP2 promotes the canalicular efflux of conjugated bilirubin. Consistent with the pattern of gene regulation, activation of PXR in transgenic mice has been shown to prevent experimental hyperbilirubinemia [25].

13.4.2.3 PXR in Adrenal Steroid Homeostasis and Drug–Hormone Interactions

PXR plays an important endobiotic role in adrenal steroid homeostasis. Our recent study showed that genetic (VP-hPXR transgene) and pharmacological (ligand and rifampicin) activation of hPXR in mice markedly increased plasma concentrations of corticosterone and aldosterone, the respective primary glucocorticoid and mineralocorticoid in rodents. The increased levels of corticosterone and aldosterone were associated with activation of adrenal steroidogenic enzymes, including CYP11a1, CYP11b1, CYP11b2 and 3β-Hsd. The PXR-activating transgenic mice also exhibited hypertrophy of the adrenal cortex, a loss of glucocorticoid circadian rhythm and a lack of glucocorticoid responses to psychogenic stress [65].

Interestingly, the VP-hPXR transgenic mice had normal pituitary secretion of adrenocorticotropic hormone (ACTH) and the corticosterone suppressing effect of dexamethasone (DEX) was intact, suggesting a functional hypothalamus–pituitary–adrenal (HPA) axis despite a severe disruption of adrenal steroid homeostasis. The ACTH-independent hypercortisolism in the PXR-activating transgenic mice is reminiscent of the pseudo-Cushing's syndrome in patients, the clinical hallmark of which is the normal DEX suppression despite a high circulation level of glucocorticoid. Pseudo-Cushing's syndrome is mostly seen in alcoholic, depressed or obese subjects. It is of interest to know whether or not these susceptible patients are associated with increased expression and/or activity of PXR. The glucocorticoid effect appeared to be PXR specific, as the activation of CAR in transgenic mice had little effect. We propose that PXR is a potential endocrine disrupting factor that may have broad implications in steroid homeostasis and drug–hormone interactions [65].

13.4.2.4 PXR in Lipid Metabolism

PXR has also recently been shown to play an endobiotic role by impacting lipid homeostasis [66]. Expression of an activated PXR in the livers of transgenic mice resulted in an increased hepatic deposit of triglycerides. This PXR-mediated lipid accumulation was independent of the activation of the lipogenic transcriptional factor sterol regulatory element binding protein 1c (SREBP-1c) and its primary lipogenic target enzymes, including fatty acid synthase (FAS) and acetyl CoA carboxylase 1 (ACC-1). Instead, the lipid accumulation in transgenic mice was associated with an increased expression of the free fatty acid transporter CD36 and several accessory lipogenic enzymes such as stearoyl CoA desaturase-1 (SCD-1) and long chain free fatty acid elongase (FAE). Studies using transgenic and knockout mice showed that PXR is both necessary and sufficient for *Cd36* activation. Promoter analyses revealed a DR-3 type of PXR response element in the mouse *Cd36* gene promoter, establishing *Cd36* as a direct transcriptional target of PXR. The hepatic lipid accumulation and *Cd36* induction was also seen in the hPXR 'humanized' mice treated with the hPXR agonist RIF. The activation of PXR was also associated with an inhibition of pro-β-oxidative genes, such as PPARα and thiolase, and an upregulation of PPARγ, a positive regulator of CD36. The cross-regulation of CD36 by PXR and PPARγ suggests that this fatty acid transporter may function as a common target of orphan nuclear receptors in their regulation of lipid homeostasis.

13.5 Species Specificity of PXR and the Creation of "Humanized" Mice

13.5.1 Challenges of using Rodents as Drug Metabolism Models

Primary hepatocytes are a common *in vitro* system to evaluate metabolism, toxicity and enzyme induction [67–69]. Human hepatocytes are considered the most relevant system to evaluate or predict human metabolism or effects of a new drug. A significant disadvantage of this system is the lack of routine availability of good quality human liver tissues or cells. Other model systems that may provide more consistent access include immortalized hepatocytes or humanized animal models [10,70]. Over the years, it has been perceived that rodent models have limited utility in predicting drug-related human effects due to significant species differences in drug-metabolizing enzymes, transporters and nuclear hormone receptors. For example, PCN is an effective CYP3A inducer in rodents but not in humans, and rifampicin induces CYP3A in humans but not in rats [71]. These findings have been attributed to the species differences in the effect of several drugs on CYP3A expression mediated by PXR [72]. These differences across species exemplify the need to develop humanized animal models to evaluate the potential effect of a chemical in humans using an animal model.

13.5.2
Species Specificity of the Rodent and Human PXR

Both hPXR/SXR and mPXR are highly expressed in the liver and small intestine and share many functional properties, in particular, the regulation of *CYP3A* genes. However, as discussed earlier, these two orthologs pharmacologically distinct as strong activators of one receptor are often poor activators of the other. This species-specific ligand profile is reflected by the sequence divergence in the LBDs of mouse and human receptors. The crystal structure of the hPXR LBD has been solved [73]. The hydrophobic ligand-binding cavity of hPXR is composed of a large, smooth surface containing only a small number of polar residues, suggesting that it is not necessary for activators to conform to a restricted orientation. Based on site-directed mutational analysis, the position and nature of these polar residues were found to be crucial to establishing the precise pharmacological activation profile of hPXR [73]. Indeed, conversion of four amino acids of mouse PXR that correspond to the hPXR-specific activator SR12813-interacting residues in hPXR produces a hybrid mouse–human PXR that was no longer activated by PCN and was only weakly activated by SR12813 in reporter assays [73]. The structural and pharmacological differences between hPXR and mPXR and that of other species might reflect the difference in the diets of rodents and primates and the evolutionary need to respond to a different set of ingested nutrients and xenobiotics.

13.5.3
Creation and Characterization of the hPXR "Humanized" Mice

On the basis of the fact that the species origin of the receptor determines the ligand specificity between species, we have created transgenic mouse models to determine whether the human receptor is sufficient to establish a human response profile [10] (Figure 13.2). First, hPXR transgenic mice were generated by causing the hPXR expression in their livers. The liver-specific expression of the transgene was accomplished by using the mouse albumin promoter. Because the resulting mice harbor both mPXR and hPXR in their livers, the transgenic mice exhibited a chimeric or combined CYP3A response to both the rodent-specific inducer PCN and the human-specific

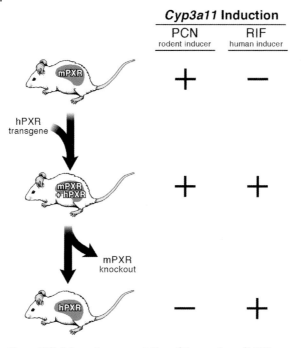

Figure 13.2 Schematic representation of the creation of hPXR 'humanized' mice. The humanization was achieved in the liver only when the liver-specific albumin promoter was used to direct the transgene expression, or in both the liver and the intestine when the fatty acid binding protein promoter was used. PCN, pregnenolone-16α-carbonitrile; RIF, rifampicin. '+' and '−' mean induction and lack of induction, respectively.

inducer rifampicin. These results imply that the mice expressing hPXR only could be fully humanized for the xenobiotic response. These animal models were created by breeding the hPXR transgene into the mPXR knockout background. In contrast to the null mice that are devoid of CYP3A induction by steroids, replacement of mPXR with transgenic hPXR restores xenobiotic regulation with a humanized response profile. These mice readily responded to human inducers, such as rifampicin, in the equivalent range of the standard oral dosing regimen in humans (300–600 mg per 70 kg man) and exhibited similar pharmacokinetics of CYP3A regulation [10,74]. A 'fully' human profile of CYP3A inducibility is obtained in the mPXR null/hPXR transgenic mice. Therefore, these experiments provide compelling evidence that PXR functions as a species-specific xenosensor mediating the adaptive hepatic response. This is also one of the rare examples in which replacing a single transcriptional regulator enables conversion of species-specific gene regulation.

The original 'humanized' mice bear the expression hPXR exclusively in the liver [10]. As both drug-metabolizing enzymes and xenobiotic receptors are also highly expressed in the intestinal tracts, it is conceivable that mouse models with the humanized receptors expressed in both the liver and the intestine would represent a more complete humanized mouse model. The liver and intestine dual humanization has been achieved by using the fatty acid binding protein (FABP) gene promoter that targets the expression of hPXR transgene to both the liver and the intestine [66]. An alternative strategy is to 'knock-in' hPXR in the mouse locus. This would not only direct expression of hPXR in both the liver and the intestine, but also normalize expression levels and tissue patterns to the endogenous gene.

13.5.4
Significance of Humanized Mice in Drug Metabolism Studies and Drug Development

The creation of mouse models with humanized xenobiotic response may aid pharmaceutical development by predicting potential drug–drug interactions [5–7,75]. For decades, rodent models have been standard components in the assessment of potential toxicity for the development of candidate human drugs. However, their reliability as predictors of the human xenobiotic response is limited because of the species specificity of the xenobiotic response. To date, there has been no reliable system outside of humans to assess drug–drug interactions directly and quantitatively. Primary cultures of human hepatocytes are valuable. However, since the human hepatocytes are drawn from individual patients, their utility is compromised by interindividual variability, limited tissue resources and high cost. The humanized mice exhibited a 'humanized' hepatic xenobiotic response profile, readily responding to the human-specific inducer RIF in a concentration range equivalent to the standard oral dosing regimen in humans [10]. The creation of these mice represents a major step toward generating a humanized rodent toxicological model that is continuously renewable and completely standardized. In addition, a PXR-mediated and mechanism-based transfection and reporter gene system has also been shown to be an effective *in vitro* approach to screen for drugs that may be precocious hPXR activators. Although the *in vitro* screening is fast, the availability of hPXR 'humanized' mice offers a unique screening tool to evaluate drug–drug interactions *in vivo*. The humanized mouse models represent important steps in the development of safer human drugs.

Another benefit of the hPXR humanized mice is its utility as a pharmacological model to dissect the function of PXR *in vivo*. The PXR ligand effect on gene expression and/or pathophysiological outcome may require long-term drug treatment. Although PCN is a potent rodent PXR agonist, the wild-type mice cannot tolerate a long-term treatment of PCN because of its toxicity. In this case, the use of humanized mice becomes necessary since mice can tolerate chronic treatment of RIF, the hPXR agonist. We have successfully used the humanized mice in our recent studies of PXR effect on lipogenesis and adrenal steroid homeostasis [65,66].

13.6
Conclusions

PXR has been extensively studied in the past 7–8 years. It has become clear that PXR can function as a master xenosensor to regulate the expression of drug-metabolizing enzymes and transporters. Another notable development is the appreciation of PXR as an 'endobiotic receptor'. PXR not only can sense the endobiotics, but also control the homeostasis of many endobiotic chemicals and thus the receptor has many potential pathophysiological implications. Moreover, there is significant crosstalk among xenobiotic receptors that are manifested by overlap in both ligands and target gene regulations. The receptor crosstalk is believed to be the molecular basis for the fail–safe xenobiotic regulatory networks that facilitate host protection.

It also appears that the PXR-controlled xenobiotic regulation is a double-edged sword. One of the remaining outstanding challenges is whether the biological actions of PXR make this receptor a drug target for treatment of human diseases such as hyperbilirubinemia and bile acid-associated cholestasis and colon cancers. Although a PXR-activating drug has potential for adverse drug–drug interactions, we have also begun to appreciate that activation of PXR does not always lead to drug–drug interactions [55]. Moreover, the potential of drug–drug interaction alone should not exclude PXR as a therapeutic target. The latter notion is supported by the facts that many of the PXR-activating agents, including rifampicin and phenobarbital, have been used successfully as clinical drugs.

Since xenobiotic receptors mediate pharmacological and genetic control of the expression of drug-metabolizing enzymes and transporters, the identification of PXR and other xenobiotic receptors has also opened up a new perspective in pharmacogenetics and pharmacogenomics. Pharmacogenetics has traditionally focused on the polymorphism within the coding sequences of genes that encode various enzymes and transporters. After enhancing our understanding of pharmacogenetics, the cDNA polymorphisms may not explain all of the interindividual and interrace variations in enzyme activity. The identification of xenobiotic nuclear receptors leads to several important questions from a pharmacogenomic perspective: (1) Are there natural allelic variants of PXR or other xenobiotic receptors that exhibit differential transactivation potency to induce enzymes and transporters? (2) Whether there are polymorphisms in the promoter regions of target enzyme or transporter genes that may alter the binding affinity of xenobiotic receptors? Recent reports appear to support the implications of PXR in pharmacogenetics [6,76,77]. However, we believe many more comprehensive studies are necessary before this pharmacogenomic information can be applied to practice true 'personalized' medicine.

On a final note, it is important to appreciate that the expression of drug-metabolizing enzymes and transporters has also been shown to be under the control of transcriptional factors other than nuclear hormone receptors. One such non-NR transcriptional factor is aryl hydrocarbon receptor (AhR), a member of the basic helix–loop–helix PER-ARNT-SIM (PAS) transcription factor family. AhR regulates gene expression by heterodimerizing with the aryl hydrocarbon receptor nuclear translocator (Arnt). Like PXR and CAR, activation of AhR has been reported to simultaneously induce both phase I and phase II enzymes [78,79]. The mammalian CYP1A1, CYP1A2 and CYP1B1 genes are AR target genes identified initially [80,81]. More recently, it was found that AhR is also implicated in the regulation of UGTs, including UGT1A1 and 1A6 [82–86]. AhR-binding xenobiotic responsive elements (XREs) have been identified in the human and rodent UGT1A1 gene promoter [82]. An XRE was reported for the rat

Ugt1a6 gene and found to be both necessary and sufficient to mediate *Ugt1a6* transactivation by AhR [86]. Interestingly, the PXR-, CAR- and AhR-responsive elements are closely localized in the UGT1A1 gene promoter [87]. It remains to be determined whether there is coordination or crosstalk between AhR and CAR/PXR in xenobiotic regulation. The existing literature does support crosstalk between AhR and nuclear receptor signaling. For example, AhR has been shown to interact with a number of NRs such as the estrogen receptor α (ERα), chicken ovalbumin upstream promoter-transcription factor I (COUP-TFI) and estrogen-related receptor α (ERRα). Moreover, the AhR/Arnt heterodimers can utilize certain NR response elements such as the naturally occurring estrogen response elements [83].

Abbreviations

ACC	acetyl CoA carboxylase
AF-2	activation function 2
AhR	aryl hydrocarbon receptor
CAR	constitutive androstane receptor
CN	Crigler–Najjar
COUP-TFI	chicken ovalbumin upstream promoter-transcription factor I
CYP	cytochrome P450
DBD	DNA-binding domain
DR-3	direct repeat of the consensus sequence AG(G/T)TCA separated by three nucleotides
ER	estrogen receptor
ER-6	everted repeat spaced by six nucleotides
ERR	estrogen-related receptor
FABP	fatty acid binding protein
FAE	fatty acid elongase
FAS	fatty acid synthase
FXR	farnesoid X receptor
GST	glutathione *S*-transferase
HREs	hormone response elements
IR-0	inverted repeat without a spacing nucleotide
LBD	ligand-binding domain
LXR	liver X receptor
MRP	multidrug resistance-associated protein
MDR	multidrug resistance protein
NR	nuclear receptor
OATP	organic anion-transporting polypeptide
PAR	pregnane-activated receptor
PBRE	phenobarbital response element
PGP	P-glycoprotein
PPAR	peroxisome proliferator activated receptor
PXR	pregnane xenobiotic receptor

PXRE	pregnane xenobiotic response element
RIF	rifampicin
RXR	retinoid X receptor
SCD	stearoyl CoA desaturase
SJW	St John's wort
SREBP	sterol regulatory element binding protein
SULT	sulfotransferase
SXR	steroid and xenobiotic receptor
TCM	traditional Chinese medicine
UGT	UDP-glucuronosyltransferase
XREs	xenobiotic response elements

Acknowledgments

We apologize to authors and laboratories whose original research is not cited in this chapter due to space limitation. Our original research described here was supported in part by NIH grants ES012479, ES014626 and CA107011, the Department of Defense Breast Cancer Research Program grant BC23189, the Susan G. Komen Breast Cancer Foundation grant PDF0503458, and the Competitive Medical Research Fund and the Central Research Development Fund from the University of Pittsburgh.

References

1 Kliewer, S.A., Moore, J.T., Wade, L., Staudinger, J.L., Watson, M.A., Jones, S. A., McKee, D.D., Oliver, B.B., Willson, T. M., Zetterstrom, R.H., Perlmann, T. and Lehmann, J.M. (1998) An orphan nuclear receptor activated by pregnanes defines a novel steroid signaling pathway. *Cell*, **92**, 73–82.

2 Blumberg, B., Sabbagh, W., Jr, Juguilon, H., Bolado, J., Jr, van Meter, C.M., Ong, E. S. and Evans, R.M. (1998) SXR, a novel steroid and xenobiotic-sensing nuclear receptor. *Genes & Development*, **12**, 3195–3205.

3 Bertilsson, G., Heidrich, J., Svensson, K., Asman, M., Jendeberg, L., Sydow-Backman, M., Ohlsson, R., Postlind, H., Blomquist, P. and Berkenstam, A. (1998) Identification of a human nuclear receptor defines a new signaling pathway for CYP3A induction. *Proceedings of the National Academy of Sciences of the United States of America*, **95**, 12208–12213.

4 Maglich, J.M., Sluder, A., Guan, X., Shi, Y., McKee, D.D., Carrick, K., Kamdar, K., Willson, T.M. and Moore, J.T. (2001) Comparison of complete nuclear receptor sets from the human, *Caenorhabditis elegans* and *Drosophila* genomes. *Genome Biology*, **2**, RESEARCH0029. 1-0029.7.

5 Kliewer, S.A., Goodwin, B. and Willson, T. M. (2002) The nuclear pregnane X receptor: a key regulator of xenobiotic metabolism. *Endocrine Reviews*, **23**, 687–702.

6 Sonoda, J., Rosenfeld, J.M., Xu, L., Evans, R.M. and Xie, W. (2003) A nuclear receptor-mediated xenobiotic response and its implication in drug metabolism

and host protection. *Current Drug Metabolism*, **4**, 59–72.

7 Xie, W., Uppal, H., Saini, S.P., Mu, Y., Little, J.M., Radominska-Pandya, A. and Zemaitis, M.A. (2004) Orphan nuclear receptor-mediated xenobiotic regulation in drug metabolism. *Drug Discovery Today*, **9**, 442–449.

8 Maurel, P. (1996) The CYP3A family, in *Cytochrome P450: Metabolic and Toxicological Aspects*, (ed. Ioannides C.) CRC Press, Boca Raton, FL, pp. 241–270.

9 Lehmann, J.M., McKee, D.D., Watson, M. A., Willson, T.M., Moore, J.T., and Kliewer, S.A. (1998) The human orphan nuclear receptor PXR is activated by compounds that regulate CYP3A4 gene expression and cause drug interactions. *Journal of Clinical Investigation*, **102**, 1016–1023.

10 Xie, W., Barwick, J.L., Downes, M., Blumberg, B., Simon, C.M., Nelson, M. C., Neuschwander-Tetri, B.A., Brunt, E. M., Guzelian, P.S. and Evans, R.M. (2000) Humanized xenobiotic response in mice expressing nuclear receptor SXR. *Nature*, **406**, 435–439.

11 Staudinger, J.L., Goodwin, B., Jones, S.A., Hawkins-Brown, D., MacKenzie, K.I., LaTour, A., Liu, Y., Klaassen, C.D., Brown, K.K., Reinhard, J., Willson, T.M., Koller, B.H. and Kliewer, S.A. (2001) The nuclear receptor PXR is a lithocholic acid sensor that protects against liver toxicity. *Proceedings of the National Academy of Sciences of the United States of America*, **98**, 3369–3374.

12 Wang, H. and Negishi, M. (2003) Transcriptional regulation of cytochrome p450 2B genes by nuclear receptors. *Current Drug Metabolism*, **4**, 515–525.

13 Honkakoski, P., Zelko, I., Sueyoshi, T. and Negishi, M. (1998) The nuclear orphan receptor CAR–retinoid X receptor heterodimer activates the phenobarbital-responsive enhancer module of the CYP2B gene. *Molecular and Cellular Biology*, **18**, 5652–5658.

14 Goodwin, B., Moore, L.B., Stoltz, C.M., McKee, D.D. and Kliewer, S.A. (2001) Regulation of the human CYP2B6 gene by the nuclear pregnane X receptor. *Molecular Pharmacology*, **60**, 427–431.

15 Dvorak, Z., Modriansky, M., Pichard-Garcia, L., Balaguer, P., Vilarem, M.J., Ulrichova, J., Maurel, P. and Pascussi, J.M. (2003) Colchicine down-regulates cytochrome P450 2B6, 2C8, 2C9, and 3A4 in human hepatocytes by affecting their glucocorticoid receptor-mediated regulation. *Molecular Pharmacology*, **64**, 160–169.

16 Ferguson, S.S., Chen, Y., LeCluyse, E.L., Negishi, M. and Goldstein, J.A. (2005) Human CYP2C8 is transcriptionally regulated by the nuclear receptors constitutive androstane receptor, pregnane X receptor, glucocorticoid receptor, and hepatic nuclear factor 4alpha. *Molecular Pharmacology*, **68**, 747–757.

17 Hayes, J.D. and Pulford, D.J. (1995) The glutathione S-transferase supergene family: regulation of GST and the contribution of the isoenzymes to cancer chemoprotection and drug resistance. *Critical Reviews in Biochemistry and Molecular Biology*, **30**, 445–600.

18 Salinas, A.E. and Wong, M.G. (1999) Glutathione S-transferases – a review. *Current Medicinal Chemistry*, **6**, 279–309.

19 Nagata, K. and Yamazoe, Y. (2000) Pharmacogenetics of sulfotransferase. *Annual Review of Pharmacology and Toxicology*, **40**, 159–176.

20 Tukey, R.H. and Strassburg, C.P. (2000) Human UDP-glucuronosyltransferases: metabolism, expression, and disease. *Annual Review of Pharmacology and Toxicology*, **40**, 581–616.

21 Sheweita, S.A. (2000) Drug-metabolizing enzymes: mechanisms and functions. *Current Drug Metabolism*, **1**, 107–132.

22 Mackenzie, P.I., Owens, I.S., Burchell, B., Bock, K.W., Bairoch, A., Belanger, A., Fournel-Gigleux, S., Green, M., Hum, D. W., Iyanagi, T., Lancet, D., Louisot, P., Magdalou, J., Chowdhury, J.R., Ritter, J.K., Schachter, H., Tephly, T.R., Tipton, K.F. and Nebert, D.W. (1997) The UDP glycosyltransferase gene superfamily: recommended nomenclature update based on evolutionary divergence. *Pharmacogenetics*, **7**, 255–269.

23 Radominska-Pandya, A., Czernik, P.J., Little, J.M., Battaglia, E. and Mackenzie, P.I. (1999) Structural and functional studies of UDP-glucuronosyltransferases. *Drug Metabolism Reviews*, **31**, 817–899.

24 Mackenzie, P.I., Gregory, P.A., Gardner-Stephen, D.A., Lewinsky, R.H., Jorgensen, B.R., Nishiyama, T., Xie, W. and Radominska-Pandya, A. (2003) Regulation of UDP glucuronosyltransferase genes. *Current Drug Metabolism*, **4**, 249–257.

25 Xie, W., Yeuh, M.F., Radominska-Pandya, A., Saini, S.P., Negishi, Y., Bottroff, B.S., Cabrera, G.Y., Tukey, R.H. and Evans, R.M. (2003) Control of steroid, heme, and carcinogen metabolism by nuclear pregnane X receptor and constitutive androstane receptor. *Proceedings of the National Academy of Sciences of the United States of America*, **100**, 4150–4155.

26 Chen, C., Staudinger, J.L. and Klaassen, C.D. (2003) Nuclear receptor, pregnane X receptor, is required for induction of UDP-glucuronosyltransferases in mouse liver by pregnenolone-16 alpha-carbonitrile. *Drug Metabolism and Disposition*, **31**, 908–915.

27 Huang, W., Zhang, J., Chua, S.S., Qatanani, M., Han, Y., Granata, R. and Moore, D.D. (2003) Induction of bilirubin clearance by the constitutive androstane receptor (CAR). *Proceedings of the National Academy of Sciences of the United States of America*, **100**, 4156–4161.

28 Sugatani, J., Kojima, H., Ueda, A., Kakizaki, S., Yoshinari, K., Gong, Q.H., Owens, I.S., Negishi, M. and Sueyoshi, T. (2001) The phenobarbital response enhancer module in the human bilirubin UDP-glucuronosyltransferase UGT1A1 gene and regulation by the nuclear receptor CAR. *Hepatology*, **33**, 1232–1238.

29 Barbier, O., Villeneuve, L., Bocher, V., Fontaine, C., Torra, I.P., Duhem, C., Kosykh, V., Fruchart, J.C., Guillemette, C. and Staels, B. (2003) The UDP-glucuronosyltransferase 1A9 enzyme is a peroxisome proliferator-activated receptor alpha and gamma target gene. *Journal of Biological Chemistry*, **278**, 13975–13983.

30 Kota, B.P., Huang, T.H. and Roufogalis, B.D. (2005) An overview on biological mechanisms of PPARs. *Pharmacological Research*, **51**, 85–94.

31 Morris, M.E. and Pang, K.S. (1987) Competition between two enzymes for substrate removal in liver: modulating effects due to substrate recruitment of hepatocyte activity. *Journal of Pharmacokinetics and Biopharmaceutics*, **15**, 473–496.

32 Qian, Y.M., Sun, X.J., Tong, M.H., Li, X.P., Richa, J. and Song, W.C. (2001) Targeted disruption of the mouse estrogen sulfotransferase gene reveals a role of estrogen metabolism in intracrine and paracrine estrogen regulation. *Endocrinology*, **142**, 5342–5350.

33 Sonoda, J., Xie, W., Rosenfeld, J.M., Barwick, J.L., Guzelian, P.S. and Evans, R.M. (2002) Regulation of a xenobiotic sulfonation cascade by nuclear pregnane X receptor (PXR). *Proceedings of the National Academy of Sciences of the United States of America*, **99**, 13801–13806.

34 Song, C.S., Echchgadda, I., Baek, B.S., Ahn, S.C., Oh, T., Roy, A.K. and Chatterjee, B. (2001) Dehydroepiandrosterone sulfotransferase gene induction by bile acid activated farnesoid X receptor. *Journal of Biological Chemistry*, **276**, 42549–42556.

35 Saini, S.P., Sonoda, J., Xu, L., Toma, D., Uppal, H., Mu, Y., Ren, S., Moore, D.D., Evans, R.M. and Xie, W. (2004) A novel constitutive androstane receptor-mediated and CYP3A-independent pathway of bile acid detoxification. *Molecular Pharmacology*, **65**, 292–300.

36 Uppal, H., Saini, S.P.S., Moschetta, A., Mu, Y., Zhou, J., Gong, H., Zhai, Y., Ren, S., Michalopoulos, G.K., Mangelsdorf, D.J. and Xie, W. (2007) Activation of LXRs prevents bile acid toxicity and cholestasis in female mice. *Hepatology*, **45**, 422–432.

37 Falkner, K.C., Pinaire, J.A., Xiao, G.H., Geoghegan, T.E. and Prough, R.A. (2001) Regulation of the rat glutathione S-transferase A2 gene by glucocorticoids: involvement of both the glucocorticoid and pregnane X receptors. *Molecular Pharmacology*, **60**, 611–619.

38 Townsend, D.M. and Tew, K.D. (2003) The role of glutathione-S-transferase in anti-cancer drug resistance. *Oncogene*, **22**, 7369–7375.

39 Rosenfeld, J.M., Vargas, R., Jr, Xie, W. and Evans, R.M. (2003) Genetic profiling defines the xenobiotic gene network controlled by the nuclear receptor pregnane X receptor. *Molecular Endocrinology*, **17**, 1268–1282.

40 Maglich, J.M., Stoltz, C.M., Goodwin, B., Hawkins-Brown, D., Moore, J.T. and Kliewer, S.A. (2002) Nuclear pregnane X receptor and constitutive androstane receptor regulate overlapping but distinct sets of genes involved in xenobiotic detoxification. *Molecular Pharmacology*, **62**, 638–646.

41 Gong, H., Singh, S.V., Singh, S.P., Mu, Y., Lee, J.H., Saini, S.P., Toma, D., Ren, S., Kagan, V.E., Day, B.W., Zimniak, P. and Xie, W. (2006) Orphan nuclear receptor pregnane X receptor sensitizes oxidative stress responses in transgenic mice and cancerous cells. *Molecular Endocrinology*, **20**, 279–290.

42 Klaassen, C.D. and Slitt, A.L. (2005) Regulation of hepatic transporters by xenobiotic receptors. *Current Drug Metabolism*, **6**, 309–328.

43 Hagenbuch, B. and Meier, P.J. (2003) The superfamily of organic anion transporting polypeptides. *Biochimica et Biophysica Acta*, **1609**, 1–18.

44 Staudinger, J.L., Madan, A., Carol, K.M. and Parkinson, A. (2003) Regulation of drug transporter gene expression by nuclear receptors. *Drug Metabolism and Disposition*, **31**, 523–527.

45 Guo, G.L., Choudhuri, S. and Klaassen, C.D. (2002) Induction profile of rat organic anion transporting polypeptide 2 (oatp2) by prototypical drug-metabolizing enzyme inducers that activate gene expression through ligand-activated transcription factor pathways. *The Journal of Pharmacology and Experimental Therapeutics*, **300**, 206–212.

46 Kruh, G.D. and Belinsky, M.G. (2003) The MRP family of drug efflux pumps. *Oncogene*, **22**, 7537–7552.

47 Kast, H.R., Goodwin, B., Tarr, P.T., Jones, S.A., Anisfeld, A.M., Stoltz, C.M., Tontonoz, P., Kliewer, S., Willson, T.M. and Edwards, P.A. (2002) Regulation of multidrug resistance-associated protein 2 (ABCC2) by the nuclear receptors pregnane X receptor, farnesoid X-activated receptor, and constitutive androstane receptor. *Journal of Biological Chemistry*, **277**, 2908–2915.

48 Teng, S., Jekerle, V. and Piquette-Miller, M. (2003) Induction of ABCC3 (MRP3) by pregnane X receptor activators. *Drug Metabolism and Disposition*, **31**, 1296–1299.

49 Geick, A., Eichelbaum, M. and Burk, O. (2001) Nuclear receptor response elements mediate induction of intestinal MDR1 by rifampin. *Journal of Biological Chemistry*, **276**, 14581–14587.

50 Xie, W., Barwick, J.L., Simon, C.M., Pierce, A.M., Safe, S., Blumberg, B., Guzelian, P.S. and Evans, R.M. (2000) Reciprocal activation of xenobiotic response genes by nuclear receptors SXR/PXR and CAR. *Genes & Development*, **14**, 3014–3023.

51 Saini, S.P., Mu, Y., Gong, H., Toma, D., Uppal, H., Ren, S., Li, S., Poloyac, S.M. and Xie, W. (2005) Dual role of orphan nuclear receptor pregnane X receptor in bilirubin detoxification in mice. *Hepatology*, **41**, 497–505.

52 Synold, T.W., Dussault, I. and Forman, B.M. (2001) The orphan nuclear receptor SXR coordinately regulates drug metabolism and efflux. *Nature Medicine*, **7**, 584–590.

53 Dussault, I., Lin, M., Hollister, K., Wang, E.H., Synold, T.W. and Forman, B.M. (2001) Peptide mimetic HIV protease inhibitors are ligands for the orphan receptor SXR. *Journal of Biological Chemistry*, **276**, 33309–33312.

54 Kliewer, S.A., Lehmann, J.M., Milburn, M.V. and Willson, T.M. (1999) The PPARs and PXRs: nuclear xenobiotic receptors that define novel hormone signaling pathways. *Recent Progress in Hormone Research*, **54**, 345–367.

55 Sinz, M., Kim, S., Zhu, Z., Chen, T., Anthony, M., Dickinson, K. and Rodrigues, A.D. (2006) Evaluation of 170 xenobiotics as transactivators of human pregnane X receptor (hPXR) and correlation to known CYP3A4 drug interactions. *Current Drug Metabolism*, **7**, 375–388.

56 Moore, L.B., Goodwin, B., Jones, S.A., Wisely, G.B., Serabjit-Singh, C.J., Willson, T.M., Collins, J.L. and Kliewer, S.A. (2000) St. John's wort induces hepatic drug metabolism through activation of the pregnane X receptor. *Proceedings of the National Academy of Sciences of the United States of America*, **97**, 7500–7502.

57 Mu, Y., Zhang, J., Zhang, S., Zhou, H.H., Toma, D., Ren, S., Huang, L., Yaramus, M., Baum, A., Venkataramanan, R. and Xie, W. (2006) Traditional Chinese medicines Wu Wei Zi (*Schisandra chinensis* Baill) and Gan Cao (*Glycyrrhiza uralensis* Fisch) activate pregnane X receptor and increase warfarin clearance in rats. *The Journal of Pharmacology and Experimental Therapeutics*, **316**, 1369–1377.

58 Goldstein, J.A. (2001) Clinical relevance of genetic polymorphisms in the human CYP2C subfamily. *British Journal of Clinical Pharmacology*, **52**, 349–355.

59 Mu, Y., Stephenson, C.R., Kendall, C., Saini, S.P., Toma, D., Ren, S., Cai, H., Strom, S.C., Day, B.W., Wipf, P. and Xie, W. (2005) A pregnane X receptor agonist with unique species-dependent stereoselectivity and its implications in drug development. *Molecular Pharmacology*, **68**, 403–413.

60 Gong, H., Sinz, M.W., Feng, Y., Chen, T., Venkataramanan, R. and Xie, W. (2005) Animal models of xenobiotic receptors in drug metabolism and diseases. *Methods in Enzymology*, **400**, 598–618.

61 Xie, W., Radominska-Pandya, A., Shi, Y., Simon, C.M., Nelson, M.C., Ong, E.S., Waxman, D.J. and Evans, R.M. (2001) An essential role for nuclear receptors SXR/PXR in detoxification of cholestatic bile acids. *Proceedings of the National Academy of Sciences of the United States of America*, **98**, 3375–3380.

62 Stedman, C.A., Liddle, C., Coulter, S.A., Sonoda, J., Alvarez, J.G., Moore, D.D., Evans, R.M. and Downes, M. (2005) Nuclear receptors constitutive androstane receptor and pregnane X receptor ameliorate cholestatic liver injury. *Proceedings of the National Academy of Sciences of the United States of America*, **102**, 2063–2068.

63 Zhang, J., Huang, W., Qatanani, M., Evans, R.M. and Moore, D.D. (2004) The constitutive androstane receptor and pregnane X receptor function coordinately to prevent bile acid-induced hepatotoxicity. *Journal of Biological Chemistry*, **279**, 49517–49522.

64 Uppal, H., Toma, D., Saini, S.P., Ren, S., Jones, T.J. and Xie, W. (2005) Combined loss of orphan receptors PXR and CAR heightens sensitivity to toxic bile acids in mice. *Hepatology*, **41**, 168–176.

65 Zhai, Y., Pai, H.V., Zhou, J., Amico, J.A., Vollmer, R.R. and Xie, W. (2007)

Activation of pregnane X receptor disrupts glucocorticoid and mineralocorticoid homeostasis. *Molecular Endocrinology*, **21**, 138–147.

66 Zhou, J., Zhai, Y., Mu, Y., Gong, H., Uppal, H., Toma, D., Ren, S., Evans, R.M. and Xie, W. (2006) A novel pregnane X receptor-mediated and sterol regulatory element-binding protein-independent lipogenic pathway. *Journal of Biological Chemistry*, **281**, 15013–15020.

67 Tucker, C.M., Petersen, S., Herman, K.C., Fennell, R.S., Bowling, B., Pedersen, T. and Vosmik, J.R. (2001) Self-regulation predictors of medication adherence among ethnically different pediatric patients with renal transplants. *Journal of Pediatric Psychology*, **26**, 455–464.

68 Weaver, R.J. (2001) Assessment of drug–drug interactions: concepts and approaches. *Xenobiotica*, **31**, 499–538.

69 Weaver, S.A., Russo, M.P., Wright, K.L., Kolios, G., Jobin, C., Robertson, D.A. and Ward, S.G. (2001) Regulatory role of phosphatidylinositol 3-kinase on TNF-alpha-induced cyclooxygenase 2 expression in colonic epithelial cells. *Gastroenterology*, **120**, 1117–1127.

70 Mills, J.B., Rose, K.A., Sadagopan, N., Sahi, J. and de Morais, S.M. (2004) Induction of drug metabolism enzymes and MDR1 using a novel human hepatocyte cell line. *The Journal of Pharmacology and Experimental Therapeutics*, **309**, 303–309.

71 Kocarek, T.A., Schuetz, E.G., Strom, S.C., Fisher, R.A. and Guzelian, P.S. (1995) Comparative analysis of cytochrome P4503A induction in primary cultures of rat, rabbit, and human hepatocytes. *Drug Metabolism and Disposition*, **23**, 415–421.

72 Jones, S.A., Moore, L.B., Shenk, J.L., Wisely, G.B., Hamilton, G.A., McKee, D.D., Tomkinson, N.C., LeCluyse, E.L., Lambert, M.H., Willson, T.M., Kliewer, S.A. and Moore, J.T. (2000) The pregnane X receptor: a promiscuous xenobiotic receptor that has diverged during evolution. *Molecular Endocrinology*, **14**, 27–39.

73 Watkins, R.E., Wisely, G.B., Moore, L.B., Collins, J.L., Lambert, M.H., Williams, S.P., Willson, T.M., Kliewer, S.A. and Redinbo, M.R. (2001) The human nuclear xenobiotic receptor PXR: structural determinants of directed promiscuity. *Science*, **292**, 2329–2333.

74 Kolars, J.C., Schmiedlin-Ren, P., Schuetz, J.D., Fang, C. and Watkins, P.B. (1992) Identification of rifampin-inducible P450IIIA4 (CYP3A4) in human small bowel enterocytes. *Journal of Clinical Investigation*, **90**, 1871–1878.

75 Moore, J.T. and Kliewer, S.A. (2000) Use of the nuclear receptor PXR to predict drug interactions. *Toxicology*, **153**, 1–10.

76 Zhang, J., Kuehl, P., Green, E.D., Touchman, J.W., Watkins, P.B., Daly, A., Hall, S.D., Maurel, P., Relling, M., Brimer, C., Yasuda, K., Wrighton, S.A., Hancock, M., Kim, R.B., Strom, S., Thummel, K., Russell, C.G., Hudson, J.R., Jr, Schuetz, E.G. and Boguski, M.S. (2001) The human pregnane X receptor: genomic structure and identification and functional characterization of natural allelic variants. *Pharmacogenetics*, **11** 555–572.

77 Burk, O., Tegude, H., Koch, I., Hustert, E., Wolbold, R., Glaeser, H., Klein, K., Fromm, M.F., Nuessler, A.K., Neuhaus, P., Zanger, U.M., Eichelbaum, M. and Wojnowski, L. (2002) Molecular mechanisms of polymorphic CYP3A7 expression in adult human liver and intestine. *Journal of Biological Chemistry*, **277**, 24280–24288.

78 Rushmore, T.H. and Pickett, C.B. (1990) Transcriptional regulation of the rat glutathione S-transferase Ya subunit gene. Characterization of a xenobiotic-responsive element controlling inducible expression by phenolic antioxidants. *Journal of Biological Chemistry*, **265**, 14648–14653.

79 Ma, Q. (2001) Induction of CYP1A1. The AhR/DRE paradigm: transcription,

receptor regulation, and expanding biological roles. *Current Drug Metabolism*, 2, 149–164.

80 Hoffman, E.C., Reyes, H., Chu, F.F., Sander, F., Conley, L.H., Brooks, B.A. and Hankinson, O. (1991) Cloning of a factor required for activity of the Ah (dioxin) receptor. *Science*, **252**, 954–958.

81 Gu, Y.Z., Hogenesch, J.B. and Bradfield, C.A. (2000) The PAS superfamily: sensors of environmental and developmental signals. *Annual Review of Pharmacology and Toxicology*, **40**, 519–561.

82 Yueh, M.F., Huang, Y.H., Hiller, A., Chen, S., Nguyen, N. and Tukey, R.H. (2003) Involvement of the xenobiotic response element (XRE) in Ah receptor-mediated induction of human UDP-glucuronosyltransferase 1A1. *Journal of Biological Chemistry*, **278**, 15001–15006.

83 Klinge, C.M., Bowers, J.L., Kulakosky, P.C., Kamboj, K.K. and Swanson, H.I. (1999) The aryl hydrocarbon receptor (AHR)/AHR nuclear translocator (ARNT) heterodimer interacts with naturally occurring estrogen response elements. *Molecular and Cellular Endocrinology*, **157**, 105–119.

84 Emi, Y., Ikushiro, S. and Iyanagi, T. (1996) Xenobiotic responsive element-mediated transcriptional activation in the UDP-glucuronosyltransferase family 1 gene complex. *Journal of Biological Chemistry*, **271**, 3952–3958.

85 Munzel, P.A., Lehmkoster, T., Bruck, M., Ritter, J.K. and Bock, K.W. (1998) Aryl hydrocarbon receptor-inducible or constitutive expression of human UDP glucuronosyltransferase UGT1A6. *Archives of Biochemistry and Biophysics*, **350**, 72–78.

86 Shelby, M.K., Cherrington, N.J., Vansell, N.R. and Klaassen, C.D. (2003) Tissue mRNA expression of the rat UDP-glucuronosyltransferase gene family. *Drug Metabolism and Disposition*, **31**, 326–333.

87 Zhou, J., Zhang, J. and Xie, W. (2005) Xenobiotic nuclear receptor-mediated regulation of UDP-glucuronosyl-transferases. *Current Drug Metabolism*, **6**, 289–298.

14
Ligand Features Essential for Cytochrome P450 Induction
Daniela Schuster, Theodora M. Steindl, Thierry Langer

14.1
Introduction

A well-known difficulty in drug therapy is that administration of the same medication does not always lead to constant effects in the human body. Variability in drug response can be observed among different individuals as well as within the same person at different times. Many parameters concerning the state and condition of a patient's body are known to play a role and should be considered when setting up a therapeutical strategy. These influences include age, sex, weight, nature of disease, nutritional status and organ function. More recently, genetic variability of metabolizing enzymes, transporters and drug targets among different individuals has been investigated, which explains varying responses to drugs.

Apart from patient-specific parameters, external factors – most importantly the concomitant uptake of certain other chemicals present in diet, environment and especially other drugs – influence drug actions. Possible effects are manifold and can affect all stages of pharmacokinetic and pharmacodynamic processes in the body. Also direct interaction and inactivation of concomitantly administered substances are possible. Drug–drug interactions via modulation of metabolism present a very hot topic in pharmaceutical research and drug design.

The fate of a drug in the human body is determined by chemical processes aimed at enhancing water solubility of the resulting metabolites and easier removal. These processes involve metabolism via phase I and phase II metabolizing enzymes and transport via phase III transporters and are vulnerable to interference by drugs [1]. Metabolic reactions involving oxidative, reductive or hydrolytic alteration of a substance are considered as phase I reactions [2]. Members of the ubiquitously distributed cytochrome P450 superfamily constitute the most important phase I drug-metabolizing enzymes – P450 3A alone is responsible for the degradation of approximately 50% of all therapeutic agents that undergo oxidation.

Therefore, altered P450 activity is critical in drug therapy and can change the metabolism and, as a consequence, the plasma concentration and pharmacological

Antitargets. Edited by R. J. Vaz and T. Klabunde
Copyright © 2008 WILEY-VCH Verlag GmbH & Co. KGaA, Weinheim
ISBN: 978-3-527-31821-6

effect of a drug significantly. Several reasons exist for alterations in the performance of P450: apart from genetic polymorphisms, coadministered substances can influence the elimination of a xenobiotic, for example, via competition or even inhibition of P450 enzymes. This frequently observed phenomenon increases the blood level of the drug that is now hindered in its metabolism. However, many examples exist where coadministration of compounds augments the metabolization rate of other drugs, reducing their bioavailability and biological effect: the dose of the immunosuppressant agent cyclosporin has to be elevated often by a factor of 3 when rifamycin is taken at the same time, to achieve the same plasma level as before antibiotic therapy. In this context also, St John's wort used in the therapy of mild depressions has attracted attention, since drug failure may occur in combination with, for example, HIV-protease inhibitors or oral contraceptives [3]. These effects could be ascribed to enzyme induction: substances, the so-called inducers, provoke an upregulation of P450 proteins and stimulate the enhanced removal of substrate drugs.

Cytochrome induction, the major topic of this chapter, can be provoked by two mechanisms: direct stimulation via heteroactivation and indirect modification mediated by nuclear receptors (NRs). The first alternative means that a modulator binds within the heme-containing P450 active site, often simultaneously with other ligands (e.g. substrates) or in a nearby allosteric site [4]. The heteroactivator then induces cooperative effects, that is, an increase of P450 activity, by modulation of the active-site structure and/or increasing the catalytic activity. In contrast, NRs regulate the transcription of metabolizing enzymes and can, upon agonist binding, stimulate the production of cytochromes. Both mechanisms lead to reduced plasma levels of the drugs processed by the induced P450 enzyme and present important reasons for drug–drug interactions. The dose response of a given drug can be altered and the biological effect can be decreased dramatically, leading to therapy failure and even life-threatening conditions. Another critical aspect is the estimation of the *in vivo* clearance of investigational drugs that is commonly achieved by extrapolation from *in vitro* measurements of the P450 activity. This prediction methodology acts on the assumption that P450-mediated reactions follow simple Michaelis–Menten kinetics, which of course is not the case anymore, if cytochrome induction comes into play and leads to higher clearance [5]. There is, however, a major difference between increased P450 activity caused by heteroactivation and induction via NRs: in the first case, the duration of altered metabolism is determined by the half-life of the heteroactivator, whereas in the latter case, the rate of synthesis and turnover of the P450 enzyme is the critical parameter.

14.2
Molecular Mechanisms Leading to P450 Induction

14.2.1.1   Heteroactivation
Kinetic and modeling studies based on the X-ray crystal structures of P450 2C9 and P450 3A4 suggest substrate-binding models, where up to three ligands can occupy the spacious binding pocket concomitantly [6–8]. A ligand bound at a site distant from

the active-site heme iron atom could offer additional favorable interaction possibilities for the actual substrate and thereby accelerate its biotransformation.

14.2.1.2 Activation of Nuclear Receptors

Nuclear receptors, the stimulators for the production of metabolizing enzymes and transporters, share a common structure and mechanism of action. They can be switched between active and inactive states by binding to small lipophilic ligands in a conserved ligand-binding domain (LBD) in the C-terminal region of the receptor. Agonist binding to the LBD alters the conformation of the NR so that the activation function-2 (AF-2) helix is shifted in an active position. NRs regulate transcription via the recruitment of coregulatory proteins referred to as either coactivators or corepressors. Coregulators promote or repress transcription through a complex mechanism involving the modulation of chromatin structure. Within the LBD, a conserved coactivator-binding surface is present that can be accessed only when the binding of the agonist has stabilized the AF-2 helix to an active position in which it can participate in coactivator binding. Coactivators contain short NR-binding motifs (LXXLL, also called NR boxes). Corepressors have a related motif for binding to an overlapping region of the NR LBD. However, when the AF-2 helix is held in an active conformation, corepressor binding is prevented. Accordingly, the activity of a particular agonist depends on its ability to stabilize an active conformation, whereas antagonists hinder the AF-2 helix to occupy the active position. Partial agonists show a limited capability in promoting an active conformation [9].

The NR-coactivator or NR-corepressor complexes can interact with the promoter regions of target genes whereby the highly conserved DBD (DNA-binding domain) of the NR binds via two zinc fingers modules to the DNA. The coactivator complex can acetylate histone and thereby prepare gene promoters for transactivation by decondensation of the corresponding chromatin. In contrast, corepressor complexes affect the chromatin structure as a result of the associated histone deacetylase activity and subsequent chromatin condensation, causing gene repression [10].

In Table 14.1, a list of important NRs leading to P450 induction is given [1,11–13]. The transcription factor aryl hydrocarbon receptor (AhR) was also included because of its special role in P450 induction (especially of some P450 family 1 members).

14.2.2
Ligands Directly Inducing P450 Activity

14.2.2.1 P450 2C9 Heteroactivators

P450 2C9 can be activated by directly binding a ligand, for example, (S)-warfarin. Therefore, the transformation rate of other substrates such as 7-methoxy-4-trifluoromethylcoumarin is increased. The first published crystal structure of P450 2C9 included a cocrystallized (S)-warfarin molecule located in a large binding pocket too far away from the heme moiety to form interactions with it (pdb entry 1og5). A molecule bound at this binding site would be ideally placed to establish direct molecular interactions with another molecule interacting with the heme group [6]. To

Table 14.1 Prominent NRs and transcription factors triggering P450 gene transcription.

NR	P450s induced (examples)	Physiological function of the NR
PXR	3A4 and others	Xenobiotic metabolism regulation, antioxidant
CAR	2B, 2C, 3A4	Xenobiotic metabolism regulation
FXR	7A	Bile acid metabolism and transport
LXR	7A	Reverse cholesterol transport and absorption
AhR	1A1, 1A2, 1A6, 1B1, 2S1	Reproduction and development regulation
PPARα	4A	Fatty acid metabolism regulation
GR	2B6, 2C9, 3A4, 3A5	Immunoresponse regulation
HNF4α	2A6, 2B6, 2C9, 2D6, 3A, 7A1	Carbohydrate, lipid, protein and xenobiotic metabolism regulation
VDR	24A1	Mineral metabolism and bone growth
RXRs		Heterodimerization partners for multiple NRs

evaluate whether the (S)-warfarin-binding site as introduced in Figure 14.1 could represent a general interaction site for P450 2C9 heteroactivators, structure-based pharmacophore models were generated with the LigandScout software [14]. It was then investigated whether other P450 2C9 heteroactivators fitted into these models [15]. The models were then applied to a test set including amiodarone, niclosamide, mefenamic acid, zafirlukast and other heteroactivators with EC_{50} values in the nanomolar or low micromolar range (Chart 14.1).

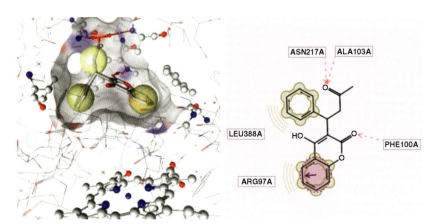

Figure 14.1 (S)-Warfarin-binding site and protein–ligand interactions as observed in the pdb entry 1og5. The ligand is located too far away from the heme moiety to establish direct interactions. However, it is well positioned to offer additional interactions for other molecules binding to the active site of P450 2C9 (left). Hydrophobic/aromatic interactions are represented as yellow spheres and hydrogen bond acceptors as red arrows. LigandScout depiction of (S)-warfarin in the putative heteroactivator-binding site of P450 2C9 (right).

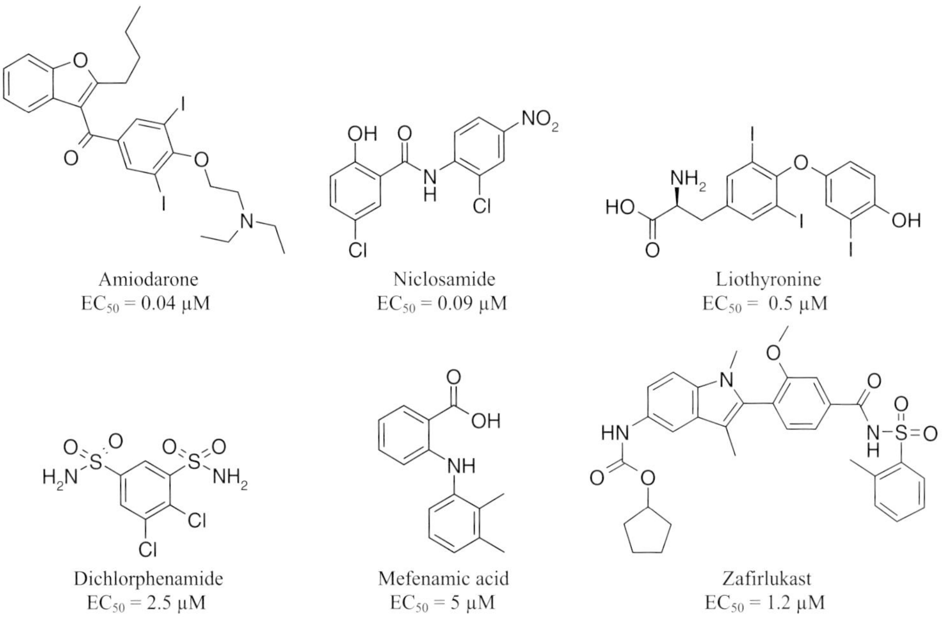

14.2 Molecular Mechanisms Leading to P450 Induction

Amiodarone	Niclosamide	Liothyronine
$EC_{50} = 0.04\ \mu M$	$EC_{50} = 0.09\ \mu M$	$EC_{50} = 0.5\ \mu M$

Dichlorphenamide	Mefenamic acid	Zafirlukast
$EC_{50} = 2.5\ \mu M$	$EC_{50} = 5\ \mu M$	$EC_{50} = 1.2\ \mu M$

Chart 14.1

The pharmacophore models derived from these experiments consisted of features representing the hydrophobic/aromatic interactions with the Phe residues present in the LBD as well as one or two hydrogen bond acceptors (HBAs) serving as anchoring sites for the warfarin molecule. As these models were able to recognize a large fraction of the test set compounds as actives, the authors concluded that the observed binding site of warfarin might well represent a general binding mode for P450 2C9 heteroactivators.

Similarly, important features for heteroactivation of 450 2C9 have been identified in a ligand-based approach. Therefore, the pharmacophoric model seeks to combine common chemical functionalities of known active ligands. From a large, 36-compound training set models with one or two HBAs, several hydrophobic (H) and sometimes one aromatic features were obtained, which is highly consistent with the knowledge from the warfarin crystal structure complex. Finally, one HBA, one ring aromatic (RA) and two H features were identified as the critical interaction determinants, giving most reliable activity predictions for the investigated P450 2C9 heteroactivators. The composition of the model suggests that it also describes heteroactivation within the active site of the enzyme [16].

P450 2C9 crystal structure information also served in docking and molecule dynamics operations to simulate the simultaneous binding of a heteroactivator and a substrate [17]. NSAID substrates were thereby positioned nearer to the heme than the heteroactivator. For the latter, limited conformational freedom was discovered because of the reduced active-site volume. The presence of an activator shortened the substrate–heme iron distances leading to an increase in the number of catalytic cycles. The metabolism rate of a substrate in combination with a particular activator further depends on substrate concentration (with low substrate concentrations yielding greater changes) and configuration. Upon structural changes caused by effector and substrate binding, a heme coordinating water is replaced transferring the iron to a high spin state, which increases its reduction potential [4].

14.2.2.2 P450 3A4 Heteroactivators

Similar to P450 2C9, a ligand-binding site located away from the active site has been observed in cocrystallization experiments (pdb entry 1w0f) [18]. The P450 3A4 heteroactivator progesterone was found rather binding to the Phe cluster on top of the active site than in a productive orientation close to the heme moiety of P450 3A4. It is speculated that this might also constitute an effector-binding site.

Attempts were made to track molecule qualities that lead to P450 3A4 heteroactivation. Common features of P450 3A4 heteroactivators identified in a ligand-based approach using six structurally diverse training set molecules include two HBAs separated by two H features. Chart 14.2 shows the three most potent P450 3A4 heteroactivators used for model building. Quantitative pharmacophore-based predictions of P450 3A4 heteroactivation as measured by metabolite formation of carbamazepin (a P450 3A4 substrate) revealed good correlation for the external test set [19].

14.2.3
P450 Inducers Acting via Nuclear Receptors

14.2.3.1 Pregnane X Receptor

Of all NRs involved in xenobiotics metabolism induction, PXR is the most prominent one. PXR functions as a xenobiotic sensor and is activated by a large variety of chemically diverse compounds, for example lovastatin, nifedipine, rifampicin, SR12813, troglitazone or hyperforin (Chart 14.3), many of them standard therapeutic agents for common diseases [20–25].

Target genes for PXR-mediated activation encode, for instance, several P450 isoenzymes (most importantly P450 3A4) as well as important phase II and III metabolizing and transporting proteins such as glutathione S-transferases (GSTs), sulfotransferase 2A1 (SULT2A1), UDP-glucuronosyltransferases (UGTs), multidrug

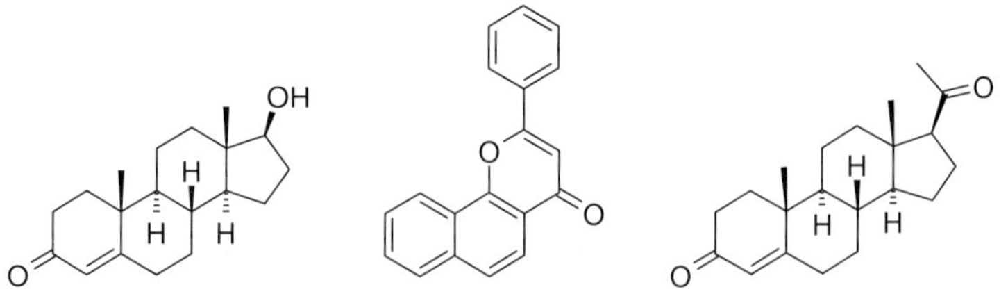

Testosterone
$V_{150\%} = 0.45\ \mu M$

-Naphthoflavone
$V_{150\%} = 1.5\ \mu M$

Progesterone
$V_{150\%} = 3\ \mu M$

Chart 14.2

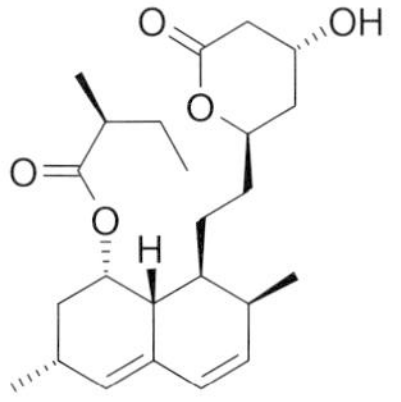

Lovastatin
five fold PXR activation

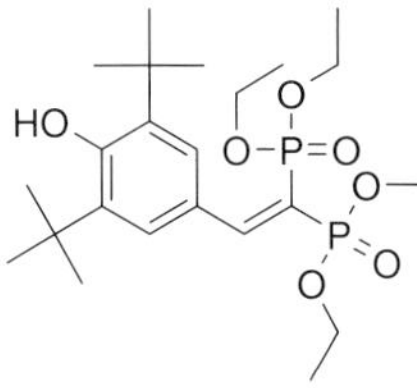

SR12813
five fold PXR activation

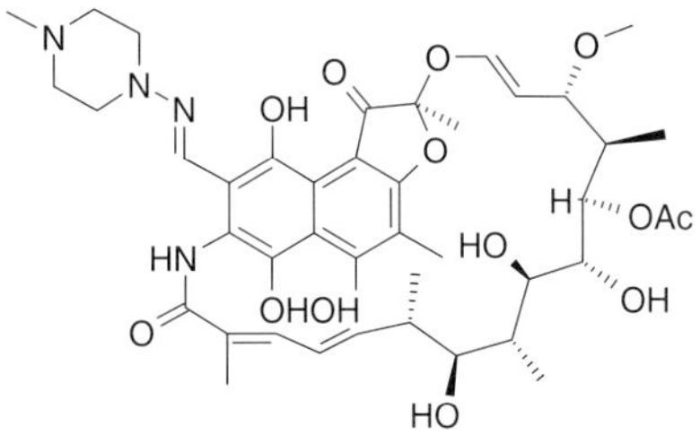

Rifampicin
9–13-fold PXR activation

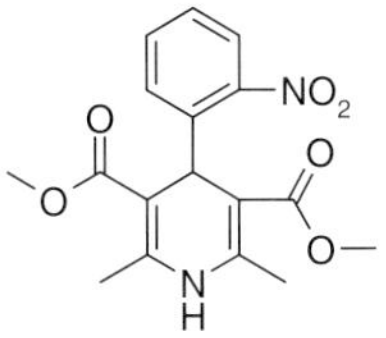

Nifedipine
nine fold PXR activation

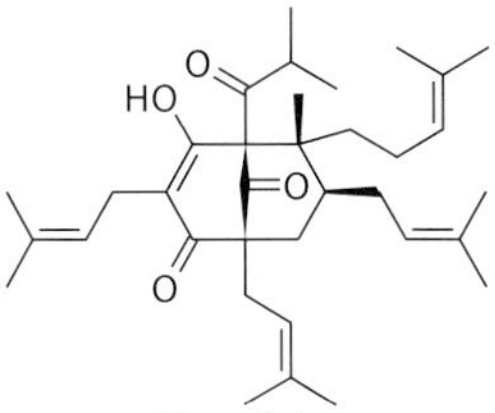

Hyperforin
seven fold PXR activation

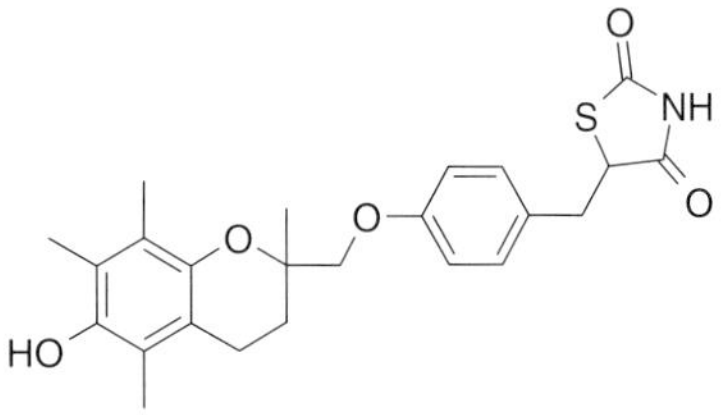

Troglitazone
Seven fold activation

Chart 14.3

resistant transporter 1 (MDR1) and organic anion transporting polypeptide 2 (OATP2) [26]. This multitude of activators, on the one hand, and their remarkable influence on systems dominating drug metabolization and transport, on the other hand, render the PXR a crucial variable to consider when planning therapeutic schemes that involve the coadministration of two or more drugs. Although PXR is known to be a very prominent xenobiotic receptor, it responds to a wide array of endogenous chemicals as well. By doing so, PXR has implications in several important physiological and pathological conditions such as bile acid induced hepatotoxicity [27].

The first ligand-based pharmacophore model for PXR ligands was based on 12 PXR agonists exhibiting EC_{50} values between 0.023 and >10 µM. Model generation was performed with Catalyst software [28,29]. The proposed model consisted of one HBA and four H features, where the HBA–H distances ranged from 3.6 to 7.6 Å. The importance of one HBA in combination with H features as well as the correctness of the proposed distances could be confirmed in quantitative structure–activity relationship (QSAR) studies on PXR activators (barbitals, hydantoins and macrolide antibiotics) [30]. To identify chemical features of a ligand essential for PXR activation, structure- and ligand-based pharmacophore modeling were combined and compared with information from X-ray crystallography [31]. According to this approach – employing the HipHop algorithm [32] of Catalyst 4.9 – highly potent PXR activators (represented by SR12813 and hyperforin) share two HBAs as well as five H spheres. A broader training set identified one HBA and three of the five H features as common chemical functionalities of PXR agonists. Less potent agonists, for example kaempferol, possess only one H region. Thus, the only feature shared by all PXR ligands was one HBA, which is supposed to direct toward Gln285 directly or via a water molecule. A second hydrogen bond (to His407) and H interactions with Tyr306 are also hypothesized to be essential for PXR activation. Additional H interactions are not considered essential for PXR activation but seem beneficial for the agonistic potency of a ligand (Figure 14.2a).

Apart from pharmacophore-based approaches, a variety of methods were applied to decipher important ligand features of PXR activation. VolSurf descriptor-based partial least squares (PLS) regression-based models pointed toward amide responsive regions that implicated good acceptor abilities as key variables [33].

Machine learning methods (including neural network, support vector machines and k-nearest neighbor) with an incorporated feature selection method gave a more detailed information on molecular descriptors relevant to PXR activation [34]. Using a set of approximately 200 activators and nonactivators and a combination of 83 descriptors led to very precise prediction tools. Important molecular descriptors specify the count of atom types, rings, rotatable bonds, molecular connectivity, geometry, flexibility and surface area, electrotopology, shape, hydrophobicity and

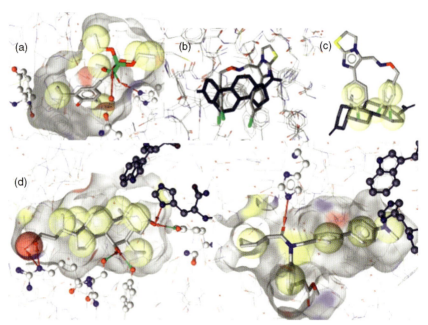

Figure 14.2 (a) Ligand features contributing to PXR activation as identified in Ref. [31]. Essential interacting amino acid residues (Tyr306, Gln285 and His407) are depicted in ball-and-stick style. (b) Alignment of hCAR pdb entries 1xv9 and 1xvp LBDs including the cocrystallized ligands (1xv9 highlighted). (c) Hydrophobic ligand features shared by 5β-pregnane-3,20-dione (highlighted) and CITCO. (d) Binding mode of 6ECDCA (left) and fexaramine (right) to the FXR LBD. Both ligands mainly form hydrophobic interactions with the receptor. The stabilized π-stacking of His444 (human: His451) and Trp466 (human: Trp473) essential for FXR activation is highlighted.

quantum chemistry. Apart from confirming the previously recognized molecular-binding features, other interesting aspects were gained from this large data set: compared to nonactivators, activating ligands from the investigated data set show more halogen and fewer nitrogen atoms, lower values for electrotopological descriptors, smaller size (smaller number of rotatable bonds, flexibility, polar surface area) and fewer possibilities for hydrogen bonds. Activators are likely to enable π–π stacking with the receptor and generally have a higher number of HBAs than HBDs. Separately applying these machine learning tools to solely human PXR activators makes allowance for the fact that PXR activation of a particular ligand is highly species specific, a fact crucial to consider, for example, when using animal models for toxicity tests [35–37].

14.2.3.2 Constitutive Androstane Receptor

Constitutive androstane receptor (CAR) reveals *in vitro* activity also in the non-liganded state and is therefore also called 'constitutive active receptor'. This receptor forms a heterodimer with RXR, which binds to retinoic acid response elements and transactivates target genes. CAR is mainly found in the liver and the intestine. *In vivo*, CAR is quiescent in the cytoplasm, where it is located in the unliganded form. Upon treatment with an inducer, CAR translocates to the nucleus where it activates the transcription of genes including members of the P450 2B and 2C families as well as P450 3A4, SULT, UGT and MRP genes, thus influencing drug response [26]. When compared with PXR, CAR regulates the transcription of many overlapping genes as well as some genes not regulated by PXR [38–40]. Endogenous CAR ligands include the two androstane metabolites, androstanol and androstenol. Other ligands influencing CAR activity are introduced in Chart 14.4 [41].

Unlike most NRs, the steroidal ligands for CAR inhibit receptor-dependent gene transcription. Both ligands act as antagonists by dissociating CAR from its coactivator and inhibiting the transactivation of CAR [1]. Physiologically, CAR plays an important role in regulating bilirubin clearance and bile acid detoxification [26].

According to an alignment of three human CAR ligands (clotrimazole, androstanol and 5β-pregnane-3,20-dione) leading to a pharmacophore model, three H features

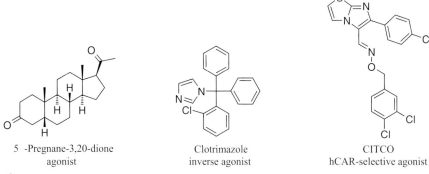

5β-Pregnane-3,20-dione   Clotrimazole   CITCO
agonist   inverse agonist   hCAR-selective agonist

Chart 14.4

and one HBA are crucial interacting chemical groups [42]. This model was found to show striking similarity with a P450 2B6 pharmacophore model. Because P450 2B6 is induced by CAR, it was speculated from this finding that ligands for an enzyme or transporter could have features in common with ligands for the receptor responsible for regulating it. Accordingly, some xenobiotics could then regulate/induce their own metabolism.

Before the elucidation of the CAR X-ray crystal structure, modeling studies mainly focused on the constitutive activity of the receptor. For example, homology modeling along with molecular dynamics simulations was combined to identify critical amino acid residues responsible for the constitutive activation [43]. Especially, the role of Tyr326 as a 'molecular mimicry' of a bound ligand in the interaction with the AF-2 helix was underscored. Mutational analyses and the later elucidation of the human CAR X-ray crystal structure confirmed an important role of this amino acid for the receptor's constitutive activity [41,44].

As the X-ray crystal structures of mouse [45] and human CAR [41] became available, the constitutive activity could be investigated more thoroughly. In CAR, the LBD is shielded from the AF-2 helix by four amino acid residues conserved over mammals. Exactly these four residues – Phe161, Asn165, Phe234 and Tyr326 – stabilize the AF-2 helix in an active conformation by direct interactions. In the sequences of CAR-related proteins such as PXR and vitamin D receptor (VDR), these residues are not conserved, which leads to their high ligand dependency. Agonist binding induces only CAR translocation from the cytoplasm (where CAR is silent) into the nucleus (where it reveals constitutive activity).

Analysis of the crystal structure of human CAR revealed that the volume of the CAR pocket is 675 Å^3. Hydrophobic contacts (enabled by Phe161, Ile164, Leu206, Phe217, Tyr224, Val232, Phe234, Phe238, Leu239, Leu242 and Phe243) are the dominating interactions for both steroids and nonsteroidal ligands. In addition, there are possibilities for hydrogen bonds (e.g. to His203) [41]. However, in the currently available hCAR crystal structures, no direct hydrogen bonds with the receptor could be observed. The cocrystallized ligands 5β-pregnane-3,20-dione and CITCO bind via hydrophobic interactions to the LBD. Both agonists occupy essentially overlapping hydrophobic areas within the CAR LBD (Figure 14.2b and c).

Apart from the X-ray crystal structures, no studies on hCAR ligand properties have been reported until January 2007.

14.2.3.3 Farnesoid X Receptor

Farnesoid X receptor (FXR) and liver X receptors (LXRs) belong to the same NR family as PXR and CAR. Their primary role lies in cholesterol and bile acid metabolism regulation. Like many NRs of this family, FXR heterodimerizes with RXR *in vivo* [46].

FXR is activated by endogenous bile acids such as chenodeoxycholic acid. Other known agonists include farnesol, GW4064 with an EC$_{50}$ of 70 nM, or AGN-31 (Chart 14.5) [12].

Its target genes regulate the secretion of bile acids and phospholipids into bile (bile salt efflux pump, MDR2 and 3), the intestinal reabsorption of bile acid (ileal bile

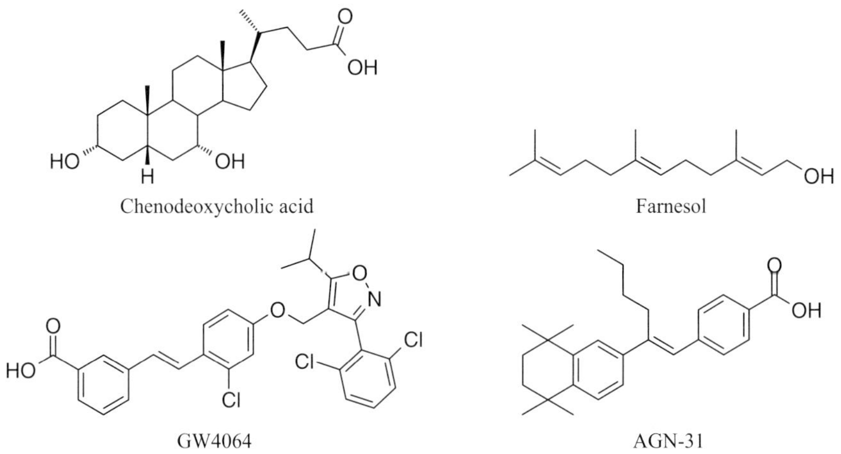

Chart 14.5

acid-binding protein) and hepatic cholesterol uptake from serum high-density lipoprotein (phospholipids transfer protein) [26]. FXR also induces a transcriptional repressor that decreases the expression of P450 7A1, the rate-limiting enzyme for bile acid synthesis from cholesterol [46]. Thus, a high bile acid level diminishes the P450 7A1-mediated transformation of cholesterol into bile acids. Further, genes essential for bile acid transport are induced. Interestingly, apart from major genes involved in bile acid regulation, not all FXR agonists induce identical gene regulation patterns [47]. Accordingly, FXR constitutes an interesting pharmacological target for liver diseases and metabolic disorders such as liver fibrosis, cholestasis or atherosclerosis [12].

A structure–activity relationship (SAR) study investigated the role of substituents at the 3-, 7- or 12-position of cholanoids. Both hydroxyl and alkyl residues on these positions diminished their ability to activate FXR [48]. Another SAR study on bile acid derivatives underlined the beneficial effect of a 6α-alkyl substitution, the need of an extended disposition of the side chain (else resulting in partial agonism or loss of activity), and the possibility for a bioisosteric substitution of the carboxylic group altering the ligand functional effects without a significant loss of potency [49]. Studies undertaken using the hologram QSAR method to generate 2D QSAR models for FXR agonists outline the importance of chirality as well as donor–acceptor properties of the ligand to activate the receptor [50].

A pharmacophore model for FXR agonists based on 14 bile acid derivatives was reported by Ekins *et al.* The hypothesis consisted of three H features and one HBA – a common functionality pattern among steroidal compounds [42].

The X-ray crystal structure of hFXR in complex with the potent and selective agonist fexaramine showed a predominantly hydrophobic LBD with water-mediated hydrogen bonds toward His298 and Ser336 [47]. Comparing the binding mode of steroidal and nonsteroidal activators represented by 6α-ethyl-chenodeoxycholic acid (6ECDCA) [51] and fexaramine (pdb entries 1osv – rFXR and 1osh – hFXR), one notices that both ligands stabilize helix H12 in an active position. However, the larger nonsteroidal ligand fexaramine occupies areas of the receptor not reached by 6ECDCA. Additionally, only a few amino acid residues that contact the ligand are in common with one another (Figure 14.2d). Although 6ECDCA is anchored in the LBD via several hydrogen bonding interactions, fexaramine is observed forming only one hydrogen bond. Interestingly, steroidal and nonsteroidal ligands lead to different gene transcription patterns induced by FXR. Accordingly, the structure of the ligand may influence the type of coactivator that is recruited by the NR, which in turn leads to diverging activation profiles [12].

3D QSAR CoMFA studies on nonsteroidal FXR ligands suggest that electropositive substitution of the aromatic rings could enhance their biological activity [52]. These results are consistent with data from X-ray crystallography, where fexaramine (in contrast to 6ECDCA) forms only one hydrogen bond with the receptor.

Thus, FXR ligands include structurally very diverse ligands displaying different binding positions within the LBD. Agonists are generally highly lipophilic compounds that stabilize the H12 in an active conformation. The formation of hydrogen bonds seems favorable for agonist activity but is not essential.

14.2.3.4 Liver X Receptors

In humans, two forms of LXRs are present – LXRα and LXRβ – which share about 80% of the sequence identity. LXRs bind oxidized cholesterol derivatives (oxysterols) such as 24(S),25-epoxycholesterol. LXRα is expressed primarily in the liver, adipose tissue, intestine, macrophage and kidney, whereas LXRβ is expressed ubiquitously [46]. These receptors are mainly located in the nucleus and must heterodimerize with RXR for activation [1]. LXRs play an important role in cholesterol regulation. Both NRs induce the expression of proteins that stimulate cholesterol efflux from macrophages, promote cholesterol transport in serum and uptake into the liver, increase cholesterol catabolism into bile acids by inducing P450 7A1 expression, increase biliary secretion of cholesterol and inhibit cholesterol absorption in the intestine [46].

Before the elucidation of X-ray crystal structures for LXRs, ligand-based studies focusing on structural requirements to activate these receptors were reported. Position-specific monooxidation of the sterol side chain was found out to be requisite for LXR high-affinity binding and activation. 24-Oxo ligands acting as HBAs showed enhanced activation. Additionally, the introduction of an oxygen on the sterol B-ring led to an LXRα-selective activation profile [53]. A combination of homology modeling of LXRα (based on retinoic acid receptor (RAR) γ) and site-directed mutagenesis studies led to the identification of Trp443 as the amino acid interacting with the epoxide oxygen at positions 24, 25 and Arg305 as interaction partner for the C3 hydroxyl group of 24(S),25-epoxycholesterol [54]. Additionally, this work rationalized the antagonistic activity of 22(S)-hydroxycholesterol. This compound can bind to LXRα without stabilizing the AF-2 helix but by forming a hydrogen bond with His421 from helix 10. In another study, Phe268 was identified as a critical residue for ligand binding in LXRβ by site-directed mutagenesis experiments combined with an LXRβ homology model based on PPARγ. In contrast, Thr272 was found to have no

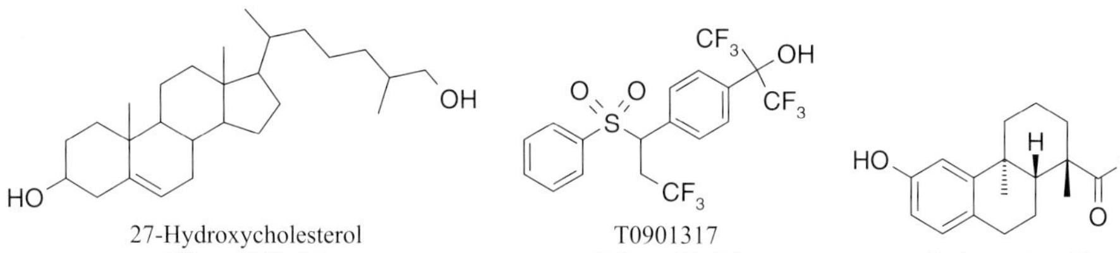

14.2 Molecular Mechanisms Leading to P450 Induction | 329

27-Hydroxycholesterol
EC_{50} = 250 nM

T0901317
EC_{50} = 60 nM

Podocarpic acid

Chart 14.6

influence on LXRβ ligand recognition [55]. Further insights into LXR activation were gained using pharmacophore modeling. Seventeen steroidal LXRα ligands and 18 LXRβ ligands were used to generate and compare respective models [42]. The previous suggestion of an additional hydrogen bond involving an oxygen on the sterol B-ring that would lead to LXRα selectivity [53] was not represented by these models. When investigating the role of sterol intermediates from the biosynthetic pathway of cholesterol on LXRs, it was observed that the presence of an unsaturated bond in the side chain of the sterol was necessary and sufficient for the activity, with the C-24 unsaturated cholesterol precursor sterols desmosterol and zymosterol exerting the largest effects [56].

Although endogenous ligands for LXRs belong to the structure class of steroids, nonsteroidal agonists are also reported (Chart 14.6) [57].

For both human LXR isoforms, X-ray crystal structures have been reported. The LXRβ LBD is hydrophobic with polar or charged residues at the two ends of the cavity. Agonists are observed to interact with either of these polar amino acids (His435 or Ser242) [58]. Upon ligand binding, His435 is positioned against the Trp457 indole ring, allowing an electrostatic interaction between Trp457 and the AF-2 helix that stabilizes an active conformation [59]. In the LXRα LBD, His421 is observed to be involved in agonist binding [60].

The crystal structures revealed that the ligand-binding pocket is quite flexible as ligands of different shapes and sizes can be accommodated. With an approximate size of 800 Å^3, the ligand-binding pocket is able to take in small ligands as well as large ones, which occupy over three-quarters of the site volume [61]. A dense network of hydrogen bonds and – in the case of ligands with acidic moieties – also charge–charge interactions to an active-site arginine (Arg305 in LXRα and Arg319 in the case of LXRβ) are the most critical features for binding apart from hydrophobic contacts. Introduction of ligand acidity usually results in an improvement of the activity [62]. The importance of this anchoring function of a hydrogen bond acceptance and also acidic ligand feature for interaction with basic site residues can be seen, for example, in SAR studies using the scaffold of the most prominent LXR ligand T0901317 (Chart 14.6). Here, activity, that is the ability to activate the LXR, stands and falls with the interacting hydroxyl group of the hexafluorocarbinol moiety, whose proper orientation and acidification requires the presence of both trifluoromethyl residues. A wider range of modifications is tolerated in the remaining part of the molecule [63]. Another successful strategy for enhancing LXR activity of small ligands is to increase their size by coupling them with appropriate, spacious, hydrophobic moieties (e.g. ring systems). In the case of podocarpic acid, for example, simple dimerization of this monomeric natural product provides considerably more active ligands [64].

14.2.4
P450 Inducers Acting via Transcription Factors

14.2.4.1 Aryl Hydrocarbon Receptor

AhR is a transcription factor and acts as a ligand-activated transcriptional regulator of various metabolizing enzymes also involved in drug disposition. Induction affects P450 1A1, 1A2, 1B1, UGT1A1, 1A6 and GSTA1, as well as the newly identified ubiquitously expressed member of the P450 superfamily 2S1. The latter has been connected with the metabolism of endogenous substrates such as retinoic acid and toxic xenobiotics (naphthalene) [65,66]. Although the inactive AhR is located in the cytoplasm, ligand binding provokes the translocation of the complex into the nucleus, where dimerization with the aryl hydrocarbon nuclear translocator results in the active transcription factor that binds to the xenobiotic response element of its specific DNA recognition site [67]. Activators include polycyclic aryl hydrocarbons such as sanguinarine, benzimidazoles, polychlorinated biphenyls, dibenzoantracene and flavonoids, as well as the most prominent ligand dioxin or TCDD (2,3,7,8-tetrachlorodibenzo-p-dioxin), leading to P450 1 induction (Chart 14.7).

The toxicity of these widespread environmental contaminants is mediated by AhR via alteration of specific genes. As AhR controls a highly regulated chain of developmental and physiological events, the harmful effects of exogenous interfering compounds are obvious: reproductive and development defects, dysregulation of female fertility, liver damage, thymic atrophy, hormone balance alteration and cancer [66–71].

To identify molecular properties that favor ligand binding at the AhR, several QSAR studies have been undertaken: a partial charge-based descriptor was identified as the main feature influencing binding affinity [72]. Furthermore, the degree and mode of ligand halogenation, as well as molecular planarity, aromaticity and a specific rectangular envelope because of possible π–π stacking interactions [73–77], proved to be critical. A CoSCoSA (comparative structural connectivity spectral analysis) modeling approach seeks to analyze quantitatively the relationship between 3D spectrometric data and biological activities of 52 AhR activators. *In vitro* binding affinities of the molecules and NMR data combined with structural connectivity information in a 3D matrix were used to generate models for AhR binding. Validation

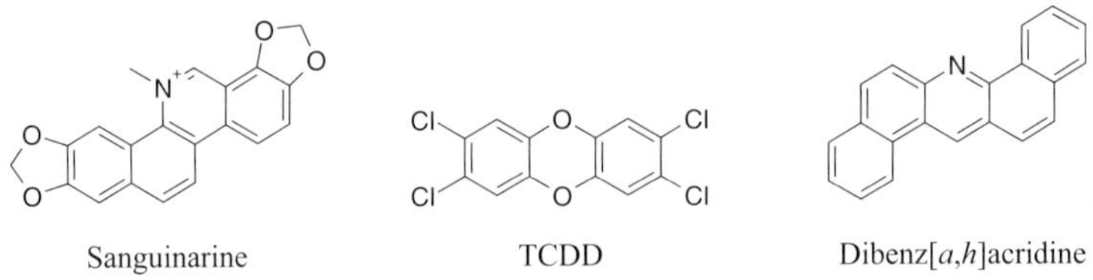

Sanguinarine   TCDD   Dibenz[*a,h*]acridine

Chart 14.7

confirmed the reliability of the models for the prediction of the test compounds [78]. When cytotoxicity and potencies to activate AhR of polycyclic aromatic hydrocarbons and their N-heterocyclic derivatives were investigated and compared, the aza-PAHs were found to be significantly more cytotoxic and more potent activators of AhR than their unsubstituted analogues with picomolar activation concentrations (comparable to TCDD, Chart 14.7). Ellipsoidal volume, molar refractivity and molecular size were the most important descriptors recognized in QSAR studies of these aza-compounds [79]. Employing a heuristic method, a large set of polybrominated diphenyl ethers and more than 400 descriptors were used to relate the compound structures with their toxicology indices (AhR relative binding affinities). Activity profiles were mainly affected by the molecular electrostatic field, nonuniformity of mass distribution, atom reactivity and conformational changes. It is assumed that the large bromyl residues enlarge the AhR-binding site [80]. Investigations on polychlorinated compounds from several structural classes revealed similar results: a high degree of branching for molecules is unfavorable for AhR binding, whereas a 'large shadow' and lower moment of inertia (mass distribution) increases the binding ability. High values for electrostatic descriptors (molecular electronegativity, charge distribution) can complicate the binding process [81]. Ligand alignment and CoMFA contour maps as well as VolSurf descriptors present other possibilities in 3D QSAR for activity predictions on the AhR. For example, hydrophobicity was related to higher AhR activity [82].

Difficulties arise from the structural diversity of AhR ligands. A good overview of synthetic and environmental chemicals and naturally occurring dietary and endogenous AhR ligands, focusing on their divergent chemical structures, is given in [83]. Also new agonists and antagonists retrieved in a large high-throughput screening bioassay were considerably diverse suggesting a very promiscuous ligand-binding pocket, which demands more sophisticated models and simulations to understand receptor–ligand interactions [84]. Investigations using wild-type and mutated proteins showed evidence that high and low affinity ligands interact with different residues in the binding site and that Tyr320 is the crucial amino acid for AhR activation [85].

To make up for the still lacking crystal structure of the AhR, several attempts have been reported to model its LBD from homologous proteins [84–86]. The hERα served as a template for homology model generation employing the Sybyl biopolymer software. Two high affinity AhR ligands, TCDD and PCB, were docked into the model. Hydrogen bonding (e.g. to arginines) appears to be significant, especially in the case of TCDD, whereas π–π stacking between the benzene rings of PCB and aromatic amino acid residues (e.g. phenylalanines) in the binding pocket seems to be the major interacting determinant for this second ligand [73]. Later, again using this homology model, these results could be confirmed and expanded by the results of QSAR runs using VolSurf descriptors and PLS and the principal component analysis as statistical evaluation tools. TCDD was identified as an outlier exhibiting different structural characteristics from the other polyhalogenated biphenyls, as well as different shapes and energies of the conformational binding site. This reflects the somewhat different binding mode, which in the case of TCDD is dominated by hydrogen bonding [33].

Summarizing, many potent AhR ligands share a highly hydrophobic, aromatic (and thus flat) structure. It will be interesting to compare the above-mentioned theoretical results with data from X-ray crystallography when they will become available in the future.

14.3
General Ligand Features Leading to NR Activation

Most NRs discussed above have endogenous steroidal ligands, so their pharmacophoric features contributing to NR activation are similar and based on the steroid scaffold. Apart from a hydrophobic core, one or multiple (optional) hydrogen bonds contribute to ligand binding. Nonsteroidal ligands often occupy other parts of the LBD and thus form different interactions. Although pharmacophoric features of these ligands are often similar to the steroidal ones – namely mostly hydrophobic – they should be treated separately in modeling studies. Regarding AhR ligands – for which no endogenous steroidal ligands are known – high hydrophobicity as well as the overall (flat) shape of the ligand is related to its potency. In this case, hydrogen bonding plays only a subsidiary role for NR activation.

Abbreviations

6ECDCA	6α-ethyl-chenodeoxycholic acid
AhR	aryl hydrocarbon receptor
AF-2	activation function-2
CAR	constitutive androstane receptor
CoMFA	comparative molecular field analysis
DBD	DNA-binding domain
FXR	farnesoid X receptor
GR	glucocorticoid receptor
GST	glutathione S-transferase
H	hydrophobic
HBA	hydrogen bond acceptor
HBD	hydrogen bond donor
HNF	hepatocytes nuclear factor
LBD	ligand-binding domain
LXR	liver X receptor
MDR1	multidrug resistant transporter 1
MRP	multidrug resistance-associated protein
NI	negatively ionizable
NR	nuclear receptor
OATP	organic anion transporting polypeptide
P450	cytochrome P450
PPAR	peroxisome proliferator-activated receptor

PXR	pregnane X receptor
QSAR	quantitative structure–activity relationship
RA	ring aromatic
RAR	retinoic acid receptor
RXR	retinoid X receptor
SAR	structure–activity relationship
SULT	sulfotransferase
UGT	UDP-glucuronosyltransferase
VDR	vitamin D receptor

Acknowledgment

We thank Gerhard Wolber from Inte:Ligand GmbH for providing the depiction of (*S*)-warfarin in P450 2C9, a functionality that will be available in a future release of LigandScout.

References

1 Xu, C., Li, C.Y.-T. and Kong, A.-N.T. (2005) Induction of phase I, II and III drug metabolism/transport by xenobiotics. *Archives of Pharmacal Research*, **28**, 249–268.

2 Mutschler, E., Geisslinger, G., Kroemer, H.K. and Schäfer-Korting, M. (eds) (2001). *Arzneimittelwirkungen*, 8th edn, Wissenschaftliche Verlagsgesellschaft mbH, Stuttgart.

3 Wilkinson, G.R. (2005) Drug metabolism and variability among patients in drug response. *The New England Journal of Medicine*, **352**, 2211–2221.

4 Locuson, C.W., Gannett, P.M. and Tracy, T.S. (2006) Heteroactivator effects on the coupling and spin state equilibrium of CYP2C9. *Archives of Biochemistry and Biophysics*, **449**, 115–129.

5 Egnell, A.-C., Houston, B. and Boyer, S. (2003) *in vivo* CYP3A4 heteroactivation is a possible mechanism for the drug interaction between felbamate and carbamazepine. *The Journal of Pharmacology and Experimental Therapeutics*, **305**, 1251–1262.

6 Williams, P.A., Cosme, J., Ward, A., Angove, H.C., Vinkovic, D.M. and Jhoti, H. (2003) Crystal structure of human cytochrome P450 2C9 with bound warfarin. *Nature*, **424**, 464–468.

7 Ekins, S., Stresser, D.M. and Andrew Williams, J. (2003) *in vitro* and pharmacophore insights into CYP3A enzymes. *Trends in Pharmacological Sciences*, **24**, 161–166.

8 Schuster, D., Laggner, C., Steindl, T.M. and Langer, T. (2006) Development and validation of an *in silico* P450 profiler based on pharmacophore models. *Current Drug Discovery Technologies*, **3**, 1–48.

9 Benoit, G., Malewicz, M. and Perlmann, T. (2004) Digging deep into the pockets of orphan nuclear receptors: insights from structural studies. *Trends in Cell Biology*, **14**, 369–376.

10 Bourguet, W., Germain, P. and Gronemeyer, H. (2000) Nuclear receptor ligand-binding domains: three-dimensional structures, molecular interactions and pharmacological implications. *Trends in Pharmacological Sciences*, **21**, 381–388.

11 Schuster, D., Steindl, T.M. and Langer, T. (2006) Predicting drug metabolism induction in silico. *Current Topics in Medicinal Chemistry*, **6**, 1627–1640.

12 Pellicciari, R., Constantino, G. and Fiorucci, S. (2005) Farnesoid X receptor: from structure to potential clinical applications. *Journal of Medicinal Chemistry*, **48**, 5383–5403.

13 Lechner, D., Kállay, E. and Cross, H.S. (2007) 1alpha,25-Dihydroxyvitamin D(3) downregulates CYP27B1 and induces CYP24A1 in colon cells. *Molecular and Cellular Endocrinology*, **263**, 55–64.

14 Wolber, G. and Langer, T. (2005) LigandScout: 3-D pharmacophores derived from protein-bound ligands and their use as virtual screening filters. *Journal of Chemical Information and Computer Sciences*, **45**, 160–169. LigandScout is available from Inte:Ligand GmbH, www.inteligand.com.

15 Schuster, D., Laggner, C., Steindl, T.M. and Langer, T. (2006) Development and validation of an in silico P450 profiler based on pharmacophore models. *Current Drug Discovery Technologies*, **3**, 1–48.

16 Egnell, A.C., Eriksson, C., Albertson, N., Houston, B. and Boyer, S. (2003) Generation and evaluation of a CYP2C9 heteroactivation pharmacophore. *The Journal of Pharmacology and Experimental Therapeutics*, **307**, 878–887.

17 Wester, M.R., Yano, J.K., Schoch, G.A., Yang, C., Griffin, J., Stout, C.D. and Johnson, E.F. (2004) The structure of human cytochrome P450 2C9 complexed with flurbiprofen at 2.0-A resolution. *The Journal of Biological Chemistry*, **279**, 35630–35637.

18 Williams, P.A., Cosme, J., Vinkovic, D. M., Ward, A., Angove, H.C., Day, P.J., Vonrhein, C., Tickle I.J., and Jhoti, H. (2004) Crystal structures of human cytochrome P450 3A4 bound to metyrapone and progesterone. *Science*, **305**, 683–686.

19 Egnell, A.-C., Houston, J.B. and Boyer, C. S. (2005) Predictive models of CYP3A4 heteroactivation: in vitro–in vivo scaling and pharmacophore modeling. *The Journal of Pharmacology and Experimental Therapeutics*, **312**, 926–937.

20 Lehmann, J.M., McKee, D.D., Watson, M. A., Willson, T.M., Moore, J.T. and Kliewer, S.A. (1998) The human orphan nuclear receptor PXR is activated by compounds that regulate CYP3A4 gene expression and cause drug interactions. *Journal of Clinical Investigation*, **102**, 1016–1023.

21 Bertilsson, G., Heidrich, J., Svensson, K., Asman, M., Jendeberg, L., Sydow-Bäckman, M., Ohlsson, R., Postling, H., Blomquist, P. and Berkenstam, A. (1998) Identification of a human nuclear receptor defines a new signaling pathway for CYP3A induction. *Proceedings of the National Academy of Sciences of the United States of America*, **95**, 12208–12213.

22 Moore, L.B., Parks, D.J., Jones, S.A., Bledsoe, R.K., Consler, T.G., Stimmel, J. B., Goodwin, B., Liddle, C. Blanchard, S. G., Willson, T.M., Collins, J.L. and Kliewer, S.A (2000) Orphan nuclear receptors constitutive androstane receptor and pregnane X receptor share xenobiotic and steroid ligands. *The Journal of Biological Chemistry*, **275**, 15122–15127.

23 Goodwin, B., Moore, L.B., Stoltz, C.M., McKee, D.D. and Kliewer, S.A. (2001) Regulation of the human CYP2B6 gene by the nuclear pregnane X receptor. *Molecular Pharmacology*, **60**, 427–431.

24 Moore, L.B., Goodwin, B., Jones, S.A., Wisely, G.B., Serabjit-Singh, C.J., Willson, T.M., Collins, J.L. and Kliewer, S. A. (2000) St. John's wort induces hepatic drug metabolism through activation of the pregnane X receptor. *Proceedings of the National Academy of Sciences of the United States of America*, **97**, 7500–7502.

25 Jones, S.A., Moore, L.B., Shenk, J.L., Wisely, G.B., Hamilton, G.A., McKee, D.D., Tomkinson, N.C., LeCluyse, E.L., Lambert, M.H., Willson, T.M., Kliewer, S.A. and Moore, J.T. (2000) The pregnane

X receptor: a promiscuous xenobiotic receptor that has diverged during evolution. *Molecular Endocrinology*, **14**, 27–39.
26 Tirona, R.G. and Kim, R.B. (2005) Nuclear receptors and drug disposition gene regulation. *Journal of Pharmaceutical Sciences*, **94**, 1169–1186.
27 Xie, W., Radominska-Pandya, A., Shi, Y., Simon, C.M., Nelson, M.C., Ong, E.S., Waxman, D.J. and Evans, R.M. (2001) An essential role for nuclear receptors SXR/PXR in detoxification of cholestatic bile acids. *Proceedings of the National Academy of Sciences of the United States of America*, **98**, 3375–3380.
28 Catalyst Version 4.5, San Diego, CA, www.accelrys.com.
29 Ekins, S. and Erickson, J.A. (2002) A pharmacophore for human pregnane X receptor ligands. *Drug Metabolism and Disposition: The Biological Fate of Chemicals*, **30**, 96–99.
30 Kobayashi, K., Yamagami, S., Higuchi, T., Hosokawa, M. and Chiba, K. (2004) Key structural features of ligands for activation of human pregnane X receptor. *Drug Metabolism and Disposition: The Biological Fate of Chemicals*, **32**, 468–472.
31 Schuster, D. and Langer, T. (2005) The identification of ligand features essential for PXR activation by pharmacophore modeling. *Journal of Chemical Information and Modeling*, **45**, 431–439.
32 Clement, O.O. and Mehl, A.T. (2000) HipHop: pharmacophores based on multiple common-feature alignments. in *Pharmacophore Perception, Development, and Use in Drug Design* (ed. O.F. Güner), International University Line, La Jolla, CA, pp.69–84.
33 Jacobs, M.N. (2004) *In silico* tools to aid risk assessment of endocrine disrupting chemicals. *Toxicology*, **205**, 43–53.
34 Ung, C.Y., Li, H., Yap, C.W. and Chen, Y.Z. (2007) *In silico* prediction of pregnane X receptor activators by machine learning approaches. *Molecular Pharmacology*, **71**, 158–168.

35 Krasowski, M.D., Yasuda, K., Hagey, L.R. and Schuetz, E.G. (2005) Evolutionary selection across the nuclear hormone receptor superfamily with a focus on the NR1I subfamily (vitamin D, pregnane X, and constitutive androstane receptors). *Nuclear Receptor*, **3**, 1–20.
36 Dai, G. and Wan, Y.-J.Y. (2005) Animal models of xenobiotic receptors. *Current Drug Metabolism*, **6**, 341–355.
37 Mu, Y., Stephenson, C.R.J., Kendall, C., Saini, S.P.S., Toma, D., Ren, S., Cai, H., Strom, S.C. Day, B.W., Wipf, P. and Xie, W. (2005) A pregnane X receptor agonist with unique species-dependent stereoselectivity and its implications in drug development. *Molecular Pharmacology*, **68**, 403–413.
38 Maglich, J.M., Stoltz, C.M., Goodwin, B., Hawkins-Brown, D., Moore, J.T. and Kliewer, S.A. (2002) Nuclear pregnane X receptor and constitutive androstane receptor regulate overlapping but distinct sets of genes involved in xenobiotic detoxification. *Molecular Pharmacology*, **62**, 638–646.
39 Wei, P., Zhang, J., Dowhan, D.H., Han, Y. and Moore, D.D. (2002) Specific and overlapping functions of the nuclear hormone receptors CAR and PXR in xenobiotic response. *The Pharmacogenomics Journal*, **2**, 117–126.
40 Moore, J.T., Moore, L.B., Maglich, J.M. and Kliewer, S.A. (2003) Functional and structural comparison of PXR and CAR. *Biochimica et Biophysica Acta*, **1619**, 235–238.
41 Xu, R.X., Lambert, M.H., Wisely, B.B., Warren, E.N., Weinert, E.E., Waitt, G.M., Williams, J.D., Collins, J.L., Moore, L.B., Willson, T.M. and Moore, J.T. (2004) A structural basis for constitutive activity in the human CAR/RXRa heterodimer. *Molecules and Cells*, **16**, 919–928.
42 Ekins, S., Mirny, L. and Schuetz, E.G. (2002) A Ligand-based approach to understanding selectivity of nuclear hormone receptors PXR, CAR, FXR,

LXRa, and LXRb. *Pharmaceutical Research*, **19**, 1788–1800.

43 Windshuegel, B., Jyrkkaerinne, J., Poso, A., Honkakoski, P. and Sippl, W. (2005) Molecular dynamics simulations of the human CAR ligand-binding domain: deciphering the molecular basis for constitutive activity. *Journal of Molecular Modelling (Online)*, **11**, 69–79.

44 Jyrkkarinne, J., Windshugel, B., Makinen, J., Ylisirnio, M., Perakyla, M., Poso, A., Sippl, W. and Honkakoski, P. (2005) Amino acids important for ligand specificity of the human constitutive androstane receptor. *The Journal of Biological Chemistry*, **280**, 5960–5971.

45 Suino, K., Peng, L., Reynolds, R., Li, Y., Cha, J.-Y., Repa, J.J., Kliewer, S.A. and Xu, H.E. (2004) The nuclear xenobiotic receptor CAR: structural determinants of constitutive activation and heterodimerization. *Molecules and Cells*, **16**, 893–905.

46 Shulman, A.I. and Mangelsdorf, D.J. (2005) Retinoid X receptor heterodimers in the metabolic syndrome. *The New England Journal of Medicine*, **353**, 604–615.

47 Downes, M., Verdecia, M.A., Roecker, A.J., Hughes, R., Hogenesch, J.B., Kast-Woelbern, H.R., Bowman, M.E., Ferrer, J.L., Anisfield, A.M., Edwards, P.A., Rosenfeld, J.M., Alvarez, J.G., Noel, J.P., Nicolaou, K.C. and Evans, R.M. (2003) A chemical, genetic, and structural analysis of the nuclear bile acid receptor FXR. *Molecules and Cells*, **11**, 1079–1092.

48 Fujino, T., Une, M., Imanaka, T., Inoue, K. and Nishimaki-Mogami, T. (2004) Structure–activity relationship of bile acids and bile acid analogs in regard to FXR activation. *Journal of Lipid Research*, **45**, 132–138.

49 Pellicciari, R., Costantino, G., Camaioni, E., Sadeghpour, B.M., Entrena, A., Willson, T.M., Fiorucci, S., Clerici, C. and Gioielli, A. (2004) Bile acid derivatives as ligands of the farnesoid X receptor. Synthesis, evaluation, and structure–activity relationship of a series of body and side chain modified analogues of chenodeoxycholic acid. *Journal of Medicinal Chemistry*, **47**, 4559–4569.

50 Honorio, K.M., Garratt, R.C. and Andricopulo, A.D. (2005) Hologram quantitative structure–activity relationships for a series of farnesoid X receptor activators. *Bioorganic & Medicinal Chemistry Letters*, **15**, 3119–3125.

51 Mi, L.Z., Devarakonda, S., Harp, J.M., Han, Q., Pellicciari, R., Willson, T.M., Khorasanizadeh, S. and Rastinejad, F. (2003) Structural basis for bile acid binding and activation of the nuclear receptor FXR. *Molecules and Cells*, **11**, 1093–1100.

52 Honório, K.M., Garratt, R.C., Polikarpov, I. and Andricopulo, A.D. (2007) 3D QSAR comparative molecular field analysis on nonsteroidal farnesoid X receptor activators. *Journal of Molecular Graphics & Modelling*, **25**, 921–927.

53 Janowski, B.A., Grogan, M.J., Jones, S.A., Wisely, G.B., Kliewer, S.A., Corey, E.J. and Mangelsdorf, D.J. (1999) Structural requirements of ligands for the oxysterol liver X receptors LXRα and LXRβ. *Proceedings of the National Academy of Sciences of the United States of America*, **96**, 266–271.

54 Spencer, T.A., Li, D., Russel, J.S., Collins, J.L., Bledsoe, R.K., Consler, T.G., Moore, L.B., Galardi, C.M., McKee, D.D., Moore, J.T., Watson, M.A., Parks, D.J., Lambert, M.H. and Willson, T.M. (2001) Pharmacophore analysis of the nuclear oxysterol receptor LXRα. *Journal of Medicinal Chemistry*, **44**, 886–897.

55 Urban, F. Jr, Cavazos, G., Dunbar, J., Tan, B., Escher, P., Tafuri, S. and Wang, M. (2000) The important role of residue F268 in ligand binding by LXRβ. *FEBS Letters*, **484**, 159–163.

56 Yang, C., McDonald, J.G., Patel, A., Zhang, Y., Umetani, M., Xu, F., Westover, E.J., Covey, D.F., Mangelsdorf, D.J., Cohen, J.C. and Hobbs, H.H. (2006) Sterol intermediates from cholesterol biosynthetic pathway as liver X receptor

ligands. *The Journal of Biological Chemistry*, **281**, 27816–27826.
57 Buijsman, R.C., Hermkens, P.H.H., van Rijn, R.D., Stock, H.T. and Teerhuis, N.M. (2005) Non-steroidal steroid receptor modulators. *Current Medicinal Chemistry*, **12**, 1017–1075.
58 Farnegardh, M., Bonn, T., Sun, S., Ljunggren, J., Ahola, H., Wilhelmsson, A., Gustafsson, J.A. and Carlquist, M. (2003) The three-dimensional structure of the liver X receptor beta reveals a flexible ligand-binding pocket that can accommodate fundamentally different ligands. *The Journal of Biological Chemistry*, **278**, 38821–38828.
59 Williams, S., Bledsoe, R.K., Collins, J.L., Boggs, S., Lambert, M.H., Miller, A.B., Moore, J., McKee, D.D., Moore, L., Nichols, J., Parks, D., Watson, M., Wisely, B. and Willson, T.M. (2003) X-ray crystal structure of the liver X receptor β ligand binding domain: regulation by a histidine–tryptophan switch. *The Journal of Biological Chemistry*, **278**, 27138–27143.
60 Svensson, S., Oestberg, T., Jacobsson, M., Norstroem, C., Stefansson, K., Hallen, D., Johansson, I.C., Zachrisson, K., Ogg, D. and Jendeberg, L. (2003) Crystal structure of the heterodimeric complex of LXRα and RXRβ ligand-binding domains in a fully agonistic conformation. *The EMBO Journal*, **22**, 4625–4633.
61 Huang, T.H.-W., Razmovski-Naumovski, V., Salam, N.K., Duke, R.K., Tran, V.H., Duke, C.C. and Roufogalis, B.D. (2005) A novel LXR-α activator identified from the natural product *Gynostemma pentaphyllum*. *Biochemical Pharmacology*, **70**, 1298–1308.
62 Hu, B., Collini, M., Unwalla, R., Miller, C., Singhaus, R., Quinet, E., Savio, D., Halpern, A., Basso, M., Keith, J., Clenin, V., Chen, L., Resmini, C., Lui, Q.-Y., Feingold, I., Huselton, C., Azam, F., Fanegardh, M., Ennoth, C., Bonn, T., Goos-Nilsson, A., Wilhelmsson, A., Nambi, P. and Wrobel, J. (2006) Discovery of phenyl acetic acid substituted quinolines as novel liver X receptor agonists for the treatment of atherosclerosis. *Journal of Medicinal Chemistry*, **49**, 6151–6154.
63 Li, L., Liu, J., Zhu, L., Cutler, S., Hasegawa, H., Shan, B. and Medina, J.C. (2006) Discovery and optimization of a novel series of liver X receptor-alpha agonists. *Bioorganic & Medicinal Chemistry Letters*, **16**, 1638–1642.
64 Liu, W., Chen, S., Dropinski, J., Colwell, L., Robins, M., Szymonifka, M., Hayes, N., Sharma, N., MacNaul, K., Hernandez, M., Burton, C., Sparrow, C.P., Menke, J.G. and Singh, S.B. (2005) Design, synthesis, and structure–activity relationship of podocarpic acid amides as liver X receptor agonists for potential treatment of atherosclerosis. *Bioorganic & Medicinal Chemistry Letters*, **15**, 4574–4578.
65 Saarikoski, S.T., Rivera, S.P., Hankinson, O. and Husgafvel-Pursiainen, K. (2005) CYP2 S1: a short review. *Toxicology and Applied Pharmacology*, **207**, S62–S69.
66 Ramadoss, P., Marcus, C. and Perdew, G.H. (2005) Role of the aryl hydrocarbon receptor in drug metabolism. *Expert Opinion on Drug Metabolism & Toxicology*, **1**, 9–21.
67 Pocar, P., Fischer, B., Klonisch, T. and Hombach-Klonisch, S. (2005) Molecular interactions of the aryl hydrocarbon receptor and its biological and toxicological relevance for reproduction. *Reproduction*, **129**, 379–389.
68 Mandal, P.K. (2005) Dioxin: a review of its environmental effects and its aryl hydrocarbon receptor biology. *Journal of Comparative Physiology B: Biochemical, Systemic, and Environmental Physiology*, **175**, 221–230.
69 Okey, A.B., Franc, M.A., Moffat, I.D., Tijet, N., Boutros, P.C., Korkalainen, M., Tuomisto, J. and Pohjanvirta, R. (2005) Toxicological implications of polymorphisms in receptors for xenobiotic chemicals: the case of the aryl hydrocarbon receptor. *Toxicology and Applied Pharmacology*, **207**, S43–S51.

70 Hombach-Klonisch, S., Pocar, P., Kietz, S. and Klonisch, T. (2005) Molecular actions of polyhalogenated arylhydrocarbons (PAHs) in female reproduction. *Current Medicinal Chemistry*, **12**, 599–616.

71 Karp, J.M., Rodrigo, K.A., Pei, P., Pavlick, M.D., Andersen, J.D., McTigue, D.J., Fields, H.W. and Mallery, S.R. (2005) Sanguinarine activates polycyclic aromatic hydrocarbon associated metabolic pathways in human oral keratinocytes and tissues. *Toxicology Letters*, **158**, 50–60.

72 Caprioara, M. and Diudea, M.V. (2003) QSAR modeling of polychlorinated aromatic compounds. *Indian Journal of Chemistry*, **42A**, 1368–1378.

73 Jacobs, M.N., Dickins, M. and Lewis, D.F.V. (2003) Homology modeling of the nuclear receptors: human estrogen receptor β (hERβ), the human pregnane-X-receptor (PXR), the Ah receptor (AhR) and the constitutive androstane receptor (CAR) ligand binding domains from the human estrogen receptor α (hERα) crystal structure, and the human peroxisome proliferator activated receptor α (PPARα) ligand binding domain from the human PPARγ crystal structure. *The Journal of Steroid Biochemistry and Molecular Biology*, **84**, 117–132.

74 Saeki, K.-I., Matsuda, T., Kato, T.-A., Yamada, K., Mizutani, T., Matsui, S., Fukuhara, K. and Miyata, N. (2003) Activation of the human Ah receptor by aza-polycyclic aromatic hydrocarbons and their halogenated derivatives. *Biological & Pharmaceutical Bulletin*, **26**, 448–452.

75 Kato, T.-A., Matsuda, T., Matsui, S., Mizutani, T. and Saeki, K.-I. (2002) Activation of the aryl hydrocarbon receptor by methyl yellow and related congeners: structure–activity relationships in halogenated derivatives. *Biological & Pharmaceutical Bulletin*, **25**, 466–471.

76 Lewis, D.F.V. and Jacobs, M.N. (1999) A QSAR study of some PCBs' ligand-binding affinity to the cytosolic Ah receptor (AhR). *Organohalogen Compounds*, **41**, 537–540.

77 Vetter, W., Hahn, M.E., Tomy, G., Ruppe, S., Vatter, S., Chahbane, N., Lenoir, D., Schramm, K.W. and Scherer, G. (2005) Biological activity and physicochemical parameters of marine halogenated natural products 2,3,3′,4,4′,5,5′-heptachloro-1′-methyl-1,2′-bipyrrole and 2,4,6-tribromoanisole. *Archives of Environmental Contamination and Toxicology*, **48**, 1–9.

78 Beger, R.D., Buzatu, D.A. and Wilkes, J.G. (2002) Combining NMR spectral and structural data to form models of polychlorinated dibenzodioxins, dibenzofurans, and biphenyls binding to the AhR. *Journal of Computer-Aided Molecular Design*, **16**, 727–740.

79 Sovadinova, I., Blaha, L., Janosek, J., Hilscherova, K., Giesy, J.P., Jones, P.D. and Holoubek, I. (2006) Cytotoxicity and aryl hydrocarbon receptor-mediated activity of N-heterocyclic polycyclic aromatic hydrocarbons: structure–activity relationships. *Environmental Toxicology and Chemistry*, **25**, 1291–1297.

80 Wang, Y., Zhao, C., Ma, W., Liu, H., Wang, T. and Jiang, G. (2006) Quantitative structure–activity relationship for prediction of the toxicity of polybrominated diphenyl ether (PBDE) congeners. *Chemosphere*, **64**, 515–524.

81 Luan, F., Ma, W.P., Zhang, X.Y., Zhang, H.X., Liu, M.C., Hu, Z.D. and Fan, B.T. (2006) QSAR study of polychlorinated dibenzodioxins, dibenzofurans, and biphenyls using the heuristic method and support vector machine. *QSAR & Combinatorial Science*, **25**, 46–55.

82 Lo Piparo, E., Koehler, K., Chana, A. and Benfenati, E. (2006) Virtual screening for aryl hydrocarbon receptor binding prediction. *Journal of Medicinal Chemistry*, **49**, 5702–5709.

83 Denison, M.S. and Nagy, S.R. (2003) Activation of the aryl hydrocarbon

receptor by structurally diverse exogenous and endogenous chemicals. *Annual Review of Pharmacology and Toxicology*, **43**, 309–334.

84 Denison, M.S. (2005) Exactly the same but different: analysis and implications of the promiscuity of ligand-dependent activation of the Ah receptor signaling pathway, in Abstracts of Papers, 230th ACS National Meeting, Washington, DC, United States, August 28–September 1, 2005, pp. AGRO-050.

85 Procopio, M., Lahm, A., Tramontano, A., Bonati, L. and Pitea, D. (1999) Homology modeling of the AhR ligand binding domain. *Organohalogen Compounds*, **42**, 405–408.

86 Pandini, A., Denison, M.S., Song, Y., Soshilov, A.A. and Bonati, L. (2007) Structural and functional characterization of the aryl hydrocarbon receptor ligand binding domain by homology modeling and mutational analysis. *Biochemistry*, **46**, 696–708.

15
Transporters and Drugs – An Overview
Hartmut Glaeser, Martin F. Fromm, Jörg König

15.1
Introduction

Drug efficacy results from the interplay of multiple processes that regulate drug response as well as drug absorption, elimination and disposition. Although in the past decades studies focused on the contribution of drug-metabolizing enzymes (e.g. cytochrome P450 enzymes) to these processes, it has become evident in the recent years that transport proteins are also key players in the field of drug treatment. Generally, transport proteins can be classified into two major groups: uptake transporters and export pumps. Uptake transporters are responsible for the translocation of substances and drugs from the outside into cells. They mostly belong to the superfamily of solute carriers (SLCs) [1]. Important transporter families within this SLC transporter superfamily are the organic anion transporting polypeptides (OATPs; gene family SLCO or SLC21A) [2,3], the organic anion and cation transporter family (OATs and OCTs, gene family SLC22A) [4] and the peptide transporter family (PEPT, gene family SLC15A) [5]. In this chapter, we will focus mostly on drug transport mediated by members of the human OATP family. For detailed information about the superfamily of solute carriers see http://www.bioparadigms.org.

Efflux transporters are mainly members of the ATP-binding cassette (ABC) transporter family [6–8]. These transporters use energy derived from ATP hydrolysis to mediate substrate translocation, basically against concentration gradients, from inside the cell to the extracellular environment. There are 48 different ABC transporters encoded in the human genome grouped into seven ABC transporter subfamilies (ABC**A** to ABC**G**) [9,10]. Important drug transporting export pumps are P-glycoprotein (P-gp), a member of the ABCB subfamily, members of the MRP (multidrug resistance protein) family belonging to the ABCC subfamily and BCRP (breast cancer resistance protein) belonging to the ABCG subfamily.

Reflecting the increasing importance of drug transporters in pharmacokinetics, we need to extend the historical two-phase concept for the metabolism of xenobiotics. As shown in Figure 15.1, the metabolic phases I (oxidation) and II (conjugation) are flanked by drug transporter phases 0 (uptake) and III (export). Phase 0 is the first step

Antitargets. Edited by R. J. Vaz and T. Klabunde
Copyright © 2008 WILEY-VCH Verlag GmbH & Co. KGaA, Weinheim
ISBN: 978-3-527-31821-6

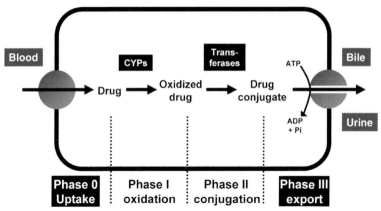

Figure 15.1 Schematic illustration of the extended phase concept during drug elimination in the kidney and liver. Phase 0 = uptake of drugs from the blood into the hepatocytes or proximal tubule epithelial cells. This uptake is mediated by transport proteins belonging to the SLC (solute carrier) transporter superfamily. Phase I and phase II = biotransformation (oxidation, mainly by cytochrome P450 enzymes followed by the conjugation of the activated drug metabolite. Phase III = efflux of the drug metabolite over the apical membrane into bile or urine. The efflux is mediated mostly by members of the ABC (ATP-binding cassette) transporter superfamily.

in drug elimination with the uptake of drug from the blood over the basolateral membrane into the cells (Figure 15.1). Phase III, the excretion step, is predominantly maintained by energy-driven ATP-binding cassette transporters, mediating the uphill transport of xenobiotics across cell membranes [6,7,11]. It is only when all components of this extended phase concept are present, the directed elimination of drugs is secured.

Important for the understanding of transporter-mediated drug disposition is the interplay between uptake transporters and efflux pumps. Epithelial cells, such as the proximal tubule epithelial cells in the kidney or the hepatocytes in the liver, contain two distinct membrane domains: the basolateral membrane domain facing the blood and the apical membrane domain facing the lumen of the tubules or the bile canaliculi. Drugs that have to be eliminated via the kidney or the liver first have to be taken up from the blood over the basolateral membrane into the cells before their subsequent elimination over the apical membrane into urine or bile. Furthermore, drug-metabolizing enzymes are located intracellularly, and therefore, the uptake of drugs into the cells is the first necessary step before drugs can be metabolized. Therefore, for many drugs, the combined actions of transporters expressed in the respective membrane domains determine the extent and the direction of drug movement across organs.

15.2
Organic Anion Transporting Polypeptides and Drug Transport

The OATP family of uptake transporters consists of 52 members comprising 12 different subfamilies across eight species. In humans, 11 different OATP family members have been identified, including 10 OATPs and the prostaglandin transporter OATP2A1 (formerly termed as PGT) [2,12]. Important members for drug transport are OATP1A2 (formerly termed as OATP or OATP-A) [13], expressed in a variety of different tissues and the hepatocellularly expressed family members OATP1B1 (formerly termed as OATP2 or OATP-C) [14–16] and OATP1B3 (formerly termed as OATP8) [17]. Because many mouse/rat OATPs have no human ortholog and trivial names for individual proteins mostly do not correspond to the continuous numbering based on the chronology of protein identification, Hagenbuch and Meier [2] introduced a new nomenclature for the OATP family. In this chapter, this new nomenclature will be used.

All OATPs share a similar transmembrane topology with 12 predicted transmembrane domains and a large fifth extracellular loop. Additional conserved features are N-glycosylation sites in the extracellular loops 2 and 5 and the so-called OATP family signature at the border between the extracellular loop 3 and the transmembrane domain 6 [2]. Several human OATP family members are expressed in multiple tissues. OATP1A2 shows the highest expression in the brain and testis [13] whereas OATP2B1 and OATP4A1 are ubiquitously expressed in all tissues investigated so far [18,19] with OATP2B1 being highly expressed in the heart [20], liver [21] and placenta [22]. Interestingly, both well-characterized family members OATP1B1 and OATP1B3 are exclusively expressed in human hepatocytes [14–17,23]. OATP1B1, OATP1B3 and OATP2B1 have been localized to the basolateral membrane of human hepatocytes (Figure 15.2) [16,17], whereas OATP1A2 has been localized to the basolateral membrane of brain endothelial cells [24–26]. In addition to the basolateral localization in hepatocytes, OATP2B1 has been detected in the apical membrane of enterocytes (Figure 15.2) and of the syncytiotrophoblast in the placenta [22,27], in the vascular endothelium of the heart [20] and in the basolateral membrane of the nonpigmented human ciliary body epithelium together with OATP1C1 and OATP3A1 [28]. OATP4C1 is apparently highly expressed in the human kidney [29] whereas the tissue distribution and the subcellular localization of OATP5A1 and OATP6A1 remain to be clarified.

Originally, OATP1A2 was cloned based on its sequence homology to rat Oatp1a1 (Oatp1) [30] and was characterized as a sodium-independent uptake transporter for organic anions and bile acids [13]. Further studies have demonstrated that OATP1A2 is capable of transporting a variety of organic anions including drugs and toxins such as fexofenadine and microcystin. Determining the substrate spectrum of other human OATP family members demonstrated that most of them have a broad substrate spectrum partially overlapping with the substrate spectrum of OATP1A2. The substrates of OATP1B1 and OATP1B3 include bilirubin and its glucuronide conjugates, bile acids, hormones and their conjugates, peptides, eicosanoids and drugs (for review, see [3]). Although OATP1B1 and OATP1B3 share a wide overlapping substrate spectrum, some substances such as the intestinal peptide cholecystokinin 8 [31] and the cardiac glycoside digoxin [18] are transported only by OATP1B3. As illustrated in Table 15.1, OATP1B1 and OATP1B3 are the best characterized family members with respect to mediating drug transport. Several classes of drugs have been identified as substrates for both transporters including

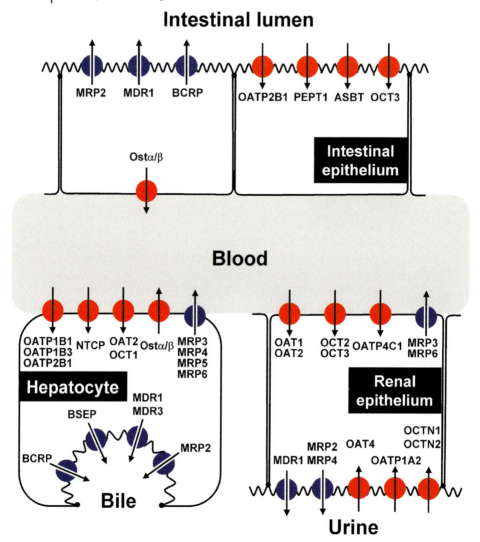

Figure 15.2 Transport proteins involved in the intestinal absorption and the renal and hepatic excretion of drugs. In the intestine, drugs are taken up from the luminal side into enterocytes before the subsequent elimination into blood. In hepatocytes, drugs are taken up from the blood over the basolateral membrane and excreted over the canalicular membrane into bile. In the renal epithelium, drugs undergo secretion (drugs are taken up from the blood and excreted into the urine) or reabsorption (drugs are taken up from the urine and are excreted back into blood). Uptake transporters belonging to the SLC transporter superfamily are shown in red and export pumps belonging to the ABC transporter superfamily are shown in blue. OATP = organic anion transporting polypeptide (SLCO), PEPT = peptide transporter (SLC15), MDR = members of the ABCB transporter family (MDR1 = ABCB1 and MDR3 = ABCB4), MRP = member of the ABCC transporter family (MRP1 – MRP6 = ABCC1 – ABCC6), ASBT = apical sodium-dependent bile acid transporter (SLC10), OCT(N) = organic cation transporter (SLC22), OAT = organic anion transporter (SLC22), Ostα/β = organic solute transporter, BSEP = bile salt export pump (ABCB11), BCRP = breast cancer resistance protein (ABCG2).

Table 15.1 Drugs identified as substrates for human OATPs.

Drug	OATP Substrate	P-Glycoprotein Substrate	MRP Substrate
Atorvastatin	OATP1B1, −2B1	+	Inhibitor
Benzylpenicillin	OATP1B1, −2B1, −3A1, −4A1	−	−
Capsofungin	OATP1B1	−	−
Deltorphin II	OATP1A2	+	−
Digoxin	OATP1B3	+	−
Docetaxel	OATP1B3	+	−
DPDPE	OATP1A2	+	−
Fexofenadine	OATP1A2	+	−
Fluvastatin	OATP1B1, −1B3	−	−
Methotrexate	OATP1A2, −1B3	+	MRP1, −2, −4, −8
Microcystin-LR	OATP1A2, −1B1, −1B3	−	−
Paclitaxel	OATP1B3	+	MRP7
Pitavastatin	OATP1B1	+	MRP2
Pravastatin	OATP1B1, −1B3	+	MRP2
Repaglinide	OATP1B1	−	−
Rifampicin	OATP1B1	+	−
Rosuvastatin	OATP1B1, −1B3, −2B1	+	MRP2

This table also indicates whether these drugs interact with P-gp or MRPs. For references see the following articles: for OATPs [3,32], for P glycoprotein [33–36] and for MRPs [7,37,38].

antibiotics such as rifampicin and HMG-CoA-reductase inhibitors (statins) such as fluvastatin [39] and pravastatin [40]. Several statins have been identified as substrates for only one or two human OATP family members. Atorvastatin acid, for example, is transported by OATP1B1 [41] and OATP2B1 [20] whereas simvastatin acid, a substrate for OATP1B1 [42], is not transported by OATP2B1.

Several drugs have been reported to inhibit the transport function of OATP1B1 and OATP1B3 resulting in the potential for transporter-mediated drug–drug interactions. Most of these interactions involve statins [43–45] that are administered to patients with hypercholesterolemia. So it has been suggested that the inhibition of OATP1B1-mediated uptake may be responsible for an observed pharmacokinetic interaction between cyclosporin and rosuvastatin [37], which undergoes limited metabolic transformation by CYP2C9. Comparing the rosuvastatin kinetics of heart-transplanted patients receiving cyclosporin with those receiving rosuvastatin alone demonstrated that the area under the curve (AUC) of rosuvastatin was 7.1-fold higher in transplant patients [46]. These data suggest that the inhibition of OATP1B1-mediated rosuvastatin uptake by coadministered cyclosporin may be, at least in part, the underlying mechanism of the observed drug–drug interaction.

A potential drug–drug interaction has also been described for the combination of gemfibrozil with several statins. Studies in healthy volunteers have demonstrated that gemfibrozil increased the AUC and C_{max} of rosuvastatin by 1.9- and 2.2-fold, respectively [43]. These results could be confirmed *in vitro* using *Xenopus* oocytes. In these experiments, gemfibrozil inhibited OATP1B1-mediated rosuvastatin transport with an IC_{50} value of 4 μM [45]. Interestingly, more than 50% of the individuals who developed rhabdomyolysis after cerivastatin therapy also took gemfibrozil [47]. Studies have demonstrated that gemfibrozil increased the AUC of coadministered cerivastatin to 560% of the values when cerivastatin therapy alone was used and that the formation of the hydroxylated cerivastatin metabolite M-23 is inhibited [48]. These data suggest that in addition to the metabolic interaction, transport inhibition may account for these drug–drug interaction effects. Our recent *in vitro* studies have demonstrated that also macrolides such as erythromycin may interact with the uptake of organic anions mediated by OATP1B1 and OATP1B3 and that this uptake inhibition may account for altered pravastatin kinetics observed after coadministration of macrolides and this statin [40].

After looking at the substrate spectrum of OATP1B1 and OATP1B3, it is obvious that uptake inhibition by one drug may not only affect a coadministered drug but also affect the hepatic uptake of endogenous substances. Especially, the inhibition of bilirubin uptake may account for drug-induced hyperbilirubinemia. This was carefully investigated by Campbell and coworkers [49] who demonstrated that rifampicin-induced hyperbilirubinemia, for example, may be due to the uptake inhibition of bilirubin into hepatocytes.

15.3
Multidrug Resistance Proteins and Drug Transport

Multidrug resistance proteins were initially named after their ability to cause drug resistance of tumor cells by ATP-dependent efflux of chemotherapeutic agents out of the cells [7,8,37,50]. However, the nomenclature 'MRP' may not be always relevant because not all the members of the MRP family are responsible for multidrug resistance. The MRP transporters belong to an evolutionarily conserved family of ATP-binding cassette proteins. In the human genome, 48 ABC proteins are encoded [9,10]. According to the amino acid alignments, the ABC transporters can be divided into seven subfamilies from A to G. In the following chapter, the multidrug resistance proteins that belong to the ABC subfamily 'C' will be discussed in more detail. The ABC proteins are characterized by a specific ABC domain that binds to ATP and hydrolyzes it [51]. The ABC unit or nucleotide binding domain (NBD) consists of various conserved motifs including the Walker A, the Walker B and the ABC signature (LSGGQ), of which the ABC signature is specific for all ABC transporters. In addition, in most cases, ABC transporters contain transmembrane domains (TMDs) composed of six membrane-spanning helices. For the MRPs, it is most likely that the substrate interaction sites are located in the TMDs. The topology of the MRPs reveals the following sequence of domains: TMD1–NBD1–TMD2–NBD2 (for MRP4, −5, −8, 9) and TMD0–TMD1–NBD1–TMD2–NBD2 (for MRP1, −2, −3, −6, −7) [7].

The first member of the ABCC subfamily ABCC1 encoding MRP1 was cloned from a doxorubicin-selected lung cancer cell line that showed multidrug resistance, yet no expression of P-glycoprotein [52,53]. MRP1 is expressed at high levels in numerous tissue and organs such as lung, testis, kidney, placenta and skeletal and cardiac muscle [52,54], indicating that MRP1 might be a key player in the defense against xenobiotics and in the elimination of endogenous metabolites in these organs. Although the expression level of MRP1 is quite low in human adult liver, MRP1 is expressed at a considerably high level in liver tumor cell lines such as HepG2 [55]. In polarized cells, MRP1 is found to be localized in either apical or basolateral membrane depending on the cell types. The mechanisms of cell-specific localization and trafficking remain to be clarified. For example, MRP1 localizes in the apical membrane of cells in the placental syncytiotrophoblast and in the endothelial cells of brain capillaries [56,57], whereas it localizes in the basolateral membrane in the proximal tubule epithelial cells of the kidney and in enterocytes [58,59]. Further *in vitro* experiments using a MRP1–GFP fusion protein (green fluorescent protein) and the fluorescent drug doxorubicin revealed that the drug accumulates in the MRP1-containing compartment, highlighting that MRP1 also plays an important role in the maintenance of intracellular compartments [60].

Several *in vitro* studies demonstrated that MRP1 mediates the transport of a vast array of organic compounds and/or conjugates. Cytotoxic agents such as anthracyclines, epipodophyllotoxins and vinca alkaloids are substrates of MRP1 [58,59,61]. Moreover, MRP1 and MRP2 can mediate the drug resistance to doxorubicin and vincristine [62–64]. This is of particular interest because both drugs can be used for the treatment of small cell lung cancer (SCLC). The expression of MRP1 in untreated SCLC is lower than that in untreated nonsmall cell lung cancer (NSCLC), which is characterized by moderate and high levels of MRP1. This may account in part for the inherent multidrug resistance of NSCLC. In prostate cancer, MRP1 may also contribute to hormonally refractory cancer types since MRP1 overexpressing cells show increased transport of the antiandrogen flutamide and its active metabolite hydroxyflutamide [65]. Other clinically relevant substrates of MRP1 include irinotecan, conjugated and unconjugated SN-38 (an active metabolite of irinotecan), methotrexate and potentially the HIV-protease inhibitor saquinavir [66–68]. MRP1 also contributes to the efflux of several endogenous compounds such as leukotriene C_4 (LTC_4) and estradiol-17β-glucuronide ($E_2 17\beta G$), indicating that MRP1 might influence inflammation and immune response and might as well protect the testis from feminization by estrogen [63,64,69–72]. Furthermore, glutathione (GSH) and oxidized glutathione (GSSG) are transported by MRP1, suggesting a potential role of MRP1 in decreasing the intracellular concentration of these reactive metabolites [71,73]. In contrast to P-glycoprotein, the MRP1- and MRP2-mediated transport has been shown to involve additional mechanism(s) other than the hydrolysis of ATP. Indeed, the transport of several unmodified drugs such as vincristine, doxorubicin and verapamil seems to be dependent on or enhanced by GSH [64,74,75].

MRP2/Mrp2, another important member of the ABCC/MRP subfamily, is encoded by the *ABCC2* gene and it was first cloned from rat liver [76–78]. In contrast to MRP1, MRP2 shows a more restricted tissue expression and appears exclusively on apical membranes. In humans, MRP2 is expressed in the canalicular membrane of hepatocytes and in the apical membrane of renal tubule cells and enterocytes (Figure 15.2) [76,79,80]. Similar to MRP1, MPR2 contributes to the efflux of multiple endogenous and exogenous compounds [37,81]. The data regarding the clinical relevance for MRP2-mediated multidrug resistance during chemotherapy are limited. It has been shown that an increased expression of MRP2 in colorectal carcinoma was associated with resistance to cisplatin [82]. The physiological and pathophysiological importance of MRP2 for endogenous compounds was discovered using various Mrp2-deficient animal models such as the GY/TR$^-$ rat and the EHBR rat [76–78]. These animals were characterized by a decreased biliary transport of bilirubin glucuronides and LTC$_4$ due to mutations in the *Abcc2* gene leading to the absence of Mrp2 from the canalicular membrane of hepatocytes. In humans, a variety of mutations (for review see [37,83]) in the *ABCC2* gene can cause the Dubin–Johnson syndrome, which is also characterized by a complete loss of immunodetectable MRP2 in the liver [84]. The clinical symptoms of these patients are hyperbilirubinemia and jaundice. Divalent bile salts are also substrates of MRP2 [81,85,86]. In addition to the typical physiological substrates such as glucuronides, glutathione conjugates, and sulfates, several drugs have been identified as substrates of MRP2. For example, pravastatin, methotrexate, irinotecan and the HIV-protease inhibitors such as saquinavir and indinavir are transported by MRP2 [62,87].

MRP3, encoded by the *ABCC3* gene, is expressed in the intestine, adrenal gland, pancreas, gut, gall bladder and placenta. MRP3 expression at a low level was also reported in the prostate, liver and kidney [88–90]. In polarized cells such as hepatocytes or enterocytes, MRP3 localizes in the basolateral membrane [90,91]. The induction of hepatic MRP3 expression under cholestatic conditions highlights an important physiological role of MRP3 [92]. Since it is known that bile acids such as glycocholate and taurocholate are transported by MRP3, it is speculated that MRP3 may protect the hepatocytes by extruding these toxic bile acids out of the cells. Similar to MRP2, the role of MRP3 for drug resistance has not yet been fully elucidated. MRP3 contributes, in part, to cellular resistance against etoposide and teniposide, whereas the underlying mechanism seems to be GSH independent [93,94].

Additional MRPs such as MRP4 (ABCC4), MRP5 (ABCC5) and MRP8 (ABCC11) are characterized by their ability to transport nucleotide analogues [95–98]. However, the three transporters show overlapping but not identical substrate specificities. For example, the nucleoside analogue 9-(-2-phosphonylmethoxyethyl)adenine (PMEA) is transported by MRP4, MRP5 and MRP8. In addition, MRP4 also confers a resistance to the antiviral drugs such as azidothymidine monophosphate (AZT) and lamivudine. In contrast to MRP4, MRP5 and MRP8 confer a resistance to 5-fluorouracil [95,99]. As these three transporters are able to transport cyclic nucleotides such as cGMP and cAMP, their potential physiological roles in nucleotide and cyclic nucleotide homeostasis have been discussed. Interestingly, it was shown that the transport activity of MRP5 is inhibited by the drug sildenafil (Viagra) [96]. MRP4 mediates the transport of prostaglandins such as PGE$_1$ and PGE$_2$ that can be inhibited by nonsteroidal anti-inflammatory drugs, indicating an important role for inflammatory processes [100,101]. Regarding the expression, both MRP4 and MRP5 are expressed in various tissues. MRP4 is expressed in the prostate, lung, adrenals, ovary, testis, pancreas, platelets [102] and small intestine (for references see [7]). Interestingly, in most tissues MRP4 is localized in basolateral membrane, whereas in proximal tubule cells of kidney MRP4 is localized in the apical membrane [103]. MRP5 shows ubiquitous expression with high levels in skeletal muscle, heart, brain and smooth muscle of the genitourinary tract [7]. Notably, the cellular localization of MRP5 does not seem to be restricted to either the apical or the basolateral side. For MRP8, the knowledge regarding its expression is yet not fully complete. It is expressed in axons of the human CNS and the peripheral nervous system and mediates the transport of steroid sulfates [104]. However, mRNA analysis revealed that MRP8 may also be expressed in other tissues with exception of kidney, spleen and colon [7].

The less characterized MRPs are MRP6 (ABCC6), MRP7 (ABCC10) and MRP9 (ABCC12). MRP6 has been associated with pseudoxanthoma elasticum (PXE) [105]. MRP6 is highly expressed in the liver and kidney, but not in the tissues that are affected by PXE, indicating that PXE might be caused by the absence of substances secreted by kidney and liver [106]. Moreover, MRP6 confers a low resistance against etoposide, teniposide, anthracyclines and cisplatin [107,108]. Even though little is known about MRP7, it should be noted that MRP7 appears to be the only MRP that mediates a resistance to taxanes such as paclitaxel and docetaxel [109]. This might be of particular interest since these drugs are used for the chemotherapy of breast cancer, ovary cancer and NSCLC.

15.4
Role of P-Glycoprotein for Drug Disposition

The importance of drug efflux transporters in drug disposition has been well recognized since the cloning of the *MDR1* gene (encoding P-glycoprotein) [110–112]. P-glycoprotein is an efflux pump responsible for developing drug resistance for multiple chemotherapeutic agents [113]. The overexpression of P-glycoprotein in tumor cells increases the ATP-dependent efflux of chemotherapeutic drugs out of cells, resulting in the multidrug resistance phenotype. In addition, P-glycoprotein plays an important physiological role in the defense against drugs and xenobiotics. The expression of P-glycoprotein is found in various tissues and organs such as the intestine (apical membrane of enterocytes), liver (canalicular membrane of hepatocytes), kidney (proximal tubule epithelial cells [Figure 15.2]), brain (luminal membrane of endothelial cells in brain capillaries), pancreas (apical membrane of secretarial duct), placenta (microvilli of syncytiotrophoblast), testis (luminal membrane of endothelial cells) and lymphocytes [114–118]. Schinkel *et al.* [119] demonstrated the relevance of P-glycoprotein for drug elimination and disposition using an *mdr1* knockout mouse model. Following intravenous and oral administration of the P-glycoprotein substrate digoxin, the *mdr1* knockout mice showed significantly higher concentrations of digoxin in plasma (2–4-fold) and brain (30–50-fold) than wild-type mice [119,120]. Additional differences included substantial secretion of intravenously administered digoxin into the intestine of wild-type mice in contrast to that of minimal secretion in the *mdr1* knockout mice [120]. These findings highlighted the importance of P-glycoprotein in drug disposition *in vivo*, and the results have been further confirmed in a number of studies in humans (e.g. [121]).

The significance of intestinal P-glycoprotein in humans was demonstrated in a drug interaction study using digoxin and rifampicin [122]. This study reported that the intestinal P-glycoprotein is a major determinant for the bioavailability of orally administered digoxin and that the induction of P-glycoprotein is a mechanism for drug interactions between rifampicin and digoxin. A study by Drescher et al. [121] has further demonstrated the relevance of intestinal P-glycoprotein for the elimination of intravenously administered digoxin in healthy volunteers by using a multilumen perfusion catheter. Subsequent studies in healthy volunteers underlined the importance of P-glycoprotein by determining the pharmacokinetics of several P-glycoprotein substrates such as talinolol, loperamide and fexofenadine. One of the earlier studies using the β-adrenoceptor blocker talinolol showed that simultaneous administration of talinolol with erythromycin results in an increased AUC (area under the plasma concentration–time curve) of talinolol in humans [123]. Further drug interaction studies combining talinolol with digoxin and rifampicin confirmed the role of intestinal P-glycoprotein for the disposition and elimination of talinolol [124,125]. Another example of P-glycoprotein-mediated drug interactions includes fexofenadine. Fexofenadine undergoes negligible metabolism and exhibits marginal side effects, and because of these features fexofenadine is extensively used for pharmacokinetic studies in humans. Although it is known that fexofenadine is also a substrate for the uptake transporter OATP1A2, most studies have examined the impact of P-glycoprotein on fexofenadine disposition [32,126]. A drug interaction study between fexofenadine and rifampicin revealed an increased clearance of orally administered fexofenadine after rifampicin intake (600 mg/days for 6 days) due to an induction of intestinal P-glycoprotein [127]. Other studies used fexofenadine as a P-glycoprotein probe to investigate the influence of polymorphisms in the *ABCB1* gene on the pharmacokinetics in humans [128,129]. However, the results of these studies were conflicting, suggesting that there may be additional determinants, such as OATP functions [130], for the pharmacokinetics of fexofenadine in humans.

In addition to the intestine, P-glycoprotein also plays an important role in the blood–brain barrier (BBB), limiting the entry of drugs into the brain. P-glycoprotein-mediated exclusion of drug molecules from the CNS (central nervous system) endothelial cells may provide protection from central side effects or result in a failure of desired pharmacological and therapeutic effects. For example, the μ-opioid receptor agonist loperamide is a substrate of P-glycoprotein [131]. Loperamide is extensively used as an antidiarrheal drug without central side effects because the entry of loperamide into the CNS compartment is limited by P-glycoprotein. In a study with healthy volunteers, the simultaneous administration of loperamide with the P-glycoprotein inhibitor quinidine led to respiratory depression due to a central opioid effect of loperamide [132]. In children whose BBB function is not fully developed, loperamide can cause serious central side effects such as constipation and respiratory depression [133]. Further studies tried to clarify the influence of polymorphisms in the *ABCB1* gene on the central side effects of loperamide [134,135]. However, the investigated polymorphisms in the *ABCB1* gene had only limited impact on the pharmacokinetics and central pharmacodynamics of loperamide in humans. Using PET (positron emission tomography) imaging these results were also confirmed with the P-glycoprotein substrate verapamil [136,137].

The impact of P-glycoprotein on the HIV therapy was well demonstrated by a study that investigated the brain entry of HIV-protease inhibitors using *mdr1* knockout mice [138,139]. These data indicate that P-glycoprotein expressed in the blood–brain barrier limits the access of HIV-protease inhibitors to the brain, thereby possibly contributing to the persistence of virus in this sanctuary site.

As discussed earlier, P-glycoprotein is also expressed in the syncytiotrophoblast of the human placenta transporting toxins and drugs from the fetal to the maternal circulation. The expression of P-glycoprotein in the human placenta was extensively investigated [114,140–142]. The fact that P-glycoprotein is expressed in the trophoblast facing the maternal side supports the role of P-glycoprotein in protecting the fetus from toxic substances. In fact, a study investigating the P-glycoprotein expression in human placenta during pregnancy revealed a twofold lower P-glycoprotein expression at late term than at earlier time points during pregnancy, suggesting that the ability of fetal protection decreases during pregnancy [142]. The consideration of the placental P-glycoprotein expression in the context of the fetal accessibility to HIV-protease inhibitors adds additional clinical relevance of placental P-glycoprotein in affecting the drug disposition and effectiveness of HIV drugs [138,139,143]. This notion was confirmed by investigating the transplacental passage of protease inhibitors at delivery [144]. Measurement of the HIV-protease inhibitor concentration in the peripheral maternal vein and the umbilical cord at the time of delivery revealed that lopinavir, nelfinavir, ritonavir and saquinavir did not cross the placenta–blood barrier. In contrast, the nonnucleoside reverse transcriptase inhibitor nevirapine showed a sufficient passage across the placental barrier, indicating that the fetus would benefit more from this treatment. Indeed, the analysis revealed that the time point of nevirapine intake during labor pains correlates well with the nevirapine concentration in the cord blood that also showed an association with the HIV-1 transmission to infants [145].

The role of other ABC-efflux transporters (ATP-binding cassette transporters) such as MRPs and BCRP was also intensively investigated over the past decade. However, most studies investigating the importance of ABC transporters for drug disposition were performed with different cell lines and animal models. In comparison to P-glycoprotein, only limited human data are available about the role of ABC transporters on drug disposition. In this chapter mainly human data will be discussed. For more detailed description and *in vitro* data about MRPs see Section 15.3.

A study investigating the impact of rifampicin treatment on the expression of intestinal MRPs revealed that rifampicin leads to the induction of P-glycoprotein and intestinal MRP2 [79]. Subsequently, several drug–drug interaction studies highlighted the relevance of intestinal MRP2 for the pharmacokinetics of drugs. For example a study investigating the pharmacokinetics of the β-adrenoceptor blocker carvedilol demonstrated that CYP2D6, MRP2 and P-glycoprotein were the major determinants for the variable pharmacokinetics [146]. Similar results were obtained in a drug interaction study using talinolol and carbamazepine [147]. Accordingly, rifampicin and carbamazepine induced MRP2 and P-glycoprotein, and both transporters were involved in the elimination of talinolol. Recent findings also showed the impact of polymorphisms in the *ABCC2* gene on the pharmacokinetics and toxicity of high dose of methotrexate (MTX), indicating that MRP2 may be one of the major determinants of

MTX elimination [148,149]. Similar studies have demonstrated the role of genetic variations in the *ABCC2* gene on doxorubicin-induced cardiotoxicity [150].

For other efflux transporters such as BCRP (ABCG2), human pharmacokinetic and pharmacodynamic data are currently rare. However, an investigation of the influence of polymorphisms in ABC-transporter genes on the accumulation of nelfinavir in peripheral blood mononuclear cells (PBMCs) revealed no associations between the polymorphisms in the transporters analyzed and the accumulation of nelfinavir in the PBMCs [151]. A second study in patients clearly demonstrated an increase in the AUC of the orally and intravenously administered BCRP substrate topotecan when it is given with GF120918, an inhibitor of P-glycoprotein and BCRP [152].

15.5
Vectorial Drug Transport

Polarized tissues directly involved in drug absorption (intestine) or excretion (liver and kidney) and restricted drug disposition (blood–tissue barriers) asymmetrically express a variety of different drug transporters in the apical or basolateral membrane resulting in vectorial drug transport. This vectorial drug transport is characterized by two transport processes: the uptake into the cell and subsequently the directed elimination out of the cell (Figure 15.3). Because the uptake of substances

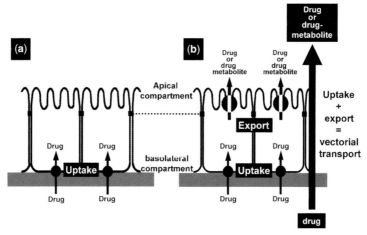

Figure 15.3 Cell systems used for the analysis of transport processes. (a) Single-transfected MDCKII cell stably expressing an uptake transporter. These cell systems are useful for the analysis of single transport processes. A substance (drug) can be added into the basolateral compartment and the uptake can be measured by analyzing the radioactivity in the cell at respective time points. (b) Double-transfected MDCKII cells stably expressing an uptake transporter in the basolateral membrane and an export pump in the apical membrane. These cell systems can be used for the analysis of vectorial transport processes. A substance (drug) can be added into the basolateral compartment, and the appearance of the drug in the apical compartment can be measured. Therefore, the detailed analysis of both transport steps (uptake and export) under defined conditions is possible.

limits the intracellular substrate concentration, it is clear that subsequent metabolic conversion as well as the final excretion step directly depend on the first transport process. Therefore, both transport processes can be characterized as one vectorial transport step.

In the epithelium of the small intestine, the uptake transporters such as PEPT1 (peptide transporter 1), the Na^+-dependent bile salt transporter (ASBT) and the uptake transporter OATP2B1 for organic anions are localized in the apical (luminal) membrane (Figure 15.2). Especially OATP2B1, transporting drugs such as fluvastatin [39], atorvastatin [20] and fexofenadine [153], is important for the absorption of substances into the body. Primary active efflux transporters such as P-glycoprotein, MRP2 (multidrug resistance protein 2; ABCC2) and BCRP are also expressed in the apical membrane for the purpose of extruding xenobiotics (for details, see Sections 15.3 and 15.4). These ABC transporters play an important role in determining the bioavailability of drugs. If a drug is a preferred substrate for one of these efflux pumps, it will be transported back in the gut lumen before it can reach the systemic circulation. Therefore, such a drug has a low bioavailability. Interestingly, the role of basolaterally localized intestinal efflux transporters is not yet fully understood. Studies in rat have demonstrated that the Mrp3 protein is localized in the basolateral membrane of enterocytes and that this protein is responsible for the extrusion of bile acids and other organic anions into blood. In humans, the role of MRP3 has to be clarified.

Hepatocytes are highly polarized cells responsible for the vectorial efflux of endogenous substances and drugs that are hepatobiliary eliminated. Therefore, several uptake transporters belonging to different SLC families are located in the basolateral hepatocytes membrane and several export pumps of the ABC transporter superfamily are localized in the canalicular membrane (for review see [154]) (Figure 15.2). Interestingly, at least four members of the MRP family are also localized in the basolateral hepatocyte membrane transporting substances out of the hepatocytes back into the blood [37]. This localization can be important under cholestatic conditions when the canalicular efflux into bile is impaired. It has been demonstrated that under these conditions the apically localized efflux pumps are downregulated and the basolaterally localized export pumps may account for the transport of substances out of the hepatocytes back into the blood followed by an excretion into the urine [155–158].

In the renal epithelium, many uptake transporters are localized in the basolateral membrane and efflux transporters are localized in the apical membrane (Figure 15.2). As a result, vectorial transport of endogenous substances and of drugs from the blood into the urine is achieved. The important uptake transporters are members of the SLC22 family of solute carriers, especially the family members OCT2 and OAT2, which are highly expressed in human kidney.

Generally, the *in vitro* analysis of transport proteins is performed using the so-called single-transfected cells that recombinantly express an uptake transporter (Figure 15.3a) or an export pump. Using these cells systems substances and drugs can be identified as substrates for a specific transport protein and kinetic parameters of a substance/drug can be determined. The disadvantage of these single transfectants is that only one transport process, the uptake into cells or the export out of cells, can be analyzed. The vectorial transport as the combination of uptake and efflux can be analyzed in double-transfected cell lines. The bases of these cell models are, for example, MDCKII cells, which grow in a polarized fashion with distinct apical and basolateral membrane domains (Figure 15.3). Recombinant MDCKII cells stably expressing a human uptake transporter, expressed in the basolateral membrane, and an export pump in the apical membrane have been established in recent years [159–163]. Using these double-transfected cell lines (Figure 15.3b) the vectorial transport of substrates and drugs can be studied under defined conditions. Substances can be added into the basolateral compartment, and in addition to the uptake of the substance into the cell, mediated by the uptake transporter, the appearance of the substance in the apical compartment as the combination of uptake and efflux can be analyzed. Therefore, kinetic parameters of both transport processes can be determined. Furthermore, these cell systems are the bases for the detailed analysis of uptake transporters and export pumps during the hepatobiliary or renal elimination of drugs. Recently, Kopplow and coworkers [39] established a quadruple-transfected cell line that simultaneously expresses the three major hepatic OATP family members OATP1B1, OATP1B3 and OATP2B1 together with the apically localized export pump MRP2 (ABCC2). Using this cell line the vectorial transport of endogenous substances and drugs mediated by hepatic transport proteins could be analyzed. These studies have demonstrated that *in vitro* cell systems are useful tools to study the kinetic properties of transport proteins not only in isolation but also in combination of uptake transporters and efflux pumps. Furthermore, these systems may be useful in screening of drugs and drug candidates with respect to their substrate or inhibitor properties. In the future, these systems can be used as the bases for further cell lines, containing in addition drug-metabolizing enzymes.

Abbreviations

ABC transporter	ATP binding cassette transporter
ABCB1	gene encoding human P-glycoprotein
ABCC2	gene encoding human MRP2
AUC	area under the curve
BCRP	breast cancer resistance protein
MRP	human multidrug resistance protein
OAT	organic anion transporter
OATP	organic anion transport polypetide
OCT	organic cation transporter
P-gp	P-glycoprotein
SLC	solute carrier
SLCO	genes encoding human OATP family members
TMD	transmembrane domains

Acknowledgment

This work was supported by grants DFG Ko 2120/1-3, DFG Fr 1298/2-4 and DFG Gl 588/2-1 of the Deutsche Forschungsgemeinschaft.

References

1 Hediger, M.A., Romero, M.F., Peng, J.B., Rolfs, A., Takanaga, H. and Bruford, E.A. (2004) The ABCs of solute carriers: physiological, pathological and therapeutic implications of human membrane transport proteins. *Pflugers Archiv*, **447**, 465–468.

2 Hagenbuch, B. and Meier, P.J. (2004) Organic anion transporting polypeptides of the OATP/SLC21 family: phylogenetic classification as OATP/SLCO superfamily, new nomenclature and molecular/functional properties. *Pflugers Archiv*, **447**, 653–665.

3 König, J., Seithel, A., Gradhand, U. and Fromm, M.F. (2006) Pharmacogenomics of human OATP transporters. *Naunyn-Schmiedeberg's Archives of Pharmacology*, **372**, 432–443.

4 Koepsell, H. and Endou, H. (2004) The SLC22 drug transporter family. *Pflugers Archiv*, **447**, 666–676.

5 Daniel, H. and Kottra, G. (2004) The proton oligopeptide cotransporter family SLC15 in physiology and pharmacology. *Pflugers Archiv*, **447**, 610–618.

6 Borst, P. and Elferink, R.O. (2002) Mammalian ABC transporters in health and disease. *Annual Review of Biochemistry*, **71**, 537–592.

7 Deeley, R.G., Westlake, C. and Cole, S.P. (2006) Transmembrane transport of endo- and xenobiotics by mammalian ATP-binding cassette multidrug resistance proteins. *Physiological Reviews*, **86**, 849–899.

8 Kruh, G.D. and Belinsky, M.G. (2003) The MRP family of drug efflux pumps. *Oncogene*, **22**, 7537–7552.

9 Dean, M. and Allikmets, R. (2001) Complete characterization of the human ABC gene family. *Journal of Bioenergetics and Biomembranes*, **33**, 475–479.

10 Dean, M., Rzhetsky, A. and Allikmets, R. (2001) The human ATP-binding cassette (ABC) transporter superfamily. *Genome Research*, **11**, 1156–1166.

11 Chan, L.M., Lowes, S. and Hirst, B.H. (2004) The ABCs of drug transport in intestine and liver: efflux proteins limiting drug absorption and bioavailability. *European Journal of Pharmaceutical Sciences*, **21**, 25–51.

12 Hagenbuch, B. and Meier, P.J. (2003) The superfamily of organic anion transporting polypeptides. *Biochimica et Biophysica Acta*, **1609**, 1–18.

13 Kullak-Ublick, G.A., Hagenbuch, B., Stieger, B., Schteingart, C.D., Hofmann, A.F., Wolkoff, A.W. and Meier, P.J. (1995) Molecular and functional characterization of an organic anion transporting polypeptide cloned from human liver. *Gastroenterology*, **109**, 1274–1282.

14 Abe, T., Kakyo, M., Tokui, T., Nakagomi, R., Nishio, T., Nakai, D., Nomura, H., Unno, M., Suzuki, M., Naitoh, T., Matsuno, S. and Yawo, H. (1999) Identification of a novel gene family encoding human liver-specific organic anion transporter LST-1. *The Journal of Biological Chemistry*, **274**, 17159–17163.

15 Hsiang, B., Zhu, Y., Wang, Z., Wu, Y., Sasseville, V., Yang, W.P. and Kirchgessner, T.G. (1999) A novel human hepatic organic anion transporting polypeptide (OATP2).

Identification of a liver-specific human organic anion transporting polypeptide and identification of rat and human hydroxymethylglutaryl-CoA reductase inhibitor transporters. *The Journal of Biological Chemistry*, **274**, 37161–37168.

16 König, J., Cui, Y., Nies, A.T. and Keppler, D. (2000) A novel human organic anion transporting polypeptide localized to the basolateral hepatocyte membrane. *American Journal of Physiology. Gastrointestinal and Liver Physiology*, **278**, G156–G164.

17 König, J., Cui, Y., Nies, A.T. and Keppler, D. (2000) Localization and genomic organization of a new hepatocellular organic anion transporting polypeptide. *The Journal of Biological Chemistry*, **275**, 23161–23168.

18 Kullak-Ublick, G.A., Ismair, M.G., Stieger, B., Landmann, L., Huber, R., Pizzagalli, F., Fattinger, K., Meier, P.J. and Hagenbuch, B. (2001) Organic anion-transporting polypeptide B (OATP-B) and its functional comparison with three other OATPs of human liver. *Gastroenterology*, **120**, 525–533.

19 Tamai, I., Nezu, J., Uchino, H., Sai, Y., Oku, A., Shimane, M. and Tsuji, A. (2000) Molecular identification and characterization of novel members of the human organic anion transporter (OATP) family. *Biochemical and Biophysical Research Communications*, **273**, 251–260.

20 Grube, M., Kock, K., Oswald, S., Draber, K., Meissner, K., Eckel, L., Bohm, M., Felix, S.B., Vogelgesang, S., Jedlitschky, G., Siegmund, W., Warzok, R. and Kroemer, H.K. (2006) Organic anion transporting polypeptide 2B1 is a high-affinity transporter for atorvastatin and is expressed in the human heart. *Clinical Pharmacology and Therapeutics*, **80**, 607–620.

21 Ismair, M.G., Stieger, B., Cattori, V., Hagenbuch, B., Fried, M., Meier, P.J. and Kullak-Ublick, G.A. (2001) Hepatic uptake of cholecystokinin octapeptide by organic anion-transporting polypeptides OATP4 and OATP8 of rat and human liver. *Gastroenterology*, **121**, 1185–1190.

22 St-Pierre, M.V., Hagenbuch, B., Ugele, B., Meier, P.J. and Stallmach, T. (2002) Characterization of an organic anion-transporting polypeptide (OATP-B) in human placenta. *The Journal of Clinical Endocrinology and Metabolism*, **87**, 1856–1863.

23 Abe, T., Unno, M., Onogawa, T., Tokui, T., Kondo, T.N., Nakagomi, R., Adachi, H., Fujiwara, K., Okabe, M., Suzuki, T., Nuoki, K., Sato, E., Kakyo, M., Nishio, T., Sugita, J., Asano, N., Tanemoto, M., Seki, M., Date, F., Ono, K., Kondo, Y., Shiiba, K., Suzuki, M., Ohtani, H., Shimosegawa, T., Iinuma, K., Nagura, H., Ito, S., and Matsuno, S. (2001) LST-2, a human liver-specific organic anion transporter, determines methotrexate sensitivity in gastrointestinal cancers. *Gastroenterology*, **120**, 1689–1699.

24 Bronger, H., König, J., Kopplow, K., Steiner, H.H., Ahmadi, R., Herold-Mende, C., Keppler, D. and Nies, A.T. (2005) ABCC drug efflux pumps and organic anion uptake transporters in human gliomas and the blood–tumor barrier. *Cancer Research*, **65**, 11419–11428.

25 Gao, B., Hagenbuch, B., Kullak-Ublick, G.A., Benke, D., Aguzzi, A. and Meier, P.J. (2000) Organic anion-transporting polypeptides mediate transport of opioid peptides across blood–brain barrier. *Journal of Pharmacology and Experimental Therapeutics*, **294**, 73–79.

26 Lee, W., Glaeser, H., Smith, L.H., Roberts, R.L., Moeckel, G.W., Gervasini, G., Leake, B.F. and Kim, R.B. (2005) Polymorphisms in human organic anion transporting polypeptide 1A2: implications for altered drug disposition and CNS drug entry. *Journal of Biological Chemistry*, **280**, 9610–9617.

27 Grube, M., Reuther, S., Meyer Zu Schwabedissen, H., Kock, K., Draber, K., Ritter, C.A., Fusch, C., Jedlitschky, G.

and Kroemer, H.K. (2007) Organic anion transporting polypeptide 2B1 and breast cancer resistance protein interact in the transepithelial transport of steroid sulfates in human placenta. *Drug Metabolism and Disposition*, **35**, 30–35.
28. Gao, B., Huber, R.D., Wenzel, A., Vavricka, S.R., Ismair, M.G., Reme, C. and Meier, P.J. (2005) Localization of organic anion transporting polypeptides in the rat and human ciliary body epithelium. *Experimental Eye Research*, **80**, 61–72.
29. Mikkaichi, T., Suzuki, T., Onogawa, T., Tanemoto, M., Mizutamari, H., Okada, M., Chaki, T., Masuda, S., Tokui, T., Eto, N., Abe, M., Satoh, F., Unno, M., Hishinuma, T., Inui, K., Ito, S., Goto, J. and Abe, T. (2004) Isolation and characterization of a digoxin transporter and its rat homologue expressed in the kidney. *Proceedings of the National Academy of Sciences of the United States of America*, **101**, 3569–3574.
30. Jacquemin, E., Hagenbuch, B., Stieger, B., Wolkoff, A.W. and Meier, P.J. (1994) Expression cloning of a rat liver Na(+)-independent organic anion transporter. *Proceedings of the National Academy of Sciences of the United States of America*, **91**, 133–137.
31. Letschert, K., Komatsu, M., Hummel-Eisenbeiss, J. and Keppler, D. (2005) Vectorial transport of the peptide CCK-8 by double-transfected MDCKII cells stably expressing the organic anion transporter OATP1B3 (OATP8) and the export pump ABCC2. *Journal of Pharmacology and Experimental Therapeutics*, **313**, 549–556.
32. Glaeser, H., Bailey, D.G., Dresser, G.K., Gregor, J.C., Schwarz, U.I., McGrath, J.S., Jolicoeur, E., Lee, W. et al. (2007) Intestinal drug transporter expression and the impact of grapefruit juice in humans. *Clinical Pharmacology and Therapeutics*, **81**, 362–370.
33. Fromm, M.F. (2004) Importance of P-glycoprotein at blood–tissue barriers. *Trends in Pharmacological Sciences*, **25**, 423–429.
34. Dagenais, C., Graff, C.L. and Pollack, G.M. (2004) Variable modulation of opioid brain uptake by P-glycoprotein in mice. *Biochemical Pharmacology*, **67**, 269–276.
35. de Graaf, D., Sharma, R.C., Mechetner, E.B., Schimke, R.T. and Roninson, I.B. (1996) P-glycoprotein confers methotrexate resistance in 3T6 cells with deficient carrier-mediated methotrexate uptake. *Proceedings of the National Academy of Sciences of the United States of America*, **93**, 1238–1242.
36. Neuvonen, P.J., Niemi, M. and Backman, J.T. (2006) Drug interactions with lipid-lowering drugs: mechanisms and clinical relevance. *Clinical Pharmacology and Therapeutics*, **80**, 565–581.
37. König, J., Nies, A.T., Cui, Y., Leier, I. and Keppler, D. (1999) Conjugate export pumps of the multidrug resistance protein (MRP) family: localization, substrate specificity, and MRP2-mediated drug resistance. *Biochimica et Biophysica Acta*, **1461**, 377–394.
38. Lau, Y.Y., Okochi, H., Huang, Y. and Benet, L.Z. (2006) Multiple transporters affect the disposition of atorvastatin and its two active hydroxy metabolites: application of *in vitro* and *ex situ* systems. *Journal of Pharmacology and Experimental Therapeutics*, **316**, 762–771.
39. Kopplow, K., Letschert, K., König, J., Walter, B. and Keppler, D. (2005) Human hepatobiliary transport of organic anions analyzed by quadruple-transfected cells. *Molecular Pharmacology*, **68**, 1031–1038.
40. Seithel, A., Eberl, S., Singer, K., Auge, D., Heinkele, G., Wolf, N.B., Dorje, F., Fromm, M.F. and König, J. (2007) The influence of macrolide antibiotics on the uptake of organic anions and drugs mediated by OATP1B1 and OATP1B3. *Drug Metabolism and Disposition*, **35**, 1–8.
41. Kameyama, Y., Yamashita, K., Kobayashi, K., Hosokawa, M. and Chiba, K. (2005)

Functional characterization of SLCO1B1 (OATP-C) variants, SLCO1B1*5, SLCO1B1*15 and SLCO1B1*15+C1007G, by using transient expression systems of HeLa and HEK293 cells. *Pharmacogenetics and Genomics*, **15**, 513–522.

42 Pasanen, M.K., Neuvonen, M., Neuvonen, P.J. and Niemi, M. (2006) SLCO1B1 polymorphism markedly affects the pharmacokinetics of simvastatin acid. *Pharmacogenetics and Genomics*, **16**, 873–879.

43 Schneck, D.W., Birmingham, B.K., Zalikowski, J.A., Mitchell, P.D., Wang, Y., Martin, P.D., Lasseter, K.C., Brown, C.D., Windass, A.S. and Raza, A. (2004) The effect of gemfibrozil on the pharmacokinetics of rosuvastatin. *Clinical Pharmacology and Therapeutics*, **75**, 455–463.

44 Shitara, Y., Hirano, M., Adachi, Y., Itoh, T., Sato, H. and Sugiyama, Y. (2004) in vitro and in vivo correlation of the inhibitory effect of cyclosporin A on the transporter-mediated hepatic uptake of cerivastatin in rats. *Drug Metabolism and Disposition*, **32**, 1468–1475.

45 Shitara, Y., Hirano, M., Sato, H. and Sugiyama, Y. (2004) Gemfibrozil and its glucuronide inhibit the organic anion transporting polypeptide 2 (OATP2/OATP1 B1:SLC21A6)-mediated hepatic uptake and CYP2C8-mediated metabolism of cerivastatin: analysis of the mechanism of the clinically relevant drug–drug interaction between cerivastatin and gemfibrozil. *Journal of Pharmacology and Experimental Therapeutics*, **311**, 228–236.

46 Simonson, S.G., Raza, A., Martin, P.D., Mitchell, P.D., Jarcho, J.A., Brown, C.D., Windass, A.S. and Schneck, D.W. (2004) Rosuvastatin pharmacokinetics in heart transplant recipients administered an antirejection regimen including cyclosporine. *Clinical Pharmacology and Therapeutics*, **76**, 167–177.

47 Psaty, B.M., Furberg, C.D., Ray, W.A. and Weiss, N.S. (2004) Potential for conflict of interest in the evaluation of suspected adverse drug reactions: use of cerivastatin and risk of rhabdomyolysis. *Journal of the American Medical Association*, **292**, 2622–2631.

48 Backman, J.T., Kyrklund, C., Neuvonen, M. and Neuvonen, P.J. (2002) Gemfibrozil greatly increases plasma concentrations of cerivastatin. *Clinical Pharmacology and Therapeutics*, **72**, 685–691.

49 Campbell, S.D., de Morais, S.M. and Xu, J.J. (2004) Inhibition of human organic anion transporting polypeptide OATP 1B1 as a mechanism of drug-induced hyperbilirubinemia. *Chemico-Biological Interactions*, **150**, 179–187.

50 Borst, P., Evers, R., Kool, M. and Wijnholds, J. (1999) The multidrug resistance protein family. *Biochimica et Biophysica Acta*, **1461**, 347–357.

51 Hyde, S.C., Emsley, P., Hartshorn, M.J., Mimmack, M.M., Gileadi, U., Pearce, S.R., Gallagher, M.P., Gill, D.R., Hubbard, R.E. and Higgins, C.F. (1990) Structural model of ATP-binding proteins associated with cystic fibrosis, multidrug resistance and bacterial transport. *Nature*, **346**, 362–365.

52 Cole, S.P., Bhardwaj, G., Gerlach, J.H., Mackie, J.E., Grant, C.E., Almquist, K.C., Stewart, A.J., Kurz, E.U., Duncan, A.M. and Deeley, R.G. (1992) Overexpression of a transporter gene in a multidrug-resistant human lung cancer cell line. *Science*, **258**, 1650–1654.

53 Mirski, S.E., Gerlach, J.H. and Cole, S.P. (1987) Multidrug resistance in a human small cell lung cancer cell line selected in adriamycin. *Cancer Research*, **47**, 2594–2598.

54 Flens, M.J., Zaman, G.J., van der Valk, P., Izquierdo, M.A., Schroeijers, A.B., Scheffer, G.L., van der Groep, P., de Haas, M., Meijer, C.J., and Scheper, R.J. (1996) Tissue distribution of the

multidrug resistance protein. *The American Journal of Pathology*, **148**, 1237–1247.
55 Roelofsen, H., Vos, T.A., Schippers, I.J., Kuipers, F., Koning, H., Moshage, H., Jansen, P.L. and Muller, M. (1997) Increased levels of the multidrug resistance protein in lateral membranes of proliferating hepatocyte-derived cells. *Gastroenterology*, **112**, 511–521.
56 St-Pierre, M.V., Serrano, M.A., Macias, R.I., Dubs, U., Hoechli, M., Lauper, U., Meier, P.J. and Marin, J.J. (2000) Expression of members of the multidrug resistance protein family in human term placenta. *American Journal of physiology. Regulatory, Integrative and Comparative Physiology*, **279**, R1495–R1503.
57 Zhang, Y., Schuetz, J.D., Elmquist, W.F. and Miller, D.W. (2004) Plasma membrane localization of multidrug resistance-associated protein homologs in brain capillary endothelial cells. *Journal of Pharmacology and Experimental Therapeutics*, **311**, 449–455.
58 Cole, S.P., Sparks, K.E., Fraser, K., Loe, D.W., Grant, C.E., Wilson, G.M. and Deeley, R.G. (1994) Pharmacological characterization of multidrug resistant MRP-transfected human tumor cells. *Cancer Research*, **54**, 5902–5910.
59 Zaman, G.J., Flens, M.J., van Leusden, M.R., de Haas, M., Mulder, H.S., Lankelma, J., Pinedo, H.M., Scheper, R. J., Baas, F. and Broxterman, H.J. ,(1994) The human multidrug resistance-associated protein MRP is a plasma membrane drug-efflux pump. *Proceedings of the National Academy of Sciences of the United States of America*, **91**, 8822–8826.
60 Rajagopal, A. and Simon, S.M. (2003) Subcellular localization and activity of multidrug resistance proteins. *Molecular Biology of the Cell*, **14**, 3389–3399.
61 Cole, S.P. and Deeley, R.G. (1998) Multidrug resistance mediated by the ATP-binding cassette transporter protein MRP. *Bioessays*, **20**, 931–940.
62 Cui, Y., König, J., Buchholz, J.K., Spring, H., Leier, I. and Keppler, D. (1999) Drug resistance and ATP-dependent conjugate transport mediated by the apical multidrug resistance protein, MRP2, permanently expressed in human and canine cells. *Molecular Pharmacology*, **55**, 929–937.
63 Loe, D.W., Almquist, K.C., Cole, S.P. and Deeley, R.G. (1996) ATP-dependent 17 beta-estradiol 17-(beta-D-glucuronide) transport by multidrug resistance protein (MRP). Inhibition by cholestatic steroids. *Journal of Biological Chemistry*, **271**, 9683–9689.
64 Loe, D.W., Almquist, K.C., Deeley, R.G. and Cole, S.P. (1996) Multidrug resistance protein (MRP)-mediated transport of leukotriene C4 and chemotherapeutic agents in membrane vesicles. Demonstration of glutathione-dependent vincristine transport. *Journal of Biological Chemistry*, **271**, 9675–9682.
65 Grzywacz, M.J., Yang, J.M. and Hait, W. N. (2003) Effect of the multidrug resistance protein on the transport of the antiandrogen flutamide. *Cancer Research*, **63**, 2492–2498.
66 Chu, X.Y., Suzuki, H., Ueda, K., Kato, Y., Akiyama, S. and Sugiyama, Y. (1999) Active efflux of CPT-11 and its metabolites in human KB-derived cell lines. *Journal of Pharmacology and Experimental Therapeutics*, **288**, 735–741.
67 Hooijberg, J.H., Broxterman, H.J., Kool, M., Assaraf, Y.G., Peters, G.J., Noordhuis, P., Scheper, R.J., Borst, P., Pinedo, H.M. and Jansen, G. (1999) Antifolate resistance mediated by the multidrug resistance proteins MRP1 and MRP2. *Cancer Research*, **59**, 2532–2535.
68 Huisman, M.T., Smit, J.W., Crommentuyn, K.M., Zelcer, N., Wiltshire, H.R., Beijnen, J.H. and Schinkel, A.H. (2002) Multidrug resistance protein 2 (MRP2) transports HIV protease inhibitors, and transport can be enhanced by other drugs. *AIDS*, **16**, 2295–2301.

69 Jedlitschky, G., Leier, I., Buchholz, U., Barnouin, K., Kurz, G. and Keppler, D. (1996) Transport of glutathione, glucuronate, and sulfate conjugates by the MRP gene-encoded conjugate export pump. *Cancer Research*, **56**, 988–994.

70 Jedlitschky, G., Leier, I., Buchholz, U., Center, M. and Keppler, D. (1994) ATP-dependent transport of glutathione S-conjugates by the multidrug resistance-associated protein. *Cancer Research*, **54**, 4833–4836.

71 Leier, I., Jedlitschky, G., Buchholz, U., Center, M., Cole, S.P., Deeley, R.G. and Keppler, D. (1996) ATP-dependent glutathione disulphide transport mediated by the MRP gene-encoded conjugate export pump. *Biochemical Journal*, **314** (Pt 2), 433–437.

72 Leier, I., Jedlitschky, G., Buchholz, U. and Keppler, D. (1994) Characterization of the ATP-dependent leukotriene C4 export carrier in mastocytoma cells. *European Journal of Biochemistry*, **220**, 599–606.

73 Ishikawa, T., Müller, M., Klunemann, C., Schaub, T. and Keppler, D. (1990) ATP-dependent primary active transport of cysteinyl leukotrienes across liver canalicular membrane. Role of the ATP-dependent transport system for glutathione S-conjugates. *Journal of Biological Chemistry*, **265**, 19279–19286.

74 Loe, D.W., Deeley, R.G. and Cole, S.P. (1998) Characterization of vincristine transport by the M(r) 190,000 multidrug resistance protein (MRP): evidence for cotransport with reduced glutathione. *Cancer Research*, **58**, 5130–5136.

75 Renes, J., de Vries, E.G., Nienhuis, E.F., Jansen, P.L. and Müller, M. (1999) ATP- and glutathione-dependent transport of chemotherapeutic drugs by the multidrug resistance protein MRP1. *British Journal of Pharmacology*, **126**, 681–688.

76 Büchler, M., König, J., Brom, M., Kartenbeck, J., Spring, H., Horie, T. and Keppler, D. (1996) cDNA cloning of the hepatocyte canalicular isoform of the multidrug resistance protein, cMrp, reveals a novel conjugate export pump deficient in hyperbilirubinemic mutant rats. *Journal of Biological Chemistry*, **271**, 15091–15098.

77 Ito, K., Suzuki, H., Hirohashi, T., Kume, K., Shimizu, T. and Sugiyama, Y. (1997) Molecular cloning of canalicular multispecific organic anion transporter defective in EHBR. *The American Journal of Physiology*, **272**, G16–22.

78 Paulusma, C.C., Bosma, P.J., Zaman, G. J., Bakker, C.T., Otter, M., Scheffer, G.L., Scheper, R.J., Borst, P. et al. (1996) Congenital jaundice in rats with a mutation in a multidrug resistance-associated protein gene. *Science*, **271**, 1126–1128.

79 Fromm, M.F., Kauffmann, H.M., Fritz, P., Burk, O., Kroemer, H.K., Warzok, R. W., Eichelbaum, M., Siegmund, W. and Schrenk, D. (2000) The effect of rifampin treatment on intestinal expression of human MRP transporters. *The American Journal of Pathology*, **157**, 1575–1580.

80 Schaub, T.P., Kartenbeck, J., König, J., Spring, H., Dorsam, J., Staehler, G., Storkel, S., Thon, W.F. and Keppler, D. 1999) Expression of the MRP2 gene-encoded conjugate export pump in human kidney proximal tubules and in renal cell carcinoma. *Journal of the American Society of Nephrology*, **10**, 1159–1169.

81 König, J., Nies, A.T., Cui, Y. and Keppler, D. (2003) MRP2, the apical export pump for anionic conjugates, in *ABC Proteins: From Bacteria to Man* (eds I.B. Holland, S. P.C. Cole, K. Kuchler and C.F. Higgins), Academic Press, London, pp.423–443.

82 Hinoshita, E., Uchiumi, T., Taguchi, K., Kinukawa, N., Tsuneyoshi, M., Maehara, Y., Sugimachi, K. and Kuwano, M. (2000) Increased expression of an ATP-binding cassette superfamily transporter, multidrug resistance protein 2, in human colorectal carcinomas. *Clinical Cancer Research*, **6**, 2401–2407.

83 Jedlitschky, G., Hoffmann, U. and Kroemer, H.K. (2006) Structure and function of the MRP2 (ABCC2) protein and its role in drug disposition. *Expert Opinion on Drug Metabolism and Toxicology*, **2**, 351–366.

84 Kartenbeck, J., Leuschner, U., Mayer, R. and Keppler, D. (1996) Absence of the canalicular isoform of the MRP gene-encoded conjugate export pump from the hepatocytes in Dubin–Johnson syndrome. *Hepatology*, **23**, 1061–1066.

85 Keppler, D., Kamisako, T., Leier, I., Cui, Y., Nies, A.T., Tsujii, H. and König, J. (2000) Localization, substrate specificity, and drug resistance conferred by conjugate export pumps of the MRP family. *Advances in Enzyme Regulation*, **40**, 339–349.

86 Keppler, D., König, J. and Nies, A.T. (2001) Conjugate export pumps of the multidrug resistance protein (MRP) family in liver, in *The Liver: Biology and Pathobiology* (eds I.M. Arias *et al.*), Lippincott Williams & Wilkins, New York, pp.373–382.

87 Yamazaki, M., Li, B., Louie, S.W., Pudvah, N.T., Stocco, R., Wong, W., Abramovitz, M., Demartis, A., Laufer, R., Hochman, J.H., Prueksaritanont, T. and Lin, J.H. (2005) Effects of fibrates on human organic anion-transporting polypeptide 1B1-, multidrug resistance protein 2- and P-glycoprotein-mediated transport. *Xenobiotica*, **35**, 737–753.

88 Belinsky, M.G., Bain, L.J., Balsara, B.B., Testa, J.R. and Kruh, G.D. (1998) Characterization of MOAT-C and MOAT-D, new members of the MRP/cMOAT subfamily of transporter proteins. *Journal of the National Cancer Institute*, **90**, 1735–1741.

89 Hirohashi, T., Suzuki, H. and Sugiyama, Y. (1999) Characterization of the transport properties of cloned rat multidrug resistance-associated protein 3 (MRP3). *Journal of Biological Chemistry*, **274**, 15181–15185.

90 König, J., Rost, D., Cui, Y. and Keppler, D. (1999) Characterization of the human multidrug resistance protein isoform MRP3 localized to the basolateral hepatocyte membrane. *Hepatology*, **29**, 1156–1163.

91 Rost, D., Mahner, S., Sugiyama, Y. and Stremmel, W. (2002) Expression and localization of the multidrug resistance-associated protein 3 in rat small and large intestine. *American Journal of Physiology. Gastrointestinal and Liver Physiology*, **282**, G720–726.

92 Donner, M.G. and Keppler, D. (2001) Up-regulation of basolateral multidrug resistance protein 3 (Mrp3) in cholestatic rat liver. *Hepatology*, **34**, 351–359.

93 Kool, M., van der Linden, M., de Haas, M., Scheffer, G.L., de Vree, J.M., Smith, A.J., Jansen, G., Peters, G.J., Ponne, N., Scheper, R.J., Elferink, R.P., Baas, F. and Borst, P. (1999) MRP3, an organic anion transporter able to transport anti-cancer drugs. *Proceedings of the National Academy of Sciences of the United States of America*, **96**, 6914–6919.

94 Zelcer, N., Saeki, T., Reid, G., Beijnen, J.H. and Borst, P. (2001) Characterization of drug transport by the human multidrug resistance protein 3 (ABCC3). *Journal of Biological Chemistry*, **276**, 46400–46407.

95 Guo, Y., Kotova, E., Chen, Z.S., Lee, K., Hopper-Borge, E., Belinsky, M.G. and Kruh, G.D. (2003) MRP8, ATP-binding cassette C11 (ABCC11), is a cyclic nucleotide efflux pump and a resistance factor for fluoropyrimidines $2',3'$-dideoxycytidine and $9'$-$(2'$-phosphonylmethoxyethyl)adenine. *Journal of Biological Chemistry*, **278**, 29509–29514.

96 Jedlitschky, G., Burchell, B. and Keppler, D. (2000) The multidrug resistance protein 5 functions as an ATP-dependent export pump for cyclic nucleotides. *Journal of Biological Chemistry*, **275**, 30069–30074.

97 Pastor-Anglada, M., Molina-Arcas, M., Casado, F.J., Bellosillo, B., Colomer, D. and Gil, J. (2004) Nucleoside transporters in chronic lymphocytic leukaemia. *Leukemia*, **18**, 385–393.

98 Schuetz, J.D., Connelly, M.C., Sun, D., Paibir, S.G., Flynn, P.M., Srinivas, R.V., Kumar, A. and Fridland, A. (1999) MRP4: a previously unidentified factor in resistance to nucleoside-based antiviral drugs. *Nature Medicine*, **5**, 1048–1051.

99 Pratt, S., Chen, V., Perry, W.I., 3rd, Starling, J.J. and Dantzig, A.H. (2006) Kinetic validation of the use of carboxydichlorofluorescein as a drug surrogate for MRP5-mediated transport. *European Journal of Pharmaceutical Sciences*, **27**, 524–532.

100 Reid, G., Wielinga, P., Zelcer, N., van der Heijden, I., Kuil, A., de Haas, M., Wijnholds, J. and Borst, P. (2003) The human multidrug resistance protein MRP4 functions as a prostaglandin efflux transporter and is inhibited by nonsteroidal antiinflammatory drugs. *Proceedings of the National Academy of Sciences of the United States of America*, **100**, 9244–9249.

101 Rius, M., Thon, W.F., Keppler, D. and Nies, A.T. (2005) Prostanoid transport by multidrug resistance protein 4 (MRP4/ABCC4) localized in tissues of the human urogenital tract. *The Journal of Urology*, **174**, 2409–2414.

102 Jedlitschky, G., Tirschmann, K., Lubenow, L.E., Nieuwenhuis, H.K., Akkerman, J.W., Greinacher, A. and Kroemer, H.K. (2004) The nucleotide transporter MRP4 (ABCC4) is highly expressed in human platelets and present in dense granules, indicating a role in mediator storage. *Blood*, **104**, 3603–3610.

103 van Aubel, R.A., Smeets, P.H., Peters, J.G., Bindels, R.J. and Russel, F.G. (2002) The MRP4/ABCC4 gene encodes a novel apical organic anion transporter in human kidney proximal tubules: putative efflux pump for urinary cAMP and cGMP. *Journal of the American Society of Nephrology*, **13**, 595–603.

104 Bortfeld, M., Rius, M., König, J., Herold-Mende, C., Nies, A.T. and Keppler, D. (2006) Human multidrug resistance protein 8 (MRP8/ABCC11), an apical efflux pump for steroid sulfates, is an axonal protein of the CNS and peripheral nervous system. *Neuroscience*, **137**, 1247–1257.

105 Bergen, A.A., Plomp, A.S., Hu, X., de Jong, P.T. and Gorgels, T.G. (2007) ABCC6 and pseudoxanthoma elasticum. *Pflugers Archiv*, **453**, 685–691.

106 Scheffer, G.L., Hu, X., Pijnenborg, A.C., Wijnholds, J., Bergen, A.A. and Scheper, R.J. (2002) MRP6 (ABCC6) detection in normal human tissues and tumors. *Laboratory Investigation*, **82**, 515–518.

107 Belinsky, M.G., Chen, Z.S., Shchaveleva, I., Zeng, H. and Kruh, G.D. (2002) Characterization of the drug resistance and transport properties of multidrug resistance protein 6 (MRP6, ABCC6). *Cancer Research*, **62**, 6172–6177.

108 Kool, M., van der Linden, M., de Haas, M., Baas, F. and Borst, P. (1999) Expression of human MRP6, a homologue of the multidrug resistance protein gene MRP1, in tissues and cancer cells. *Cancer Research*, **59**, 175–182.

109 Hopper-Borge, E., Chen, Z.S., Shchaveleva, I., Belinsky, M.G. and Kruh, G.D. (2004) Analysis of the drug resistance profile of multidrug resistance protein 7 (ABCC10): resistance to docetaxel. *Cancer Research*, **64**, 4927–4930.

110 Juliano, R.L. and Ling, V. (1976) A surface glycoprotein modulating drug permeability in Chinese hamster ovary cell mutants. *Biochimica et Biophysica Acta*, **455**, 152–162.

111 Bellamy, W.T., Dalton, W.S., Kailey, J.M., Gleason, M.C., McCloskey, T.M., Dorr, R.T. and Alberts, D.S. (1988) Verapamil reversal of doxorubicin resistance in multidrug-resistant human myeloma

cells and association with drug accumulation and DNA damage. *Cancer Research*, **48**, 6365–6370.

112 Gottesman, M.M., Ambudkar, S.V., Cornwell, M.M., Pastan, I. and Germann, U.A. (1996) Multidrug resistance transporter, in *Molecular Biology of Membrane Transport Disorders*, (ed. S.G. Schultz), Plenum Press, New York, NY, pp.243–257.

113 Perez-Tomas, R. (2006) Multidrug resistance: retrospect and prospects in anti-cancer drug treatment. *Current Medicinal Chemistry*, **13**, 1859–1876.

114 Cordon-Cardo, C., O'Brien, J.P., Boccia, J., Casals, D., Bertino, J.R. and Melamed, M.R. (1990) Expression of the multidrug resistance gene product (P-glycoprotein) in human normal and tumor tissues. *Journal of Histochemistry and Cytochemistry*, **38**, 1277–1287.

115 Cordon-Cardo, C., O'Brien, J.P., Casals, D., Rittman-Grauer, L., Biedler, J.L., Melamed, M.R. and Bertino, J.R. (1989) Multidrug-resistance gene (P-glycoprotein) is expressed by endothelial cells at blood–brain barrier sites. *Proceedings of the National Academy of Sciences of the United States of America*, **86**, 695–698.

116 Klimecki, W.T., Futscher, B.W., Grogan, T.M. and Dalton, W.S. (1994) P-glycoprotein expression and function in circulating blood cells from normal volunteers. *Blood*, **83**, 2451–2458.

117 MacFarland, A., Abramovich, D.R., Ewen, S.W. and Pearson, C.K. (1994) Stage-specific distribution of P-glycoprotein in first-trimester and full-term human placenta. *The Histochemical Journal*, **26**, 417–423.

118 Thiebaut, F., Tsuruo, T., Hamada, H., Gottesman, M.M., Pastan, I. and Willingham, M.C. (1987) Cellular localization of the multidrug-resistance gene product P-glycoprotein in normal human tissues. *Proceedings of the National Academy of Sciences of the United States of America*, **84**, 7735–7738.

119 Schinkel, A.H., Mayer, U., Wagenaar, E., Mol, C.A., van Deemter, L., Smit, J.J., van der Valk, M.A., Voordouw, A.C., Spits, H., Van Tellingen, O., Zijlmans, J.M., Fibbe, W.E. and Borst, P. (1997) Normal viability and altered pharmacokinetics in mice lacking mdr1-type (drug-transporting) P-glycoproteins. *Proceedings of the National Academy of Sciences of the United States of America*, **94**, 4028–4033.

120 Mayer, U., Wagenaar, E., Beijnen, J.H., Smit, J.W., Meijer, D.K., van Asperen, J., Borst, P. and Schinkel, A.H. (1996) Substantial excretion of digoxin via the intestinal mucosa and prevention of long-term digoxin accumulation in the brain by the mdr 1a P-glycoprotein. *British Journal of Pharmacology*, **119**, 1038–1044.

121 Drescher, S., Glaeser, H., Murdter, T., Hitzl, M., Eichelbaum, M. and Fromm, M.F. (2003) P-glycoprotein-mediated intestinal and biliary digoxin transport in humans. *Clinical Pharmacology and Therapeutics*, **73**, 223–231.

122 Dilger, K., Greiner, B., Fromm, M.F., Hofmann, U., Kroemer, H.K. and Eichelbaum, M. (1999) Consequences of rifampicin treatment on propafenone disposition in extensive and poor metabolizers of CYP2 D6. *Pharmacogenetics*, **9**, 551–559.

123 Schwarz, U.I., Gramatte, T., Krappweis, J., Oertel, R. and Kirch, W. (2000) P-glycoprotein inhibitor erythromycin increases oral bioavailability of talinolol in humans. *International Journal of Clinical Pharmacology and Therapeutics*, **38**, 161–167.

124 Westphal, K., Weinbrenner, A., Zschiesche, M., Franke, G., Knoke, M., Oertel, R., Fritz, P., von Richter, O., Warzok, R., Hachenberg, T., Kauffmann, H.M., Schrenk, D., Terhaag, B., Kroemer, H.K. and Siegmund, W. (2000) Induction of P-glycoprotein by rifampin increases intestinal secretion of talinolol in human beings: a new type of

drug/drug interaction. *Clinical Pharmacology and Therapeutics*, **68**, 345–355.

125 Westphal, K., Weinbrenner, A., Giessmann, T., Stuhr, M., Franke, G., Zschiesche, M., Oertel, R., Terhaag, B., Kroemer, H.K. and Siegmund, W. (2000) Oral bioavailability of digoxin is enhanced by talinolol: evidence for involvement of intestinal P-glycoprotein. *Clinical Pharmacology and Therapeutics*, **68**, 6–12.

126 Cvetkovic, M., Leake, B., Fromm, M.F., Wilkinson, G.R. and Kim, R.B. (1999) OATP and P-glycoprotein transporters mediate the cellular uptake and excretion of fexofenadine. *Drug Metabolism and Disposition*, **27**, 866–871.

127 Hamman, M.A., Bruce, M.A., Haehner-Daniels, B.D. and Hall, S.D. (2001) The effect of rifampin administration on the disposition of fexofenadine. *Clinical Pharmacology and Therapeutics*, **69**, 114–121.

128 Tirona, R.G., Leake, B.F., Merino, G. and Kim, R.B. (2001) Polymorphisms in OATP-C: identification of multiple allelic variants associated with altered transport activity among European- and African-Americans. *Journal of Biological Chemistry*, **276**, 35669–35675.

129 Drescher, S., Schaeffeler, E., Hitzl, M., Hofmann, U., Schwab, M., Brinkmann, U., Eichelbaum, M. and Fromm, M.F. (2002) MDR1 gene polymorphisms and disposition of the P-glycoprotein substrate fexofenadine. *British Journal of Clinical Pharmacology*, **53**, 526–534.

130 Niemi, M., Kivisto, K.T., Hofmann, U., Schwab, M., Eichelbaum, M. and Fromm, M.F. (2005) Fexofenadine pharmacokinetics are associated with a polymorphism of the SLCO1B1 gene (encoding OATP1B1). *British Journal of Clinical Pharmacology*, **59**, 602–604.

131 Schinkel, A.H., Wagenaar, E., Mol, C.A. and van Deemter, L. (1996) P-glycoprotein in the blood–brain barrier of mice influences the brain penetration and pharmacological activity of many drugs. *Journal of Clinical Investigation*, **97**, 2517–2524.

132 Sadeque, A.J., Wandel, C., He, H., Shah, S. and Wood, A.J. (2000) Increased drug delivery to the brain by P-glycoprotein inhibition. *Clinical Pharmacology and Therapeutics*, **68**, 231–237.

133 Turck, D., Berard, H., Fretault, N. and Lecomte, J.M. (1999) Comparison of racecadotril and loperamide in children with acute diarrhoea. *Alimentary Pharmacology & Therapeutics*, **13** (Suppl. 6), 27–32.

134 Skarke, C., Jarrar, M., Schmidt, H., Kauert, G., Langer, M., Geisslinger, G. and Lotsch, J. (2003) Effects of ABCB1 (multidrug resistance transporter) gene mutations on disposition and central nervous effects of loperamide in healthy volunteers. *Pharmacogenetics*, **13**, 651–660.

135 Pauli-Magnus, C., Feiner, J., Brett, C., Lin, E. and Kroetz, D.L. (2003) No effect of MDR1 C3435T variant on loperamide disposition and central nervous system effects. *Clinical Pharmacology and Therapeutics*, **74**, 487–498.

136 Takano, A., Kusuhara, H., Suhara, T., Ieiri, I., Morimoto, T., Lee, Y.J., Maeda, J., Ikoma, Y., Ito, H., Suzuki, K. and Sugiyama, Y. (2006) Evaluation of in vivo P-glycoprotein function at the blood–brain barrier among MDR1 gene polymorphisms by using 11C-verapamil. *Journal of Nuclear Medicine*, **47**, 1427–1433.

137 Brunner, M., Langer, O., Sunder-Plassmann, R., Dobrozemsky, G., Muller, U., Wadsak, W., Krcal, A., Karch, R., Mannhalter, C., Dudczak, R., Kletter, K., Steiner, I., Baumgartner, C. and Muller, M. (2005) Influence of functional haplotypes in the drug transporter gene ABCB1 on central nervous system drug distribution in humans. *Clinical Pharmacology and Therapeutics*, **78**, 182–190.

138 Kim, R.B., Fromm, M.F., Wandel, C., Leake, B., Wood, A.J., Roden, D.M. and

Wilkinson, G.R. (1998) The drug transporter P-glycoprotein limits oral absorption and brain entry of HIV-1 protease inhibitors. *Journal of Clinical Investigation*, **101**, 289–294.

139 Nestorowicz, A., Wilson, B.A., Schoor, K. P., Inoue, H., Glaser, B., Landau, H., Stanley, C.A., Thornton, P.S., Clement, J. P., Bryan, J., Aguilar-Bryan, L. and Permutt, M.A. (1996) Mutations in the sulonylurea receptor gene are associated with familial hyperinsulinism in Ashkenazi Jews. *Human Molecular Genetics*, **5**, 1813–1822.

140 Atkinson, D.E., Greenwood, S.L., Sibley, C.P., Glazier, J.D. and Fairbairn, L.J. (2003) Role of MDR1 and MRP1 in trophoblast cells, elucidated using retroviral gene transfer. *American Journal of Physiology. Cell Physiology*, **285**, C584–C591.

141 Hitzl, M., Schaeffeler, E., Hocher, B., Slowinski, T., Halle, H., Eichelbaum, M., Kaufmann, P., Fritz, P., Fromm, M.F. and Schwab, M. (2004) Variable expression of P-glycoprotein in the human placenta and its association with mutations of the multidrug resistance 1 gene (MDR1, ABCB1). *Pharmacogenetics*, **14**, 309–318.

142 Gil, S., Saura, R., Forestier, F. and Farinotti, R. (2005) P-glycoprotein expression of the human placenta during pregnancy. *Placenta*, **26**, 268–270.

143 Lee, C.G., Gottesman, M.M., Cardarelli, C.O., Ramachandra, M., Jeang, K.T., Ambudkar, S.V., Pastan, I. and Dey, S. (1998) HIV-1 protease inhibitors are substrates for the MDR1 multidrug transporter. *Biochemistry*, **37**, 3594–3601.

144 Marzolini, C., Tirona, R.G. and Kim, R.B. (2004) Pharmacogenomics of the OATP and OAT families. *Pharmacogenomics*, **5**, 273–282.

145 Jackson, J.B., Parsons, T., Musoke, P., Nakabiito, C., Donnell, D., Fleming, T., Mirochnick, M., Mofenson, L., Fowler, M.G., Mmiro, F. and Guay, L. (2006) Association of cord blood nevirapine concentration with reported timing of dose and HIV-1 transmission. *AIDS*, **20**, 217–222.

146 Giessmann, T., Modess, C., Hecker, U., Zschiesche, M., Dazert, P., Kunert-Keil, C., Warzok, R., Engel, G. Weitschies, W., Cascorbi, I., Kroemer, H.K. and Siegmund, W. (2004) CYP2D6 genotype and induction of intestinal drug transporters by rifampin predict presystemic clearance of carvedilol in healthy subjects. *Clinical Pharmacology and Therapeutics*, **75**, 213–222.

147 Giessmann, T., May, K., Modess, C., Wegner, D., Hecker, U., Zschiesche, M., Dazert, P., Grube, M. Schroeder, E., Warzok, R., Cascorbi, I., Kroemer, H.K. and Siegmund, W. (2004) Carbamazepine regulates intestinal P-glycoprotein and multidrug resistance protein MRP2 and influences disposition of talinolol in humans. *Clinical Pharmacology and Therapeutics*, **76**, 192–200.

148 Hulot, J.S., Villard, E., Maguy, A., Morel, V., Mir, L., Tostivint, I., William-Faltaos, D., Fernandez, C., Hatem, S., Deray, G., Komajda, M., Leblond, V. and Lechat, P. (2005) A mutation in the drug transporter gene ABCC2 associated with impaired methotrexate elimination. *Pharmacogenet Genomics*, **15**, 277–285.

149 Rau, T., Erney, B., Gores, R., Eschenhagen, T., Beck, J. and Langer, T. (2006) High-dose methotrexate in pediatric acute lymphoblastic leukemia: impact of ABCC2 polymorphisms on plasma concentrations. *Clinical Pharmacology and Therapeutics*, **80**, 468–476.

150 Wojnowski, L., Kulle, B., Schirmer, M., Schluter, G., Schmidt, A., Rosenberger, A., Vonhof, S., Bickeboller, H., Toliat, M. R., Suk, E.K., Tzvetkov, M., Kruger, A., Seifert, S., Kloess, M., Hahn, H., Loeffler, M., Nürnberg, P., Pfreundschuh, M., Trumper, L., Brockmoller, J., and Hasenfuss, G. (2005) NAD(P)H oxidase and multidrug resistance protein genetic

polymorphisms are associated with doxorubicin-induced cardiotoxicity. *Circulation*, **112**, 3754–3762.

151 Colombo, S., Soranzo, N., Rotger, M., Sprenger, R., Bleiber, G., Furrer, H., Buclin, T., Goldstein, D., Decosterd, L. and Telenti, A. (2005) Influence of ABCB1, ABCC1, ABCC2, and ABCG2 haplotypes on the cellular exposure of nelfinavir *in vivo*. *Pharmacogenetics and Genomics*, **15**, 599–608.

152 Kruijtzer, C.M., Beijnen, J.H., Rosing, H., ten Bokkel Huinink, W.W., Schot, M., Jewell, R.C., Paul, E.M. and Schellens, J.H. (2002) Increased oral bioavailability of topotecan in combination with the breast cancer resistance protein and P-glycoprotein inhibitor GF120918. *Journal of Clinical Oncology*, **20**, 2943–2950.

153 Nozawa, T., Imai, K., Nezu, J., Tsuji, A. and Tamai, I. (2004) Functional characterization of pH-sensitive organic anion transporting polypeptide OATP-B in human. *Journal of Pharmacology and Experimental Therapeutics*, **308**, 438–445.

154 Shitara, Y., Horie, T. and Sugiyama, Y. (2006) Transporters as a determinant of drug clearance and tissue distribution. *European Journal of Pharmaceutical Sciences*, **27**, 425–446.

155 Kojima, H., Nies, A.T., König, J., Hagmann, W., Spring, H., Uemura, M., Fukui, H. and Keppler, D. (2003) Changes in the expression and localization of hepatocellular transporters and radixin in primary biliary cirrhosis. *Journal of Hepatology*, **39**, 693–702.

156 Trauner, M., Arrese, M., Lee, H., Boyer, J.L. and Karpen, S.J. (1998) Endotoxin downregulates rat hepatic ntcp gene expression via decreased activity of critical transcription factors. *Journal of Clinical Investigation*, **101**, 2092–2100.

157 Trauner, M., Arrese, M., Soroka, C.J., Ananthanarayanan, M., Koeppel, T.A., Schlosser, S.F., Suchy, F.J., Keppler, D. and Boyer, J.L. (1997) The rat canalicular conjugate export pump (Mrp2) is down-regulated in intrahepatic and obstructive cholestasis. *Gastroenterology*, **113**, 255–264.

158 Trauner, M., Meier, P.J. and Boyer, J.L. (1998) Molecular pathogenesis of cholestasis. *New England Journal of Medicine*, **339**, 1217–1227.

159 Cui, Y., König, J. and Keppler, D. (2001) Vectorial transport by double-transfected cells expressing the human uptake transporter SLC21A8 and the apical export pump ABCC2. *Molecular Pharmacology*, **60**, 934–943.

160 Liu, L., Cui, Y., Chung, A.Y., Shitara, Y., Sugiyama, Y., Keppler, D. and Pang, K.S. (2006) Vectorial transport of enalapril by Oatp1a1/Mrp2 and OATP1B1 and OATP1B3/MRP2 in rat and human livers. *Journal of Pharmacology and Experimental Therapeutics*, **318**, 395–402.

161 Mita, S., Suzuki, H., Akita, H., Hayashi, H., Onuki, R., Hofmann, A.F. and Sugiyama, Y. (2006) Vectorial transport of unconjugated and conjugated bile salts by monolayers of LLC-PK1 cells doubly transfected with human NTCP and BSEP or with rat Ntcp and Bsep. *American Journal of Physiology. Gastrointestinal and Liver Physiology*, **290**, G550–G556.

162 Mita, S., Suzuki, H., Akita, H., Stieger, B., Meier, P.J., Hofmann, A.F. and Sugiyama, Y. (2005) Vectorial transport of bile salts across MDCK cells expressing both rat Na^+-taurocholate cotransporting polypeptide and rat bile salt export pump. *American Journal of Physiology. Gastrointestinal and Liver Physiology*, **288**, G159–167.

163 Sasaki, M., Suzuki, H., Ito, K., Abe, T. and Sugiyama, Y. (2002) Transcellular transport of organic anions across a double-transfected Madin–Darby canine kidney II cell monolayer expressing both human organic anion-transporting polypeptide (OATP2/SLC21A6) and multidrug resistance-associated protein 2 (MRP2/ABCC2). *Journal of Biological Chemistry*, **277**, 6497–6503.

16
Computational Models for P-Glycoprotein Substrates and Inhibitors
Patrizia Crivori

16.1
P-Glycoprotein Structure, Expression, Mechanism of Transport and Role on Drug Pharmacokinetics

P-glycoprotein (P-gp), encoded by human MDR1 gene, is a 170-kDa member of the large ATP-binding cassette (ABC) superfamily of transport proteins. This transporter recognizes and extrudes a wide variety of structurally unrelated xenobiotics out of the cells. The overexpression of P-gp is associated with a multidrug resistance phenotype in various forms of cancer that causes a suboptimal efficacy of many cytotoxic chemotherapeutic agents [1–7]. In addition to the expression in tumor cells, human P-gp is also highly expressed in normal tissues. This transporter is localized on the canalicular surface of hepatocytes in liver, the apical surface of epithelial cells of proximal tubules in kidney, the columnar epithelial cells of intestine, the epithelial cells of placenta and luminal surface of capillary endothelial cells in brain [5]. Studies on P-gp knockout mice show that the protein is not essential for viability or fertility and does not confer any phenotypic abnormality other than hypersensitivity to drugs [7]. These observations as well as the human anatomical localization of P-gp suggest that the efflux transporter can functionally protect the body against toxic xenobiotics by excreting them into bile, urine and the intestinal lumen, and by preventing their accumulation in brain [5,6]. Consequently, P-gp also plays a significant role in the absorption, distribution, metabolism and excretion processes of a wide range of drugs. P-gp has been shown to limit oral absorption (e.g. paclitaxel and docetaxel), modulate hepatic, renal or intestinal elimination and restrict central nervous system entry (e.g. cyclosporin A, digoxin, indinavir, ritonavir and saquinavir) of certain drugs [5,6,8–11]. Moreover, analogous to drug–drug interactions mediated by CYP450 enzymes, coadministration of a drug that is substrate, inhibitor or an inducer of the P-gp may effect the kinetics of drugs that are substrates for the same transporter [5,6]. For example, inhibition of P-gp may be partially responsible for quinidine–digoxin, verapamil–digoxin, ketoconazole–ritonavir and ketoconazole–saquinavir interactions [6].

Antitargets. Edited by R. J. Vaz and T. Klabunde
Copyright © 2008 WILEY-VCH Verlag GmbH & Co. KGaA, Weinheim
ISBN: 978-3-527-31821-6

Although P-gp has been one of the most extensively investigated transporters, the nature of its binding site(s) and the mechanism of solute translocation still remain the center of considerable controversy [1–19]. P-gp is a single polypeptide of 1280 residues, organized as two homologous halves (which are only 43% identical in human P-gp). Each half consists of six transmembrane (TM) segments (which form a transmembrane domain, TMD) followed by a nucleotide-binding domain (NBD) located at the cytoplasmic face of the membrane. The two halves of P-gp are essential for the activity of the transporter. Several TMs have been suggested to contribute to the drug-binding pocket(s) [1,2,4,12,13]. Low-resolution structures of human P-gp have been obtained by electron microscopy (EM) [14–16]. However, in the absence of high-resolution crystal structures of the mammalian P-gp alone and in complex with interacting molecules, it is unlikely to ascertain whether these binding sites form a single large binding pocket or whether there are multiple binding sites scattered along the whole protein [13].

Several mechanistic models have also been proposed to elucidate how the P-gp transporter translocates substrates across cellular membranes. In the 'vacuum cleaner' model, which seems the most probable, the compounds partition into the membrane bilayer and then P-gp extracts the substrates from the inner (cytosolic) membrane leaflet and releases them through a protein channel into the extracellular medium (Figure 16.1) [7,14,17].

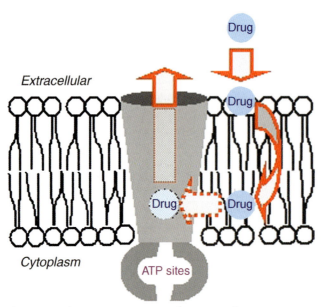

Figure 16.1 Schematic representation of a 'vacuum cleaner' model: drug partitions into the membrane bilayer, flips across the lipid core, accesses the TMDs from the inner membrane leaflet and is released through a protein channel into the extracellular medium.

Briefly, the transport cycle is initiated by substrate binding in the TMDs of P-gp, which increases the ATP affinity for the protein [1,4,7,17–19]. The binding and/or hydrolysis of a first ATP molecule in one of the NBDs seems to cause the rearrangements of the TM segments that facilitate the release of the substrate to the extracellular medium [19]. After hydrolysis of a second ATP molecule, P-gp returns to its original conformation and is ready for another cycle. Thus, the inhibition of P-gp drug transport could potentially result from competition for drug-binding sites or other P-gp modulator sites as well as from blockage of ATP binding and/or ATP hydrolysis or interaction with the cell membrane [1,5,6]. Although the competition of two substrates for the same P-gp site usually results in an inhibitory effect on the substrate transport, activation of P-glycoprotein efflux has been reported in some cases [5,6]. This evidence suggests a cooperative interaction between two substrates that are able to bind to P-gp simultaneously at different sites similar to CYP3A4 [5].

The interaction of compounds with P-gp is clearly a complex mechanism and our understanding of such a process is still at an early stage. However, it has become apparent that transport proteins are potential sources for variability in drug pharmacokinetics and pharmacodynamics. Therefore, the only way to circumvent these problems is both to identify the potential for P-gp activity in compounds early in the drug discovery process and to design candidate drugs that are unlikely to interact with P-gp. This chapter will summarize the computational models developed for identifying potential P-gp substrates and inhibitors. As the majority of the quantitative structure–property relationships and pharmacophores have been obtained using *in vitro* data, for a better understanding of the mechanisms underlying these models, a brief overview of the most used *in vitro* systems will be discussed.

16.2
In Vitro Models for Studying P-gp Interacting Compounds

A variety of *in vitro* assays are available to identify compounds as substrates and inhibitors of P-gp. These assays, which have been reviewed elsewhere in great detail [20–24], can be classified into three general categories: (1) transport, (2) accumulation/efflux and (3) ATPase activity [20–28]. It is important to note that these *in vitro* model systems can be adapted for measuring the interaction of drugs with other important drug transporter systems [22].

In the transport assays, the permeability of a compound in both absorption and secretion directions is measured using polarized epithelial cells that constitutively express high levels of P-gp (e.g. Caco-2) or have been transfected with the gene for a specific P-gp (e.g. MDR1-transfected MDCK or LLC-PK1 cells). Since P-gp is expressed on the apical membrane, ratios of basolateral-to-apical (B → A) permeability versus apical-to-basolateral (A → B) permeability greater than 1 may indicate an active efflux transport process. Bidirectional permeability measurements can also be performed in the presence of a specific P-gp inhibitor. Thus, apical-to-basolateral permeability increases and basolateral-to-apical permeability decreases such that permeability in both directions becomes similar. Although the transport assays from a drug-screening standpoint are very informative and can also be designed to distinguish P-gp substrates from inhibitors [22], a careful interpretation of the results should always be done [20]. For example, Caco-2, one of the most commonly used cell model systems, can express variable levels of P-gp and other efflux/uptake transporters (depending on the subclones used within each laboratory). Hence, a high efflux ratio (ER) could be a result of multiple transport mechanisms or, on the contrary, lack of efflux could be due to insufficient P-gp expression for allowing an efficient polarized transport. It has also been reported that the transport assays tend to be insensitive to highly permeable compounds (Papp A → B > 300 nm/s), providing false negative results [20]. This is probably due to fast movement of highly permeable molecules across the membrane, which could lead to either insufficient compound concentrations within the inner membrane leaflet or a saturation of P-gp activity [29].

Accumulation/efflux studies can be performed on different cell systems or membrane vesicle preparations. In the accumulation assays, uptake of a probe over time, typically either fluorescent (e.g. calcein-AM (CAM) [25–27]) or radiolabeled, into the cell or membrane vesicles is measured in the presence or absence of a known P-gp inhibitor. As P-gp transports substrates out of the cells, the inhibition of the protein would result in an increase in the amount of the probe in the cell. Accumulation studies in cells that overexpress P-gp can be compared to those obtained in the parental cell line that does not have as high a level of P-gp expression. The probe in the absence of inhibitors shows lower accumulation in P-gp expressing cells than in P-gp deficient cells. Similarly, probe accumulation is increased under conditions where P-gp is inhibited such that the difference in accumulation in P-gp deficient and overexpressing cells, respectively, becomes smaller. Accumulation assays poorly distinguish substrates and inhibitors of P-gp and, as far as transport assays are concerned, are also influenced by a passive diffusion property of molecules [20]. In contrast to transport assays, both accumulation (i.e. calcein-AM assay) and ATPase assays tend to fail in the identification of relatively low permeable compounds as P-gp active compounds [20].

The transport of compounds by P-gp and other efflux transporters is driven by the hydrolysis of ATP to ADP and inorganic phosphate. Thus, monitoring the release of inorganic phosphate in cell membrane preparations or purified membrane proteins represents a method to evaluate P-gp interacting compounds [28]. Sodium orthovanadate is used to inhibit P-gp ATPase activity, and the extent of stimulated P-gp ATPase activity mediated by the substrate is determined as the difference between the inorganic phosphate released in the presence and absence of sodium orthovanadate, respectively. Although many substrates have been shown to stimulate P-gp ATPase activity, other known substrates such as colchicines and etoposide did not increase basal ATPase activity [21]. In addition, some compounds such as daunomycin and vinblastine have been observed to inhibit ATPase activity in some studies, but increase in others, suggesting that modulation of ATPase activity highly depends on experimental conditions and should not be used as the single criterion to discriminate substrates and inhibitors [20–23].

16.3
Computational Models for Predicting P-gp Interacting Compounds

The ability to predict the likelihood of a compound to interact with P-gp or other relevant transporters on the basis of its molecular features should be extremely useful in drug discovery. This allows a most efficient design of potential drugs not interacting with undesirable targets such as P-gp. Although a variety of models have been proposed in the literature, the structural requirements for interaction with P-gp are not definitively resolved [30,31]. In the case of other proteins characterized by high flexibility, a low substrate affinity and broad specificity (e.g. CYP450 enzymes and other transporters), P-gp represents a challenging 'antitarget'. Furthermore, the experimental data used to develop computational models are often not fully mechanistically interpretable. In addition, depending on the type of *in vitro* assay (e.g. ATPase assay, efflux assay, etc.), conditions (e.g. cell lines, probe substrates/inhibitors, compound concentrations, etc.) and criteria for substrate identification (e.g. minimum efflux ratio to define a substrate), the compounds employed in the different studies are not absolutely defined as substrates, nonsubstrates or inhibitors. The use of different molecular data sets and computational methods (e.g. descriptors, statistical analysis, etc.) makes the comparison of the proposed models more difficult. An additional source of complexity is the lack of high-resolution X-ray P-gp structures. Crystallographic P-gp structures in complex with various substrates and inhibitors would be very helpful to confirm or refine the different pharmacophore hypotheses proposed so far.

In principle, the modeling techniques employed for characterizing P-gp or other transporter interacting molecules can be classified as ligand, protein or ligand–protein interaction-based approaches. Ligand-based approaches, which include quantitative structure–activity/property relationship (QSAR or QSPR) studies, three-dimensional QSAR (3D-QSAR) and pharmacophore modeling, use the structural information of the molecules interacting with the 'antitarget' of interest. The QSAR is a mathematical relationship linking the chemical structure (encoded by calculated descriptors) and biological activities on a target or antitarget in a quantitative manner. Various linear and nonlinear statistical techniques in combination with different molecular descriptors can be used for such purposes. 3D-QSAR involves the analysis of the quantitative relationship between the biological activity of a set of aligned compounds and their 3D electronic, steric and hydrophobic properties. Pharmacophore modeling is a procedure to identify, from a series of molecules characterized by a similar property, the common structural features and their 3D spatial arrangements assumed to be responsible for that activity.

Protein-based methods rely upon the structural information extracted from the X-ray crystallographic and/or homology protein structures. These also include docking techniques for the exploration of possible binding modes of a ligand to a given transporter protein.

Although, it is well known that not every inhibitor is a substrate, it is generally believed that every substrate should show inhibitory effects on other substrates. However, many substrates show a relatively low affinity for the protein and do not significantly inhibit P-gp at reasonable concentration [21]. Therefore, a classification of the computational models as specifically related to substrates or inhibitors is difficult to follow. Also, the experimental data used in the majority of the modeling studies are unable to give this information. Hence, when not clearly indicated in the text, the models described will be for P-gp interacting compounds (also indicated as modulators); that is, molecules that could behave as either substrate or inhibitor or both.

16.3.1
Ligand-Based Approach

A plethora of structure–property relationship (SPR) studies, attempting to define a set of common molecular features characterizing compounds interacting with P-gp, have been published over recent years [30,32–54]. This also included SPR analysis derived from a congeneric series of compounds, which are useful in predicting and optimizing specific chemical series [32–38]. Studies mainly dealing with a relatively large number of chemically heterogeneous compounds, which should provide more general insights on molecular P-gp interaction requirements, will be revisited (Table 16.1).

Litman et al. [39] analyzed a set of 34 diverse drugs and their effect on the kinetics of the ATPase activity of the microsomal membrane fraction of a P-gp overexpressing CHO cell line. They found a correlation ($r = 0.75$) between the affinity of the modulators and their van der Waals surface area, while the affinity data did not correlate with the lipophilicity of the compounds, expressed as calculated octanol/water partition coefficients.

Seelig [40] carefully analyzed the three-dimensional structures of 100 compounds previously classified on the basis of literature data as substrates, nonsubstrates and inducers. A set of well-defined structural elements based entirely on specific arrangements of hydrogen bonding groups was identified for each class of compounds. Two major pharmacophores (called recognition elements) were derived: type I recognition element that contained two electron donor groups separated by 2.5 ± 0.3 Å and type II consisted of two or three electron donor groups with a spatial separation of the outer two groups of 4.6 ± 0.6 Å (Figure 16.2). All molecules that contained at least one type I or type II element were predicted to be P-gp substrates. However, in some substrates up to eight of these elements were found. By a semiquantitative estimation of the hydrogen bonding energy, it was found that for substrates with comparable membrane partition coefficients, the binding data to P-gp (based on competition studies) were proportional not only to the number but also to the strength of hydrogen bonds. Although the number of inducers analyzed was relatively low, a set of hydrogen bonding acceptor elements was also identified for this class of compounds. It was found that at least one type II element was required for compounds inducing P-gp expression. Furthermore, as for the substrates, a correlation was found between the predicted hydrogen bonding energy and P-gp overexpression property. Further, Seelig and Landwojtowicz [41] found a correlation between measured air–water partition coefficients, which in turn were highly correlated with the corresponding lipid–water partition coefficients, and the ATPase data for 11 drugs. They suggested that membrane partitioning is the rate-limiting step

Table 16.1 QSPR and pharmacophore models for P-gp interacting compounds.

End point	No. of compounds in the training set	No. of compounds in the test set	Model performance on the test set	Assay type	Cell line	Major structural findings	Reference
P-gp modulator	34 drugs			ATPase	Cell membrane from CHO CR1R12	P-gp activity correlated with van der Waals surface area. No correlation found with calculated log $P_{o/w}$	[39]
P-gp substrate	64 substrates			Different sources		P-gp substrates characterized by well-defined sets of H-bond acceptor elements (type I and type II units). Strength and number of H-bond acceptor groups were also found important for P-gp activity	[40]
P-gp inducers	18 inducers			Different sources		P-gp inducers characterized by type II unit	[40]
P-gp modulator	11 drugs			ATPase	Vesicles from CHO CR1R12 [39]	P-gp activity correlated with measured log $K_{a/l}$	[41]
P-gp modulator	14 drugs [39]	7 drugs [39]	RMSE = 0.59	[39]		P-gp activity correlated with molecular surface area, H-bonding and polarizability	[42]
P-gp substrate	76 substrates and 68 nonsubstrates	32 substrates and 19 nonsubstrates	53% substrates and 79% nonsubstrates correctly predicted. Overall accuracy 63%	Different sources		Ensemble of pharmacophores: combinations of hydrophobic or aromatic groups, H-bond donors and H-bond acceptors	[43]

Table 16.1 (Continued)

End point	No. of compounds in the training set	No. of compounds in the test set	Model performance on the test set	Assay type	Cell line	Major structural findings	Reference
P-gp inhibitor	27 inhibitors of digoxin transport	9 digoxin inhibitors, 19 inhibitors of vinblastine binding, 17 inhibitors of vinblastine accumulation and 18 inhibitors of calcein accumulation	Except for the 18 inhibitors of calcein accumulation, the model produced a reasonable rank ordering of the tested molecules	Transport	Caco-2 cell line	Pharmacophore consisted of four hydrophobic regions and one H-bond acceptor group	[44]
P-gp inhibitor	21 inhibitors of vinblastine binding	27 inhibitors of digoxin transport, 17 inhibitors of vinblastine accumulation and 18 inhibitors of calcein accumulation	The model produced a reasonable rank ordering of the 27 inhibitors of digoxin transport	Binding to vinblastine site	Vesicles of CEM/VLB$_{100}$	Pharmacophore consisted of three aromatic rings and one hydrophobic feature	[44]
P-gp inhibitor	17 inhibitors of vinblastine accumulation	27 inhibitors of digoxin transport and 17 inhibitors of vinblastine accumulation	Poor performance	Accumulation	LLC-PK1 cells	Pharmacophore consisted of four hydrophobic features and one H-bond acceptor group	[44]
P-gp inhibitor	18 inhibitors of calcein accumulation data	27 inhibitors of digoxin transport and 21 inhibitors of vinblastine binding	Poor performance	Accumulation	LLC-PK1 cells	Pharmacophore consisted of two hydrophobic features, one H-bond acceptor group and one H-bond donor group	[44]

16.3 Computational Models for Predicting P-gp Interacting Compounds | 375

Table 16.1 (Continued)

End point	No. of compounds in the training set	No. of compounds in the test set	Model performance on the test set	Assay type	Cell line	Major structural findings	Reference
P-gp inhibitor	16 inhibitors of verapamil–P-gp binding [46]	Test sets in [44]	The model produced a reasonable rank ordering of the test sets	Binding to the verapamil site [46]	Vinblastine-induced Caco-2 cells [46]	Pharmacophore consisted of two hydrophobic, one H-bond acceptor and one ring aromatic features	[45]
P-gp modulator	25 compounds			Binding to the verapamil site and ATPase [46,48]		The most often observed combination of pharmacophoric elements: H-bond acceptor group and two hydrophobic centers	[47]
P-gp modulator	7 compounds			ATPase	DC-3F/ADX	Identified two partially overlapping pharmacophores consisting of an electron donor group, an aromatic group and alkyl areas	[49]
P-gp substrate	63 substrates and 32 nonsubstrates	35 substrates and 23 nonsubstrates	94% substrates and 74% nonsubstrates correctly predicted. Overall accuracy 86%	Transport	MDR1-MDCK cell line	High probability to be a nonsubstrate: high ES value of a nitrogen atom with zero hydrogen atoms and high number of the O–C hydrides. P-gp substrate showed MoIES > 110	[50]

Table 16.1 (Continued)

End point	No. of compounds in the training set	No. of compounds in the test set	Model performance on the test set	Assay type	Cell line	Major structural findings	Reference
P-gp substrate	116 substrates and 85 nonsubstrates	19 substrates and 6 nonsubstrates	84.2% substrates and 66.7% nonsubstrates correctly predicted. Overall accuracy 86%	Different sources		Descriptors selected in the final models: electrotopological E-state values, connectivity indices, number of H-bond donor and number of chlorine atoms	[51]
P-gp substrate and inhibitor	95 substrates, 78 inhibitors and 32 overlapping		83.3% substrates and 80.8% inhibitors correctly predicted			Descriptors selected: electrotopological E-state values, connectivity indices and physicochemical properties	[52]
P-gp substrate	109 compounds	20 compounds	Predicted values in good agreement with experimental data	Transport and accumulation	Caco-2 cell line	Pharmacophoric descriptors as well as size, shape (e.g. MW, molecular volume and rugosity) and H-bonding related descriptors correlated positively with P-gp activity. log $P_{o/w}$ not significant in the model	[53]
P-gp substrate	109 compounds	20 compounds	Predicted values in good agreement with experimental data	Transport and accumulation	Caco-2 cell line	The pharmacophore hypothesis consisted of two H-bond acceptor groups and two hydrophobic areas	[53]

16.3 Computational Models for Predicting P-gp Interacting Compounds

Table 16.1 (Continued)

End point	No. of compounds in the training set	No. of compounds in the test set	Model performance on the test set	Assay type	Cell line	Major structural findings	Reference
P-gp substrate	22 substrates and 31 nonsubstrates	115 substrates and 157 nonsubstrates	61% substrates and 81% nonsubstrates correctly predicted. Overall accuracy 72.4%	Transport	Caco-2 cell line	Size, shape (e.g. molecular surface and globularity), hydrophilic and H-bonding related descriptors correlated positively with P-gp activity. log $P_{o/w}$ not significant	[54]
P-gp substrate and inhibitor	9 substrates and 14 inhibitors	69 substrates and 56 inhibitors from literature source [52]	88.4% substrates and 75% inhibitors correctly predicted. Overall accuracy 82.4%	Transport and accumulation	Caco-2 (transport) and LoVoDx (accumulation) cell lines	Pharmacophore (substrate) hypothesis consisted of two H-bond acceptor groups and two H-bond donor groups. Pharmacophore (inhibitor) hypothesis consisted of hydrophobic areas and H-bond acceptor groups	[54]

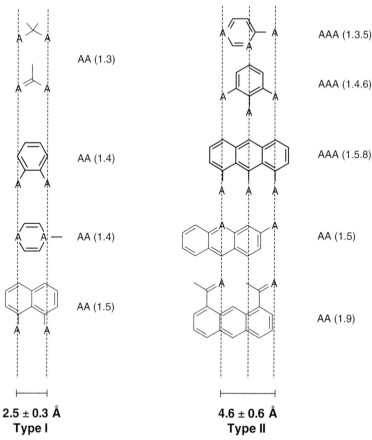

Figure 16.2 Proposed P-gp recognition elements: type I contains two electron donor groups separated by 2.5 ± 0.3 Å and type II consists of two or three electron donor groups with a spatial separation of the outer two groups of 4.6 ± 0.6 Å. A donates a hydrogen bonding acceptor group (electron donor group). Reprinted with permission from ref. [40]. Copyright 1998 Blackwell Publishing Ltd.

for interaction with P-gp, while the number and strength of the hydrogen bonds determine the dissociation of the complex.

Osterberg and Norinder [42] further analyzed a subset of the ATPase data published by Litman et al. [39]. The ATPase activity values were correlated by multivariate statistics with calculated descriptors (MolSurf descriptors) related to physicochemical properties such as lipophilicity, polarity, polarizability and hydrogen bonding. After exclusion of one outlier and large molecules such as valinomycin, gramicidin S and so on, which were not handled by the MolSurf software, only 21 compounds were included in the study. Two models were derived: model 1, based on a training set of 14 compounds, and model 2, based on the entire data set. Both derived models were statistically significant and predictive. In line with the observation of Litman et al. [39], they found a similar relationship between ATPase activity and van der Waals surface area. This was interpreted as an indication that the binding between modulators and P-gp occurs across a wide interaction surface on the protein. As already underlined in Seelig's study [40], it was also found that the strength and number of hydrogen bonds have an important effect on the P-gp-associated ATPase activity.

Using the Seelig study [40] as the starting point, Penzotti et al. [43] assembled a data set of 195 chemically diverse compounds, which included 108 P-gp substrates (i.e. compounds transported by P-gp or reported to induce P-gp overexpression) and 87 nonsubstrates (i.e. compounds not transported by P-gp), from literature sources. A subset of 144 molecules was used to generate an ensemble of 100 two-to-four-point pharmacophores to discriminate between P-gp substrates and nonsubstrates. The remaining 32 substrates and 19 nonsubstrates, were used to test the pharmacophore hypotheses. Compounds matching at least 20 of the 100 pharmacophores were predicted as potential P-gp substrates. The predominant structural features identified for the pharmacophoric ensemble, which included combinations of hydrophobic or aromatic groups, hydrogen bond donors and hydrogen bond acceptors, further supported the importance of hydrogen bonding capacity in P-gp substrate recognition. Although this approach employed a broad number and types of pharmacophores that should account for different binding modes/sites, its predictive performance was not so high. The method predicted correctly only 53 and 79% of the P-gp substrates and nonsubstrates, respectively, included in the test set, with an overall accuracy of 63%.

Four different 3D-QSAR models [44] were generated using *in vitro* data associated with inhibition of P-gp function and Catalyst software. Each of the four data sets collected was employed for developing a pharmacophore that was then validated using the remaining data sets. The first pharmacophore generated using 27 inhibitors of digoxin transport measured in Caco-2 cells, consisted of four hydrophobic regions and one hydrogen bond acceptor (nonbasic amines that have a lone pair, sp or sp^2 nitrogens, sp^3 oxygens or sulfurs and sp^2 oxygens). The model that showed an r^2 value of 0.77 reasonably categorized a test set of nine molecules. The digoxin inhibition pharmacophore was further evaluated using two additional test sets, which included 19 and 17 inhibitors of vinblastine binding and accumulation data generated in vesicles of CEM/VLB_{100} and in LLC-PK1 cells, respectively. Although four and nine molecules of the two additional test sets were predicted outside one log-unit cutoff, the model was able to produce a significant rank ordering of the tested molecules. On the contrary, the model poorly predicted the calcein accumulation data set (that contains the same molecules of the vinblastine accumulation data set) generated in LLC-PK1 cells. The second pharmacophore, generated with 21 inhibitors of vinblastine binding in vesicles of CEM/VLB_{100}, consisted of three aromatic rings and one hydrophobic feature, and had an r^2 value of 0.88. The model reasonably rank ordered the digoxin transport inhibition data, which were used for generating the first pharmacophore, while it did not appropriately rank vinblastine accumulation and calcein accumulation data. The third P-gp model generated from 17 inhibitors of vinblastine accumulation in L-MDR-1 cells, consisted of four hydrophobic features and one hydrogen acceptor group, and had an r^2 value of 0.88. The model showed poorly predictive performance for the digoxin transport data set and did not rank significantly the data for inhibition of vinblastine binding. A fourth model, generated from 18 inhibitors of calcein accumulation, consisted of two hydrophobic features, one hydrogen bond acceptor group and one hydrogen bond donor group. Similar to the previous model, it also predicted poorly the digoxin transport inhibition data set and did not significantly rank the data for inhibition of vinblastine binding. Ekins *et al.* [45] also developed a fifth pharmacophore using available literature data on verapamil–P-gp binding in vinblastine-induced Caco-2 cells for 16 compounds [46]. These data produced a pharmacophore consisting of two hydrophobic features, one hydrogen bond acceptor and one ring aromatic features (Figure 16.3), which was able to rank order the data sets previously used to generate the four pharmacophores. On the basis of the results obtained, the authors suggested a possible overlap in the sites to which the three P-gp substrates (i.e. verapamil, vinblastine and digoxin) bind.

Pajeva and Wiese [47] also proposed a general pharmacophore model of substrates and modulators that bind to the verapamil-binding site of the P-gp. The affinity data obtained from competition experiments with a radiolabeled [^{3}H]-verapamil at the verapamil-binding site of P-gp induced in Caco-2 cells, were taken from the literature [46,48]. In addition to the compounds from the verapamil-binding data set, seven drugs with ATPase effects were studied [39]. The model was developed by using a Genetic Algorithm Similarity Program (GASP). Different molecular alignments, conformations and interacting forms (i.e. neutral or protonated) were considered in the analysis. The pharmacophoric points were identified by overlays of the most active and rigid compounds in the data set (i.e. vinblastine and rhodamine 123). Up to six pharmacophoric points were identified, including two hydrophobic regions (coded as H_1 and H_2), three hydrogen bond acceptor groups (coded as A_1, A_2 and A_D; A_D group can act as both an acceptor and a donor simultaneously) and one

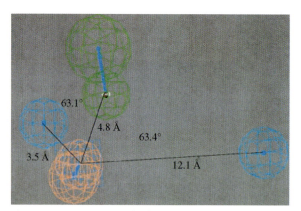

Figure 16.3 P-gp–verapamil binding inhibition pharmacophore (orange, ring aromatic feature; green, hydrogen bond acceptor; cyan, hydrophobic features). Reprinted with permission from ref. [45]. Copyright 2002 American Society for Pharmacology and Experimental Therapeutics.

16.3 Computational Models for Predicting P-gp Interacting Compounds | 381

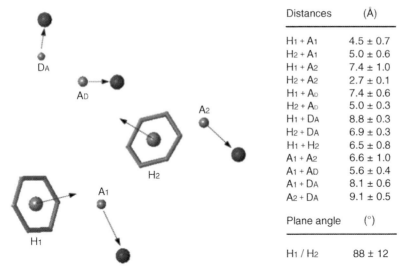

Distances	(Å)
$H_1 + A_1$	4.5 ± 0.7
$H_2 + A_1$	5.0 ± 0.6
$H_1 + A_2$	7.4 ± 1.0
$H_2 + A_2$	2.7 ± 0.1
$H_1 + A_D$	7.4 ± 0.6
$H_2 + A_D$	5.0 ± 0.3
$H_1 + D_A$	8.8 ± 0.3
$H_2 + D_A$	6.9 ± 0.3
$H_1 + H_2$	6.5 ± 0.8
$A_1 + A_2$	6.6 ± 1.0
$A_1 + A_D$	5.6 ± 0.4
$A_1 + D_A$	8.1 ± 0.6
$A_2 + D_A$	9.1 ± 0.5

Plane angle	(°)
H_1 / H_2	88 ± 12

Figure 16.4 General pharmacophoric pattern of drugs at the verapamil-binding site of P-gp. H_1 and H_2 are hydrophobic points (shown as spheres) located around the centers of the aromatic rings. A_1, A_2 and A_D are hydrogen bond (HB) acceptor points. D_A is an HB donor point. The complementary HB donor and acceptor points of the receptor are also presented as spheres at a distance of about 2.9 Å from the acceptor or donor atom of the drug in the direction of the lone pair or hydrogen. The arrows show directions of the hydrophobic and HB interactions. The values of distances between the pharmacophoric points and of the angle between the aromatic planes are the average ± SD of the first several overlays with the highest fitness scores of vinblastine (on rhodamine 123), rhodamine 123 (on vinblastine), and S- and R-verapamil (on vinblastine). Reprinted with permission from ref. [47]. Copyright 2002 American Chemical Society.

hydrogen bond donor point (coded as D_A) (Figure 16.4). The A_1 hydrogen bond acceptor group together with the two hydrophobic centers H_1 and H_2 was the most often observed combination of pharmacophoric elements present in the active P-gp drugs. The authors attempted to compare their findings with other published results. For example, the average plane angle between the aromatic rings (i.e. H_1 and H_2) was found to be similar to those proposed by Suzuki et al. [32] (i.e. 88° versus 100°), who stressed the importance of such features for highly active MDR modulators. In agreement with Seelig and Landwojtowicz [40,41], P-gp recognition elements formed by electron donor groups (oxygen, tertiary amino group, cyano group, halides, sulfides or π-electron system) with around 4.6 and 2.5 Å spatial separation, respectively, were found in their proposed pharmacophore (Figure 16.4). In addition, other pharmacophoric elements such as two hydrophobes, one hydrogen bond acceptor and one hydrogen bond donor were found similar to those proposed by Penzotti et al. [43] and Ekins et al. [45].

Garrigues et al. [49] characterized two different but partially overlapping P-gp pharmacophores. Initially, they determined affinities and mutual relationships from the changes in P-gp ATPase activity induced by a series of cyclic peptides, peptide-like compounds used alone or in combination. Because verapamil binding to P-gp excluded the binding of the majority of the molecules tested (i.e. cyclosporin A, actinomycin D, pristinamycin IA and bromocriptine), they investigated whether the studied molecules had structural elements in common with verapamil. All five molecules displayed three common pharmacophoric features: an electron donor, an aromatic group and alkyl areas. An additional consensus alkyl area was found for three of the five molecules analyzed. A second but partially overlapping pharmacophore, which included an electron donor, an aromatic group and three alkyl areas, was identified by alignment of only two molecules (i.e. vinblastine and tentoxin). The authors suggested that the binding pocket of P-gp binds the various substrates at defined sites, according to the shape, size and distribution of the hydrophobic and polar elements of the substrates, making possible the recognition of various chemical structures. The authors concluded that the pharmacophores identified were qualitatively in good agreement with other published models [40,42,44,45].

A QSAR model was developed to predict whether a given compound is a P-gp substrate or not [50]. The training set consisted of 95 compounds classified as substrates or nonsubstrates based on the results from *in vitro* monolayer transport assays performed using an MDR1-MDCK cell line. A compound was considered a P-gp substrate if the ratio of its apparent permeability in the B → A direction to that in the A → B direction was greater than 2.0, and a nonsubstrate if this ratio was less than 1.5. For compounds with efflux ratios between 1.5 and 2.0, further experiments were conducted in the presence of a potent P-gp inhibitor. These molecules were labeled as P-gp substrate if the efflux ratio collapsed to 1 in the presence of the inhibitor. The final two-group linear discriminant model used a set of 27 descriptors, which included electrotopological state (ES) values, topological shape indices, number of H-bond donors and acceptors, octanol/water partition coefficient, molecular weight and so on, calculated from the bidimensional structures of the training set. Typically, the interpretation of the models based on electrotopologial and tolopogical values is not straightforward. However, the authors found that a higher ES value of a nitrogen atom with no hydrogen atoms and a higher number of the O–C hydrides increased the probability of a compound to be a nonsubstrate. They also identified that the larger the molecule, the higher the chance of that compound to be a P-gp substrate. The QSAR model was able to correctly predict 33 and 17 compounds out of 35 and 23 P-gp substrates and nonsubstrates, respectively, with an overall accuracy of 86%. The authors analyzed the eight 'mispredicted' compounds to define the limitations of their model. Five false positives were found to have high permeability. As mentioned above, with a class of such molecules it is difficult to accurately determine if a compound is a P-gp substrate or not by using the monolayer transport assay. Therefore, it is possible that the model correctly classified these compounds as substrates. The remaining three mispredicted compounds were chemically diverse compared to the training molecules. This aspect should always be taken into account when QSPRs are applied to predict molecules that belong to chemical classes not included in the models. The authors observed that molecules with MolES (i.e. an E-state descriptor representing the molecular bulk of a compound) greater than 110 were predominately P-gp substrates and those with MolES less than 49 were nonsubstrates. Therefore, these simple rules to identify potential P-gp substrates and nonsubstrates should be useful to rapidly screen a large number of molecules.

A machine learning method, SVM, was explored for the prediction of P-gp substrates [51]. A total of 116 P-gp substrates and 85 nonsubstrates collected from the literature and 22 final molecular descriptors, including electrotopological E-state values, connectivity indices, number of H-bond donors and the number of chlorine atoms, were used to develop the model. Although the prediction accuracies for an independent evaluation set of 6 nonsubstrates and 19 substrates were reasonably good, accounting for 66.7 and 84.2%, respectively, the structural features responsible for discriminating between the two groups of compounds were not identified.

Using seven descriptors (including molecular connectivity indices, physicochemical and electrotopological state descriptors) and a data set of 95 substrates, 78 inhibitors and 32 overlapping compounds collected from the literature, an unsupervised machine learning approach based on the Kohonen self-organizing maps (SOMs) was explored to classify potential drugs as P-gp substrates or inhibitors [52]. The method classified correctly 83.3 and 80.8% of the substrates and inhibitors, respectively. As for the SVM model [51], this approach appears valuable for a fast screening of large libraries but it seems unable to provide hints on structural modifications that should be pursued to modify the P-gp interacting characteristic of a compound.

Two 3D-QSAR models were proposed for a set of 129 molecules characterized using both Caco-2 permeability measurements and CAM assay [53]. The P-gp CAM inhibition values of 109 compounds were correlated by partial least squares analysis (PLS) with a set of physicochemical descriptors combined with pharmacophoric features calculated using Volsurf and Almond software, respectively. The robustness and quality of the three-latent variable (LV) model (r^2 value of 0.82 and q^2 value, after leave-one-out cross validation, of 0.73) was confirmed by predicting the activity of 20 molecules not included in the training set. Some pharmacophoric descriptors mostly contributed to explain the pharmacological activities, while the physicochemical properties were relatively less important in the model. The authors suggested that substrate–protein recognition played a greater role in explaining P-gp activity data than the partitioning of the compound into the membrane. This is likely due to the relatively high lipophilicity of the compounds investigated that can easily cross the membrane and accumulate in the bilayer. In agreement with other studies [39,42,50], size and shape-related descriptors such as molecular weight, molecular volume, polarizability and rugosity, as well as hydrogen bonding properties, correlated positively with P-gp activity, while octanol/water partition coefficient (log P) showed a marginal role in the model. A statistically significant and validated PLS model (three LVs, r^2 value of 0.83 and q^2 value, after leave-one-out cross validation, of 0.74) was generated using only pharmacophoric descriptors (GRIND descriptors) extracted from the calculated molecular interaction fields (MIFs). The pharmacophore hypothesis derived, which consisted of two H-bond acceptor groups and two hydrophobic areas (Figure 16.5), was found to be similar to those developed by other groups [43–45,49]. Although the structural requirements identified for P-gp interaction were the same for the data set analyzed, the authors underlined that the experimental data used were unable to provide any indication on the binding site(s) involved, and therefore, to assign the pharmacophore found to one or more of the binding sites that have been described in the literature.

In our group, we developed a computational model for discriminating P-gp substrates from nonsubstrates, based on Caco-2 ER values [54]. A partial least

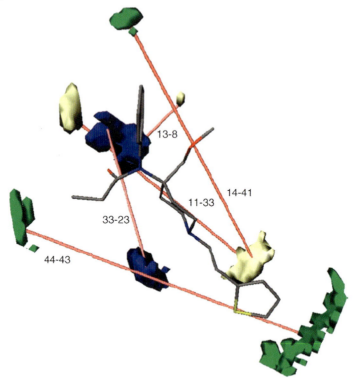

Figure 16.5 Resulting pharmacophore for P-gp actively transported molecules. The depicted molecule is the analgesic (narcotic) sufentanil. The colored areas around the molecules are the GRID fields produced by the molecule: yellow for DRY probe, green for TIP probe and blue for N1 probe. Reprinted with permission from ref. [53]. Copyright 2005 American Chemical Society.

squares discriminant (PLSD) model using physicochemical descriptors (Volsurf) was generated. The training set consisted of 22 P-gp and 31 non-P-gp substrates. The number of compounds utilized to develop the model was too small to be divided into training and test sets; therefore, 115 P-gp substrates and 157 non-P-gp substrates characterized in terms of ER from Caco-2 permeability measurements were extracted from our internal database to be used as an external test set. Although the expression of P-gp protein was monitored by the flow cytometry analysis, the Caco-2 cell line can also express different levels of other ABC transporters. Consequently, it is likely that some of the efflux ratios analyzed could be a result of multiple transport mechanisms. The model correctly classified 72% of the entire evaluation set with an interclass accuracy of 61% for the P-gp substrates and 81% for the nonsubstrates. By model interpretation, we gained more insight on the most relevant physicochemical properties characterizing the P-gp substrates of this study. In agreement with other findings [39–42,53], descriptors related to the size, shape and flexibility of the molecule such as molecular surface, globularity and elongation were directly correlated with P-gp substrates. Wide hydrophilic regions and their relative spatial arrangements as well as the presence of H-bond donor and acceptor regions were the characteristics of the P-gp substrates. As observed by other authors [39,53], log P had a very marginal role in discriminating between P-gp substrates and nonsubstrates. However, an appropriate hydrophobic–hydrophilic balance of P-gp substrates is probably a prerequisite for membrane partitioning and binding to P-gp as it was already underlined by Seelig and Landwojtowicz [41]. To further explore the 3D pharmacophoric features that mainly differentiated the substrates with poor or no inhibitory activity in the range of the therapeutically relevant concentrations from inhibitors having no evidence of significant transport, a PLSD model based on GRIND descriptors was developed combining the experimental results of a CAM assay with the Caco-2 efflux ratio data. The model that included 9 P-gp substrates and 14 inhibitors was validated by predicting an external set of 69 substrates and 56 inhibitors taken from the literature [52]. The overall accuracy was 82%, and the accuracy in predicting the substrates and inhibitors was 88 and 75%, respectively. The most important descriptors for P-gp substrates were mainly related to H-bonding potential properties. Compounds thus analyzed were characterized by favorable interacting energy regions around two H-bond donor groups (GRIND 22-16 descriptor in Figure 16.6) and two H-bond acceptor groups (GRIND 33-38 descriptor in Figure 16.6) placed, respectively, 6.5 and 15 Å apart. Favorable interacting regions

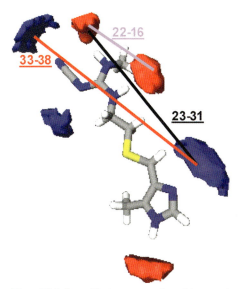

Figure 16.6 A graphical representation of the most important 3D pharmacophoric GRIND features for the cimetidine substrate as a reference compound. The colored areas around the molecule are the GRID MIFs calculated with the O (red) and N1 (blue) probes. Reprinted with permission from ref. [54]. Copyright 2006 American Chemical Society.

around an H-bond donor and an H-bond acceptor group (GRIND 23-31 descriptor in Figure 16.6) separated by a distance of 12.5 Å also differentiated P-gp substrates from inhibitors. A picture summarizing the relevant pharmacophoric features for the P-gp substrate cimetidine is reported in Figure 16.6. As mentioned above, Cianchetta *et al.* [53] proposed a quantitative model, based on a combination of GRIND and Volsurf descriptors developed using a different set of P-gp substrates. Although the two models were not directly comparable, a key recognition element, that is, two H-bond acceptor groups around 11.5–15 Å apart [53], was found in both studies. The 3D pharmacophoric elements that differentiate inhibitors from substrates, which are graphically represented for astemizole inhibitor as reference compound in Figure 16.7I, were also identified. Two hydrophobic regions placed at an optimal distance of around 11.5 Å (GRIND 11-28 descriptor in Figure 16.7I) as well as the presence of favorable interacting regions placed 8.0 Å apart around two H-bond acceptor groups (GRIND 33-20 descriptor in Figure 16.7I) characterized P-gp inhibitors. Favorable interacting regions around a hydrophobic moiety and a H-bond acceptor group separated by a distance of 4.0 Å also differentiated P-gp inhibitors (GRIND 13-7 descriptor in Figure 16.7I) from substrates. The key elements of this pharmacophore (Figure 16.7I) are similar, to some extent, to those proposed by Ekins *et al.* [44,45]. The model was also able to correctly predict the P-gp inhibitory properties of three additional compounds (Figure 16.8) recently described in the literature [55]. In fact, these molecules were characterized by P-gp recognition elements similar to our studied inhibitors (Figure 16.7II). It is important to underline that the experimental data used for developing our models were unable to provide strong indications on the recognition site and mechanism (i.e. competitive, noncompetitive inhibition, etc.) of molecular P-gp interaction. However, we identified for both P-gp substrates and P-gp

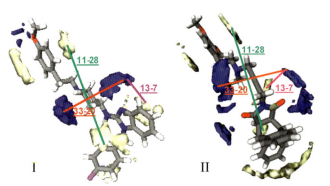

Figure 16.7 (I) A graphical representation of the most important 3D pharmacophoric GRIND features for the astemizole inhibitor as a reference compound [54]. (II) Similar pharmacophoric features found for three potent P-gp inhibitors [55]; as an example, these features are reported for compound **3a**. The colored areas around the molecule are the GRID MIFs calculated with the DRY (yellow) and N1 (blue) probes.

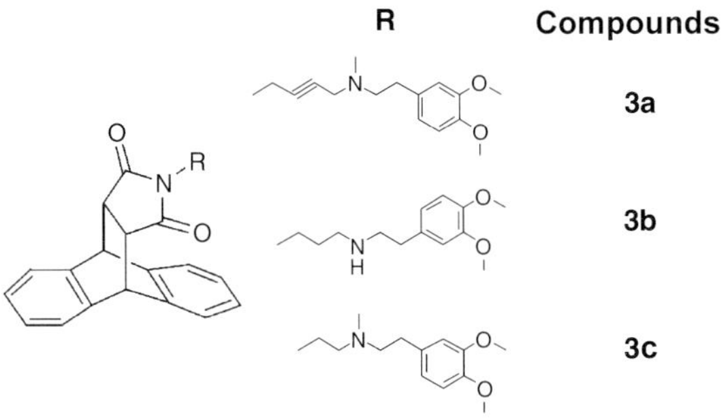

Figure 16.8 Structures of three potent P-gp inhibitors taken from the literature [55].

inhibitors, respectively, a common set of pharmacophoric features. Thus, a common mechanism of action was assumed to characterize each group of the analyzed compounds. Nevertheless, without any additional experimental information, any hypothesis on the molecular recognition cannot be made.

16.3.2
Protein and Ligand–Protein Interaction-Based Approaches

Detailed three-dimensional models of P-gp–substrate complexes representing the various steps of the catalytic cycle would be significantly helpful in elucidating the molecular mechanism for substrate binding and release. The relatively poor 3D structural information available [14–16] and the complex mechanism for compounds undergoing P-gp–compound interactions explain why only a few groups have attempted to build and study 3D homology models for P-gp [56,58,60,70].

The P-gp structure was modeled by Seigneuret and Garnier-Suillerot [56] using comparative modeling and the dimer structure of *E. coli* MsbA [57] as a template. The model was characterized by a large inner chamber open to the cytoplasmic site. The presence of a bilayer accessible chamber was in agreement with an EM study [14]. However, this was the only point of similarity because the EM studies [14–17] suggested an inverted V-form chamber, with a large opening to the extracellular side that is closed at the cytoplasmic face of the membrane. The authors suggested that the model proposed could represent an open structure of P-gp that during the catalytic cycle could switch to a closed conformation.

An atomic scale model for the human P-gp was derived from disulfide cross-linking and homology model [58]. Each half of P-gp was modeled using the crystal structures of two prokaryotic ABC transporters: MsbA [57] and BtuCD [59]. Next, the two halves were translated and rotated to reflect the close interactions of the two NBDs. Energy minimization was then performed to optimize the packing of the two TMDs. The TM domains formed a compact helix bundle surrounding a chamber that was open at the extracellular surface, entirely consistent with the low-resolution EM data for P-gp [14,15,17]. The resulting model was in relatively good agreement with generated cross-linking data.

Pajeva et al. [60] proposed two partial models of the TM domains of P-gp based on the E. coli MsbA crystal structure [57] and the cross-linking results of Loo and Clarke [61–69], respectively. The models were used to identify the potential binding sites of the Hoechst 33342 (H-site) and rhodamine 123 (R-site), respectively. A pharmacophoric pattern for the H-site including a chain of hydrophobic centers and two hydrogen bond interaction points was proposed. After analysis of both models, the authors suggested that the H-binding site involved amino acids belonging to either TM5 or TM11. As for the H-site, the opposite orientation of the amino acids involved in the R-site in both models (i.e. TM6 and TM12) reinforced the hypothesis that the TM domains undergo rotation exposing the substrate bound from the membrane to the pore. The authors concluded that the homology and the cross-linking models represented a 'nucleotide-free', that is, an inactive, form and a 'nucleotide-bound', that is, an active, form of P-gp structures, respectively.

Recently, Vandevuer et al. [70] proposed a 3D atomic model of P-glycoprotein in the absence of ATP. The model, obtained using a combination of different computational methods including comparative modeling and rigid body dynamics, was further validated by docking a number of P-gp ligands. Each P-gp half was modeled independently using the structure of E. coli MsbA monomers. Then, an initial model of the P-gp NBD dimer was built using the NBD dimer of E. coli BtuCD [59] and each NBD monomer of E. coli MsbA [57] as templates. The authors followed this approach since there was considerable support that the NBDs in P-gp form an interface in its resting state, as also shown by cysteine disulfide cross-linking experiments [71,72] in the absence of ATP. Each P-gp half was then fitted on the modeled P-gp NBD dimer. Rigid body dynamics simulations, embodying all available cross-linking data characterizing the whole protein in the absence of ATP, were performed on the model. Finally, the loops connecting the TM helices to the NBDs were added by comparative modeling using as additional template segments from the initial model of each P-gp half. The authors reported that the interresidue distances derived in both the TM domains and the NBDs were in quite good agreement with the experimental data. The model was also in good agreement with EM data [16], particularly in terms of size and topology. Some openings were observed between the TM helices 5 and 8 on one the hand and TM helices 2 and 11 on the other hand. They both lead to a large cavity open toward the extracellular side. As suggested by other studies [17], this lateral opening should allow the P-gp substrates to move from the inner leaflet bilayer into the protein cavity. Seven P-gp compounds were docked inside this large pocket and different binding modes involving H-bonds, π–π and cation–π interactions were identified for all compounds. This corroborated the experimental evidence indicating the existence of multiple drug-binding sites. The six predicted affinities, calculated for each molecule as an average of 25 docking runs, roughly correlated with the experimental values.

Although these aforementioned studies represent interesting results, at present it is unlikely that docking approaches applied to homology models of proteins like P-gp, which are characterized by a large, highly flexible cavity able to accommodate a wide variety of chemically different compounds, is able to differentiate between P-gp interacting and noninteracting compounds. However, the 3D models of P-gp could in some extent help to interpret and refine pharmacophore models obtained from the analysis of P-gp interacting molecules.

16.4
Computational Models for Other Important Drug Transporters

Although a variety of uptake and efflux transporters are expressed in humans, only a handful of them appear to be broadly relevant in the disposition of a majority of compounds in discovery or in clinical use [8]. Several reviews describing the expression and function of other drug transporters are available [73–75]. However, compared to P-gp, the knowledge currently available is relatively poor. Moreover, their clinical relevance is not fully understood [9]. For these reasons, *in silico/in vitro/in vivo* tools for identifying compounds interacting with other transporters in drug research and development process are less well established. To our knowledge, only a very few computational models for other drug transporters such as ABCG2, hOCT1 and OATP have been described in the literature [76–78]. The most important findings of these studies will be briefly discussed here.

The overexpression of the human ATP-binding cassette transporter ABCG2 (BCRP/ABCP/MXR) confers cancer cell resistance to anticancer drugs such as mitoxantrone and anthracycline. In normal tissues, ABCG2 is expressed in the placental trophoblast cells, hepatocyte canalicular membrane, apical intestinal epithelia and vascular endothelial cells. There appears to be some overlap in the substrates and tissue distribution between BCRP and P-gp. Saito *et al.* [76] proposed a quantitative structure–activity relationship to evaluate human ABCG2–drug interactions. Forty-nine different therapeutic drugs and natural products were tested for their ability to inhibit ABCG2-mediated [^{3}H]methotrexate (MTX) transport in plasma membrane vesicles prepared from ABCG2-expressing Sf9 cells. Based on their inhibition profiles, a QSAR model was developed using chemical fragmentation codes deduced from the structure of the analyzed compounds. This analysis allowed the identification of a well-defined set of substructural elements responsible for the inhibition of ABCG2-mediated MTX transport. In summary, it was suggested that one amine bonded to one carbon of a heterocyclic ring, fused heterocyclic ring(s) and two substituents on a carbocyclic ring of the fused heterocyclic ring(s) are important moieties for the interaction with ABCG2.

The human OCT1 is a member of organic cation transporter family (OCT). This transporter appears to be expressed predominantly in liver, where it is presumed to play a role in the hepatic metabolism and elimination of cationic drugs. OCT1 has been shown to be capable of mediating the uptake of small protonated molecules such as 1-methyl 1,4-phenylpyridinium and tetraethylammonium as well as larger cations including *N*-methyl-quinine, *N*-methyl-quinidine and quinidine. Bednarczyk *et al.* [77] evaluated quantitatively the ability of a set of 30 molecules to inhibit [^{3}H]tetraethylammonium uptake in HeLa cells stably expressing hOCT1. By analyzing the IC_{50} values of a subset of 22 inhibitors, a pharmacophore consisting of three hydrophobic features at the distances of 5.12, 4.19 and

5.32 Å from the center of a positive ionizable group was identified. However, its predictive performance on a test set consisted of eight structurally similar compounds was relatively poor. To achieve a computational model with greater predictability, a descriptor-based QSPR model was also developed. Descriptors related to molecular hydrophobicity as well as hydrogen bond donor, shape and charge features contributed to explain hOCT1 inhibitor properties of the analyzed compounds.

The organic anion transporter polypeptides (OATPs) expressed in organs such as the intestine, liver and blood–brain barrier are key determinants in the cellular uptake of many endogenous and exogenous chemicals, including drugs in clinical use. Chang *et al.* [78] generated pharmacophore models for rat Oatp1a1 and human OATP1B1, respectively, using published K_m values from different experimental systems. Both human and rat pharmacophores, obtained by combining the data from multiple cell types, consisted of two hydrogen bond acceptors and three hydrophobes. The difference between the two models was only related to the 3D spatial arrangement of the five pharmacophoric elements. Thus, the authors suggested that a certain degree of similarity characterizes the two proteins.

16.5
Conclusions

In recent years, drug transporters have received increasing attention, owing to the recognition that drug–transporter interactions could both lead to important drug pharmacokinetic variability and cause serious drug–drug interactions. Although progress is being made, *in vitro* and *in silico* models for identifying and evaluating substrates and inhibitors of drug transporters are much less advanced than those for other antitargets such as enzyme CYP450s. Undoubtedly, our knowledge of P-gp and the tools to study it are more advanced than other transporters.

The availability of an increased amount of P-gp-related experimental data of compounds prompted the efforts in developing structure-based computational tools that may be used at lower cost and even before the actual drug candidate is synthesized. From the modeling studies, general pictures, often in reasonable agreement, of the major structural requirements for interaction with P-gp have been derived. Almost all models proposed so far provide a qualitative estimation of compounds as potential P-gp interacting and noninteracting molecules. Although useful in the very early phases of drug discovery as first filters on large libraries of compounds, the lack of valuable quantitative computational models addressing all possible sites and mechanisms of P-gp interactions makes the rational drug design for circumventing P-gp hard to pursue. Undoubtedly, considerable progress has been made in the development of new computational models for identifying P-gp interacting compounds; however, there is still much to do. In some cases, the computational models reported in the literature are suboptimal as they are derived from small data sets and/or their validation is not always of adequate quality. Also, the experimental measurements that are the key elements for deriving relevant computational models are often affected by high variability and complexity (i.e. they address multiple mechanisms), or their relevance with respect to the *in vivo* situation is not fully exploited.

A clear understanding of drug-binding sites and the mechanisms of transport/inhibition as well as the resolution of P-gp crystal structures in complex with different substrates/inhibitors will definitively aid in the development of better models. However, considering the complexity underlying these antitargets, we strongly believe that this will not be reached in the near future.

Abbreviations

3D	three-dimensional
ABC	ATP-binding cassette
ADP	adenosine diphosphate
ATP	adenosine triphosphate
BCRP	breast cancer resistance protein
Caco-2	human Caucasian colon adenocarcinoma cells
CAM	calcein acetoxymethyl ester
CEM/VLB$_{100}$	multidrug-resistant human leukemia cells
CYP450	cytochrome P450
EM	electron microscopy
ER	efflux ratio
ES	electrotopological state
GASP	genetic algorithm similarity program
IC$_{50}$	half maximal inhibitory concentration
Km	Michaelis-Menten constant
LLC-PK1	pig kidney epithelial cells
LV	latent variable
MDCK	Madin-Darby canine kidney cells
MDR	multidrug resistance
MIF	molecular interaction field
Mo1ES	E-state descriptor representing the molecular bulk
MTX	methotrexate
MXR	mitoxantrone resistance protein
NBD	nucleotide-binding domain
OATP	organic anion transporter polypeptide
OCT	organic cation transporter
P-gp	P-glycoprotein
PLS	partial least squares
PLSD	partial least squares discriminant
QSAR	quantitative structure–activity relationship
QSPR	quantitative structure–property relationship
SOM	self-organizing map
TM	transmembrane
TMD	transmembrane domain

References

1 Ambudkar, S.V., Dey, S., Hrycyna, C.A., Ramachandra, M., Pastan, I. and Gottesman, M.M. (1999) Biochemical, cellular, and pharmacological aspects of the multidrug transporter. *Annual Review of Pharmacology and Toxicology*, **39**, 361–398.

2 Ambudkar, S.V., Kimchi-Sarfaty, C., Sauna, Z.E. and Gottesman, M.M. (2003) P-glycoprotein: from genomics to mechanism. *Oncogene*, **22**, 7468–7485.

3 Schinkel, A.H. and Jonker, J.W. (2003) Mammalian drug efflux transporters of the ATP binding cassette (ABC) family: an overview. *Advanced Drug Delivery Reviews*, **55**, 3–29.

4 Ambudkar, S.V., Kim, I.-W. and Sauna, Z.E. (2006) The power of the pump: mechanisms of action of P-glycoprotein (ABCB1). *European Journal of Pharmaceutical Sciences*, **27**, 392–400.

5 Lin, J.H. and Yamazaki, M. (2003) Role of P-glycoprotein in pharmacokinetics. Clinical implications. *Clinical Pharmacokinetics*, **42**, 59–98.

6 Lin, J.H. (2003) Drug–drug interaction mediated by inhibition and induction of P-glycoprotein. *Advanced Drug Delivery Reviews*, **55**, 53–81.

7 Loo, T.W. and Clarke, D.M. (2005) Recent progress in understanding the mechanism of P-glycoprotein-mediated drug efflux. *Journal of Membrane Biology*, **206**, 173–185.

8 Kim, R.B. (2006) Transporters and drug discovery: why, when, and how. *Molecular Pharmaceutics*, **3**, 26–32.

9 Zhang, L., Strong, J.M., Qiu, W., Lesko, L.J. and Huang, S.-M. (2006) Scientific perspectives on drug transporters and their role in drug interactions. *Molecular Pharmaceutics*, **3**, 62–69.

10 Benet, L.Z., Izumi, T., Zhang, Y., Silverman, J.A. and Wacher, V.J. (1999) Intestinal MDR transport proteins and P-450 enzymes as barriers to oral drug delivery. *Journal of Controlled Release*, **62**, 25–31.

11 Varma, M.V.S., Sateesh, K. and Panchagnula, R. (2005) Functional role of P-glycoprotein in limiting intestinal absorption of drugs: contribution of passive permeability to P-glycoprotein meditated efflux transport. *Molecular Pharmaceutics*, **2**, 12–21.

12 Dey, S., Ramachandra, M., Pastan, I., Gottesman, M.M. and Ambudkar, S.V. (1997) Evidence for two nonidentical drug-interaction sites in the human P-glycoprotein. *Proceedings of the National Academy of Sciences of the United States of America*, **94**, 10594–10599.

13 Martin, C., Berridge, G., Higgins, C.F., Mistry, P., Charlton, P. and Callaghan, R. (2000) Communication between multiple drug binding sites on P-glycoprotein. *Molecular Pharmacology*, **58**, 624–632.

14 Rosenberg, M.F., Callaghan, R., Ford, R.C. and Higgins, C.F. (1997) Structure of the multidrug resistance P-glycoprotein to 2.5 nm resolution determined by electron microscopy and image analysis. *Journal of Biological Chemistry*, **272**, 10685–10694.

15 Lee, J.Y., Urbatsch, I.L., Senior, A.E. and Wilkens, S. (2002) Projection structure of P-glycoprotein by electron microscopy. Evidence for a closed conformation of the nucleotide binding domains. *Journal of Biological Chemistry*, **277**, 40125–40131.

16 Rosenberg, M.F., Callaghan, R., Modok, S., Higgins, C.F. and Ford, R.C. (2005) Three-dimensional structure of P-glycoprotein. *Journal of Biological Chemistry*, **280**, 2857–2862.

17 Rosenberg, M.F., Velarde, G., Ford, R.C., Martin, C., Berridge, G., Kerr, I.D., Callaghan, R., Schmidlin, A., Wooding, C., Linton, K.J. and Higgins, C.F. (2001)

Repacking of the transmembrane domains of P-glycoprotein during the transport ATPase cycle. *EMBO Journal*, **20**, 5615–5625.

18 Higgins, C.F. and Linton, K.J. (2004) The ATP switch model for ABC transporters. *Nature Structural and Molecular Biology*, **11**, 918–926.

19 Rothnie, A., Storm, J., Campbell, J., Linton, K.J., Kerr, I.D. and Callaghan, R. (2004) The topography of transmembrane segment six is altered during the catalytic cycle of P-glycoprotein. *Journal of Biological Chemistry*, **279**, 34913–34921.

20 Polli, J.W., Wring, S.A., Humphreys, J.E., Huang, L., Morgan, J.B., Webster, L.O. and Serabjit-Singh, C.S. (2001) Rational use of in vitro P-glycoprotein assays in drug discovery. *Journal of Pharmacology and Experimental Therapeutics*, **299**, 620–628.

21 Hochman, J.H., Yamazaki, M., Ohe, T. and Lin, J.H. (2002) Evaluation of drug interactions with P-glycoprotein in drug discovery: in vitro assessment of the potential for drug–drug interactions with P-glycoprotein. *Current Drug Metabolism*, **3**, 257–273.

22 Zhang, Y., Bachmeier, C. and Miller, D.W. (2003) In vitro and in vivo models for assessing drug efflux transporter activity. *Advanced Drug Delivery Reviews*, **55**, 31–51.

23 Schwab, D., Fischer, H., Tabatabaei, A., Poli, S. and Huwyler, J. (2003) Comparison of in vitro P-glycoprotein screening assays: recommendations for their use in drug discovery. *Journal of Medicinal Chemistry*, **46**, 1716–1725.

24 Rautio, J., Humphreys, J.E., Webster, L.O., Balakrishnan, A., Keogh, J.P., Kunta, J.R., Serabjit-Singh, C.J. and Polli, J.W. (2006) In vitro P-glycoprotein inhibition assays for assessment of clinical drug interaction potential of new drug candidates: a recommendation for probe substrates. *Drug Metabolism and Disposition*, **34**, 786–792.

25 Homolya, L., Hollo, Z., Germann, U.A., Pastan, I., Gottesman, M.M. and Sarkadi, B. (1993) Fluorescent cellular indicators are extruded by the multidrug resistance protein. *Journal of Biological Chemistry*, **268**, 21493–21496.

26 Hollo, Z., Homolya, L., Davis, C.W. and Sarkadi, B. (1994) Calcein accumulation as a fluorometric functional assay of the multidrug resistance transporter. *Biochimica et Biophysica Acta*, **1191**, 384–388.

27 Tiberghien, F. and Loor, F. (1996) Ranking of P-glycoprotein substrates and inhibitors by a calcein-AM fluorometry screening assay. *Anti-Cancer Drugs*, **7**, 568–578.

28 Scarborough, G.A. (1995) Drug-stimulated ATPase activity of the human P-glycoprotein. *Journal of Bioenergetics and Biomembranes*, **27**, 37–41.

29 Eytan, G.D., Regev, R., Oren, G. and Assaraf, Y.G. (1996) The role of passive transbilayer drug movement in multidrug resistance and its modulation. *Journal of Biological Chemistry*, **271**, 12897–12902.

30 Stouch, T.R. and Gudmundsson, O. (2002) Progress in understanding the structure–activity relationships of P-glycoprotein. *Advanced Drug Delivery Reviews*, **54**, 315–328.

31 Raub, T.J. (2006) P-glycoprotein recognition of substrates and circumvention through rational drug design. *Molecular Pharmaceutics*, **3**, 3–25.

32 Suzuki, T., Fukazawa, N., Sannohe, K., Sato, W., Yano, O. and Tsuruo, T. (1997) Structure–activity relationship of newly synthesized quinoline derivatives for reversal of multidrug resistance in cancer. *Journal of Medicinal Chemistry*, **40**, 2047–2052.

33 Ford, J.M., Prozialeck, W.C. and Hait, W.N. (1989) Structural features determining activity of phenothiazines and related drugs for inhibition of cell growth and reversal of multidrug

resistance. *Molecular Pharmacology*, **35**, 105–115.

34 Ford, J.M., Bruggemann, E.P., Pastan, I., Gottesman, M.M. and Hait, W.N. (1990) Cellular and biochemical characterization of thioxanthenes for reversal of multidrug resistance in human and murine cell lines. *Cancer Research*, **50**, 1748–1756.

35 Thimmaiah, K.N., Horton, J.K., Qian, X.D., Beck, W.T., Houghton, J.A. and Houghton, J. (1990) Structural determinants of phenoxazine type compounds required to modulate the accumulation of vinblastine and vincristine in multidrug-resistant cell lines. *Cancer Communications*, **2**, 249–259.

36 Dhainaut, A., Regnier, G., Atassi, G., Pierre, A., Leonce, S., Kraus-Berthier, L. and Prost, J.F. (1992) New triazine derivatives as potent modulators of multidrug resistance. *Journal of Medicinal Chemistry*, **35**, 2481–2496.

37 Dhainaut, A., Regnier, G., Tizot, A., Pierre, A., Leonce, S., Guilbaud, N., Kraus-Berthier, L. and Atassi, G. (1996) New purines and purine analogs as modulators of multidrug resistance. *Journal of Medicinal Chemistry*, **39**, 4099–4108.

38 Ferte, J., Kuhnel, J.M., Chapuis, G., Rolland, Y., Lewin, G. and Schwaller, M.A. (1999) Flavonoid-related modulators of multidrug resistance: synthesis, pharmacological activity, and structure–activity relationships. *Journal of Medicinal Chemistry*, **42**, 478–489.

39 Litman, T., Zeuthen, T., Skovsgaard, T. and Stein, W.D. (1997) Structure–activity relationships of P-glycoprotein interacting drugs: kinetic characterization of their effects on ATPase activity. *Biochimica et Biophysica Acta*, **1361**, 159–168.

40 Seelig, A. (1998) A general pattern for substrate recognition by P-glycoprotein. *European Journal of Biochemistry*, **251**, 252–261.

41 Seelig, A. and Landwojtowicz, E. (2000) Structure–activity relationship of P-glycoprotein substrates and modifiers. *European Journal of Pharmaceutical Sciences*, **12**, 31–40.

42 Osterberg, T. and Norinder, U. (2000) Theoretical calculation and prediction of P-glycoprotein interacting drugs using MolSurf parametrization and PLS statistics. *European Journal of Pharmaceutical Sciences*, **10**, 295–303.

43 Penzotti, J.E., Lamb, M.L., Evensen, E. and Grootenhuis, P.D.J. (2002) A computational ensemble pharmacophore model for identifying substrates of P-glycoprotein. *Journal of Medicinal Chemistry*, **45**, 1737–1740.

44 Ekins, S., Kim, R.B., Leake, B.F., Dantzig, A.H., Schuetz, E.G., Lan, L.-B., Yasuda, K., Shepard, R.L., Winter, M.A., Schuetz, J.D., Wikel, J.H. and Wrighton, S.A. (2002) Three-dimensional quantitative structure–activity relationships of inhibitors of P-glycoprotein. *Molecular Pharmacology*, **61**, 964–973.

45 Ekins, S., Kim, R.B., Leake, B.F., Dantzig, A.H., Schuetz, E.G., Lan, L.-B., Yasuda, K., Shepard, R.L., Winter, M.A., Schuetz, J.D., Wikel, J.H. and Wrighton, S.A. (2002) Application of three-dimensional quantitative structure–activity relationships of P-glycoprotein inhibitors and substrates. *Molecular Pharmacology*, **61**, 974–981.

46 Neuhoff, S., Langguth, P., Dressler, C., Andersson, T.B., Regardh, C.G. and Spahn-Langguth, H. (2000) Affinities at the verapamil binding site of MDR1-encoded P-glycoprotein: drugs and analogs, stereoisomers and metabolites. *International Journal of Clinical Pharmacology and Therapeutics*, **38**, 168–179.

47 Pajeva, I.K. and Wiese, M. (2002) Pharmacophore model of drugs involved in P-glycoprotein multidrug resistance: explanation of structural variety

(hypothesis). *Journal of Medicinal Chemistry*, **45**, 5671–5686.
48 Doeppenschmitt, S., Spahn-Langguth, H., Regard, C.G. and Langguth, P. (1998) Radioligand-binding assay employing P-glycoprotein-overexpressing cells: testing drug affinities to the secretory intestinal multidrug transporter. *Pharmaceutical Research*, **15**, 1001–1006.
49 Garrigues, A., Loiseau, N., Delaforge, M., Ferte, J., Garrigos, M., Andre, F. and Orlowski, S. (2002) Characterization of two pharmacophores on the multidrug transporter P-glycoprotein. *Molecular Pharmacology*, **62**, 1288–1298.
50 Gombar, V.K., Polli, J.W., Humphreys, J.E., Wring, S.A. and Serabjit-Singh, C.S. (2004) Predicting P-glycoprotein substrates by a quantitative structure–activity relationship model. *Journal of Pharmaceutical Sciences*, **93**, 957–968.
51 Xue, Y., Yap, C.W., Sun, L.Z., Cao, Z.W., Wang, J.F. and Chen, Y.Z. (2004) Prediction of P-glycoprotein substrates by a support vector machine approach. *Journal of Chemical Information and Computer Sciences*, **44**, 1497–1505.
52 Wang, Y.-H., Li, Y., Yang, S.-L. and Yang, L. (2005) Classification of substrates and inhibitors of P-glycoprotein using unsupervised machine learning approach. *Journal of Chemical Information and Computer Sciences*, **45**, 750–757.
53 Cianchetta, G., Singleton, R.W., Zhang, M., Wildgoose, M., Giesing, D., Fravolini, A., Cruciani, G. and Vaz, R.J. (2005) A pharmacophore hypothesis for P-glycoprotein substrate recognition using GRIND-based 3D-QSAR. *Journal of Medicinal Chemistry*, **48**, 2927–2935.
54 Crivori, P., Reinach, B., Pezzetta, D. and Poggesi, I. (2006) Computational models for identifying potential P-glycoprotein substrates and inhibitors. *Molecular Pharmaceutics*, **3**, 33–44.
55 Bisi, A., Gobbi, S., Rampa, A., Belluti, F., Piazzi, L., Valenti, P., Gyemant, N. and Molnar, J. (2006) New potent P-glycoprotein inhibitors carrying a polycyclic scaffold. *Journal of Medicinal Chemistry*, **49**, 3049–3051.
56 Seigneuret, M. and Garnier-Suillerot, A. (2003) A structural model for the open conformation of the mdr1 P-glycoprotein based on the MsbA crystal structure. *Journal of Biological Chemistry*, **278**, 30115–30124.
57 Chang, G. and Roth, C.B. (2001) Structure of MsbA from *E. coli*: a homolog of the multidrug resistance ATP binding cassette (ABC) transporters. *Science*, **293**, 1793–1800.
58 Stenham, D.R., Campbell, J.D., Sansom, M.S.P., Higgins, C.F., Kerr, I.D. and Linton, K.J. (2003) An atomic detail model for the human ATP binding cassette transporter P-glycoprotein derived from disulphide cross-linking and homology modelling. *FASEB Journal*, **17**, 2287–2289.
59 Locher, K.P., Lee, A.T. and Rees, D.C. (2002) The *E. coli* BtuCD structure: a framework for ABC transporter architecture and mechanism. *Science*, **296**, 1091–1098.
60 Pajeva, I.K., Globisch, C. and Wiese, M. (2004) Structure–function relationships of multidrug resistance P-glycoprotein. *Journal of Medicinal Chemistry*, **47**, 2523–2533.
61 Loo, T.W. and Clarke, D.M. (1999) Determining the structure and mechanism of the human multidrug resistance P-glycoprotein using cysteine-scanning mutagenesis and thiol-modification techniques. *Biochimica et Biophysica Acta*, **1461**, 315–325.
62 Loo, T.W. and Clarke, D.M. (1999) Identification of residues in the drug binding domain of human P-glycoprotein. *Journal of Biological Chemistry*, **274**, 35388–35392.

63 Loo, T.W. and Clarke, D.M. (2000) Identification of residues within the drug-binding domain of the human multidrug resistance P-glycoprotein by cysteine-scanning mutagenesis and reaction with dibromobimane. *Journal of Biological Chemistry*, **275**, 39272–39278.

64 Loo, T.W. and Clarke, D.M. (2001) Determining the dimensions of the drug-binding domain of human P-glycoprotein using thiol crosslinking compounds as molecular rulers. *Journal of Biological Chemistry*, **276**, 36877–36880.

65 Loo, T.W. and Clarke, D.M. (2001) Defining the drug-binding site in the human multidrug resistance P-glycoprotein using a methanethiosulfonate analog of verapamil, MTS-verapamil. *Journal of Biological Chemistry*, **276**, 14972–14979.

66 Loo, T.W. and Clarke, D.M. (2002) Vanadate trapping of nucleotide at the ATP-binding sites of human multidrug resistance P-glycoprotein exposes different residues to the drug-binding site. *Proceedings of the National Academy of Sciences of the United States of America*, **99**, 3511–3516.

67 Loo, T.W. and Clarke, D.M. (2002) Location of the rhodamine-binding site in the human multidrug resistance P-glycoprotein. *Journal of Biological Chemistry*, **277**, 44332–44338.

68 Loo, T.W., Bartlett, M.C. and Clarke, D.M. (2004) Val133 and Cys137 in transmembrane segment 2 are close to Arg935 and Gly939 in transmembrane segment 11 of human P-glycoprotein. *Journal of Biological Chemistry*, **279**, 18232–18238.

69 Loo, T.W., Bartlett, M.C. and Clarke, D.M. (2003) Substrate-induced conformational changes in the transmembrane segments of human P-glycoprotein. *Journal of Biological Chemistry*, **278**, 13603–13606.

70 Vandevuer, S., Van Bambeke, F., Tulkens, P.M. and Prevost, M. (2006) Predicting the three-dimensional structure of human P-glycoprotein in absence of ATP by computational techniques embodying crosslinking data: insight into the mechanism of ligand migration and binding sites. *Proteins*, **63**, 466–478.

71 Loo, T.W., Bartlett, M.C. and Clarke, D.M. (2002) The LSGGQ motif in each nucleotide-binding domain of human P-glycoprotein is adjacent to the opposing walker A sequence. *Journal of Biological Chemistry*, **277**, 41303–41306.

72 Urbatsch, I.L., Gimi, K., Wilke-Mounts, S., Lerner-Marmarosh, N., Rousseau, M.E., Gros, P. and Senior, A.E. (2001) Cysteines 431 and 1074 are responsible for inhibitory disulfide cross-linking between the two nucleotide-binding sites in human P-glycoprotein. *Journal of Biological Chemistry*, **276**, 26980–26987.

73 Ho, R.H. and Kim, R.B. (2005) Transporters and drug therapy: implications for drug disposition and disease. *Clinical Pharmacology and Therapeutics*, **78**, 260–277.

74 Lee, W. and Kim, R.B. (2004) Transporters and renal drug elimination. *Annual Review of Pharmacology and Toxicology*, **44**, 137–166.

75 Mizuno, N., Niwa, T., Yotsumoto, Y. and Sugiyama, Y. (2003) Impact of drug transporter studies on drug discovery and development. *Pharmacological Reviews*, **55**, 425–461.

76 Saito, H., Hirano, H., Nakagawa, H., Fukami, T., Oosumi, K., Murakami, K., Kimura, H., Kouchi, T., Konomi, M., Tao, E., Tsujikawa, N., Tarui, S., Nagakura, M., Osumi, M. and Ishikawa, T. (2006) A new strategy of high-speed screening and quantitative structure–activity relationship analysis to evaluate human ATP-binding cassette transporter ABCG2–drug interactions. *Journal of Pharmacology*

and Experimental Therapeutics, **317**, 1114–1124.

77 Bednarczyk, D., Ekins, S., Wikel, J.H. and Wright, S.H. (2003) Influence of molecular structure on substrate binding to the human organic cation transporter, hOCT1. *Molecular Pharmacology*, **63**, 489–498.

78 Chang, C., Pang, K.S., Swaan, P.W. and Ekins, S. (2005) Comparative pharmacophore modeling of organic anion transporting polypeptides: a meta-analysis of rat Oatp1a1 and human OATP1B1. *Journal of Pharmacology and Experimental Therapeutics*, **314**, 533–541.

IV
Case Studies of Drug Optimization Against Antitargets

Antitargets. Edited by R. J. Vaz and T. Klabunde
Copyright © 2008 WILEY-VCH Verlag GmbH & Co. KGaA, Weinheim
ISBN: 978-3-527-31821-6

17
Selective Dipeptidyl Peptidase IV Inhibitors for the Treatment of Type 2 Diabetes: The Discovery of JANUVIA™ (Sitagliptin)
Scott D. Edmondson, Dooseop Kim

17.1
Introduction

Type 2 diabetes is often associated with metabolic syndrome and is characterized by chronic hyperglycemia. It can lead to serious vascular complications, significant comorbidities, such as stroke or coronary heart disease, and mortality. The rapid growth in the incidence of diabetes worldwide prompted the United Nations to recognize the threat of diabetes as a worldwide epidemic in December 2006 [1]. Although current type 2 diabetes therapies have proven therapeutically beneficial, they often possess undesirable effects such as hypoglycemia, weight gain and inadequate durability [2]. Accordingly, there is a significant unmet medical need for the treatment of this disease. Inhibition of dipeptidyl peptidase IV (DPP-4), a serine protease, has recently emerged as a potential approach for the treatment of type 2 diabetes [3]. DPP-4 inhibitors indirectly stimulate insulin secretion by enhancing the action of the incretin hormones glucagon-like peptide 1 (GLP-1) and glucose-dependent insulinotropic polypeptide (GIP) [4,5]. GLP-1 and GIP are secreted by the gut in response to the ingestion of nutrients, and each stimulates the secretion of insulin in a glucose-dependent manner, thus posing little or no risk of hypoglycemia. Additionally, GLP-1 stimulates insulin biosynthesis and inhibits the release of glucagon [5]. Other beneficial effects of GLP-1 include delayed gastric emptying [6] and reduction of appetite [7], and GLP-1 receptor agonists have been shown to induce weight loss in humans [8]. Recently, rodent data suggest a potential role for GLP-1 in the restoration of β-cell function, indicating that this mechanism might even slow down or reverse the progression of the disease [9]. GLP-1 and GIP are both rapidly degraded *in vivo* through the action of DPP-4, which cleaves a dipeptide from the N-terminus to give the inactive truncated peptides [10]. Inhibition of DPP-4 thus increases the half-life of GLP-1 and GIP and prolongs the beneficial effects of these incretins on blood glucose homeostasis. Compound **1**, JANUVIA (sitagliptin phosphate) [11], a DPP-4 inhibitor recently approved by the Food and Drug Administration (FDA), is a potent, selective and orally active antidiabetic agent

Antitargets. Edited by R. J. Vaz and T. Klabunde
Copyright © 2008 WILEY-VCH Verlag GmbH & Co. KGaA, Weinheim
ISBN: 978-3-527-31821-6

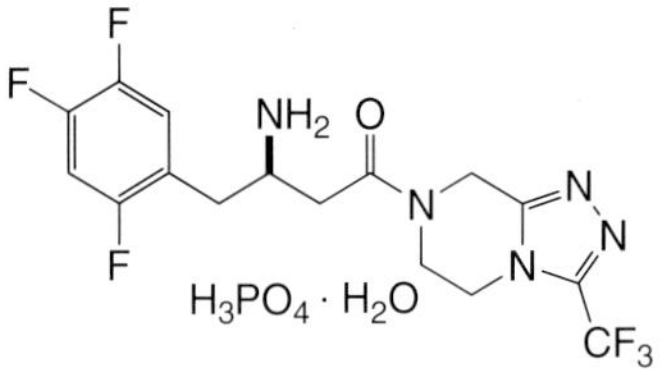

Figure 17.1 Compound **1**, JANUVIA™ (sitagliptin phosphate).

that provides a new treatment option for patients with type 2 diabetes (Figure 17.1) [12]. In this chapter, our efforts at optimization of DPP-4 activity, selectivity, pharmacokinetic profile and *in vivo* efficacy will be described for two structurally distinct classes of DPP-4 inhibitors. Ultimately, this work led to the discovery of sitagliptin (**1**).

17.2 Selectivity of DPP-4 Inhibitors

At the outset of the DPP-4 program at Merck, several potential safety concerns were identified. First, DPP-4 had been shown to be identical to the T-cell activation marker CD26, which was under investigation for potential applications in immunology [13]. Indeed, early pyrrolidide derived DPP-4 inhibitors were reported to possess immune system effects including inhibition of the proliferation of immune cells [14]. Second, DPP-4 had been shown to cleave multiple peptides *in vitro*, which possess a broad range of immunological, neurological and endocrine functions [15]. As DPP-4 is a cell surface peptidase that is ubiquitously expressed, a possibility existed that DPP-4 inhibition might affect a broad range of potential biological functions. On the contrary, DPP-4-deficient mice described in the literature were reported to develop normally and appeared to be healthy [16,17]. Moreover, two companies demonstrated sufficient safety profiles to introduce their DPP-4 inhibitors into clinical trials, including Novartis with NVP-DPP-728 (**2**) [18] and Probiodrug with P32/98 (**3**) (Figure 17.2) [19].

In 2000, Merck in-licensed *threo* (**3**, P32/98) and *allo* (**4**) isoleucine thiazolidides from Probiodrug so as to fast forward the internal program and increase our understanding of DPP-4 inhibition. These two compounds are equipotent against DPP-4 ($IC_{50} = 440$ nM for **3** and 460 nM for **4**), and each compound showed similar *in vivo* efficacies in oral glucose tolerance tests (OGTT) administered to diet-induced

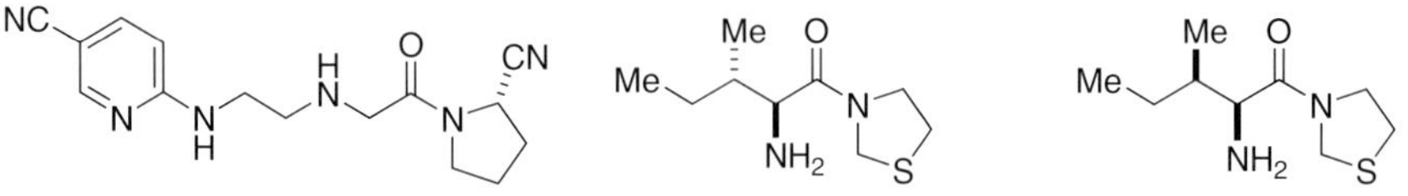

NVP-DPP728 (**2**) P32/98, *threo*-Ile-thiazolidide (**3**) *allo*-Ile-thiazolidide (**4**)

Figure 17.2 Early DPP-4 inhibitors.

obese (DIO) mice [20]. Additionally, Probiodrug demonstrated that **3** reduced glycemic excursion and was well tolerated in a single dose phase 1 clinical trial in healthy volunteers [21]. Furthermore, Probiodrug reported that single-dose administration of **3** enhanced insulin secretion and improved glucose tolerance in diabetic patients [22].

Probiodrug evaluated **3** in 4-week toxicity studies in rats and dogs to support preclinical safety studies [20]. In rats, lung histiocytosis and thrombocytopenia were observed at relatively high doses (77.5 and 698 mg/kg, respectively); while in dogs, acute CNS toxicities such as ataxia, seizures, convulsions and tremors were observed at the dose of 75 mg/kg. Furthermore, bloody diarrhea was observed in dogs treated with 225 mg/kg of **3** after acute dosing. Longer chronic toxicity studies were performed by Merck; however, **3** caused mortality and profound toxicities in dogs at doses $\geq$25 mg/kg/day after 5–6 weeks of dosing. Parallel toxicity studies with the *allo* diastereomer **4** revealed bloody diarrhea in dogs after a single dose that was more than 10-fold lower than that of **3**. In a 4-week rat toxicity study with **4**, lung histiocytosis and thrombocytopenia were also observed, but again the *allo* compound was toxic at more than 10-fold lower dosage than the *threo* diastereomer **3**. Moreover, the other toxicities observed in dogs with **3** (anemia, splenomegaly and mortality with multiple organ pathology) were observed with **4** in rats in the 4-week toxicity study. As a result of these toxicities, Merck discontinued the development of both compounds in 2001.

The observation that **4** was more than 10-fold more toxic than **3** suggested that some factor other than DPP-4 inhibition was responsible for these effects. Each of these compounds is equipotent against DPP-4 *in vitro* and each compound afforded a similar pharmacokinetics, drug metabolism and *in vivo* efficacy profile. Taken together, these data did not support a hypothesis that DPP-4 was responsible for the above-observed toxicities in rats and dogs. Instead, a hypothesis that better explains the above data is that **3** and **4** possess some off-target activities that might lead to toxicity. Support for this hypothesis was initially obtained from studies using tissue extracts from DPP-4-deficient mice. Using a Gly-Pro-AMC fluorogenic substrate, low levels of proline-specific dipeptidyl peptidase activity were identified in kidney, liver, lung and gastrointestinal tract tissue extracts from these DPP-4-deficient mice. Unlike DPP-4, however, the ability of these tissue extracts to cleave Gly-Pro-AMC was differentially inhibited by compound **3** relative to compound **4** (IC_{50} = 726 and 86 nM, respectively) [20]. As **3** and **4** exhibit virtually identical DPP-4 IC_{50}'s using the same fluorogenic substrate *in vitro*, off-target enzymes (especially DPP-4 like proline selective peptidases) were sought that might help explain the above discrepancies between these two compounds.

Off-target activities of **3** and **4** were initially assessed by screening against an in-house series of proteases and a panel of 170 receptor, enzyme and ion channel assays (MDS Pharma Services). With the exception of a weak binding interaction of **4** at the sigma σ_1 receptor (K_i = 42 µM), the two compounds showed no significant activity (IC_{50}'s > 100 µM) against the in-house protease panel or the MDS Pharma Services assays. Recently, however, a series of DPP-4-like activity and/or structural homologue (DASH) proteins were described, which are defined as serine dipeptidyl peptidases unified by a common mechanism of postproline cleavage [23]. Enzymes that belong to the DASH protein family include quiescent cell proline dipeptidase (QPP, also known as DPP-2 and DPP-7) [24], DPP-8 [25], DPP-9 [26] and fibroblast activation protein (FAP) [27]. Since the functions of many of these enzymes are not known, in-house assays were developed that provided a method to counterscreen against these potential 'antitargets'.

When compounds **3** and **4** were screened against the DASH protein panel, inhibition was observed against QPP, DPP-8 and DPP-9 in addition to DPP-4 [20]. While both **3** and **4** possess a similar level of potency against QPP (IC$_{50}$'s = 18 and 14 µM, respectively), they displayed divergent activities against DPP-8 and DPP-9. Indeed, the *allo* compound **4** is 10-fold more potent than the *threo* compound **3** at DPP-8 (IC$_{50}$ = 220 nM for **4** versus 2180 nM for **3**) and fivefold more potent at DPP-9 (IC$_{50}$'s = 320 nM for **4** versus 1600 nM for **3**). The greater potency of **4** at DPP-8/9 relative to **3** correlated well with the differences in dose (and exposure) necessary to produce toxicity by these two compounds. We therefore hypothesized that inhibition of DPP-8 and/or DPP-9 was responsible for the observed toxicities of these compounds in preclinical species.

To gather support for this hypothesis, a series of compounds that are potent and specific inhibitors of DPP-4 (**5**), QPP (**6**) and DPP-8/9 (**7**) were identified. Each of these compounds was evaluated in 2-week rat toxicity studies and acute dog tolerability studies (Table 17.1) [20]. Toxicity observed with the potent and selective DPP-8/9 inhibitor **7** was remarkably similar to toxicity noted with the *allo* diastereomer **4** from

Table 17.1 Comparative toxicities in rats (2 weeks of treatment at doses of 10, 30 and 100 mg/kg/day) and dogs (single dose, 10 mg/kg PO) with selective inhibitors **5**, **6** and **7**.

5, DPP-4 selective
DPP-4 IC$_{50}$ = 27 nM

6, QPP selective
QPP IC$_{50}$ = 19 nM

7, DPP-8/9 selective
DPP-8 IC$_{50}$ = 38 nM
DPP-9 IC$_{50}$ = 55 nM

Species	Toxicity	DPP-4 (5)	QPP (6)	DPP-8/9 (7)	threo (3)	allo (4)
Rat	Alopecia		√			√
	Thrombocytopenia		√	√	√	√
	Anemia					√
	Reticulocytopenia	√	√	n.d.^{a}		n.d.
	Splenomegaly			√		√
	Mortality			√		√
Dog	Bloody diarrhea			√	√	√

A check (√) indicates that toxicity was observed. Historical data from safety studies with **3** and **4** are shown for comparison.
^{a}Not determined.

previous studies. Administration of DPP-8/9 inhibitor **7** to rats produced alopecia, thrombocytopenia, reticulocytopenia, enlarged spleen, multiorgan histopathological changes and mortality. Furthermore, the DPP-8/9 inhibitor produced gastrointestinal toxicity in dogs. In contrast, the selective QPP inhibitor **6** produced reticulocytopenia in rats only, and the selective DPP-4 inhibitor **5** produced no toxic effects in either species. Further evidence that DPP-4 was not responsible for toxicity was acquired from a separate 2-week toxicity study in which administration of the DPP-8/9 inhibitor **7** to wild-type and DPP-4-deficient mice resulted in similar toxicities in both species [20].

The finding that similar multiorgan toxicities in preclinical species were linked to two structurally diverse dual inhibitors of DPP-8 and DPP-9 provided a compelling argument that inhibition of one or both of these enzymes was responsible for toxicities in preclinical species. This explanation was reinforced by the demonstration that **7** is selective for DPP-8/9 inhibition over a panel of proline-specific enzymes and that the only identifiable biological activity of **4** is limited to DPP-4, DPP-8/9 and weak QPP inhibition. Furthermore, the difference in the dosage and drug exposure levels of **3** and **4** that were required for toxicity in dogs and rats correlated with the differences between the relative potency of inhibition against DPP-8/9. Finally, DPP-4 could be ruled out as a causative agent for the toxicities because DPP-4 inhibitor **5** did not induce toxic effects in rats or dogs and further because the DPP-8/9 inhibitor **7** afforded similar toxicities in both DPP-4-deficient and wild-type mice. Taken together, the above lines of evidence provided a strong rationale for the identification of DPP-8 and/or DPP-9 as the 'antitargets' responsible for the multiorgan toxicities observed with nonselective DPP-4 inhibitors **3** and **4**.

As some of the toxicity observed with DPP-8/9 inhibitors **4** and **7** in preclinical species suggested a potential immune system role, we hypothesized that the immunological effects observed with historical DPP-4 inhibitors [14] might be due to inhibition of DPP-8/9 instead of DPP-4 as initially reported. Sure enough, when these historical compounds were assayed at DPP-8/9 and DPP-4, they possessed more potent intrinsic inhibition at DPP-8/9 than at DPP-4 [20]. Furthermore, we demonstrated that DPP-8/9 inhibitor **7** is able to attenuate proliferation and IL-2 release in human *in vitro* models of T-cell activation, while a selective DPP-4 inhibitor does not. Recent tissue distribution studies also suggest a role for DPP-8 in the immune system [25].

In summary, off-target inhibition of DPP-8 and/or DPP-9 with compounds **3**, **4** and **7** led to significant multiorgan toxicity in preclinical species. Although not all of the biological functions of DPP-8/9 are known, these enzymes have been linked to cell adhesion, migration and apoptosis [28]. The relevance of DPP-8/9 inhibition to humans is also not yet known, and it is important to point out that both DPP-8 and DPP-9 are reported to be intracellular enzymes [25,29]. In contrast, DPP-4 is a ubiquitous integral membrane protein that is also expressed as a catalytically active extracellular soluble form that lacks a transmembrane domain. Although it is not clear what degree of inhibition of DPP-8/9 might contribute to the above-described toxicities, Merck made a strategic decision to avoid inhibiting these antitargets. Consequently, the objective of Merck's DPP-4 inhibitor program focused on the identification of compounds that are at least 1000-fold more selective for DPP-4 inhibition than for DPP-8 and DPP-9.

In addition to the DASH proteins, a number of other off-target screens were monitored during Merck's DPP-4 medicinal chemistry program. Specifically, binding at cardiac ion channels was routinely tested and compounds were optimized for selectivity over these ion channels. The hERG potassium channel was viewed as particularly important because binding to hERG has been linked to an increased risk for ventricular arrhythmia (i.e. *torsades de pointes*), which can lead to sudden death [30]. Accordingly, the medicinal chemistry program sought to optimize for DPP-4 inhibition over hERG ion channel binding in addition to DPP-8, DPP-9 and the other DASH proteins.

17.3
α-Amino Acid Amide Series

Medicinal chemistry was initiated on the DPP-4 program prior to learning about the toxicity related to DPP-8/9 inhibition and before completion of a high-throughput screen (HTS) for DPP-4 inhibitors. The earliest leads on the program were therefore derived from literature reports. As DPP-4 preferentially cleaves peptides with proline at the P-1 position, a common motif among early DPP-4 inhibitors was a pyrrolidine-bearing electrophile at the P-1 site linked to an α-amino acid at the P-2 site. Some of these electrophiles are irreversible covalent inhibitors (phosphonates), while others are reversible (nitriles, boronic acids). These covalent inhibitors pose formidable challenges, including the potential for poor chemical stability due to the propensity of the free amine group to cyclize with the electrophile to form a six-membered ring [31]. Even though covalent inhibitors were generally reported to be more potent than their noncovalent counterparts, we pursued an early strategy to focus on optimization of noncovalent DPP-4 inhibitors. One way to assess potent and stable noncovalent inhibitors would be to optimize for increased potency at the P-2 substituent of α-amino acid inhibitors that lack reactive electrophiles.

17.3.1
Cyclohexylglycines and Related Derivatives

(S)-Cyclohexylglycine pyrrolidide **8**, which did not possess a reactive electrophile, was an early DPP-4 inhibitor reported by Ferring [32]. A strategy to modify the P-2 position of this inhibitor lead led to the introduction of nitrogen-linked amides, carbamates and sulfonamides at the 4 position of the cyclohexane ring (Table 17.2). In particular, aryl sulfonamides such as **9** and **10** increased DPP-4 potency relative to **8**, showed good selectivity over hERG and QPP, but possessed moderate bioavailability in rats (F_{rat} = 36 and 38%, respectively) [33].

To further increase bioavailability in this series, modification of the pyrrolidine moiety was investigated. Incorporation of fluorine into the pyrrolidine ring led to the discovery that the (S)-3-fluoropyrrolidide and 3,3-difluoropyrrolidide derivatives

17.3 α-Amino Acid Amide Series

Table 17.2 Off-target activities of cyclohexylglycine DPP-4 inhibitors (IC$_{50}$'s, nM).

Compound **8**: DPP-4 IC$_{50}$ = 320 nM

Compound	R	X	DPP-4	QPP	DPP-8	DPP-9	hERG
9	2,4-diF	H	88	8800	750	3300	35 000
10	4-CF$_3$O	H	89	6400	1900	2300	5700
11	2,4-diF	F	48	12 000	990	2700	49 000
12	4-CF$_3$O	F	35	8300	1400	1700	5500

afforded compounds with equivalent or increased potency against DPP-4 compared to the free pyrrolidide [34]. When compared to **9** and **10**, the (S)-3-fluoropyrrolidide derivatives **11** and **12** showed increased potency at DPP-4 and similar bioavailabilities in rats (F_{rat} = 53 and 37%, respectively). Compound **12** was further found to possess an excellent pharmacokinetic profile in dogs (F = 89%, $t_{1/2}$ = 12 h) and rhesus monkeys (F = 64%, $t_{1/2}$ = 5 h). Unfortunately, when assays for DPP-8 and DPP-9 became available, all of the cyclohexylglycine derivatives were found to display unacceptable (<1000-fold) selectivity over these enzymes.

Further efforts in related series afforded structurally similar DPP-4 inhibitors such as acid **13** (E. Parmee, unpublished results), thiazole **14** [35] and cyclopentylglycine **15** (Figure 17.3) [36]. Although **15** did not show improved selectivity over DPP-8/9, **13** and **14** showed some improvement in selectivity over these counterscreens. Nevertheless, none of these compounds exhibited sufficient selectivity to merit further pursuit; zwitterion **13** also possesses low oral bioavailability in rats (F < 1%). Consequently, focus in the α-amino acid series shifted to acyclic derivatives.

17.3.2
β-Substituted Phenylalanines as Potent and Selective DPP-4 Inhibitors

Although the development of diastereomers **3** and **4** was discontinued, these compounds nevertheless provided a valuable hint on how to improve selectivity over DPP-8/9 in the α-amino acid series. By switching the β-chiral center adjacent to the amine group from a 'syn' to an 'anti' configuration, the selectivity for DPP-8 and DPP-9 improved 5–10-fold. Since modification of the P-2 position succeeded in improving DPP-4 potency in the cyclohexylglycines, we wondered if modification of the ethyl group of **3** would also lead to improved potency and/or selectivity. Substitution of the ethyl with a phenyl (**16**) resulted in a loss of potency at DPP-4, DPP-8 and DPP-9, but 4-arylation of the phenyl group selectively improved DPP-4 potency (Table 17.3) [37]. Fluorophenyl derivative **17** possessed a promising rat pharmacokinetic profile (F = 85%, $t_{1/2}$ = 2.2 h), but activity at the hERG potassium

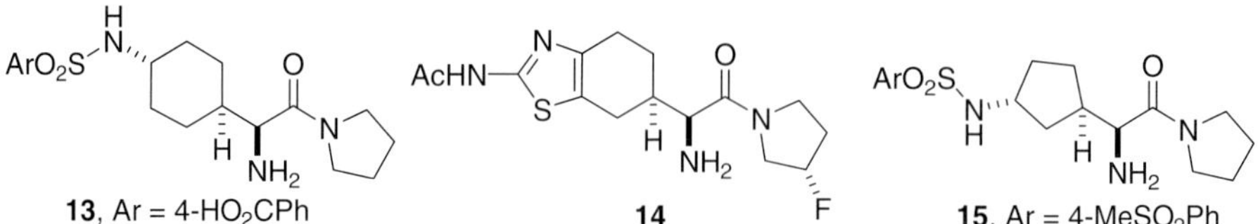

Figure 17.3 Potent cyclohexylglycine and cyclopentylglycine DPP-4 inhibitors.

13, Ar = 4-HO$_2$CPh
DPP-4 IC$_{50}$ = 15 nM
DPP-8 IC$_{50}$ = 2800 nM
DPP-9 IC$_{50}$ = 6200 nM

14
DPP-4 IC$_{50}$ = 13 nM
DPP-8 IC$_{50}$ = 3900 nM
DPP-9 IC$_{50}$ = 930 nM

15, Ar = 4-MeSO$_2$Ph
DPP-4 IC$_{50}$ = 13 nM
DPP-8 IC$_{50}$ = 570 nM
DPP-9 IC$_{50}$ = 880 nM

channel and moderate DPP-4 potency prohibited further development of this compound. To improve DPP-4 potency and hERG selectivity, we sought to replace the fluorophenyl moiety with a more polar heterocycle.

The SAR of fluorophenyl substitutes revealed that a hydrogen bond acceptor at the 4 position was important for maintaining or improving DPP-4 potency and selectivity, but the introduction of polar substituents often resulted in inferior rat pharmacokinetic properties [37]. Pyridone **18** showed promise, however, with modest improvements in DPP-4 potency and hERG selectivity [38]. As **18** was still suboptimal with respect to hERG binding and rat oral bioavailability (F_{rat} = 10%), further modification of the pyridone ring was pursued and ultimately led to the identification of methyl pyridone **19**. This compound showed improved selectivity over hERG and an improved pharmacokinetic profile in rats, dogs and rhesus monkeys ($t_{1/2} \geq 1.4$ h, $F \geq 34\%$). Unfortunately, in vitro and in vivo metabolism studies of **19** revealed that N-demethylation occurs at the pyridone ring to give the free pyridone **18** [38]. The formation of an active metabolite that possessed an unacceptable level of hERG binding thus led to the suspension of any further development of **19**.

As the introduction of polar aryl groups into phenylalanine analogues afforded improvements in DPP-4 potency and selectivity, alternative locations to introduce polarity into this series were examined. Replacement of the β-methyl group with a

Table 17.3 Potency and off-target activities of β-methyl phenylalanine-based DPP-4 inhibitors (IC$_{50}$'s, nM).

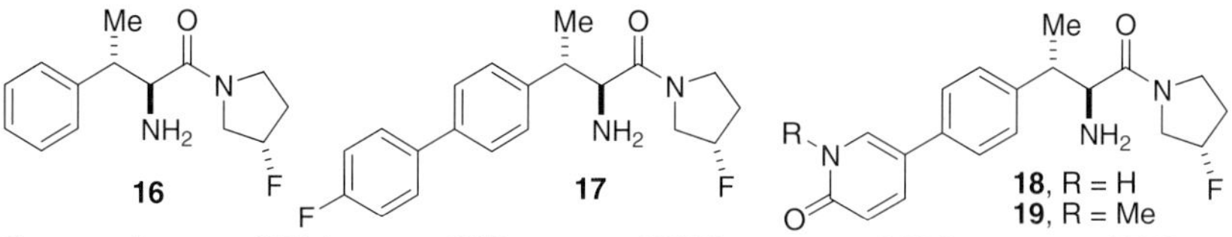

Compound	DPP-4	QPP	DPP-8	DPP-9	hERG
16	1400	59 000	>100 000	>100 000	—
17	64	2700	88 000	86 000	1100
18	25	5000	>100 000	>100 000	8300
19	34	8000	>100 000	>100 000	>100 000

17.3 α-Amino Acid Amide Series

Table 17.4 Potency and off-target activities of phenylalanine-based DPP-4 inhibitors (IC$_{50}$'s, nM).

Compound	DPP-4	QPP	DPP-8	DPP-9	hERG
20	6.6	>100 000	>100 000	>100 000	76 000
21	12	45 000	>100 000	69 000	4600
22	8.0	>100 000	>100 000	>100 000	83 000
23	4.3	>100 000	>100 000	>100 000	86 000
24	8.8	100 000	>100 000	>100 000	>100 000

Structures: **20**, X = OH; **21**, X = NMe$_2$; **22**; **23**, X = H; **24**, X = F.

carboxylic acid (**20**) or a dimethylamide (**21**) resulted in greater than fivefold improvements both in potency at DPP-4 and in selectivity over DPP-8/9 and hERG (Table 17.4) [39]. While acid **20** showed an excellent overall off-target selectivity profile relative to **21**, the oral bioavailability in rats ($F = 16\%$) and dogs ($F = 45\%$) was still suboptimal. Conversely, dimethylamide **21** exhibited an excellent pharmacokinetic profile in rats, dogs and rhesus monkeys ($t_{1/2} \geq 3.5$ h, $F \geq 56\%$) but possessed suboptimal selectivity over hERG binding. Further introduction of polarity by replacing the fluorophenyl group with a methylpyridone (**22**) increased hERG selectivity but had a negative impact on pharmacokinetic properties (rat $t_{1/2} = 1.2$ h, $F = 2\%$) [40]. Moreover, **22** still possessed the metabolically labile N-methylpyridone moiety. By fusing five-membered ring heterocycles to a pyridine ring, we were able to generate a series of metabolically stable bioisosteres to the N-methylpyridone moiety to ultimately lead to triazole **23**. Not only did **23** exhibit excellent DPP-4 potency and off-target selectivity, but this triazole also showed significant improvements in the rat pharmacokinetic profile ($t_{1/2} = 2.0$ h, $F = 43\%$) relative to **22**.

An X-ray crystal structure of triazole **23** bound to the active site of DPP-4 shows significant overlap with a previously reported DPP-4 inhibitor, valine pyrrolidide (DPP-4: IC$_{50}$ = 1580 nM; DPP-8/9: IC$_{50}$'s = 3300 nM; Figure 17.4a) [41]. Each pyrrolidine moiety from these inhibitors occupies the S1 hydrophobic pocket and the pyrrolidide carbonyls and primary amines of these two compounds interact with Asn710, Glu205 and Glu206 in the binding pocket. By extending into the S2 pocket, however, **23** can interact with Tyr547 via a hydrogen bond between the tyrosine ring and the dimethylamide carbonyl. This particular interaction offers an improvement of roughly fivefold in DPP-4 potency. The fused triazole ring stacks against the side chain of Phe357 and hydrogen bonds with the side chain of Arg358, also enhancing inhibition of DPP-4. Indeed, inhibitors that possess functional groups within hydrogen bonding distance of Arg358 appear to offer a substantial boost to intrinsic DPP-4 inhibition. In contrast, SAR shows that DPP-8 and DPP-9 inhibition is not enhanced when a compound is able to interact with Tyr547, Phe357 and Arg358, thus

(a)

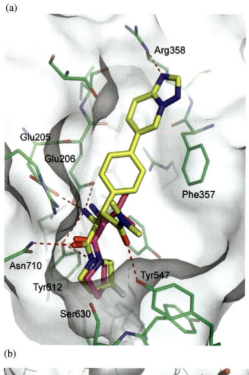

(b)

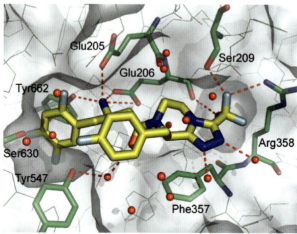

Figure 17.4 (a) Compound **23** bound to DPP-4. The overlay of compound **23** (yellow) and the substrate analogue valine pyrrolidide (magenta) shows a similar orientation between the two compounds. Interactions of compound **23** with DPP-4 are shown as red dotted lines. (b) Compound **48** bound to DPP-4. Interactions of compound **48** with DPP-4 are shown as red dotted lines.

suggesting that one method for building selectivity for DPP-4 inhibition over DPP-8 and DPP-9 is to incorporate interactions with these residues in the binding pocket of DPP-4 [42,43].

Further evaluation of **23** revealed a potential for metabolic activation at the fluoropyrrolidine ring, which was solved by adding a second fluorine atom to the pyrrolidine ring to furnish compound **24** (Table 17.4). Difluoropyrrolidide **24** exhibited excellent selectivity (IC_{50}'s $\geq$ 100 000 nM) over the DASH proteins, hERG binding, a panel of in-house proteases, and a series of cytochrome P450 enzymes. Furthermore, **24** showed high selectivity (IC_{50}'s > 10 000 nM) over other cardiac ion channels and an extensive panel of receptors, enzymes and ion channels (MDS Pharma Services). In four separate preclinical species, **24** displayed an excellent pharmacokinetic profile ($t_{1/2} \geq 1.6$ h, $F \geq 56\%$ in rats, mice, dogs and monkeys), and this compound was subsequently evaluated in a series of *in vivo* efficacy models [40]. In lean mice, for example, **24** was effective in lowering blood glucose levels at oral doses as low as 0.1 mg/kg (relative to vehicle) after a glucose challenge was administered. In summary, compound **24** represented the optimal balance of DPP-4 potency, off-target selectivity, drug metabolism profile, pharmacokinetic properties and *in vivo* efficacy in the phenylalanine class of DPP-4 inhibitors.

17.4
Early β-Amino Acid Amide DPP-4 Inhibitors

Efforts to find potent and selective DPP-4 inhibitors in the α-amino acid amide series were made in parallel with those in the β-amino acid amides series. The structural origin of the earliest β-amino acid amide DPP-4 inhibitors traces back to two Merck HTS hits: proline derivative **25** and piperazine derivative **26**. These two screening leads were further progressed to β-amino acid amide DPP-4 inhibitors incorporating thiazolidine, proline and piperazine amide moieties (Figure 17.5).

Thiazolidides **27–29** and proline amides **30** and **31** derive from the proline amide HTS hit **25**. In the course of the SAR development of the thiazolidides, an interesting fluorine effect on DPP-4 activity was observed [44]. The addition of a fluorine atom at the 2-position of the phenyl ring of **27** increased DPP-4 potency roughly threefold (Table 17.5). Sequential addition of two more fluorine atoms at the 4- and 5-positions of the phenyl ring further increased potency to give rise to **29**, the most potent thiazolidide in this series. The 2,4,5-trifluorophenyl ring later became one of the key features incorporated into the design of sitagliptin. In general, most of the thiazolidides exhibited good selectivity against QPP, but selectivity over DPP-8 was suboptimal. Moreover, compound **29** suffered from high clearance, a short half-life and poor oral bioavailability ($F = 3\%$) in the rat due to oxidative metabolism of the thiazolidine ring. Extensive optimization of the proline hit **25** provided potent zwitterion **30** with high DPP-4 selectivity over the DASH protein counterscreens [45]. Incorporation of fluorine atoms into the left side phenyl again increased DPP-4 potency to give the most potent DPP-4 inhibitor (**31**) in this series. Although these potent zwitterionic DPP-4 inhibitors exhibited excellent DPP-4 potency together with

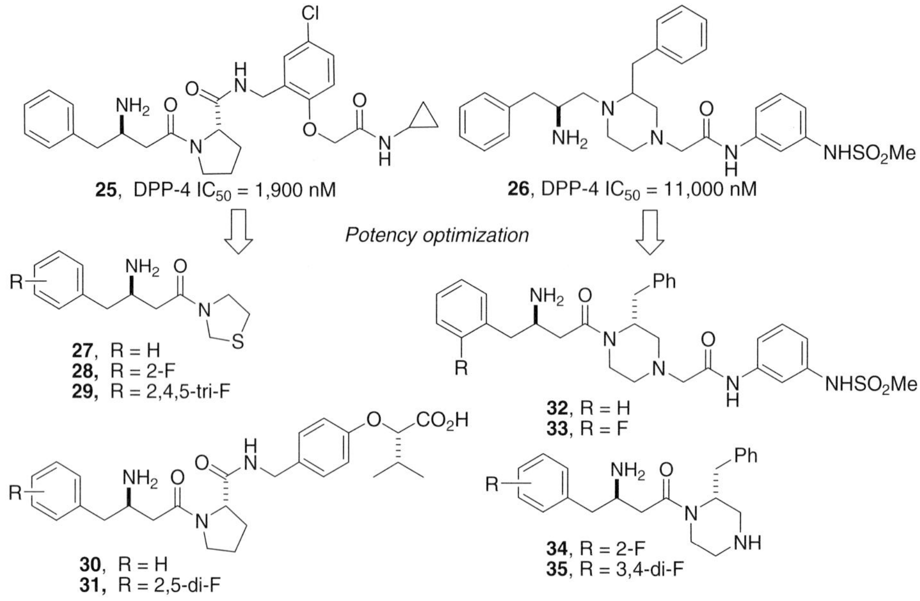

Figure 17.5 HTS hits and early β-amino acid acid derived DPP-4 inhibitors.

high selectivity over DPP-8, DPP-9 and cardiac ion channels, they typically possessed poor oral absorption ($F = 1\%$), which precluded further development.

Incorporation of the β-amino acid amide from **25** into piperazine hit **26** improved DPP-4 inhibition by more than 80-fold and led to the discovery of **32** (Table 17.5) [46]. The 2-fluorophenyl effect on DPP-4 potency was again observed with compound **33**; however, this compound showed very high clearance and no oral bioavailability in the rat. More simplified truncated compounds **34** and **35** showed decreased DPP-4 potency by 10-fold compared to **33**, and the removal of the benzyl group resulted in a more significant loss of DPP-4 activity (**36**). Unfortunately, while **34** showed good

Table 17.5 Off-target activities of early β-amino acid DPP-4 inhibitors (IC_{50}'s, nM).

Compound	DPP-4	QPP	DPP-8	DPP-9	hERG
27	3000	>100 000	2000	23 200	—
28	932	>100 000	2000	25 500	—
29	119	25 400	2700	19 400	22 400
30	18	>100 000	>100 000	>100 000	—
31	0.48	>100 000	>100 000	>100 000	76 000
32	134	7700	6300	8200	—
33	14	18 600	7600	12 500	—
34	139	>100 000	>100 000	41 700	—
35	206	12 500	>100 000	34 900	—

DPP-8/9 selectivity, it exhibited poor oral bioavailability in the rat due to metabolic instability of the piperazine ring.

17.4.1
Discovery of JANUVIA™ (Sitagliptin) and Related Structure–Activity Relationships

To stabilize the piperazine ring toward metabolism, we evaluated a number of fused heterocycles as piperazine replacements. A variety of fused heterocycles were found to be effective in improving metabolic stability and pharmacokinetic properties, as well as increasing DPP-4 potency. Interestingly, imidazopiperazine analogue **37** showed a fivefold increase in DPP-4 potency over its piperazine counterpart **36** (Table 17.6) [47]. Incorporation of another nitrogen into compound **37** afforded triazolopiperazine analogue **39** with similar potency [11]. Although the effect of an ethyl substituent was insignificant in the case of imidazopiperazine **38**, it was promising in the triazolopiperazine series (**40**) [11]. It was worth noting that ethyl analogue **40** was stable to metabolism in rat hepatocytes, even though it showed poor oral bioavailability in rats (2%). Although the enhancement of DPP-4 potency due to fluorine atoms on phenyl was again observed with triazolopiperazine analogues **41** and **42**, each of these compounds showed poor oral bioavailability in rats ($F = 2\%$). Nevertheless, it was encouraging to note that both **41** and **42** showed improvements in DPP-4 potency, while maintaining excellent selectivity over DPP-8, DPP-9 and hERG in the triazolopiperazine series.

To improve DPP-4 activity as well as pharmacokinetic properties, the effects of substituents on triazolopiperazine moieties were examined (Table 17.7) [48]. Aryl and heteroaryl substituents were found to be as effective as alkyl substituents at improving DPP-4 activity. Unlike most of the previous β-amino acid amide analogues in this series, these analogues (e.g. **43** and **44**) exhibited unacceptable DPP-8 and hERG activities (IC_{50}'s $= 1$–$4\,\mu M$). The effects of perfluoroalkyl substituents were

Table 17.6 Preliminary SAR of triazolopiperazines.

Compound	DPP-4	QPP	DPP-8	DPP-9	hERG
36	3056	>100 000	>100 000	>100 000	—
37	635	51 000	—	>100 000	—
38	560	31 000	24 000	>100 000	—
39	455	66 000	38 000	>100 000	—
40	231	64 000	45 000	>100 000	38 800
41	43	>100 000	53 900	>100 000	77 200
42	37	>100 000	39 200	>100 000	24 800

Table 17.7 Potency and off-target activities of fused triazole derivatives.

Compound	DPP-4	QPP	DPP-8	DPP-9	hERG
43	60	38 100	980	36 600	1200
44	65	>100 000	4200	84 000	—
1	18	>100 000	48 000	>100 000	33 000
45	71	78 000	54 000	80 000	51 000
46	59	54 000	62 000	>100 000	30 000

next examined using two different triazolopiperazine isomers. Not only did trifluoromethyl analogue **1** afford a twofold increase in DPP-4 potency over ethyl analogue **42**, but also it showed a dramatic improvement in oral bioavailability over **42**, from 2% to 76%. While pentafluoroethyl analogue **45** showed decreased DPP-4 potency relative to **1**, it exhibited acceptable oral bioavailability in rats ($F = 61\%$) and slightly reduced off-target activity at hERG. Although compound **46**, a regioisomer of **1**, exhibited a threefold decrease in DPP-4 potency, it showed an interesting rat pharmacokinetic profile ($F = 100\%$, $t_{1/2} = 3.6$ h). Since the off-target activity at hERG was not considered significant, this series merited further pursuit [49].

Overall, trifluorophenyl analogue **1** was superior to all triazolopiperazine derivatives with respect to DPP-4 potency, off-target selectivity, pharmacokinetic profile and *in vivo* efficacy in preclinical species and was selected for further development. *In vitro* potency and selectivity of compound **1** (sitagliptin) are illustrated in Table 17.8. Sitagliptin is a very selective DPP-4 inhibitor, showing 2700-fold selectivity over DPP-8 and more than 5000-fold selectivity over other DASH proteins. Sitagliptin was further profiled in an extensive panel of over 170 enzymes, receptors and ion

Table 17.8 *In vitro* potency and selectivity of compound **1** (sitagliptin).

	Enzyme	IC$_{50}$ (nM)	Selectivity
DPP-4 gene family	DPP-4	18	1
	DPP-8	48 000	2700
	DPP-9	>100 000	>5000
	FAP	>100 000	>5000
	PEP	>100 000	>5000
Other proline-specific enzymes	QPP/DPP-7	>100 000	>5000
	APP	>100 000	>5000
	Prolidase	>100 000	>5000

channels (MDS Pharma Services) and no significant binding or activity was observed at 10 000 nM. Furthermore, no cytochrome P450 enzyme inhibition or induction was observed and **1** was not a time-dependent inhibitor of CYP3A4.

A significant increase in potency (~18-fold) was previously observed with the incorporation of a benzyl substituent into the piperazine moiety as exemplified by **35** (DPP-4: $IC_{50} = 206$ nM) compared to **36** (DPP-4: $IC_{50} = 3700$ nM) [46]. Incorporation of a benzyl substituent onto the piperazine moiety might likewise increase the DPP-4 potency in the triazolopiperazine series. As expected, the introduction of a 4-fluorobenzyl group at the 8-position of triazolopiperazine resulted in a dramatic 100-fold boost in DPP-4 potency (**48**, $IC_{50} = 0.18$ nM; Table 17.9) [50]. Unfortunately, many of these extremely potent compounds displayed unacceptable levels of DPP-8 inhibition (e.g. **47–49**). Fine-tuning of the benzyl substituent revealed that incorporation of trifluoromethyl groups onto the phenyl ring improved selectivity over DPP-8, especially with derivative **51**. Oral bioavailability for this compound in the rat was, however, only 19% and hERG binding was suboptimal.

X-ray crystal structure determination for the potent compound **48** in complex with DPP-4 (Figure 17.4b) revealed that the absolute stereochemistry at the 8-position of the triazolopiperazine in the more potent diastereomer was (R). Similar to compound **1**, the 2,4,5-trifluorophenyl moiety fully occupies the S1 hydrophobic pocket and the (R)-β-amino group forms four hydrogen bonding interactions with the side chains of a Tyr662, Glu205 and Glu206. A water molecule bridges the carboxylic oxygen and the hydroxyl of Tyr547. Several other water-mediated interactions are also present between the nitrogen atoms of the triazolopiperazine and the protein. The triazolopiperazine is stacked against the side chain of Phe357 and the trifluoromethyl substituent interacts with the side chains of Arg358 and Ser209. The 4-fluorobenzyl-group of **48** fully occupies the hydrophobic pocket, suggesting that relatively large groups are well tolerated at this position, as exemplified by compounds **50** and **51**. The superior DPP-4 potency of compound **48** may be attributed to an additional water

Table 17.9 Potency and off-target activities of benzyl analogues.

Compound	R	DPP-4	QPP	DPP-8	DPP-9	hERG
47	H	0.66	52 000	622	24 000	—
48	4-F	0.18	33 000	332	20 000	13 300
49	2-F	0.46	47 000	1103	39 000	19 800
50	2-CF$_3$	0.31	41 000	8011	>100 000	—
51	bis-3,5-CF$_3$	6.3	>100 000	>10 000	>100 000	8000

molecule-bridged interaction between 4-fluorophenyl and Ser630 in addition to the hydrophobic interaction mentioned above.

17.5
Conclusions

For most medicinal chemistry programs, the importance of developing highly selective compounds is recognized from the outset of a project, and the relevant counterscreens are developed early on in a project and on a regular basis. From the outset of the DPP-4 inhibitor program, it was clear that high selectivity over cardiac ion channels such as hERG would be critical for a potential clinical candidate. It was, however, only after medicinal chemistry was initiated that the Merck team identified that DPP-8 and DPP-9 inhibition can lead to toxicity in preclinical species. As a result of this discovery, an early objective for the DPP-4 inhibitor program was to achieve more than 1000-fold selectivity over DPP-8 and DPP-9 before extensive preclinical evaluation of any drug candidate. The importance of achieving highly selective DPP-4 inhibition is underscored by the finding that a nonselective DPP-4, DPP-8 and DPP-9 inhibitor caused treatment-related necrotic skin lesions in monkeys during a 3-month study of toxicity at all doses [51]. Meanwhile, the potent and selective DPP-4 inhibitor sitagliptin (1) did not show any dermatological toxicity at concentrations more than 40-fold that of the highest human dose (C_{max}) where more than 90% DPP-4 inhibition was sustained throughout the dosing interval of the study [51]. The 3-month monkey skin toxicity study with sitagliptin was requested by the FDA prior to approval and was spurred by reports of dermatological toxicity with other DPP-4 inhibitors in development.

In conclusion, α-amino acid amide derivative 24 and β-amino acid amide derivative 1 demonstrate that extending into the P-2 binding pocket of DPP-4 can be an effective strategy to optimize for DPP-4 inhibition of noncovalent inhibitors while minimizing activity against DPP-8 and DPP-9. It should, however, be noted that extension into the P-2 pocket alone may not be sufficient to impart substantial selectivity over DPP-8 and DPP-9 (e.g. compounds 9–15, 43, 44 and 47–49). Recently, other groups have demonstrated that extending noncovalent DPP-4 inhibitors into the P-2 binding pocket may be used as a means to introduce selectivity over DPP-8 and DPP-9 [52]. Although the compounds reviewed in this chapter were optimized using empirical methods, reports are beginning to emerge that describe the active sites and SAR specific to DPP-8 and DPP-9 in more detail [42,43,53]. The introduction of polar functional groups into DPP-4 inhibitors (e.g. carboxylic acids and polar heterocycles) proved a successful strategy to increase selectivity over hERG binding. Nevertheless, optimization of DMPK properties must also be considered in a drug discovery effort, and often the best solution to one problem (e.g. selectivity) might detract from properties in another area (e.g. pharmacokinetics or drug metabolism). In this regard, the optimization of the α- and β-amino acid amide series both demonstrate that a reevaluation of structural features over the course of a medicinal chemistry program is often necessary to identify the top drug candidate from a particular structure class.

Abbreviations

CNS	central nervous system
DASH proteins	DPP-4-like activity and/or structural homologue proteins
DIO	diet-induced obese
DPP II	dipeptidyl peptidase II (same as QPP)
DPP-4	dipeptidyl peptidase IV
DPP-8	dipeptidyl peptidase 8
DPP-9	dipeptidyl peptidase 9
F	oral bioavailability
FAP	fibroblast activation protein
FDA	Food and Drug Administration
GIP	glucose-dependent insulinotropic polypeptide
GLP-1	glucagon-like peptide 1
hERG	human ether-a-go-go
HTS	high-throughput screen
OGTT	oral glucose tolerance test
QPP	quiescent cell proline peptidase
SAR	structure–activity relationships
$t_{1/2}$	half-life

References

1 For more information about the prevalence of diabetes, see URLs by the International Diabetes Federation at www.idf.org; the World Health Organization at www.who.int; the Center for Disease Control at www.cdc.gov/diabetes; and the American Diabetes Association at www.diabetes.org.

2 (a) Wagman, A.S. and Nuss, J.M. (2001) Current therapies and emerging targets for the treatment of diabetes. *Current Pharmaceutical Design*, 7, 417–450; (b) Inzucchi, S.E. (2002) Oral antihyperglycemic therapy for type 2 diabetes. *Journal of the American Medical Association*, 287, 360–372.

3 For recent reviews see Weber, A.E. (2004) Dipeptidyl peptidase IV inhibitors for the treatment of diabetes. *Journal of Medicinal Chemistry*, 47, 4135–4141; (b) Augustyns, K., der Veken, P.V., Senten, K. and Haemers, A. (2005) The therapeutic potential of inhibitors of dipeptidyl peptidase IV (DPP-IV) and related proline-specific dipeptidyl aminopeptidases. *Current Medicinal Chemistry*, 12, 971–998; (c) Augustyns, K., der Veken, P.V. and Haemers, A. (2005) Inhibitors of proline-specific dipeptidyl peptidases: DPP-4 inhibitors as a novel approach for the treatment of type 2 diabetes. *Expert Opinion on Therapeutic Patents*, 15, 1387–1407; (d) von Geldern, T.W. and Trevillyan, J.M. (2006) "The next big thing" in diabetes: clinical progress on DPP-IV inhibitors. *Drug Development Research*, 67, 627–642.

4 (a) Orsakov, C. (1992) Glucagon-like peptide 1, a new hormone of the enteroinsular axis. *Diabetologia*, 35, 701–711; (b) Zander, M., Madsbad, S.,

Madsen, J.L. and Holst, J.J. (2002) Effect of 6-week course of glucagon-like peptide 1 on glycaemic control, insulin sensitivity, and β-cell function in type 2 diabetes: a parallel-group study. *Lancet*, **359**, 824–830.

5 For recent reviews see Knudsen, L.B. (2004) Glucagon-like peptide 1: a basis of a new class of treatment for type 2 diabetes. *Journal of Medicinal Chemistry*, **47**, 4128–4134; (b) Vahl, T.P. and D'Alessio, D.A. (2004) Gut peptides in the treatment of diabetes mellitus. *Expert Opinion on Investigational Drugs*, **13**, 177–188; (c) Meier, J.J. and Nauck, M.A. (2005) Glucagon-like peptide 1 (GLP-1) in biology and pathology. *Diabetes/Metabolism Research and Reviews*, **21**, 91–117; (d) Drucker, D.J. and Nauck, M. A. (2006) The incretin system: glucagon-like peptide-1 receptor agonists and dipeptidyl peptidase-4 inhibitors in type 2 diabetes. *Lancet*, **368**, 1696–1705; (e) Combettes, M.M.J. (2006) GLP-1 and type 2 diabetes: physiology and new clinical advances. *Current Opinion in Pharmacology*, **6**, 598–605.

6 (a) Wettergren, A., Schjoldager, B., Martensen, P.E., Myhre, J., Christiansen, J. and Holst, J.J. (1993) Truncated GLP-1 (proglucagon 72–107 amide) inhibits gastric and pancreatic functions in man. *Digestive Diseases and Sciences*, **38**, 665–673; (b) Nauck, M.A., Niedereichholz, U., Ettler, R., Holst, J.J., Orskov, C., Ritzel, R. and Schmigel, W. H. (1997) Glucagon-like peptide-1 inhibition of gastric-emptying outweighs its insulinotropic effects in healthy humans. *American Journal of Physiology*, **273**, E981–E988.

7 Flint, A., Raben, A., Ersboll, A.K., Holst, J.J. and Astrup, A. (2001) The effect of physiological levels of glucagons-like peptide-1 on appetite, gastric emptying, energy and substrate metabolism in obesity. *International Journal of Obesity*, **25**, 781–792.

8 (a) Holst, J.J. (2005) Glucagon-like peptide-1: physiology and therapeutic potential. *Current Opinion in Endocrinology and Diabetes*, **12**, 56–62; (b) Edwards, C.M.B. (2005) The GLP-1 system as a therapeutic target. *Annals of Medicine*, **37**, 314–322; (c) Gautier, J.F., Fetita, S., Sobngwi, E. and Salaun-Martin, C. (2005) Biological actions of the incretins GIP and GLP-1 and therapeutic perspectives in patients with type 2 diabetes. *Diabetes and Metabolism*, **31**, 233–242.

9 (a) Reimer, M.K., Holst, J.J. and Ahren, B. (2002) Long-term inhibition of dipeptidyl peptidase IV improves glucose tolerance and preserves islet function in mice. *European Journal of Endocrinology*, **146**, 717–727; (b) Pospisilik, J.A., Martin, J., Doty, T., Ehses, J.A., Pamir, N., Lynn, F. C., Piteau, S., Demuth, H.-U., McIntosh, C.H.S. and Pederson, R. (2003) Dipeptidyl peptidase IV inhibitor treatment stimulates β-cell survival and islet neogenesis in streptozotocin-induced diabetic rats. *Diabetes*, **52**, 741–750.

10 (a) Kieffer, T.J., McIntosh, C.H.S. and Pederson, T.A. (1995) Degradation of glucose-dependent insulinotropic polypeptide and truncated glucagons-like peptide 1 *in vitro* and *in vivo* by dipeptidyl peptidase IV. *Endocrinology*, **136**, 3585–3596; (b) Deacon, C.F., Nauck, M.A., Toft-Nielson, M., Pridal, L., Willms, B. and Holst, J.J. (1995) Both subcutaneously and intravenously administered glucagon-like peptide-1 are rapidly degraded from the NH_2-terminus in type II diabetic patients and in healthy subjects. *Diabetes*, **44**, 1126–1131.

11 (a) Kim, D., Wang, L., Beconi, M., Eiermann, G.J., Fisher, M.H., He, H., Hickey, G.J., Kowalchick, J.E., Leiting, B., Lyons, K., Marsilio, F., McCann, M.E., Patel, R.A., Petrov, A., Scapin, G., Patel, S.B., Sinha Roy, R., Wu, J.K., Wyvratt,

M.J., Zhang, B.B., Zhu, L., Thornberry, N.A. and Weber, A.E. (2005) (2R)-4-Oxo-4-[3-(trifluoromethyl)-5,6-dihydro[1,2,4]triazolo[4,3-a]pyrazin-7(8H)-yl]-1-(2,4,5-trifluorophenyl)butan-2-amine: a potent, orally active dipeptidyl peptidase IV inhibitor for the treatment of type 2 diabetes. *Journal of Medicinal Chemistry*, **48**, 141–151; (b) U.S. Patent 6,699,871 B2,Mar. 2, 2004.

12 For more information about JANUVIA™ (sitagliptin phosphate), seewww.januvia.com.

13 De Meester, I., Korom, S., Van Damme, J. and Scharpe, S. (1999) CD26, let it cut or cut it down. *Immunology Today*, **20**, 367–375.

14 Mentlein, R. (1999) Dipeptidyl-peptidase IV (CD26)-role in the inactivation of regulatory peptides. *Regulatory Peptides*, **85**, 9–24.

15 Reinhold, D., Kahne, T., Steinbrecher, A., Wrenger, S., Neubert, K., Ansorge, S. and Brocke, S. (2002) The role of dipeptidyl peptidase IV (DP IV) enzymatic activity in T cell activation and autoimmunity. *Biological Chemistry*, **383**, 1133–1138.

16 Marguet, D., Baggio, L., Kobayashi, T., Bernard, A.M., Pierres, M., Nielsen, P.F., Ribel, U., Watanabe, T., Drucker, D.J. and Wagtmann, N. (2000) Enhanced insulin secretion and improved glucose tolerance in mice lacking CD26. *Proceedings of the National Academy of Sciences of the United States of America*, **97**, 6874–6879.

17 Conarello, S.L., Li, Z., Ronan, J., Roy, R.S., Zhu, L., Jiang, G., Liu, F., Woods, J., Zycband, E., Moller, D.E., Thornberry, N.A. and Zhang, B.B. (2003) Mice lacking dipeptidyl peptidase IV are protected against obesity and insulin resistance. *Proceedings of the National Academy of Sciences of the United States of America*, **100**, 6825–6830.

18 Ahre?n, B., Simonsson, E., Larsson, H., Landin-Olsson, M., Torgeirsson, H., Jansson, P.-A., Sandqvist, M., Ba?venholm, P., Efendic, S., Eriksson, J.W., Dickinson, S. and Holmes, D. (2002) Inhibition of dipeptidyl peptidase IV improves metabolic control over a 4-week study period in type 2 diabetes. *Diabetes Care*, **25**, 869–875.

19 Sorbera, L.A., Revel, L. and Castan?er, J. (2001) MK-0431. *Drugs Future*, **26**, 859–864.

20 Lankas, G.R., Leiting, B., Sinha Roy, R., Eiermann, G.J., Beconi, M.G., Biftu, T., Chan, C.-C., Edmondson, S., Feeney, W.P., He, H., Ippolito, D.E., Kim, D., Lyons, K.A., Ok, H.O., Patel, R.A., Petrov, A.N., Pryor, K.A., Qian, X., Reigle, L., Woods, A., Wu, J.K., Zaller, D., Zhang, X., Zhu, L., Weber, A.E. and Thornberry, N.A. (2005) Dipeptidyl peptidase IV inhibition for the treatment of type 2 diabetes: potential importance of selectivity over dipeptidyl peptidases 8 and 9. *Diabetes*, **54**, 2988–2994.

21 Glund, K., Hoffmann, T., Demuth, H.-U., Banke-Bochita, J., Rost, K.-L. and Fuder, H. (2000) Single dose-escalation study to investigate the safety and tolerability of the DP IV inhibitor P32/98 in healthy volunteers. *Experimental and Clinical Endocrinology and Diabetes*, **108**, 159.

22 Demuth, H.-U., Hoffmann, T., Glund, K., McIntosh, C.H.S., Pederson, R.A., Fuecker, K., Fischer, S. and Hanefeld, M. (2000) Single dose treatment of diabetic patients by the DP IV inhibitor P32/98. *Diabetes*, **49**(S1), A102.

23 (a) Sedo, A. and Malik, R. (2001) Dipeptidyl peptidase IV-like molecules: homologous proteins or homologous activities? *Biochimica et Biophysica Acta*, **1550**, 107–116; (b) Rosenblum, J.S. and Kozarich, J.W. (2003) Prolyl peptidases: a serine protease subfamily with high potential for drug discovery. *Current Opinion in Chemical Biology*, **7**, 1–9.

24 (a) Underwood, R., Chiravuri, M., Lee, H., Schmitz, T., Kabcenell, A.K., Yardley, K. and Huber, B.T. (1999) Sequence, purification, and cloning of an

intracellular serine protease, quiescent cell proline dipeptidase. *Journal of Biological Chemistry*, **274**, 34053–34058; (b) Maes, M.-B., Scharpe, S. and De Meester, I. (2007) Dipeptidyl peptidase II (DPPII), a review. *Clinica Chimica Acta*, **380**, 31–49.

25 Abbott, C.A., Yu, D.M.T., Woollatt, E., Sutherland, G.R., McCaughan, G.W. and Gorrell, M.D. (2000) Cloning, expression and chromosomal localization of a novel human dipeptidyl peptidase (DPP) IV homolog, DPP8. *European Journal of Biochemistry*, **267**, 6140–6150.

26 Olsen, C. and Wagtmann, N. (2002) Identification and characterization of human *DPP9*, a novel homologue of dipeptidyl peptidase IV. *Gene*, **299**, 185–193.

27 Scallan, M.J., Raj, B.K.M., Calvo, B., Garin-Chesa, P., Sanz-Moncasi, M.P., Healey, J.H., Old, L.J. and Rettig, W.J. (1994) Molecular cloning of fibroblast activation protein alpha, a member of the serine protease family selectively expressed in stromal fibroblast of epithelial cancers. *Proceedings of the National Academy of Sciences of the United States of America*, **91**, 5657–5661.

28 Yu, D.M.T., Wang, X.M., McCaughan, G.W. and Gorrell, M.D. (2006) Extraenzymatic functions of the dipeptidase IV-related proteins DP8 and DP9 in cell adhesion, migration and apoptosis. *FEBS*, **273**, 2447–2460.

29 Ajami, K., Abbott, C.A., McCaughan, G.W. and Gorrell, M.D. (2004) Dipeptidyl peptidase 9 has two forms, a broad distribution, cytoplasmic localization and DPIV-like peptidase activity. *Biochimica et Biophysica Acta*, **1679**, 18–28.

30 (a) Recanatini, M., Poluzzi, E., Masetti, M., Cavalli, A. and Ponti, F.D. (2005) QT prolongation through hERG K^+ channel blockade: current knowledge and strategies for the early prediction during drug development. *Medicinal Research Reviews*, **25**, 133–166; (b) Jamieson, C., Moir, E.M., Rankovic, Z. and Wishart, G. (2006) Medicinal chemistry of hERG optimization: highlights and hang-ups. *Journal of Medicinal Chemistry*, **49**, 5029–5046.

31 (a) Wiedeman, P.E. and Trevillyan, J.M. (2003) Dipeptidyl peptidase IV inhibitors for the treatment of impaired glucose tolerance and type 2 diabetes. *Current Opinion in Investigational Drugs*, **4**, 412–420; (b) Augustyns, K., Van der Veken, P., Senten, K. and Haemers, A. (2003) Dipeptidyl peptidase IV inhibitors as new therapeutic agents for the treatment of type 2 diabetes. *Expert Opinion on Therapeutic Patents*, **13**, 499–510.

32 Ashworth, D.M., Atrash, B., Baker, G.R., Baxter, A.J., Jenkins, P.D., Jones, D.M. and Szelke, M. (1996) 2-Cyanopyrrolidines as potent, stable inhibitors of dipeptidyl peptidase IV. *Bioorganic & Medicinal Chemistry Letters*, **6**, 1163–1166.

33 Parmee, E.R., He, J., Mastracchio, A., Edmondson, S.D., Colwell, L., Eiermann, G., Feeney, W.P., Habulihaz, B., He, H., Kilburn, R., Leiting, B., Lyons, K., Marsilio, F., Patel, R., Petrov, A., Di Salvo, J., Wu, J.K., Thornberry, N.A. and Weber, A.E. (2004) 4-Aminocyclohexylglycine analogs as potent dipeptidyl peptidase IV inhibitors. *Bioorganic & Medicinal Chemistry Letters*, **14**, 43–46.

34 Caldwell, C.G., Chen, P., He, J., Parmee, E.R., Leiting, B., Marsilio, F., Patel, R.A., Wu, J.K., Eiermann, G.J., Petrov, A., He, H., Lyons, K.A., Thornberry, N.A. and Weber, A.E. (2004) Fluoropyrrolidine amides as dipeptidyl peptidase IV inhibitors. *Bioorganic & Medicinal Chemistry Letters*, **14**, 1265–1268.

35 Mastracchio, A., Parmee, E.R., Leiting, B., Marsilio, F., Patel, R., Thornberry, N.A. and Edmondson, S.D. (2004) Heterocycle fused cyclohexylglycine derivatives as novel dipeptidyl peptidase-IV inhibitors. *Heterocycles*, **62**, 203–206.

36 Ashton, W.T., Dong, H., Sisco, R.M., Doss, G.A., Leiting, B., Patel, R.A., Wu, J.K., Marsilio, F., Thornberry, N.A. and Weber, A.E. (2004) Diastereoselective synthesis and configuration-dependent activity of (3-substituted-cycloalkyl) glycine pyrrolidides and thiazolidides as dipeptidyl peptidase IV inhibitors. *Bioorganic & Medicinal Chemistry Letters*, **14**, 859–863.

37 Xu, J., Wei, L., Mathvink, R., He, J., Park, Y.-J., He, H., Leiting, B., Lyons, K.A., Marsilio, F., Patel, R.A., Wu, J.K., Thornberry, N.A. and Weber, A.E. (2005) Discovery of potent and selective phenylalanine based dipeptidyl peptidase IV inhibitors. *Bioorganic & Medicinal Chemistry Letters*, **15**, 2533–2536.

38 Xu, J., Wei, L., Mathvink, R., Edmondson, S.D., Mastracchio, A., Eiermann, G.J., He, H., Leone, J.F., Leiting, B., Lyons, K.A., Marsilio, F., Patel, R.A., Petrov, A., Wu, J.K., Thornberry, N.A. and Weber, A.E. (2006) Discovery of potent, selective and orally bioavailable pyridone based dipeptidyl peptidase IV inhibitors. *Bioorganic & Medicinal Chemistry Letters*, **16**, 1346–1349.

39 Edmondson, S.D., Mastracchio, A., Duffy, J.L., Eiermann, G.J., He, H., Ita, I., Leiting, B., Leone, J.F., Lyons, K.A., Makarewicz, A.M., Patel, R.A., Petrov, A., Wu, J.K., Thornberry, N.A. and Weber, A.E. (2005) Discovery of potent and selective orally bioavailable β-substituted phenylalanine derived dipeptidyl peptidase IV inhibitors. *Bioorganic & Medicinal Chemistry Letters*, **15**, 3048–3052.

40 Edmondson, S.D., Mastracchio, A., Mathvink, R.J., He, J., Harper, B., Park, J.-J., Beconi, M., Di Salvo, J., Eiermann, G.J., He, H., Leiting, B., Leone, J.F., Levorse, D.A., Lyons, K., Patel, R.A., Patel, S.B., Petrov, A., Scapin, G., Shang, J., Sinha Roy, R., Smith, A., Wu, J.K., Xu, S., Zhu, B., Thornberry, N.A. and Weber, A.E. (2006) (2S,3S)-3-Amino-4-(3,3-difluoropyrrolidin-1-yl)-N,N-dimethyl-4-oxo-2-(4-[1,2,4]triazolo[1,5-a]pyridin-6-ylphenyl)butanamide: a selective α-amino amide dipeptidyl peptidase IV inhibitor for the treatment of type 2 diabetes. *Journal of Medicinal Chemistry*, **49**, 3614–3627.

41 Rasmussen, H.B., Branner, S., Wiberg, F.C. and Wagtmann, N. (2003) Crystal structure of human dipeptidyl peptidase IV/CD26 in complex with a substrate analog. *Nature Structural Biology*, **10**, 19–25.

42 For a reference supporting the importance of Arg358 and Phe357 for DPP-4 selectivity, see Rummy, C. and Metz, G. (2007) Homology models of dipeptidyl peptidases 8 and 9 with a focus on lop predictions near the active site. *Proteins: Structure, Function, and Bioinformatics*, **66**, 160–171.

43 For a reference that describes the importance of Tyr547 in human DPP-4, see Longnecker, K.L., Stewart, K.D., Madar, D.J., Jakob, C.G., Fry, E.H., Wilk, S., Lin, C.W., Ballaron, S.J., Stashko, M.A., Lubben, T.H., Yong, H., Pireh, D., Pei, Z., Basha, F., Weideman, P.E., von Geldern, T.W., Trevillyan, J.M. and Stoll, V.S. (2006) Crystal structures of DPP-IV (CD26) from rat kidney exhibit flexible accommodation of peptidase-selective inhibitors. *Biochemical*, **45**, 7474–7482.

44 Xu, J., Ok, H.O., Gonzalez, E.J., Colwell, L.F., Jr, Habulihaz, B., He, H., Leiting, B., Lyona, K.A., Marsilio, F., Patel, R.A., Wu, J.K., Thornberry, N.A., Weber, A.E. and Parmee, E.R. (2004) Discovery of potent and selective β-homophenylalanine based dipeptidyl peptidase IV inhibitors. *Bioorganic & Medicinal Chemistry Letters*, **14**, 4759–4762.

45 Edmondson, S.D., Mastracchio, A., Beconi, M., Colwell, L., Jr, Habulihaz, H., He, H., Kumar, S., Leiting, B., Lyons, K., Mao, A., Marsilio, F., Patel, R.A., Wu, J. K., Zhu, L., Thornberry, N.A., Weber, A. E. and Parmee, E.R. (2004) Potent and

selective proline derived dipeptidyl peptidase IV inhibitors. *Bioorganic & Medicinal Chemistry Letters*, **14**, 5151–5155.

46 Brockunier, L., He, J., Colwell, L.F., Jr, Habulihaz, B., He, H., Leiting, B., Lyons, K.A., Marsilio, F., Patel, R., Teffera, Y., Wu, J.K., Thornberry, N.A., Weber, A.E. and Parmee, E.R. (2004) Substituted piperazines as novel dipeptidyl peptidase IV inhibitors. *Bioorganic & Medicinal Chemistry Letters*, **14**, 4763–4766.

47 Manuscript in preparation.

48 Kim, D., Kowalchick, J.E., Edmondson, S.E., Mastracchio, A., Xu, J., Eiermann, G.J., Leiting, B., Wu, J.K., Pryor, K.D., Patel, R.A., He, H., Lyons, K.A., Thornberry, N.A. and Weber, A.E. (2007) Triazolopiperazine-amides as dipeptidyl peptidase IV inhibitors: close analogs of JANUVIA (sitagliptin phosphate). *Bioorganic & Medicinal Chemistry Letters*, **17**, 3373–3377.

49 Kowalchick, J.E., Leiting, B., Pryor, K.D., Marsilio, F., Wu, J.K., He, H., Lyons, K., Eiermann, G.J., Petrov, A., Scapin, G., Patel, R.A., Thornberry, N.A., Weber, A.E. and Kim, D. (2007) Design, synthesis, and biological evaluation of triazolopiperazine-based β-amino amides as potent, orally active dipeptidyl peptidase IV (DPP-4) inhibitors. *Bioorganic & Medicinal chemistry Letters*, **17**, 5934–5939.

50 Kim, D., Kowalchick, J.E., Brockunier, L. L., Parmee, E.R., Beconi, M.G., Eiermann, G.J., Fisher, M.H., He, H., Leiting, B., Lyons, K., Scapin, G., Patel, S. B., Petrov, A., Pryor, K.A., Sinha Roy, R., Wu, J.K., Zhang, X., Wyvratt, M.J., Zhang, B.B., Zhu, L., Thornberry, N.A. and Weber, A.E. Discovery of Potent and Selective DPP-4 Inhibitors Derived from β-Aminoamides Bearing Substituted Triazolopiperazines. *Journal of Medicinal Chemisty* (In press).

51 Sami, T. (2007) Skin toxicity seen in other DPP-4 inhibitors missing in *Januvia* – FDA review. *Pharmaceutical Approvals Monthly: The Pink Sheet DAILY*, **12**, 29–30.

52 For lead references see (a) Sakashita, H., Akahoshi, F., Yoshida, T., Kitajima, H., Hayashi, Y., Ishii, S., Takashina, Y., Tsutsumiuchi, R. and Ono, S. (2007) Lead optimization of [(S)-γ-(arylamino) prolyl]thiazolidine focused on γ-substituent: indoline compounds as potent DPP-IV inhibitors. *Bioorganic and Medicinal Chemistry*, **15**, 641–655; (b) Pei, Z., Li, X., von Geldern, T.W., Madar, D.J., Longenecker, K., Yong, H., Lubben, T.H., Stewart, K.D., Zinker, B.A., Backes, B.J., Judd, A.S., Mulhern, M., Ballaron, S.J., Stashko, M.A., Mika, A.K., Beno, D.W. A., Reinhart, G.A., Fryer, R.M., Preusser, L.C., Kempf-Grote, A.J., Sham, H.L. and Trevillyan, J.M. (2006) Discovery of ((4R,5S)-5-amino-4-(2,4,5-trifluoromethyl)-5,6-dihydro[1,2,4] triazolo[4,3-a]pyrazin-7(8H)-yl) methanone (ABT-341), a highly potent, selective, orally efficacious, and safe dipeptidyl peptidase IV inhibitor for the treatment of type 2 diabetes. *Journal of Medicinal Chemistry*, **49**, 6439–6442.

53 Lee, H.-J., Chen, Y.-S., Chou, C.-Y., Chien, C.-H., Lin, C.-H., Chang, G.-G. and Chen, X. (2006) Investigation of the dimer interface and substrate specificity of prolyl dipeptidase DPP8. *Journal of Biological Chemistry*, **281**, 38653–38662.

18
Strategy and Tactics for hERG Optimizations

Craig Jamieson, Elizabeth M. Moir, Zoran Rankovic, Grant Wishart

18.1
Introduction

In recent years, a number of marketed drugs have been withdrawn because of adverse effects on the QT interval. Prolongation of the QT interval by a chemical entity, which may lead to the cardiac arrhythmia *torsades de pointes* (TdP), has been linked to interaction with a cardiac potassium channel, a product of the human *ether-a-go-go* related gene (hERG) [1–4]. Consequently, throughout the pharmaceutical industry, efforts to predict risk of QT prolongation have been focused on assays testing *in vitro* hERG channel activity in mammalian cell lines expressing the hERG channel. Over time, an increasing body of information is being accumulated by medicinal chemistry groups on the strategy and tactics for overcoming blockade of the hERG channel.

The purpose of the following discussion is to summarize the approaches used to circumvent activity at hERG, identified through an extensive survey of the medicinal chemistry literature until the end of 2006. Optimizations are recorded as pairs of compounds, which have been categorized in terms of the tactics employed to diminish hERG activity. In contrast to global computational hERG models, where heterogeneous data sets are often used, this approach has the advantage of dealing with optimization pairs from the same chemical series, with data generated under identical assay conditions. Therefore, there exists a direct relationship between the start and end points of the optimization, which has enabled the determination of the most appropriate method to employ depending on the nature of the starting compound. The resulting analysis is aimed at producing a set of simple, empirical guidelines for attenuating hERG activity, which will be of use to the practicing medicinal chemist.

18.1.1
Classification of Optimization Literature

The reported optimizations were grouped into the following classification categories on the basis of common descriptors: discrete structural modifications (DSM), formation of zwitterions (ZI), control of log P and attenuation of pK_a. Although this review attempts to focus on examples where there is only one clear site of modification, it is not possible to alter a parameter in isolation, and this may confound interpretation of the controlling factor mitigating hERG activity. Therefore, the categorization applied here is best regarded a mnemonic rather than a rigorous classification. In addition, the primary focus of the study is on efforts to remove hERG activity that, in some cases, may result in concomitant variation of the primary target potency, further underlining the multifactorial nature of hERG optimizations. Furthermore, many of the groups that encounter hERG issues employ several tactics to diminish this unwanted activity.

18.1.1.1 Discrete Structural Modifications

Several homology models of the open and closed states of the hERG channel, derived from bacterial potassium channels, have been published [5,6]. Most drugs that inhibit hERG are believed to do so by binding to residues lining the large lipophilic central cavity of the channel. Early mutagenesis studies indicated that the aromatic residues F656 and Y652 of the channel S6 domain are important for interactions with the hERG blocker MK-499 (**1**, Figure 18.1) [5]. More detailed analysis of the roles of F656 and Y652 indicates that at position 652 an aromatic residue is required, leading to

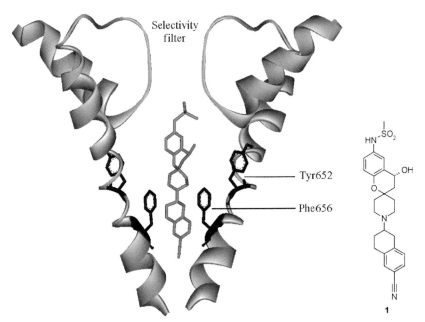

Figure 18.1 Schematic depiction of the putative interactions between a spiropiperidine antiarrhythmic agent (MK-499, **1**) and the hERG channel based on homology modeling of the closed state [5]. The inner helices (S6) and loops extending from the pore helices to the selectivity filter form the inner cavity and the ligand-binding site. For clarity, S6 and pore helix domains of only two subunits of the tetrameric channel are shown (longitudinal view with respect to the cell membrane).

the hypothesis that Y652 participates in π–cation type interactions with basic amine containing hERG ligands and presumably π-stacking interactions for neutral hERG ligands. In the case of position 656, hydrophobic surface area was found to be the key determinant, suggesting F656 participates in hydrophobic interactions [7]. Strategies that disrupt the ability of a ligand to interact with F656 and Y652 have been proposed as a possible means to remove hERG activity [8].

Discrete, often peripheral modifications to a drug molecule can have a dramatic effect on hERG potency, a phenomenon potentially explained by the disruption of the putative interactions with aromatic residues (F656 and Y652) lining the hERG channel. Strategies adopted to mitigate hERG activity in this way not only are restricted to modifications to distal aryl rings, but can also include introducing constraint and variation in stereochemistry. Criteria for including optimization pairs in this category are as follows: one clear site of modification, >1 log reduction in hERG activity, no significant reduction (<1 log unit) in key physicochemical parameters such as $c\log P$ and pK_a [9] and maintained activity at the primary target.

18.1.1.2 Formation of Zwitterions

The zwitterion approach was one of the first successfully applied strategies for reduction of hERG potency (e.g. terfenadine to fexofenadine, Table 18.2). In all reported cases in this category, the starting compound is an amine with hERG activity, into which a carboxylic group (or a suitable bioisostere) is incorporated, to obtain a zwitterionic analogue with attenuated hERG potency. This effect on hERG could be rationalized by a significant lowering of lipophilicity that accompanies incorporation of a polar group such as carboxylate. It is also possible that reduced membrane permeability of zwitterionic species could prevent access of a drug molecule to the intracellular side of the channel central cavity and, therefore, minimize potential for blocking hERG. In any case, the literature evidence suggests that this approach may be of lower general utility owing to attendant issues regarding gut absorption, oral bioavailability and CNS exposure.

18.1.1.3 Control of log P

SAR data can often show series-dependent correlation between potency of hERG block and measures of lipophilicity. This is supported by the published hERG homology models and mutagenesis data [5,7], suggesting that a lipophilic ligand-binding site exists, which is accessed from the intracellular domain. By increasing the polarity of a drug molecule, interaction with the lipophilic cavity will be destabilized, resulting in decreased hERG activity. Optimization pairs included in this category have been restricted to those meeting the following criteria: one clear site of modification, >1 log unit reduction in hERG activity, >1 log unit reduction in $c\log P$ and maintained activity at the primary target.

18.1.1.4 Attenuation of pK_a

Many of the ligands that block hERG contain a basic nitrogen moiety that is likely to be protonated at physiological pH. However, a basic amine is not a prerequisite to hERG blockade, as a suitably placed aromatic or hydrophobic group in the ligand can potentially interact with the channel equally well. On the basis of mutagenesis and homology modeling, it has been proposed that π–cation interactions between basic amine containing hERG blockers and aromatic residues within the cavity of the hERG channel, contribute to ligand-binding affinity [5,6]. Therefore, lowering the pK_a of a basic nitrogen would reduce the proportion of molecules in the protonated form at physiological pH and would be expected to disrupt any putative π–cation interactions with the channel. In general, it has been observed that modification of pK_a can often increase the polarity of a compound (e.g. piperidine to piperazine) and thus can have the same net impact as reducing log P. Criteria for inclusion of optimization pairs into this category are as follows: one clear site of modification, >1 log unit reduction in hERG activity, >1 log unit reduction in calc-pK_a and maintained activity at the primary target. Data presented in the optimization tables refer to the highest calculated basic pK_a.

18.2
Survey of Strategies Used to Diminish hERG

18.2.1
Discrete Structural Modifications

A continuous thread in a series VEGFR-2 (KDR) kinase inhibitors has been off-target activity at hERG [10]. Compounds of the type **2** (Table 18.1) represent a hybrid between two previously reported series [11], designed in an attempt to reduce the hERG channel affinity in the series. Substitution on the pyrimidine ring with basic moieties in general had a beneficial effect on the primary target potency; however, this was often accompanied by undesirable activity at hERG. In the case of the piperazine system **2**, a discrete structural modification was invoked to remove a methyl on the distal nitrogen, thus introducing another hydrogen bond donor. In this case a reduction of 20-fold or above in hERG activity was achieved, despite actually increasing the pK_a that may have otherwise been anticipated to increase affinity for the hERG channel.

Walker et al. [12] have described their efforts toward identifying a series of azabicyclic aryl amides as nicotinic α7 receptor agonists. A previous lead in the series (**4**, PNU-282,987, Figure 18.2) was compromised with hERG activity and optimization ultimately led to the selection of PHA-543,613 (**5**) as a development candidate [13]. While further evaluation of PHA-543,613 was going on, the search for alternative α7 agonists with improved hERG profile was undertaken. Taking **6** as a starting point, the Pfizer group successfully enabled a discrete structural modification, leading to compound **7** that had considerably lower antagonism at the hERG channel, while retaining the desired selectivity profile against other receptors in the Cys-loop superfamily.

In the discovery of the CCR5 antagonist maraviroc, Price et al. [14] were able to generate high selectivity over hERG. Docking of compounds from the series

Table 18.1 Application of discrete structural modifications to control hERG.

hERG active	hERG[a] (µM)	IC$_{50}$[a,b] (nM)	c log P[c]	pK$_a$[c]	hERG optimized	hERG[a] (µM)	IC$_{50}$[a,b] (nM)	c log P[c]	pK$_a$[c]	Ref
2	0.92	10	1.69	7.21	**3**	21.5	16	1.12	8.05	[10]
6	71% at 20 µM	14	1.85		**7**	<1% at 20 µM	40	1.64	9.67	[12]
8	30% at 300 nM	IC$_{90}$ 8	2.76	10.25	**9**	0% at 300 nM	IC$_{90}$ 2	3.26	10.24	[14]

(*Continued*)

Table 18.1 (Continued)

hERG active	hERG^a (μM)	IC$_{50}^{a,b}$ (nM)	c log P^c	pK_a^c	hERG optimized	hERG^a (μM)	IC$_{50}^{a,b}$ (nM)	c log P^c	pK_a^c	Ref
11	0.1	1	4.55	7.33	12	1.25	9	4.46	7.81	[15]
11	0.1	1	4.55	7.33	13	1.8	0.5	4.86	7.32	[15]
15	0.3	6.8	2.16	6.89	16	~3.8	5.1	2.52	6.89	[16]

Compound					Compound					Ref
17	0.44	1.9	2.49	6.46	18	5.8	27	2.30	8.56	[17]
17	0.44	1.9	2.49	6.46	19	7.2	29	1.77	6.46	[17]
20	0.08	<1	2.48	6.64	21	4.7	3.6	2.48	6.64	[18]
20	0.08	<1	2.48	6.64	22	7	<1	3.04	6.65	[18]

(Continued)

Table 18.1 (Continued)

hERG active	hERG[a] (μM)	IC$_{50}$[a,b] (nM)	c log P[c]	pK$_a$[c]	hERG optimized	hERG[a] (μM)	IC$_{50}$[a,b] (nM)	c log P[c]	pK$_a$[c]	Ref
23	0.12	0.68	4.78	5.18	**24**	1.6	3	5.31	5.20	[19]
25	0.99	1.3	3.56	10.19	**26**	>10	1.3	4.48	10.19	[20]

27	0.4	0.32	3.50	5.51		
28				5.6	3.9	3.33	5.63 [21]

[a]Both hERG and primary target activity refer to K_i, IC_{50} or IP unless otherwise stated.
[b]IC_{50} refers to activity at the primary target.
[c]clog P and pK_a are calculated according to [9]. pK_a is reported as the highest basic pK_a. Where no value of pK_a is reported, calculation of this parameter was not possible.

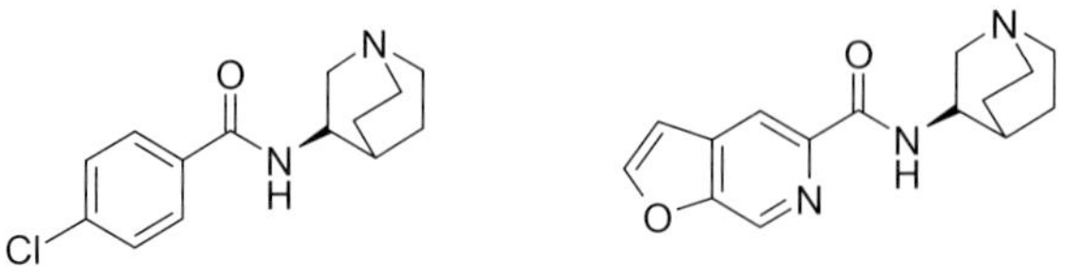

PNU-282,987 (4)
hERG = 57% at 20 µM

PHA-543,613 (5)
hERG = 29% at 20 µM

Figure 18.2 Structures of progenitor nicotinic agonists and hERG activities.

represented by **8** into a model of the hERG channel suggested that the cyclobutyl group could overlap with a lipophilic binding area, so a diverse set of cyclobutyl replacements were prepared to investigate hERG SAR in this region. Ring homologation to the cyclopentyl analogue gave a modest reduction in hERG affinity (suggested to be due to a possible steric clash with residues in the ion channel), as did the 2,2,2-trifluoroethyl analogue. Combining these two modifications in the 4,4-difluorocyclohexyl group led to Maraviroc **9**, which showed no binding to the hERG channel when screened at 300 nM. The authors suggest that this reduction may be due to a steric effect along with a possible contribution from the dipole generated by the difluoro moiety. Affinity at hERG was found to be independent of lipophilicity in this region of the molecule.

Docking of the lead compound for the series, **10** (Figure 18.3), in the model of the hERG channel suggested that the phenyl ring of the benzimidazole overlapped with a second lipophilic binding region. Thus, the benzimidazole was replaced with a range of polar heterocycles resulting in the discovery of the triazole **8**. Additional modifications of **10**, including introduction of an oxygen into the bridgehead position to form the oxogranatane, reduced the measured pK_a from 7.8 to 6.0, with negligible effect on the affinity for the hERG channel, indicating that hERG affinity is independent of the basicity of the nitrogen in this series.

Huscroft et al. [15] reported a new series of conformationally constrained hNK1 antagonists containing an 8-azabicyclo[3.2.1]octane core. The N-2 methyl tetrazole analogue **11** had an attractive profile with high hNK1 affinity (1 nM) and long lasting

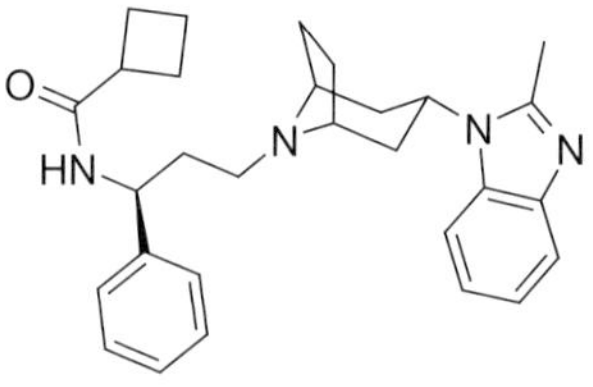

10
MIP-1 IC_{50} = 2 nM
hERG = 80% at 300 nM

Figure 18.3 Lead CCR5 antagonist and associated hERG activity.

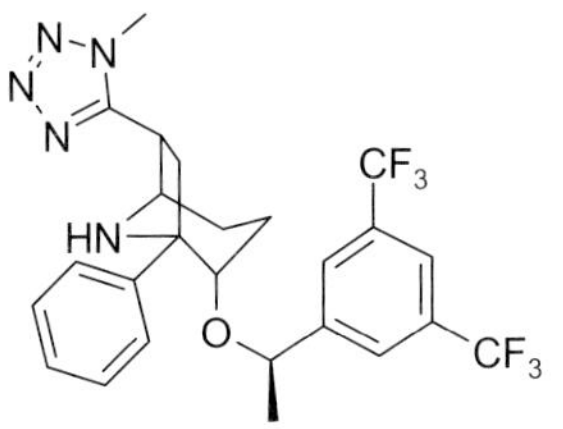

NK1 IC$_{50}$ = 1.2 nM
hERG IC$_{50}$ = 2.8 µM

Figure 18.4 Hybrid NK1 antagonist **14**.

in vivo efficacy in the foot-tapping gerbil paradigm, however, it was compromised by hERG liability (K_i 0.1 µM). A closely related *N*-1 methyl analogue **12** showed approximately 12-fold lower hERG binding (K_i 1.25 µM), indicating high sensitivity of hERG SAR to subtle structural changes in this series. However, this was accompanied by a similar reduction of hNK1 activity (IC$_{50}$ 9 nM). A more striking DSM example in this series derived from introduction of an α-methyl substituent into the pendant benzyl ether side chain to give **13**, which resulted in attenuated hERG activity (K_i 1.8 µM) while maintaining hNK1 affinity (0.5 nM) and *in vivo* efficacy. Interestingly, these two changes proved to be nonadditive, so that a hybrid analogue **14** displayed hERG activity similar to compounds **12** and **13** (Figure 18.4).

Becker *et al.* [16] used Predix's PREDICT technology to identify 5-HT$_{1A}$ agonists via virtual screening using a receptor model. Optimization to address selectivity over the α$_1$ adrenergic receptor guided by receptor models, yielded **15** which showed high affinity and potency for the 5-HT$_{1A}$ receptor (K_i 6.8 nM, EC$_{50}$ 20 nM) and selectivity greater than 500-fold over the α$_1$ adrenergic receptor. However, upon evaluation of **15** in a patch–clamp hERG assay, it was found to be a potent hERG blocker (IC$_{50}$ 0.3 µM). Optimization of the unwanted hERG activity was driven by a 3D model of the hERG pore region based upon the KcsA channel X-ray structure. Evaluation of the binding mode of **15** in the hERG model resulted in the hypothesis that the *p*-toluenesulfonyl group is contributing to the hERG activity of **15** by interacting with the important Y652 residues. Therefore, a strategy to disrupt this aromatic interaction was pursued. Replacement of the aromatic toluene moiety by the aliphatic methylcyclohexyl group to give **16** resulted in an estimated 12-fold reduction of hERG activity while maintaining high 5-HT$_{1A}$ affinity and selectivity over the α$_1$ adrenergic receptor. Compound **16** (PRX-00023) was subsequently selected as a clinical candidate from the series. Phase I clinical studies indicated that **16** (once daily and 28 day multiple doses up to 60 mg) was well tolerated with no effect on cardiovascular or ECG parameters.

Subtle structural modification was one of the strategies employed in an effort to design out hERG affinity within a series of novel pyrrolidinone-based farnesyltransferase inhibitors (FTIs) [17]. Several members of the series such as compound **17** were found to have significant hERG activity, and selectivity was, therefore, a key goal in the medicinal chemistry strategy. Migration of the pyrrolidinone carbonyl group featured in **17** out of the ring resulted in the 1-acylpyrrolidine analogue **18** with more than 10-fold reduced hERG affinity (IP =5.8 µM). However, this structural change had a similar effect on FT inhibition (IC_{50} = 27 nM), rendering analogue **18** with a similar selectivity window over hERG compared to the parent compound **17**. Similarly, deletion of a chlorine atom on the phenyl ring (compound **19**) reduced hERG activity with a concomitant reduction in potency at the primary target. In further efforts to increase the therapeutic window, a range of constrained analogues of **17** were synthesized [18]. The effect of stereochemistry on hERG activity was explored (cf. **20** and **21**) and the extension of the cyclopentyl to cyclohexyl ring was also successful, delivering tetrahydronaphthalene derivative **22**, one of the most potent and selective FTIs in the series.

Researchers at Merck have reported the optimization of a benzimidazole series of NMDA NR2B antagonists [19]. A simple change of the methyl sulfonamide of **23** to the ethyl sulfonamide in **24** resulted in a 13-fold decrease in hERG affinity. The reduction in hERG affinity may be attributed to specific structural changes to the periphery of the molecule potentially disrupting the interaction of the adjacent aromatic ring with the hERG channel-binding site.

In their CCR5 program, Kim et al. [20] adopted a number of strategies to avoid hERG activity. In this case, conversion of the *tert*-butyl group of **25** to the *sec*-butyl system of **26** was accompanied by a reduction of 10-fold or above in the hERG activity while essentially conserving potency at the primary target.

Using more wholesale structural changes and $c \log P$ to tune hERG activity, a recent report from Fish et al. [21] describes the effect of subtle structural changes in enhancing the selectivity window in a series of 5-HT_{2A} antagonists. Preparation of the isomeric fluorophenyl sulfone **28** yielded a more than 10-fold reduction of hERG activity. Given the minimal impact of this modification on the relevant physicochemical parameters, this example highlights the power that discrete structural modifications offer in terms of achieving selectivity over hERG.

18.2.2
Formation of Zwitterions

Terfenadine **29** (Table 18.2), a second-generation antihistamine was the first compound in the series to be launched in 1982. However, owing to instances of cardiac arrhythmia, terfenadine was withdrawn from the market in 1997. Interestingly, it transpired that the principal metabolite of terfenadine, carboxylate **30** (later marketed as fexofenadine), accounted for the therapeutic effect of its parent. In addition, the compound displays significantly reduced hERG affinity and no effect on QT interval [22]. The low hERG affinity of the zwitterionic fexofenadine could be attributed to the low cell membrane permeability of the compound, restricting access to the intracellular lumen of the channel, believed to be crucial to hERG blockade. The zwitterion approach has since been widely employed as one of the strategies for designing out hERG activity. In addition to destabilizing drug/channel interactions, recent molecular modeling work comparing the two compounds has also suggested that differences in conformation could explain this difference in hERG activity [6].

Table 18.2 Application of the zwitterion approach to minimize hERG activity.

hERG active	hERG[a] (μM)	IC$_{50}$[a,b] (nM)	c log P[c]	pK$_a$[c]	hERG optimized	hERG[a] (μM)	IC$_{50}$[a,b] (nM)	c log P[c]	pK$_a$[c]	Ref
29	0.056	1	6.07		30	23	15	1.96		[22]
31	0.8	62	3.81	9.99	32	>10	60	1.52	9.99	[23]

(Continued)

Table 18.2 (Continued)

hERG active	hERG[a] (μM)	IC$_{50}$[a,b] (nM)	clog P[c]	pK$_a$[c]	hERG optimized	hERG[a] (μM)	IC$_{50}$[ab] (nM)	clog P[c]	pK$_a$[c]	Ref
33	0.19		0.24	5.78	9.32	**34**	>6		0.27	5.26	[24]
33	0.19		0.24	5.78		**35**	4.8		1	2.43	9.64 [24]

0.60	2.15	10.03	**36**	>10	1.42	10.03 [25]
0.88	5.13		**38**	>10	5.34	[25]

36

37

38

39

(*Continued*)

Table 18.2 (Continued)

hERG active	hERG[a] (µM)	IC$_{50}$[a,b] (nM)	c log P[c]	pK$_a$[c]	hERG optimized	hERG[a] (µM)	IC$_{50}$[a,b] (nM)	c log P[c]	pK$_a$[c]	Ref
40	1.1	64	3.33		**41**	28	3	2.42		[29]

[a] Both hERG and primary target activity refer to K_i, IC$_{50}$ or IP unless otherwise stated.
[b] IC$_{50}$ refers to activity at the primary target.
[c] c log P and pK$_a$ are calculated according to [9]. pK$_a$ is reported as the highest basic pK$_a$. Where no value of pK$_a$ is reported, calculation of this parameter was not possible.

In their optimization of small molecule CCR8 antagonists identified by a high-throughput screen, Ghosh et al. [23] introduced polar substituents to reduce hERG activity without disrupting target potency. In one successful example of this strategy, introduction of a carboxylic acid group in the 4-position of the aminopiperidine of **31** resulted in a substantial decrease in hERG binding for compound **32**. The authors report that the introduction of polar functionality did not appreciably affect the permeability of related carboxylic acid functionalized compounds as measured by the Caco-2 assay; however, the oral bioavailability for these molecules was not discussed.

Thomson et al. [24] disclosed further hERG optimization efforts around conformationally constrained hNK1 antagonists previously reported by the same group. The best compounds in this series showed subnanomolar activity on hNK1 and also displayed significant binding to the hERG channel (e.g. compound **33**). The optimization efforts focused on the 6-*exo* position of the bicyclic core showed that acidic groups in this region maintain the high hNK1 affinity and have a marked effect on hERG binding (cf. sulfonamide **33** and acylsulfonamide **34**). Replacement of the sulfonamide group in **33** with a carboxylic functionality resulted in a zwitterionic analogue **35** displaying a similar profile, hNK1 IC_{50} 1 nM and hERG K_i 4.8 µM. However, despite their high *in vitro* potency, none of the zwitterionic analogues showed *in vivo* efficacy, which the authors attributed to a lack of brain exposure.

The concept of introducing a carboxylic acid into benzamidine containing factor Xa inhibitors as a means of reducing hERG activity has been explored by Zhu et al. [25]. Compounds were evaluated for hERG affinity using a dofetilide binding assay, and many examples were described where the introduction of a carboxylic acid group resulted in a significant decrease in hERG affinity. It was initially proposed that the effect is general for benzamidine containing factor Xa inhibitors, as a decrease in hERG was observed in several different chemical series and at different substitution points within the same chemical series. Modifying the morpholino substituent of **36** within the piperazino series to the acid **37** afforded a compound with significantly lower hERG affinity. Similar results were also observed for the pyrazole series in the transformation of nitrile **38** to carboxylic acid **39**. Unfortunately in all cases primary activity for factor Xa inhibition was not presented. It was subsequently suggested that the carboxylic acid effect may be further generalized to additionally encompass neutral hERG blockers, which could be modified in this manner. This is corroborated by recent data from Waring and Johnstone [26], who showed that out of 350 acidic compounds tested at the hERG channel, only nine entities displayed significant (<10 µM) activity.

In a series of reports on their class of DPPIV inhibitors [27–29], workers from Merck have shown that the incorporation of an acid moiety (or related bioisostere) into an amine-containing template to furnish a zwitterionic system has been of use in minimizing hERG activity. For example, incorporation of an acid bioisostere into the biaryl β-methylphenylalanine template has been shown to afford enhancements in selectivity over hERG (cf. compounds **40** and **41**) [29]. A concomitant reduction in oral bioavailability was observed between the two compounds, which underlines the main limitation in this approach.

18.2.3
Control of log P

In their series of MCHR1 antagonists, Blackburn et al. [30] screened compounds for hERG binding as a predictor of the cardiovascular liabilities that had been prevalent within the project. Replacement of the naphthyl of **42** with a quinolyl group to give **43** (Table 18.3) reduced the $c\log P$ from 4.69 to 3.41 and resulted in considerably improved hERG selectivity.

Vaz et al. [31] have used a CoMSiA QSAR model of hERG activity to enhance their selectivity profile within a series of inhibitors of human β-tryptase. By application of the model, incorporation of a carboxamide substituent was targeted as an approach to minimize hERG activity. Comparison of **44** and **45** (Table 18.3) shows a significant reduction in $c\log P$, with accompanying reduction in hERG activity. The Aventis group has also used the zwitterion approach as a means of enhancing selectivity over hERG, replacing the carboxamide moiety with a carboxylic acid.

In the series of 5-HT$_{2A}$ antagonists reported by Rowley et al. [32], a general hERG pharmacophoric model was used to seek a means of improving selectivity. The authors reasoned that the phenethyl group attached to the basic nitrogen of the lead (**46**) confers high activity at hERG ($K_i = 0.08\ \mu M$). Indeed, the truncated analogue **47** displayed significantly reduced hERG binding ($K_i = 5.7\ \mu M$) consistent with its lower $c\log P$, thus providing an attractive point for further optimization of potency and bioavailability. The group further pursued log P by targeting other parts of the molecule, which also proved successful. For example, replacement of the naphthyl moiety in **48** with a pyridyl system led to analogue **49** ($c\log P = 2.84$) that not only diminished hERG ($K_i = 2.5\ \mu M$) but also maintained 5-HT$_{2A}$ affinity (0.35 nM).

The PDE4 inhibitor **50** was identified as a potential development candidate; however, cardiovascular safety studies in anesthetized dogs at a dose of 3 mg/kg indicated an 8.9% prolongation of the QTc interval [33]. Compound **50** was optimized by modifying four key areas: tertiary alcohol substituents, 3-alkoxy substituents of the catechol, the 2,5-disubstituted pyridine and the pyridine-N-oxide. Initial SAR indicated that PDE4 inhibition, but not hERG affinity, was affected by stereochemistry. Replacement of the phenyl substituent of the tertiary alcohol by methyl resulting in the dimethyl alcohol **51** reduced hERG affinity by 19-fold (hERG $K_i = 22.7\ \mu M$), but PDE4 inhibition decreased by 17-fold (PDE4 IC$_{50} = 35$ nM). The reduction of hERG affinity by replacement of the phenyl substituent could be attributed to the loss of a key aromatic pharmacophoric interaction with the hERG channel.

Another example from the VEGFR-2 (KDR) kinase area shows how hERG blockade and QT prolongation can also be overcome through deletion of a phenyl group [11]. Replacement of the phenyl of **52** with the nitrile in **53** offered a significant reduction in activity at hERG. Reducing lipophilicity within the series, improved both the pharmacokinetics and selectivity over hERG. In this example, a reduction in $c\log P$ from 3.23 in **52** to 0.66 in **53** was observed, though hERG SAR within the series indicated that piperazine with a polar N-substituent was also important in maintaining an optimal profile. Indeed, the existence of other examples with similar $c\log P$ to **52** but lower hERG activity indicates that the structural change afforded by

Table 18.3 Control of log P to effect off-target activity at hERG.

hERG active	hERG^a (μM)	IC$_{50}$^{ab}, (nM)	c log P^c	pK$_a$^{c}	hERG optimized	hERG^a (μM)	IC$_{50}$^{ab}, (nM)	c log P^{ab},	pK$_a$^{c}	Ref
42	0.16	7	4.69	7.99	**43**	1.75	12	3.41	7.68	[30]
44	0.8	4.3	3.82		**45**	17.1	1.3	2.73		[31]

(*Continued*)

Table 18.3 (Continued)

hERG active	hERG[g] (μM)	IC$_{50}$[g,h] (nM)	clog P[i]	pK$_a$[i]	hERG optimized	hERG[g] (μM)	IC$_{50}$[g,h] (nM)	clog P[i]	pK$_a$[i]	Ref
46	0.08	0.48	6.50	9.60	47	5.7	12	4.13	10.93	[32]
48	0.11	0.34	5.46	9.00	49	2.5	0.35	2.84	8.91	[32]

50	1.18	2.1	3.69	4.63		
51		22.7	35	2.65	4.74	[33]
52	0.24	8	3.23	4.80		
53		10.6	13	0.66	4.40	[11]

[a] Both hERG and primary target activity refer to K_i, IC_{50} or IP unless otherwise stated.
[b] IC_{50} refers to activity at the primary target.
[c] log P and pK_a are calculated according to [9]. pK_a is reported as the highest basic pK_a. Where no value of pK_a is reported, calculation of this parameter was not possible.

the deletion of the phenyl ring was potentially more relevant to hERG activity than the global lipophilicity of the compound.

18.2.4 Attenuation of pK_a

The role of pK_a in the most recent series of VEGFR-2 (KDR) kinase inhibitors is somewhat confounding [10]. Comparison of **2** and **3** (Table 18.1) illustrates that hERG activity can be reduced despite a slight increase in pK_a. From consideration of compounds **54** and **55** (Table 18.4), the opposite appears to be the case. Exchange of the terminal pyrrolidine system with a morpholine unit significantly reduces pK_a and provides an improvement in hERG activity, with slight reduction in $c\log P$. Indeed, the authors conclude that it was difficult to correlate hERG activity with basicity and lipophilicity, suggesting that the overall properties of the molecule were more important.

Another series of VEGFR-2 kinase inhibitors that displayed significant activity at the hERG channel were the pyrrolotriazine-based systems exemplified by **56** [34]. The authors quickly determined that the basic amine side chain was responsible for the antitarget activity. Compounds such as **57**, where the basic amine was replaced with a sulfone derivative, exhibited considerably reduced activity at hERG, as well as reduced inhibition of CYP3A4. The alkyloxy region of the molecule proved to be highly tolerant of substitution. This enabled the use of alternative solubilizing groups, with the secondary alcohol **58** (Figure 18.5) exhibiting a similar profile. Compound **58** was then derivatized as a prodrug, which then became the development candidate.

In addition to DSM and zwitterion-based strategies, Merck scientists also employed a pK_a lowering approach in an effort to improve hERG selectivity within a series of conformationally constrained hNK1 antagonists [15]. Introduction of a fluorine atom at the C-6 position of the azabicyclic ring to give **59** (calc p$K_a = 5.00$) maintained the hNK1 affinity observed in the parent des-fluoro analogue **11** (calc p$K_a = 7.3$), but dramatically increased the selectivity over hERG ($K_i > 10\,\mu M$).

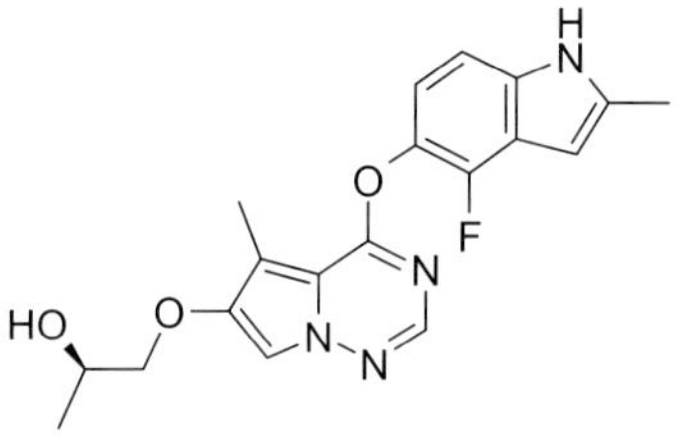

58
VEGFR IC_{50} = 25 nM
hERG IC_{50} = 18 μM

Figure 18.5 Alcohol containing VEGFR-2 inhibitor used for a prodrug approach.

Table 18.4 Reduction in pKa to facilitate selectivity over hERG.

hERG active	hERG^i (μM)	IC$_{50}^{j,k}$ (nM)	c log P^l	pK$_a^l$	hERG optimized	hERG^i (μM)	IC$_{50}^{j,k}$ (nM)	c log P^l	pK$_a^l$	Ref
54	0.38	9	1.70	8.60	55	5	12	1.05	7.41	[10]
56	66% at 1 μM	20	4.99	10.14	57	20	70	3.00		[34]
11	0.1	1	4.34	7.3	59	>10	1.5	5.12	5	[15]

(Continued)

Table 18.4 (Continued)

hERG active	hERG[j] (μM)	IC$_{50}$[j,k] (nM)	c log P[l]	pK$_a$[l]	hERG optimized	hERG[j] (μM)	IC$_{50}$[j,k] (nM)	c log P[l]	pK$_a$[l]	Ref
60	1	0.33	2.94	7.52	**62**	2.5	0.68	3.23	5.81	[21]
63	0.025	5	6.61	9.41	**64**	0.85	18	6.54	7.45	[36]

[a] Both hERG and primary target activity refer to K$_i$, IC$_{50}$ or IP unless otherwise stated.
[b] IC$_{50}$ refers to activity at the primary target.
[c] log P and pK$_a$ are calculated according to [9]. pK$_a$ is reported as the highest basic pK$_a$. Where no value of pK$_a$ is reported, calculation of this parameter was not possible.

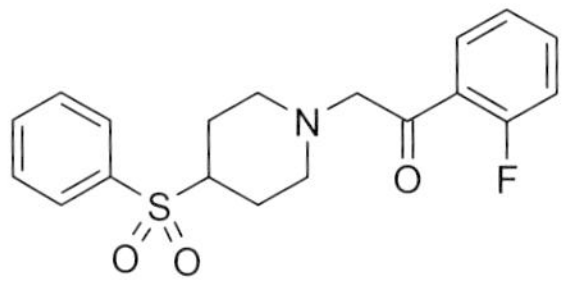

61
pKa = 6.3
hERG IP = 6.8 µM
5-HT$_{2A}$ IC$_{50}$ = 0.68 nM

Figure 18.6 hERG optimization in the 5-HT$_{2A}$ program through attenuation of pK_a.

However, this change was accompanied by a reduction in the duration of action in vivo.

Returning to the 5-HT$_{2A}$ area, Fletcher et al. [35] demonstrate how the control of pK_a can be used to gain additional selectivity over hERG. Compound **60** was found to prolong the QT interval in the anesthetized ferret by more than 10% at doses of 3 and 10 mg/kg/h. A range of modifications were made to **60** to achieve selectivity over hERG. Some success was achieved through the introduction of a ketone at the β-position to the amine (compound **61**, Figure 18.6). Such attenuation of the pK_a of the piperidine system yields a commensurate reduction in activity at hERG. The optimized compound did not show QT prolongation in the anesthetized ferret model at doses up to 10 mg/kg/h and exhibited acceptable pharmacokinetic properties. Despite these improvements, compound **61** was found to degrade in polar solvents such as methanol because of the presence of the ketone moiety and therefore could not be developed further [21]. As control of pK_a was found useful in governing activity at hERG, the group sought an additional approach to modulating pK_a. In this case, a series of 4-fluorosulfones were targeted. These were found to provide acceptable levels of selectivity over hERG, with compound **62** (Table 18.4) showing no increase in QT prolongation in anesthetized dogs at plasma concentrations up to 148 µM. Further refinement of hERG activity was also possible within this series by judicious positioning of a fluorine atom on the phenyl sulfone system (cf. **27** and **28**, Table 18.1).

A series of highly potent and selective indolylindazoylmaleimide inhibitors of protein kinase C-β (PKC-β), reported by Johnson & Johnson, were shown to have hERG activity [36]. Compound **63**, the most potent kinase inhibitor in the series (IC$_{50}$ = 5 nM), was found to have substantial activity at hERG (K_i = 25 nM). Intravenous infusion of this compound into anesthetized pigs resulted in a dose-dependent prolongation in the QTc interval by 7, 11 and 15% at 1, 3 and 10 mg/kg, respectively. According to the authors, the SAR in this series (not all reported) suggests that the basic amino group is pivotal not only for high PKC-β potency but also for hERG affinity control, which raised a concern about tractability of this class of PKC inhibitors. Interestingly, the less basic morpholine analogue **64** showed significantly attenuated hERG activity (K_i = 850 nM), compared to the parent dimethyl amine **63** while essentially conserving activity at PKC.

18.3
Summary and Analysis

18.3.1
Summary of Optimization Literature

This analysis suggests that the most regularly employed method of achieving selectivity over hERG is to make discrete structural modifications to the drug molecule. The effect of this change may result in loss of putative π-stacking interactions with the channel. DSM can be achieved by subtle changes in the template (e.g. replacement of aryl systems with a cyclohexyl, increasing steric bulk), which has minimal overall impact on molecular properties, or by making more significant alterations to the compound. Control of log P and pK_a are both highly effective as a means of controlling hERG activity. Indeed, these two approaches would appear to be closely related as a reduction in pK_a very often results in a reduction of log P. Reducing log P is clearly a useful approach owing to concomitant benefits in, for example, microsomal clearance and other crucial DMPK parameters. Out of the four categories, it is suggested that the zwitterion approach is of lowest general use owing to the attendant issues with lower permeability and hence reduced oral bioavailability/CNS exposure that are often observed.

18.3.2
Global Analysis of Reported Optimizations

A comprehensive survey covering the literature up to 2006 has yielded a data set of 58 optimizations comprising 610 hERG data points. The analysis of the hERG optimization data presented both here and in a previous publication [37] did not show any consistent trends across different optimization series with descriptors intuitive to medicinal chemists (e.g. polar surface area, $c\log P$, CMR and highest calculated basic pK_a). However, the relationship between the starting point for hERG optimization and the strategy adopted can give valuable insight into the most appropriate approach to take to reduce hERG activity. On the basis of the optimization data, a simple decision tree mnemonic can be derived (Figure 18.7). Assuming that a higher frequency of optimization category from a given starting point indicates a higher probability of success of a particular approach, then some basic guidelines can be proposed for the optimization of the undesirable hERG activity. When $c\log P < 3.0$ for the molecule to be optimized, then DSM is the most frequently observed category. For cases when $c\log P \geq 3.0$, then the correlation between $c\log P$ and hERG activity in the chemical series should be examined. For each optimization from a chemical series with a minimum of four hERG data points, the correlation coefficient R was calculated between $c\log P$ and hERG activity for the series. For $R \geq 0.5$, the $c\log P$ category dominates, and for $R < 0.5$, there is a more even distribution of categories with DSM and ZI most frequent. It is difficult to draw any conclusions related to the ZI and pK_a categories as both categories are observed to some degree across all

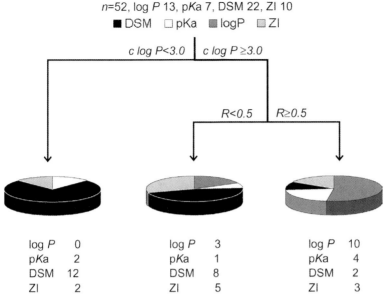

Figure 18.7 Optimization categories classified by $c \log P$ of the compound to be optimized and the correlation coefficient R between $c \log P$ and $-\log[\text{hERG activity}]$ for the series.

three terminal nodes of the decision tree. The calculated highest basic pK_a of the compound for optimization did not show any relationship with the pK_a or ZI categories, though trends may become apparent as more optimization examples are collected. This simple framework is sensible from a medicinal chemistry perspective and is consistent with the previously published findings derived from a smaller data set [37].

18.3.3
Summary and Recommendations

On the basis of the literature data summarized here and in an earlier publication [37], we can propose the following recommendations:

- Four strategies for the removal of hERG have been identified: discrete structural modifications, control of log P, formation of zwitterions and control of pK_a.
- All four strategies were found to be equally efficacious in diminishing hERG activity. Literature evidence indicates that the zwitterion approach is of lower general applicability owing to issues with membrane permeability and oral bioavailability.
- Consideration of $c \log P$ of the starting compound enables selection of the most appropriate strategy for lowering hERG activity.

- Where $c\log P \geq 3.0$: seek to reduce this by, for example, incorporation of heteroatoms, polar groups or removal of lipophilic moieties. On average 1 log unit reduction in $c\log P$ leads to 0.8 log unit reduction in hERG activity. If no correlation between hERG activity and $\log P$ can be established in the series then pursue the DSM strategy.
- If $c\log P < 3.0$: discrete structural modifications afford the highest probability of success

- A corollary to the above is to remove or modify aryl moieties in the target molecule. This may have the effect of reducing $c\log P$ and potentially disrupting π-stacking with the channel.
- Reducing pK_a of a basic nitrogen is a frequently employed strategy in attenuating hERG. This is often associated with reduction in $\log P$, which may be the more relevant parameter.
- A basic nitrogen (although not a prerequisite) is often associated with hERG activity; therefore, alternative solubilizing groups should be considered.
  - hERG data should be interpreted in the context of measured solubility. Compounds with low aqueous solubility (<5 mg/L) may have significantly underestimated activity in the hERG assay.

- Only optically pure material should be tested in the hERG assay as the effects due to stereochemistry can be dramatic.
- To ensure that the effects of discrete structural modifications can be identified and to establish reliable SAR, test a significant number of analogues within each chemical series against hERG.
- Where the amount of data available permits, consider developing a local (series specific) *in silico* model to guide medicinal chemistry efforts away from hERG activity.
  - Application of global *in silico* models is best reserved for prioritization of compounds for synthesis/acquisition from larger arrays.

- To ascertain the effects of non-hERG-mediated QT prolongation, test key compounds in relevant *ex vivo* (e.g. Langendorff [38], dog purkinje fibers [39]) or *in vivo* [40] models as early as possible.

18.3.4
Future Directions

There now exists an ever-burgeoning body of information on the tactics for optimizing against the hERG channel, which may ultimately be used to refine further the classification system discussed here. In particular, new SAR information could be used to create subzones within, for example, the DSM class of optimizations or refine the heuristics based on the nature of the starting compound or the target family it belongs to. It is anticipated that the expanding number of ion channel crystal structures, including mammalian channels [41], will have a significant influence on our understanding of drug–hERG interactions. Increasing attention is being paid to the identification of additional ion channels or mechanisms implicated in drug-induced cardiac toxicity. For example, there is evidence to suggest that drugs such as sevoflurane can exert cardiac toxicity not through hERG but via reduction in KvLQT1/minK potassium channel currents [42]. Similarly, compounds such as pentamidine may induce QT prolongation via inhibition of hERG trafficking rather than through inhibition of hERG [43]. Future optimization trajectories are likely to pay increasing attention to these and other CV-related antitargets.

Abbreviations

CMR	calculated molar refractivity
CNS	central nervous system
CV	cardiovascular
DSM	discrete structural modifications
FTI	farnesyl transferase inhibitor
hERG	human ether-a-go-go gene related product
IP	inflection point
KcsA	*Streptomyces lividans* potassium channel
KDR	kinase insert domain protein receptor
K_i	equilibrium dissociation constant
NMDA	*N*-methyl-D-aspartate
PKC	protein kinase C
QT interval	time between end of Q wave and start of T wave
TdP	*torsades de pointes*
ZI	zwitterion

References

1 Abbott, G.W., Sesti, F., Splawski, I., Buck, M.E., Lehmann, M.H., Timothy, K.W., Keating, M.T. and Goldstein, S.A. (1999) MiRP1 forms IKr potassium channels with HERG and is associated with cardiac arrhythmia. *Cell*, **97**, 175–187.

2 Fermini, B. and Fossa, A.A. (2003) The impact of drug-induced QT interval prolongation on drug discovery and development. *Nature Reviews Drug Discovery*, **2**, 439–447.

3 Keating, M.T. and Sanguinetti, M.C. (2001) Molecular and cellular mechanisms of cardiac arrhythmias. *Cell*, **104**, 569–580.

4 Pearlstein, R., Vaz, R. and Rampe, D. (2003) Understanding the structure–activity relationship of the human ether-a-go-go-related gene cardiac K^+ channel. A model for bad behavior. *Journal of Medicinal Chemistry*, **46**, 2017–2022.

5 Mitcheson, J.S., Chen, J., Lin, M., Culberson, C. and Sanguinetti, M.C. (2000) A structural basis for drug-induced long QT syndrome. *Proceedings of the National Academy of Sciences of the United States of America*, **97**, 12329–12333.

6 Pearlstein, R.A., Vaz, R.J., Kang, J., Chen, X.-L., Preobrazhenskaya, M., Shchekotikhin, A.E., Korolev, A.M., Lysenkova, L.N., Miroshnikova, O.V., Hendrix, J. and Rampe, D. (2003) Characterization of HERG potassium channel inhibition using CoMSiA 3D QSAR and homology modeling approaches. *Bioorganic & Medicinal Chemistry Letters*, **13**, 1829–1835.

7 Fernandez, D., Ghanta, A., Kauffman, G.W. and Sanguinetti, M.C. (2004) Physicochemical features of the hERG channel drug binding site. *Journal of Biological Chemistry*, **279**, 10120–10127.

8 Mitcheson, J.S. and Perry, M.D. (2003) Molecular determinants of high-affinity drug binding to HERG channels. *Current Opinion in Drug Discovery and Development*, **6**, 667–674.

9 $c \log P$'s calculated with clog P 4.10 or clog P 4.3, BioByte Corp. 201 W. 4th St. #204 Claremont, CA 91711-4707 USA, pK_a's calculated with ACD PhysChem Batch, version 4.76 or 9.03, Advanced Chemistry Development, Inc., Toronto ON, Canada.

10 Sisko, J.T., Tucker, T.J., Bilodeau, M.T., Buser, C.A., Ciecko, P.A., Coll, K.E., Fernades, C., Gibbs, J.B., Koester, T.J., Kohl, N., Lynch, J.J., Mao, X., McLoughlin, D., Miller-Stein, C.M., Rodman, L.D., Rickert, K.W., Sepp-Lorrenzino, L., Shipman, J.M., Thomas, K.A., Wong, B.K. and Hartman, G.D. (2006) Potent 2-[(pyrimidin-4-yl)amine]-1,3-thiazole-5-carbonitrile-based inhibitors of VEGFR-2 (KDR) kinase. *Bioorganic & Medicinal Chemistry Letters*, **16**, 1146–1150.

11 Bilodeau, M.T., Balitza, A.E., Koester, T.J., Manley, P.J., Rodman, L.D., Buser-Doepner, C., Coll, K.E., Fernandes, C., Gibbs, J.B., Heimbrook, D.C., Huckle, W.R., Kohl, N., Lynch, J.J., Mao, X., McFall, R.C., McLoughlin, D., Miller-Stein, C., Rickert, K.W., Sepp-Lorenzino, L., Shipman, J.M., Subramanian, R., Thomas, K.A., Wong, B.K., Yu, S. and Hartman, G.D. (2004) Potent N-(1,3-thiazoyl-2-yl)pyridine-2-amine vascular endothelial growth factor receptor tyrosine kinase inhibitors with excellent pharmacokinetics and low affinity for the hERG ion channel. *Journal of Medicinal Chemistry*, **47**, 6363–6372.

12 Walker, D.P., Wishka, D.G., Piotrowski, D.W., Jia, S., Reitz, S.C., Yates, K.M., Myers, J.K., Vetman, T.N., Margolis, B.J., Jacobsen, E.J., Acker, B.A., Groppi, V.E., Wolfe, M.L., Thornburgh, B. A., Tinholt, P.M., Cortes-Burgos, L.A., Walters, R.R., Hester, M.R., Seest, E.P., Dolak, B.A., Han, F., Olson, B.A., Fitzgerald, L., Staton, B.A., Raub, T.J., Hajos, M., Hoffman, W.E., Li, K.S., Higdon, N.R., Wall, T.M., Hurst, R.S., Wong, E.H.F. and Rogers, B.N. (2006) Design, synthesis, structure–activity relationship, and *in vivo* activity of azabicyclic aryl amides as $\alpha 7$ nicotinic acetylcholine agonists. *Bioorganic & Medicinal Chemistry*, **14**, 8219–8248.

13 Wishka, D.G., Walker, D.P., Yates, K.M., Reitz, S.C., Jia, S., Myers, J.K., Olson, K.L., Jacobsen, E.J., Wolfe, M.L., Groppi, V.E., Hanchar, A.J., Thornburgh, B.A., Cortes-Burgos, L.A., Wong, E., Staton, B. A., Raub, T.J., Higdon, N.R., Wall, T.M., Hurst, R.S., Walters, R.R., Hoffman, W.E., Hajos, M., Franklin, S., Carey, G., Gold, L.H., Cook, K.K., Sands, S.B., Zhao, S.X., Soglia, J.R., Kalgutkar, A.S., Arneric, S.P. and Rogers, B.N. (2006) Discovery of N-[(3R)-1-azabicyclo[2.2.2]oct-3-yl]furo[2,3-c]pyridine-5-carboxamide, an agonist of the $\alpha 7$ nicotinic acetylcholine receptor, for the potential treatment of cognitive deficits in schizophrenia: synthesis and structure–activity relationship. *Journal of Medicinal Chemistry*, **49**, 4425.

14 Price, D.A., Armour, D., de Groot, M., Leishman, D., Napier, C., Perros, M., Stammen, B. and Wood, A. (2006) Overcoming HERG affinity in the discovery of the CCR5 antagonist

maraviroc. *Bioorganic & Medicinal Chemistry Letters*, **16**, 4633–4637.

15 Huscroft, I.T., Carlson, E.J., Chicchi, G.G., Kurtz, M.M., London, C., Raubo, P., Wheeldon, A. and Kulagowski, J.J. (2006) Phenyl-8-azabicyclo[3.2.1]octane ethers: a novel series of neurokinin (NK$_1$) antagonists. *Bioorganic & Medicinal Chemistry Letters*, **16**, 2008–2012.

16 Becker, O.M., Dhanoa, D.S., Marantz, Y., Chen, D., Shacham, S., Cheruku, S., Heifetz, A., Mohanty, P., Fichman, M., Sharadendu, A., Nudelman, R., Kauffman, M. and Noiman, S. (2006) An integrated *in silico* 3D model-driven discovery of a novel, potent, and selective amidosulfonamide 5-HT1A agonist (PRX-00023) for the treatment of anxiety and depression. *Journal of Medicinal Chemistry*, **49**, 3116–3135.

17 Bell, I.M., Gallicchio, S.N., Abrams, M., Beshore, D.C., Buser, C.A., Culberson, J.C., Davide, J., Ellis-Hutchings, N., Fernandes, C., Gibbs, J.B., Graham, S.L., Hartman, G.D., Heimbrook, D.C., Homnick, C.F., Huff, J.R., Hassahun, K., Koblan, K.S., Kohl, N.E., Lobell, R.B., Lynch, J.J., Miller, P.A., Omer, C.A., Rodrigues, A.D., Walsh, E.S. and Williams, T.M. (2001) Design and biological activity of (S)-4-(5-{[1-(3-chlorobenzyl)-2-oxopyrrolidin-3-ylamino]methyl}imidazol-1-ylmethyl) benzonitrile, α3-aminopyrrolidinone farnesyltransferase inhibitor with excellent cell potency. *Journal of Medicinal Chemistry*, **44**, 2933–2949.

18 Bell, I.M., Gallicchio, S.N., Abrams, M., Beese, L.S., Beshore, D.C., Bhimnathwala, H., Bogusky, M.J., Buser, C.A., Culberson, J.C., Davide, J., Ellis-Hutchings, N., Fernandes, C., Gibbs, J.B., Graham, S.L., Hamilton, K.A., Hartman, G.D., Heimbrook, D.C., Homnick, C.F., Huber, H., Huff, J.R., Hassahun, K., Koblan, K.S., Kohl, N.E., Lobell, R.B., Lynch, J.J., Robinson, R., Rodrigues, A.D., Taylor, J.S., Walsh, E.S., Williams, T.M. and Zartman, C.B. (2002) 3-Aminopyrrolidinone farnesyltransferase inhibitors: design of macrocyclic compounds with improved pharmacokinetics and excellent cell potency. *Journal of Medicinal Chemistry*, **45**, 2388–2409.

19 McCauley, J.A., Theberge, C.R., Romano, J.J., Billings, S.B., Anderson, K.D., Claremon, D.A., Freidinger, R.M., Bednar, R.A., Mosser, S.D., Gaul, S.L., Connolly, T.M., Condra, C.L., Xia, M., Cunningham, M.E., Bednar, B., Stump, G.L., Lynch, J.J., Macaulay, A., Wafford, K.A., Koblan, K.S. and Liverton, N.J. (2004) NR2B-selective *N*-methyl-D-aspartate antagonists: synthesis and evaluation of 5-substituted benzimidazoles. *Journal of Medicinal Chemistry*, **47**, 2089–2096.

20 Kim, D., Wang, L., Hale, J.J., Lynch, C.L., Budhu, R.J., MacCoss, M., Mills, S.G., Malkowitz, L., Gould, S.L., Demartino, J.A., Springer, M.S., Hazuda, D., Miller, M., Kessler, J., Hrin, R.C., Carver, G., Carella, A., Henry, K., Lineberger, J., Schleif, W.A. and Emini, E.A. (2005) Potent 1,3,4-trisubstituted pyrrolidine CCR5 receptor antagonists: effects of fused heterocycles on antiviral activity and pharmacokinetic properties. *Bioorganic & Medicinal Chemistry Letters*, **15**, 2129–2134.

21 Fish, L.R., Gilligan, M.T., Humphries, A.C., Ivarsson, M., Ladduwahetty, T., Merchant, K.J., O'Connor, D., Patel, S., Philipps, E., Vargas, H.M., Hutson, P.H. and MacLeod, A.M. (2005) 4-Fluorosulfonylpiperidines: selective 5-HT$_{2A}$ ligands for the treatment of insomnia. *Bioorganic & Medicinal Chemistry Letters*, **15**, 3665–3669.

22 Rampe, D., Wibble, B., Brown, A.M. and Dage, R.C. (1993) Effects of terfenadine and its metabolites on a delayed rectifier K^+ channel cloned from human heart. *Molecular Pharmacology*, **44**, 1240–1245.

23 Ghosh, S., Elder, A., Guo, J., Mani, U., Patane, M., Carson, K., Ye, Q., Bennett, R., Chi, S., Jenkins, T., Guan, B., Kolbeck, R.,

Smith, S., Zhang, C., LaRosa, G., Jaffee, B., Yang, H., Eddy, P., Lu, C., Uttamsingh, V., Horlick, R., Harriman, G. and Flynn, D. (2006) Design, synthesis, and progress toward optimisation of potent small molecule antagonists of CC chemokine receptor 8 (CCR8). *Journal of Medicinal Chemistry*, **49**, 2669–2672.

24 Thomson, C.G., Carlson, E., Chicchi, G.G., Kulagowski, J.J., Kurtz, M.M., Swain, C.J., Tsao, K.C. and Wheeldon, A. (2006) Synthesis and structure–activity relationships of 8-azabicyclo[3.2.1]octane benzylamine NK_1 antagonists. *Bioorganic & Medicinal Chemistry Letters*, **16**, 811–814.

25 Zhu, B.-Y., Jia, Z.J., Zhang, P., Su, T., Huang, W., Goldman, E., Tumas, D., Kadambi, V., Eddy, P., Sinha, U., Scarborough, R.M. and Song, Y. (2006) Inhibitory effect of carboxylic acid group on hERG binding. *Bioorganic & Medicinal Chemistry Letters*, **16**, 5507–5512.

26 Waring, M.J. and Johnstone, C. (2007) A quantitative assessment of hERG liability as a function of lipophilicity. *Bioorganic & Medicinal Chemistry Letters*, **17**, 1759–1764.

27 Xu, J., Mathvink, R., He, J., Park, Y.J., He, H., Leiting, B., Lyons, K.A., Marsilio, F., Patel, R.A., Wu, J.K., Thornberry, N.A. and Weber, A.E. (2005) Discovery of potent and selective phenylalanine based dipeptidyl peptidase IV inhibitors. *Bioorganic & Medicinal Chemistry Letters*, **15**, 2533–2536.

28 Edmondson, S.D., Mastracchio, A., Beconi, M., Colwell, L.F., Habulihaz, B., He, H., Kumar, S., Leiting, B., Lyons, K.A., Mao, A., Marsilio, F., Patel, R.A., Wu, J.K., Zhu, L., Thornberry, N.A., Weber, A.E. and Parmee, E.R. (2004) Potent and selective proline derived dipeptidyl peptidase IV inhibitors. *Bioorganic & Medicinal Chemistry Letters*, **14**, 5151–5155.

29 Edmondson, S.D., Mastracchio, A., Duffy, J., Eiermann, G.J., He, H., Ita, I., Leiting, B., Leone, J.F., Lyons, K.A., Makarewicz, A.M., Patel, R.A., Petrov, A., Wu, J.K., Thornberry, N.A. and Weber, A.E. (2005) Discovery of potent and selective orally bioavailable β-substituted phenylalanine derived dipeptidyl peptidase IV inhibitors. *Bioorganic & Medicinal Chemistry Letters*, **15**, 3048–3052.

30 Blackburn, C., LaMarche, M.J., Brown, J., Lee Che, J., Cullis, C.A., Lai, S., Maguire, M., Marsilje, T., Geddes, B., Govek, E., Kadambi, V., Doherty, C., Dayton, B., Brodjian, S., Marsh, K.C., Collins, C.A. and Kym, P.R. (2006) Identification and characterisation of amino-piperidinequinolones and quinazolinones as MCHr1 antagonists. *Bioorganic & Medicinal Chemistry Letters*, **16**, 2621–2627.

31 Vaz, R.J., Gao, Z., Pribish, J., Chen, X., Levell, J., Davis, L., Albert, E., Brollo, M., Ugolini, A., Cramer, D.M., Cairns, J., Sides, K., Liu, F., Kwong, J., Kang, J., Rebello, S., Elliot, M., Lim, H., Chellaraj, V., Singleton, R.W. and Li, Y. (2004) Design of bivalent ligands using hydrogen bond linkers: synthesis and evaluation of inhibitors for human β-tryptase. *Bioorganic & Medicinal Chemistry Letters*, **14**, 6053–6056.

32 Rowley, M., Hallet, D.J., Goodacre, S., Moyes, C., Crawforth, J., Sparey, T.J., Patel, S., Marwood, R., Patel, S., Thomas, S., Hitzel, L., O'Connor, D., Szeto, N., Castro, J.L., Hutson, P.H. and MacLeod, A.M. (2001) 3-(4-Fluoropiperidin-3-yl)-2-phenylindoles as high affinity, selective, orally bioavailable h5-HT_{2A} receptor antagonists. *Journal of Medicinal Chemistry*, **44**, 1603–1614.

33 Friesen, R.W., Ducharme, Y., Ball, R.G., Blouin, M., Boulet, L., CôtÕ, B., Frenette, R., Girard, M., Guay, D., Huang, Z., Jones, T.R., LalibertÕ, F., Lynch, J.J., Mancini, J., Martins, E., Masson, P., Muise, E., Pon, D.J., Siegl, P.K.S., Styhler, A., Tsou, N.N., Turner, M.J., Young, R.N. and Girard, Y. (2003) Optimization of a tertiary alcohol series of phosphodiesterase-4 (PDE4) inhibitors: structure–activity relationship related to PDE4 inhibition and human ether-a-go-go related gene potassium

channel binding affinity. *Journal of Medicinal Chemistry*, **46**, 2413–2426.

34 Bhide, R.S., Cai, Z.W., Zhang, Y.Z., Quian, L., Wei, D., Barbosa, S., Lombardo, L.J., Borzilleri, R.M., Zheng, X., Wu, L.I., Barrish, J.C., Kim, S.H., Leavitt, K., Mathur, A., Leith, L., Chao, S., Wautlet, B., Mortillo, S., Jeyaseelan, R., Kukral, D., Hunt, J.T., Kamath, A., Fura, A., Vyas, V., Marathe, P., D'Arienzo, C., Derbin, G. and Fargnoli, J. (2006) Discovery and preclinical studies of (R)-1-(4-(4-fluoro-2-methyl-1H-indol-5-yloxy)-5-methylpyrrolo[2,1-f][1,2,4]triazin-6-yloxy)propan-2-ol (BMS-540215), an *in vivo* active potent VEGFR-2 inhibitor. *Journal of Medicinal Chemistry*, **49**, 2143–2146.

35 Fletcher, S.R., Burkamp, F., Blurton, P., Cheng, S.K.F., Clarkson, R., O'Connor, D., Spinks, D., Tudge, M., van Niel, M.B., Patel, S., Chapman, K., Marwood, R., Shepheard, S., Bentley, G., Cook, G.P., Bristow, L.J., Castro, J.L., Hutson, P.H. and MacLeod, A.M. (2002) 4-(Phenylsulfonyl)piperidines: novel, selective, and bioavailable 5-HT$_{2A}$ receptor antagonists. *Journal of Medicinal Chemistry*, **45**, 492–503.

36 Zhang, H.C., Derian, C.K., McComsey, D.F., White, K.B., Ye, H., Hecker, L.R., Li, J., Addo, M.F., Croll, D., Eckardt, A.J., Smith, C.E., Li, Q., Cheung, W.M., Conway, B.R., Emanuel, S., Demarest, B.P. and Maryanoff, B.E. (2005) Novel indolylindazolylmaleimides as inhibitors of protein kinase C-β: synthesis, biological activity and cardiovascular safety. *Journal of Medicinal Chemistry*, **48**, 1725–1728.

37 Jamieson, C., Moir, E.M., Rankovic, Z. and Wishart, G. (2006) Medicinal chemistry of hERG optimizations: highlights and hang-ups. *Journal of Medicinal Chemistry*, **49**, 5029–5046.

38 Valentin, J.-P., Hoffmann, P., De Clerck, F., Hammond, T.G. and Hondeghem, L. (2004) Review of the predictive value of the Langendorff heart model (Screenit system) in assessing the proarrhythmic potential of drugs. *Journal of Pharmacological and Toxicological Methods*, **49**, 171–181.

39 Gintant, G.A., Limberis, J.T., McDermott, J.S., Wegner, C.D. and Cox, B.F. (2001) The canine purkinje fiber: an *in vitro* model system for acquired long QT syndrome and drug-induced arrhythmogenesis. *Journal of Cardiovascular Pharmacology*, **37**, 607–618.

40 Carlsson, L. (2006) *in vitro* and *in vivo* models for testing arrhythmogenesis in drugs. *Journal of Internal Medicine*, **259**, 70–80.

41 Long, S.B., Campbell, E.B. and MacKinnon, R. (2005) Crystal structure of a mammalian voltage-dependent shaker family K$^+$ channel. *Science*, **309**, 897–903.

42 Kang, J., Reynolds, W.P., Chen, X.L., Ji, J., Wang, H. and Rampe, D.E. (2006) Mechanisms underlying the QT interval–prolonging effects of sevoflurane and its interactions with other QT-prolonging drugs. *Anesthesiology*, **104**, 1015–1022.

43 Kuryshev, Y.A., Ficker, E., Wang, L., Hawryluk, P., Dennis, A.T., Wible, B.A., Brown, A.M., Kang, J., Chen, X.L., Sawamura, K., Reynolds, W. and Rampe, D. (2005) Pentamidine-induced long QT syndrome and block of HERG trafficking. *Journal of Pharmacology and Experimental Therapeutics*, **312**, 316–323.

19
Structure-Based *In Silico* Driven Optimization: Discovery of the Selective 5-HT$_{1A}$ Agonist PRX-00023

Oren M. Becker

19.1
Introduction

Target selectivity is one of the main challenges that the pharmaceutical R&D is facing, aiming at the discovery and development of effective and safe drugs. Interaction of drug candidates with off targets is a pitfall that complicates further development since it often results in undesired side effects and may pose safety hazards. Given the increased sensitivity of regulatory authorities to drug safety, it is highly desired to discover drug candidates that are selective to their designated drug targets and exhibit minimal affinity to off targets. However, achieving these selectivity goals while maintaining affinity to the target itself has proven to be challenging, especially because off-target selectivity has many possible sources. It may involve undesired interactions with subtypes of the same target class, interactions with targets from other classes or interactions with known safety hazards such as the human ether-a-go-go related gene (hERG) channel. As a consequence, a significant portion of the lead optimization process is typically devoted to optimizing selectivity and safety parameters (the other portion is devoted to optimizing pharmacokinetics). This has traditionally been addressed by medicinal chemistry using structure–activity relationship (SAR) analysis and typically requires the synthesis of many hundreds or even thousands of compounds. Traditional chemical optimization is thus a very time-consuming and expensive process. Hence, introduction of new methodologies and approaches that will enable a more efficient discovery of selective drug candidates is highly desirable. Such selective drugs are expected to have a higher chance of success in clinical studies with a lower safety risk. Structure-based molecular design approaches, especially when considering not only the target affinity but also the off-target profile and pharmacokinetic (PK) properties, are possible avenues to achieve this goal.

A specific example of the problems that arise from insufficient selectivity and affinity to off-target proteins is that many of the currently approved drugs for the treatment of anxiety and depression have troublesome side effects. According to the

Antitargets. Edited by R. J. Vaz and T. Klabunde
Copyright © 2008 WILEY-VCH Verlag GmbH & Co. KGaA, Weinheim
ISBN: 978-3-527-31821-6

US National Institute of Mental Health, anxiety (in particular generalized anxiety disorder or GAD) and depression are the most prevalent mental illnesses. In the United States alone, an estimated 4 million adults suffer from GAD and nearly 19 million adults are affected by depressive disorders. As GAD and depression are often present together, most patients suffering from these disorders are treated with the same drugs [1].

The most commonly used therapies for anxiety and depression are selective serotonin reuptake inhibitors (SSRIs) and the more recently developed serotonin noradrenaline reuptake inhibitors (SNRIs). SSRIs, which constitute 60% of the worldwide antidepressant and antianxiety market, are frequently associated with sexual dysfunction, appetite disturbances and sleep disorders. Because SSRIs and SNRIs increase 5-HT levels in the brain, they can indirectly stimulate all 14 serotonergic receptor subtypes [2,3], some of which are believed to lead to adverse side effects associated with these drugs. Common drugs for short-term relief of GAD are benzodiazepines. These sedating agents are controlled substances with addictive properties and can be lethal when used in combination with alcohol. The use of benzodiazepines is associated with addiction, dependency and cognitive impairment.

A therapeutic alternative for treatment of anxiety and depression is the use of 5-HT$_{1A}$ agonists. Azapirones comprise the major class of 5-HT$_{1A}$ agonists of which buspirone (Buspar [4]) is the only FDA-approved 5-HT$_{1A}$ selective agonist (relative to the other 13 serotonin receptor subtypes) for anxiety currently on the US market (Scheme 19.1). Buspirone has shown efficacy in randomized controlled trials of GAD for which it was approved [5–7]. Unlike benzodiazepines, buspirone is not addictive

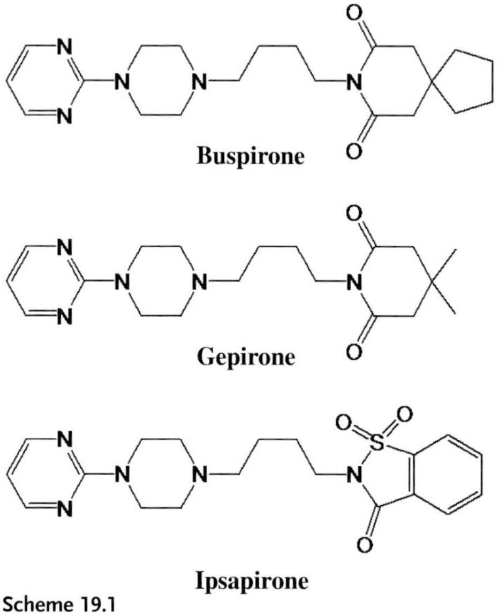

Buspirone

Gepirone

Ipsapirone

Scheme 19.1

even with long-term use and has no withdrawal effects even when the drug is discontinued quickly. Sexual dysfunction, very prominent in patients with depression and anxiety and a major side effect of other psychotropic agents, is not associated with the use of 5-HT_{1A} agonists. Buspirone has been reported to show modest activity in reducing SSRI-induced sexual side effects [8]. Other azapirones such as extended-release gepirone [9] and ipsapirone [10] (Scheme 19.1) have shown activity in major depressive disorder (MDD) studies. Gepirone, in an extended-release formulation, is now in preregistration stages for the treatment of major depressive disorder.

The main disadvantage of most azapirones is their insufficient selectivity toward adrenergic and dopaminergic receptors, mainly α_1 adrenergic and D_2 dopaminergic receptors. These result in side effects such as nausea, headache, nervousness, restlessness, dizziness, light headedness and fatigue, leading to poor tolerability. To overcome these side effects, buspirone has to be gradually titrated for several days or weeks until it reaches the therapeutic dose, taking weeks for the drug to be effective and not always reaching a fully effective dose. Its short half-life in humans necessitates three-times a day (tid) dosage, which also makes it less attractive compared to the once a day dosage of SSRIs.

Truly selective 5-HT_{1A} agonists that are selective toward other serotonergic receptor subtypes as well as to adrenergic and dopaminergic receptors remain highly desirable as potential treatment for depression and anxiety. In this report we discuss how application of a structure-based *in silico* driven optimization approach enabled us to discover a very selective and potent 5-HT_{1A} agonist (PRX-00023) for the treatment of depression, anxiety and ADHD, now in phase IIb clinical trials for depression. Applying this optimization strategy to a program in which selectivity is a key factor allowed us to discover a highly selective drug candidate within less than 6 months of lead optimization and with only 31 compounds synthesized during the entire program (and 200 compounds that were computationally evaluated).

19.2
Structure-Based *In Silico* Driven Multidimensional Optimization Paradigm

In this work we use a lead optimization paradigm, which is somewhat different from the classical structure-based approach. In general, classical 'structure-based' optimization means that the three-dimensional (3D) structural information of the protein target is used to design new analogues that will optimally bind to the target protein. This requires a 3D structure of the protein–ligand complex and involves a detailed analysis of the interaction pattern between the bound compound and the protein. On the basis of this analysis, the 'modeler' or 'structural biologist' suggests to the chemists possible modifications that will potentially enhance the interaction between the compound and the target protein. The 3D structure of the protein–ligand complex is obtained from crystallography, docking the compound into the X-ray structure of the target protein, or from modeling. However, such a structure-based approach addresses only the affinity to the target protein and does not help in the time-consuming effort of optimizing selectivity, absorption, distribution, metabolism and excretion (ADME) and pharmacokinetics.

For the structure-based approach to be effective throughout the lead optimization process, it has to be expanded beyond its regular limits and applied in the context of multidimensional compound optimization. For us, '*in silico* driven multidimensional optimization' or 'structure-based multidimensional optimization' means using structural information and *in silico* analyses to not only direct the synthesis proposals toward more active compounds but also navigate the multitude of possible pathways throughout the optimization process. We use it to optimize target affinity as well as selectivity and pharmacokinetics. The key to this approach is to pull together all the optimization proposals at each step along the way and prioritize them for synthesis according to the likelihood that they will move the program ahead toward its 'drug candidate' goal. We find it important not only to select which compounds should be synthesized but also to deprioritize proposed compounds that are likely to fail and should not be synthesized. Such prioritization and deprioritization ensure a more efficient optimization process.

As in regular structure-based molecular design, the 3D interaction patterns between the lead compounds and the target protein are analyzed, leading to new compound proposals that are expected to improve the affinity to the target protein. However, these form only a subset of the synthesis proposals at any given time. Many other proposals come from chemists' intuition, literature searches and past experience. Prioritizing and deprioritizing all the suggested compounds is at the core of our approach where only few of the suggested compounds are selected for actual synthesis. This is done by predicting *in silico* which synthesis proposals have higher probability to address specific optimization challenges and move the program toward a drug candidate that meets the 'candidate selection criteria'. For example, some of the compounds that are suggested for improving metabolic stability may be deprioritized following the *in silico* analysis if they are predicted to loose affinity to the target protein or if they are expected to increase undesired affinity to an off-target protein. This prioritization process has many sources: it makes use of 3D structures or models of the target protein, 3D structures or models of off-target proteins, various predictive ADME tools, synthetic feasibility assessments, patent protection and freedom-to-operate considerations and more. The result of this process is that all the suggested compounds are first analyzed and prioritized, resulting in only a small subset of them being actually synthesized and tested. It has been our experience from many lead optimization programs that resulted in clinical candidates, that only 10% of the suggested compounds need to be actually synthesized. This represents a 10-fold increase in efficiency for the optimization process. As lead optimization is an iterative process, this prioritization ensures that every compound is thought through prior to synthesis and that the 'hypothesis' or reason why it was selected for synthesis is clear to everyone in the team. Awareness of these hypotheses ensures that information is rapidly deduced from failed compounds as much as from successful ones, translating in turn to a quick and effective optimization process.

19.3
Clinical Candidate Selection Criteria

The first step in any of our drug discovery programs is to define the multidimensional selection criteria according to which the clinical candidate compound will be selected in due time. Having these criteria defined before the discovery program starts ensures that (1) the program is kept focused on its goal, which is obtaining a clinical candidate and moves ahead by addressing the key issues and (2) the drug candidate is recognized when it is reached, namely when most (if not all) of the selection criteria are met. In general, these selection criteria include target affinity and activity, off-target selectivity, pharmacokinetic parameters, activity in animal models, basic chemistry, manufacturing and control (CMC) parameters (e.g. chemical stability and salt forms) and basic toxicology.

Specifically, for the 5-HT$_{1A}$ agonist program, the selection criteria included a 5-HT$_{1A}$ affinity goal of <10 nM, a 5-HT$_{1A}$ *in vitro* agonist activity goal of <50 nM and a 5-HT$_{1A}/\alpha_1$ (and α_2) selectivity ratio more than 200-fold and more than 100-fold selectivity to other receptors. 5-HT$_{1A}$/hERG selectivity was set to >100-fold, with a hERG IC$_{50}$ goal of >3000 nM (patch clamp). The remaining selection criteria are not relevant to our discussion here.

19.4
Lead Identification

The lead compound was discovered through structure-based *in silico* screening using the PREDICT 3D model of the 5-HT$_{1A}$ receptor. PREDICT is a unique nonhomology-based modeling method for generating 3D models of membrane-embedded G-protein-coupled receptors (GPCRs) [11,12], which have been successfully used for *in silico* screening [13,14]. Virtual screening of a 40 000 compound library using the PREDICT 5-HT$_{1A}$ 3D model led to the selection of 78 compounds for *in vitro* testing. Sixteen of the tested compounds were confirmed as hits with binding affinities of K_i <5 µM in *in vitro* 5-HT$_{1A}$ radioligand binding assays (21% hit rate). The best hit was a novel 1.0 nM arylpiperazinylsulfonamide compound, **1**, found to be a partial agonist in a functional *in vitro* assay (intrinsic activity of 65% relative to 5-HT) with an EC$_{50}$ of 21 nM. For more details of the screening process see [14]. This activity compared favorably with buspirone and gepirone alike (buspirone: $K_i = 20$ nM, EC$_{50}$ ~ 80 nM and intrinsic activity of 50%; gepirone: $K_i = 13$ nM, EC$_{50}$ 288 nM and intrinsic activity of 80%). Compound **1** also showed drug-like pharmacokinetic properties despite its low oral bioavailability in rats and was selected as the lead compound for further optimization.

Lead, 1

As is typical of serotonergic drugs and in particular of 5-HT$_{1A}$ agonists, the main downside of compound **1** was its selectivity profile. Testing **1** *in vitro* against a panel of 50 G-protein-coupled receptors showed insufficient selectivity primarily toward α_1 and α_{2a} adrenergic receptors ($K_i = 6$ and 12 nM, respectively). Affinities below 100 nM were measured also for 5-HT$_7$, 5-HT$_{2B}$ and the D$_2$ dopamine receptor. On the other hand the hERG K$^+$ channel affinity of compound **1** was very low, with an IC$_{50} > 5000$ nM in a patch-clamp assay, which is considered safe.

19.5
Optimization Round 1: Reducing Off-Target Activities

Given the challenges of current medication for anxiety and depression, it was evident that a competitive 5-HT$_{1A}$ partial agonist with advantages over the azapirone family must be selective toward all other biogenic amine receptors, serotonergic, dopaminergic and adrenergic in a similar manner. This meant that the first priority was to improve the selectivity of the lead compound **1** toward α-adrenergic receptors. On the basis of the azapirone data, we targeted a 5-HT$_{1A}$/α_1 selectivity ratio above 200 as the highest priority for the optimization process.

Published research on several classes of 5-HT$_{1A}$ ligands has demonstrated that introducing α-adrenergic and D$_2$ selectivity while maintaining 5-HT$_{1A}$ affinity is challenging [15–22]. Most interesting is a series of studies by Lopez-Rodriguez *et al.* [23–30] who characterized a range of arylpiperarzines using SAR and QSAR analyses. These studies have shown that the pyrimidinepiperazine or arylpiperazine moiety of the compound forms the key structural motif of these series for recognition at the 5-HT$_{1A}$ receptor. These studies also showed that most of these compounds exhibit high levels of undesirable affinity to α_1 adrenergic receptor and that it is not trivial to engineer potent compounds that are selective to one or the other, though examples of selective 5-HT$_{1A}$ agonists have been reported. Homology models, which use the structure of rhodopsin as a template, have recently been used to rationalize some of this data [28,30].

Given the ability of the PREDICT technology to reliably model the 3D structure of GPCRs, we used PREDICT to model the 3D structure of not only the 5-HT$_{1A}$ receptors but also the α_1 receptor (complexed with compound **1**). With the two 3D models, 5-HT$_{1A}$ and α_1, at hand, it was now possible to compare the binding mode of **1** in the binding pocket of both receptors and look for optimization strategies that will differentiate the two. Figure 19.1 shows the lead compound **1** in the binding pocket of the 5-HT$_{1A}$ receptor and in the binding pocket of the α_1 adrenergic receptor. Not surprisingly, the compound fits very well into both binding pockets, in agreement with its high affinity to both receptors. As can be seen, the 3D model of the 5-HT$_{1A}$ binding pocket suggests that the aryl moiety of **1** interacts with Phe362 from TM6, the piperazine moiety interacts with Asp116 from TM3 and residue Ser199 interacts with the sulfonamide moiety. This binding mode is very similar to the binding mode of buspirone that was reported by us elsewhere [13]. Interestingly, residue Asn386 from TM7, which contributed to the binding of buspirone, appears not to participate in the binding of **1**.

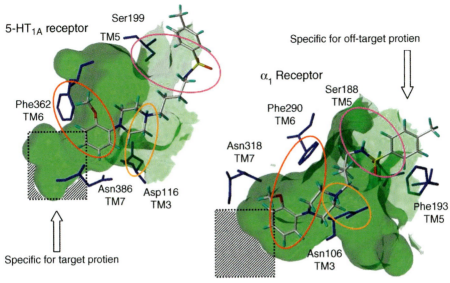

Figure 19.1 3D models of compound **1** docked into the binding pocket of 5-HT$_{1A}$ and α_1 adrenergic receptors generated by PREDICT. Interactions between the compound and key residues in the binding pocket are highlighted. The region that is present in the 5-HT$_{1A}$ pocket and is missing in the α_1 pocket is highlighted by a dashed square.

A similar binding mode is also observed for **1** in the binding pocket of the PREDICT-generated α_1 adrenergic receptor model (Figure 19.1). According to this model, the aryl moiety of **1** interacts with Phe290 from TM6, the piperazine moiety interacts with Asp106 from TM3 and residues Ser188 and Ser192 interact with the sulfonamide moiety. In addition, the *ortho*-OMe group seems to interact with Asn318 from TM7 and the phenyl ring that is attached to the sulfonamide interacts with residue Phe193 from TM5, an interaction that is not observed in the 5-HT$_{1A}$ binding site. This binding mode is in general agreement with putative binding modes suggested in a recently published ligand-supported homology model of α_1, which was successfully applied for identifying novel α_1 ligands by virtual screening [31]. Namely, the arylpiperazine (or the pyrimidinepiperazine) group is embedded within the binding pocket in a way that is similar for both 5-HT$_{1A}$ and α_1 receptors, supporting the hypothesis that this part of the molecule represents the key structural motif for receptor binding [29]. The remaining parts of the 5-HT$_{1A}$ ligands, for example the imid group of buspirone or the sulfonamide group of **1**, are oriented toward the top of the binding pocket and are less tightly bound.

By comparing the two binding sites in our PREDICT-generated 3D models (Figure 19.1), we were able to identify a subpocket near the aryl group to which we focused our optimization strategy for **1**. In the 5-HT$_{1A}$ receptor, this subpocket was empty and was not occupied by **1**, whereas in the 3D model of α_1 it was missing altogether. This subpocket has a predominantly hydrophobic character with the exception of a polar residue Asn386 lining this subpocket. The comparison of the two 3D models suggested that replacing the original *ortho*-OMe group with a bulkier moiety at the *meta*-position would lead to a more selective compound. This added moiety at the *meta*-position was expected to fit well into the 5-HT$_{1A}$ pocket but not into the α_1 site. Similar conclusions were drawn by Lopez-Rodriguez and collaborators [24] using comparative molecular field analysis (CoMFA) and artificial neural networks analysis [28] of a series of bicyclohydantoin–phenylpiperazines. Together, these observations resulted in the first two guidelines for our design strategy:

(a) Remove the *ortho*-OMe that seems to contribute to α_1 activity through an interaction with Asn318.
(b) Introduce additional substituents at the *meta*-position of the aryl ring of 4-arylpiperazine.

However, since the additional moiety at the *meta*-position is expected to bind in a very tight region of the 5-HT$_{1A}$ site, there is a risk of reducing 5-HT$_{1A}$ affinity because of the entropic penalty associated with restricting motion upon binding in a tight region. Structure-based considerations suggested that some of this entropy penalty could be offset by adding a countering stabilizing interaction such as the interaction of a polar substituent with Asn386 in the 5-HT$_{1A}$ pocket. Accordingly, two additional guidelines were added to the design strategy:

(c) Focus on relatively rigid substituents at the *meta*-position to limit the loss of flexibility upon binding (entropy).
(d) The additional substituent should have a polar moiety to interact with Asn386, thus adding a favorable interaction that could serve to compensate for the entropic loss as well.

To these we added a general consideration to avoid, if possible, functional groups that may later cause PK problems (guideline (e) in the strategy).

With these design guidelines in mind we evaluated several potential *meta*-position R groups: *O*-alkyl, *O*-aryl, esters, *N*-alkyl, amides, sulfonamides and others. The incorporation of groups such as *O*-alkyl, *N*-aryl or esters was set to a lower priority because of their lower metabolic stability, which did not agree with guideline (e). Hence, the strategy that was pursued was to introduce sulfonamides and amides at the *meta*-position. Both groups have been shown before to be tolerated at the *meta*-position of other 5-HT$_{1A}$ ligands [27,28]. Such amide and sulfonamide groups would potentially have an added value of increasing the compound's solubility and improving its metabolic stability, thus enhancing its oral bioavailability and improving the PK profile of the compound. A structure-based analysis of a virtual series of alkylamides and alkylsulfonamides docked into the 3D model of the 5-HT$_{1A}$ receptor suggested that the maximal size of the alkyl group that can be accommodated at that position was isopropyl or cyclopropyl. Furthermore, docking this series of virtual compounds, for example *meta*-^{c}PrCONH, into our 3D model of the α_1 receptor indicated that these bulkier groups disturb the binding to α_1, significantly increasing the distance between the piperazine and the aspartic acid of the receptor to more than

Table 19.1 Structure-Based Design strategy for Improving 5-HT$_{1A}$/α_1 selectivity of lead compound **1**.

(a) Remove the *ortho*-OMe substituent that seems to contribute to α_1 activity through interaction with Asn318.
(b) Introduce additional substituent at the *meta*-position of the arylpiperazine aryl group to address the 5-HT$_{1A}$ specific subpocket that is not present in the α_1 pocket.
(c) Focus on relatively rigid substituents at the *meta*-position, to limit the loss of flexibility upon binding (entropy).
(d) The new R group should have a polar moiety to interact with Asn386, thus adding a favorable interaction that could serve to compensate for the entropic loss as well.
(e) Avoid substituents with known metabolic stability liabilities.
(f) For alkylamide and alkylsulfonamide R groups, the alkyl group should not be larger than isopropyl or cyclopropyl.

5 Å and preventing the salt-bridge formation. On the basis of this analysis, we added another consideration (f) to the design strategy: for alkylamide and alkylsulfonamide substituents, the alkyl group should not be larger than isopropyl or cyclopropyl. The resulting strategy is summarized in Table 19.1.

Another outcome of the above analysis was that other parts of the molecule were rendered of lower priority for modification at this time, as these were predicted to have a smaller impact on the 5-HT$_{1A}$/α_1 selectivity ratio. For example, modifications of the arylsulfonamide side of the lead compound **1** were deferred to a later stage since this part was predicted to be bound to the open extracellular side of the binding pocket and was expected to have a smaller effect on 5-HT$_{1A}$/α_1 selectivity. Likewise, the length of the alkyl linker was not modified at this time. Structural analysis of the binding mode suggested that the relative distance between the arylpiperazine part and the arylsulfonamide part of the lead compound was adequate to form favorable interactions with the receptor. This agreed with the previously established data that maximal affinity for both 5-HT$_{1A}$ and α_1 receptors was associated with a 3-carbon or 4-carbon linker [26,27].

With these guidelines we synthesized a series of *meta*-substituted analogues of the lead compound **1** with alkylamide and alkylsulfonamide substituents. A full analysis of the SAR that was developed during this optimization process can be found in [14]. In what follows we report on the progress of the optimization process in chronological order, referring to the various compounds by their serial number, which reflects the order of their synthesis (data summarized in Table 19.2).

From the beginning it was evident that the proposed molecular design strategy was paying off, as the very first compounds to be synthesized already showed significant reduction in affinity to α_1 and α_2 while maintaining reasonable affinity to 5-HT$_{1A}$. For example, the second compound synthesized in this program (a *meta*-isopropyl-amide analogue of the lead compound (compound #2) already showed only minimal affinity to both adrenergic receptor subtypes, α_1 and α_2 (>5 µM, compared to <10 nM for the lead compound), whereas its binding affinity K_i to 5-HT$_{1A}$ was 78 nM, demonstrating a 64-fold selectivity window compared to the sixfold selectivity window of the lead compound.

Table 19.2 Key compounds is lead optimization process of cmpound (1).

Compound	5-HT$_{1A}$ K$_i$ (nM)	a$_1$ K$_i$ (nM)	a$_2$ K$_i$ (nM)	5-HT$_{1A}$ Agonist EC$_{50}$ (nM)	Rat %F	Rat t$_{1/2}$ (po) (h)	hLM t$_{1/2}$ (min)	Rat CNS (1 h, po)	hERG IC$_{50}$ (nM)	Protein binding (%)	CYP Inhibitors (nM)
Buspirone	20	367		80	1.3%	1		+		89.6	
Lead 1 from in silico screening	1	6.6	10	21	2%	1–2	9		>5000		
Compound #2	78	5000	>5000								
Compound #8	22	>5000	>5000				40				
Compound #15	6.8	3500	>5000	20	15%	3.3	39	+	300		
Compound #23 (drug candidate)	5.1	1600	>10 000	20	12%	2.5	28	+	3800	92	>30
Compound #30 (backup)	9.2	>3000	>10 000	60	50%	3.5	>90	+	3000	57	>30

19.5 Optimization Round 1: Reducing Off-Target Activities

Continuing with the same strategy, we synthesized additional alkylamide and alkylsulfonamide analogues with the goal to improve the binding affinity to 5-HT$_{1A}$ while maintaining the low affinity to both α_1 and α_2. The eighth compound synthesized (compound #8) already showed a very reasonable profile. This compound, a *meta*-cyclopropyl-amide analogue, had a K_i of 22 nM to 5-HT$_{1A}$ and a 225-fold selectivity window over both α_1 and α_2 receptors (>5000 nM). In general, the amide series showed greater 5-HT$_{1A}$/α_1 selectivity than the sulfonamide series. The SAR verified that alkyl groups ranging in size from Me to ^{n}Pr were tolerated within the 5-HT$_{1A}$ binding pocket, while the bulkier ^{t}Bu group led to reduced 5-HT$_{1A}$ affinity ($K_i = 1360$ nM). A total of 16 compounds were synthesized following this design strategy (Table 19.2). The 15th compound synthesized (*meta*-methyl-amide analogue; compound #15) was found to be the most potent and selective, with a 5-HT$_{1A}$ K_i of 6.8 nM and a 5-HT$_{1A}$/α_1 selectivity window of more than 700-fold. This compound was, as expected, conformationally very constrained in agreement with the prediction that adding rigid substituents would prevent an entropy 'penalty' upon binding. The compound also had functional activity, with an EC$_{50}$ = 20 nM in an *in vitro* cell-based assay, similar to that of the lead compound **1**.

In addition, compound #15 also had good metabolic stability in human liver microsome *in vitro* assay (hLM $t_{1/2}$ = 39 min) and in rat *in vivo* pharmacokinetic studies ($t_{1/2}$ = 3.3 h, po), with a rat oral bioavailability of 15%, showing a significant improvement in these PK parameters over the lead compound **1**. The observed improvement in PK during the optimization was another validation of the strategy discussed above. This part of the optimization process is summarized in Scheme 19.2.

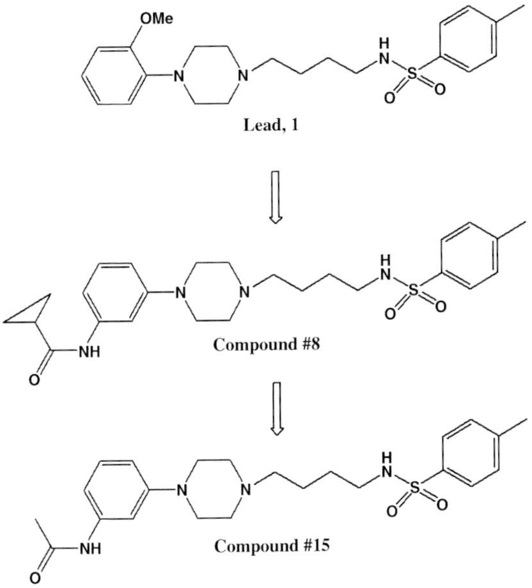

Scheme 19.2

However, when compound #15 was tested for hERG K^+ channel inhibition, a basic cardiovascular safety indicator related to the QT prolongation syndrome [32], and linked to ventricular fibrillation and sudden death [33], we were unpleasantly surprised. Although the lead compound **1** exhibited very weak binding to the hERG K^+ channel ($IC_{50} > 5000$ nM), the optimized compound #15 was a fairly potent hERG blocker with an inhibition concentration IC_{50} of 300 nM (in patch-clamp assay).

19.6
Optimization Round 2: Reducing Affinity to hERG

For the program to reach the status of development candidate, the hERG channel blocking activity of compound #15 had to be reduced by at least 10-fold. Again, we applied the structure-based *in silico* driven approach to optimize this property. As there is no experimentally determined X-ray structure of the hERG channel tetramer, we generated a 3D model of the hERG K^+ channel pore domain (segments S5-P-S6) [34] by homology modeling based on the structure of a bacterial channel KcsA ion channel (PDB code 1BL8) [35]. For details of the modeling procedure see [14]. Figure 19.2 depicts the 3D model of the hERG channel with bound astemizole.

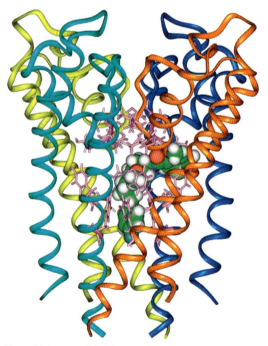

Figure 19.2 3D model of the pore-forming region (S5-P-S6) of the hERG K^+ channel with astemizole ($IC_{50} = 0.9$ nM). A side view of the hERG tetramer, showing the backbone of each monomer and a space-filling representation of the bound compound.

19.6 Optimization Round 2: Reducing Affinity to hERG

The challenge at this stage was to come up with an optimization strategy that would reduce hERG binding while maintaining the 5-HT$_{1A}$ affinity, selectivity and PK goals that were already achieved with compound #15. First, it was necessary to understand why this compound exhibited a 10-fold greater affinity to the hERG channel than the lead compound 1. Figure 19.3 shows a schematic comparison of the binding modes of the lead compound with that of compound #15 resulting from our 3D model of the hERG channel pore [14]. Given their structural similarity, both compounds are expected to adopt a similar binding mode and, according to our model, most of their interactions with the channel are similar: the p-toluenesulfonyl group interacts with Tyr652, the piperazine amine is located in the 'K$^+$ pocket' and the aryl group is sandwiched between two Phe656 residues. The only difference is that the new meta-NHCOMe moiety of compound #15, which was added to introduce selectivity and improve PK properties, is able to interact with Ser660 residues from the four monomers of the channel tetramer. This additional interaction does not exist in the binding mode of the lead compound and is most probably the reason for the higher affinity of compound #15 to the hERG channel.

As we did not want to undo the advances we achieved in compound #15, we decided to focus the efforts to reduce hERG affinity on the other side of the molecule at the p-toluenesulfonyl moiety. According to the 3D model, the p-toluenesulfonyl group contributes to the hERG binding affinity of these compounds by interacting with the aromatic Tyr562 residues (showing a π-stacking interaction). Replacing the aromatic p-toluene substituent with a nonaromatic hydrophobic group may offer an

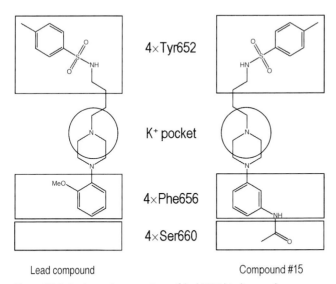

Figure 19.3 A schematic comparison of the hERG-binding modes of the lead compound 1 and compound #15, showing the four main hERG interaction regions formed by four Tyr652 residues, four Phe656 residues, four Ser660 residues and a 'K$^+$ pocket'. The greater hERG affinity of compound #15 is probably because of its interaction with the Ser660 region.

alternative strategy to reduce the hERG binding affinity. It should be noted that replacing the aromatic group by a nonaromatic group was not expected to remove hERG blocking altogether, but it was hoped that it will reduce blockage enough to obtain the required selectivity window for the drug candidate.

Various possible replacement suggestions were evaluated *in silico* in the 3D hERG model as well as in the 3D models of 5-HT$_{1A}$ and α_1, to explore whether the suggested modification would also affect the binding affinity to the primary target or the selectivity profile of compounds. According to the 5-HT$_{1A}$ 3D model there are no major aromatic residues in 5-HT$_{1A}$ that interact with the *p*-toluenesulfonyl group. This suggested that the proposed modification in this region should maintain affinity to the 5-HT$_{1A}$ target. This modification was also not expected to increase the compound's affinity for α_1 significantly. According to the 3D model of the α_1 adrenergic receptor, an aromatic Phe residue interacts with the *p*-toluene moiety, thus replacing it with a nonaromatic moiety, as suggested for reducing hERG blocking, which may even improve the 5-HT$_{1A}$/α_1 selectivity window of compounds.

Thus, the structure-based strategy that was selected for reducing the hERG affinity of compound #15 was to replace the *p*-toluene group with acyclic and cyclic alkyl groups such as isopropyl, isobutyl, cyclohexyl and cyclohexylmethyl. A few *para*-fluoroaryl analogues were also explored. A total of 15 compounds were synthesized following this strategy, which is summarized in Table 19.3.

This structure-based strategy was found to be very successful. All compounds in this series maintained excellent 5-HT$_{1A}$ affinity and a broad 5-HT$_{1A}$/α_1 selectivity window while significantly reducing hERG binding by factors ranging from fivefold to more than 10-fold.

The 23rd compound synthesized in this program (compound #23; PRX-00023) met the candidate selection criteria that were defined at the onset of the program. This cyclohexylmethylsulfonyl analogue of compound #15 has improved 5-HT$_{1A}$ activity ($K_i = 5.1$ nM) while maintaining greater than 300-fold selectivity over α_1. In the hERG patch-clamp assay, PRX-00023 was shown to be a weak hERG K$^+$ channel blocker with a low inhibitory potency of only 20% at 1 µM (estimated IC$_{50}$ = 3800 nM), indicating a 5-HT$_{1A}$/hERG selectivity window of greater than 700-fold. This compound is a full agonist with a potency of EC$_{50}$ 20 nM in an *in vitro* cell-based functional assay, shows more than a 400-fold selectivity toward other 5-HT receptors, is devoid of binding to over 50 additional protein targets tested and has a good pharmacokinetic and toxicologic profile. Hence, PRX-00023 was selected as the

Table 19.3 Structure-based design strategy for reducing hERG blocking of compound #15.

(a) The *meta*-alkylamide group interacts with Ser660 and is probably the reason for the greater hERG blocking of compound #15 relative to the lead.
(b) Removing this moiety will undo the improved selectivity and PK of compound #15, hence hERG optimization should focus on the *p*-toluenesulfonyl moiety.
(c) The *p*-toluenesulfonyl moiety interacts with the aromatic Tyr562 residues of the channel; a nonaromatic hydrophobic replacement should reduce hERG blocking.

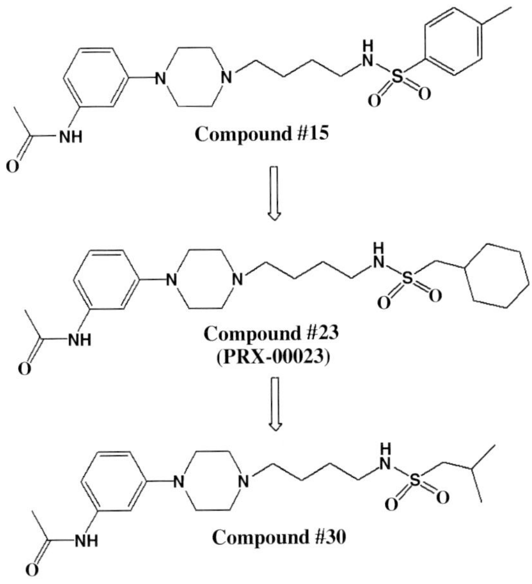

Scheme 19.3

clinical candidate. In several clinical studies this compound was found to be well tolerated in more than 250 subjects with an excellent side effect profile. Furthermore, it has good intestinal absorption, crosses the blood–brain barrier (BBB) and gets into the CNS and has a terminal half-life of more than 10 hours in humans, which supports the once-daily dosage. PRX-00023 is now in phase IIb clinical trials for depression.

The 30th compound synthesized (compound #30), an isobutyl analogue of compound #15, also showed an excellent drug candidate profile with high 5-HT$_{1A}$ binding affinity ($K_i = 9.2$ nM) and even improved 5-HT$_{1A}$/α_1 selectivity. With an EC$_{50}$ of 60 nM in the 5-HT$_{1A}$ functional assay and hERG binding IC$_{50}$ ~3000 nM, compound #30 also met all candidate selection criteria. Given its better pharmacokinetic properties, compound #30 was selected as the backup candidate for this program. This part of the optimization process is summarized in Scheme 19.3.

19.7 Conclusion

This chapter reviewed the discovery of the selective potent 5-HT$_{1A}$ agonist PRX-00023, now in phase IIb clinical studies, applying a structure-based, *in silico* driven multidimensional optimization approach. In this process, we used 3D models of the therapeutic target, the 5-HT$_{1A}$ receptor, of the main competing off-target protein, α_1-adrenergic receptor and of the hERG channel, both generated using the PREDICT modeling tool. Using these models, we were able to rapidly and effectively dial-out challenging selectivity and safety concerns in the lead compound while maintaining potency and favorable pharmacokinetic properties. The total number of compounds synthesized in this project was only 31, representing approximately 10% of all synthesis proposals that were evaluated *in silico* during the optimization process. The entire optimization process took less than 6 months from start to end, engaging only three to four medicinal chemists working together with two to three computational chemists fully dedicated to the project. This ratio of medicinal to computational chemists (close to 1 : 1) is a key aspect of our drug discovery paradigm.

Abbreviations

3D	three-dimensional
ADME	absorption, distribution, metabolism and excretion
BBB	blood–brain barrier
CMC	chemistry, manufacturing and control
CoMFA	comparative molecular field analysis
GAD	generalized anxiety disorder
GPCR	G-protein-coupled receptor
hERG	human ether-a-go-go related gene
5-HT	serotonin
IC_{50}	half maximal inhibitory concentration
K_i	binding affinity
MDD	major depressive disorder
PK	pharmacokinetic
QT	time between the start of the Q wave and the end of the T wave in hearts electrical cycle
SNRI	serotonin noradrenaline reuptake inhibitor
SSRI	selective serotonin reuptake inhibitor
TM	transmembrane

Acknowledgments

We thank the members of the Predix Pharmaceuticals (now Epix Pharmaceuticals) team who discovered PRX-00023 (computational chemists, medicinal chemists and biologists like): Dale S. Dhanoa, Yael Marantz, Dongli Chen, Sharon Shacham, Srinivasa Cheruku, Alexander Heifetz, Pradyumna Mohanty, Merav Fichman, Anurag Sharadendu, Raphael Nudelman, Michael Kauffman and Silvia Noiman.

References

1 Nemeroff, C. (2003) Anxiolytics: past, present and future agents. *Journal of Clinical Psychopharmacology,* **64**, 3–6.

2 Martin, G.R., Eglen, R.M., Hoyer, D., Hamblin, M.W. and Yocca, F. (eds)(1998) *Advances in Serotonin Receptor Research: Molecular Biology, Signal Transmission,*

and Therapeutics, Annals of the New York Academy of Sciences, New York.
3. Hoyer, D., Hannon, J.P. and Martin, G.R. (2002) Molecular, pharmacological and functional diversity of 5-HT receptors. *Pharmacology, Biochemistry, and Behavior*, **71**, 533–554.
4. Goldberg, H.L. and Finnerty, R.J. (1979) The comparative efficacy of buspirone and diazepam in the treatment of anxiety. *The American Journal of Psychiatry*, **136**, 1184–1187.
5. Sramek, J.J., Tansman, M., Suri, A., Hornig-Rohan, M., Amsterdam, J.D., Stahl, S.M., Weisler, R.H. and Cutler, N.R. (1996) Efficacy of buspirone in generalized anxiety disorder with coexisting mild depressive symptoms. *Journal of Clinical Psychopharmacology*, **57**, 287–291.
6. Delle Chiaie, R., Pancheri, P., Casacchia, M., Stratta, P., Kotzalidis, G.D. and Zibellini, M. (1995) Assessment of the efficacy of buspirone in patients affected by generalized anxiety disorder, shifting to buspirone from prior treatment with lorazepam: a placebo-controlled, double-blind study. *Journal of Clinical Psychopharmacology*, **15**, 12–19.
7. Gammans, R.E., Stringfellow, J.C., Hvizdos, A.J., Seidehamel, R.J., Cohn, J.B., Wilcox, C.S., Fabre, L.F., Pecknold, J.C., Smith, W.T. and Rickels, K. (1995) Use of buspirone in patients with generalized anxiety disorder and coexisting depressive symptoms. A meta-analysis of eight randomized, controlled studies. *Neuropsychobiology*, **25**, 193–201.
8. Landen, M., Eriksson, E., Agren, H. and Fahlen, T. (1999) Effect of buspirone on sexual dysfunction in depressed patients treated with selective serotonin reuptake inhibitors. *Journal of Clinical Psychopharmacology*, **19**, 268–271.
9. Feiger, A.D., Heiser, J.F., Shrivastava, R.K., Weiss, K.J., Smith, W.T., Sitsen, J.M. and Gibertini, M. (2003) Gepirone extended-release: new evidence for efficacy in the treatment of major depressive disorder. *Journal of Clinical Psychopharmacology*, **64**, 243–249.
10. Stahl, S.M., Kaiser, L., Roeschen, J., Keppel Hesselink, J.M. and Orazem, J. (1998) Effectiveness of ipsapirone, a 5-HT-1A partial agonist, in major depressive disorder: support for the role of 5-HT-1A receptors in the mechanism of action of serotonergic antidepressants. *International Journal of Neuropsychopharmacology*, **1**, 11–18.
11. Becker, O.M., Shacham, S., Marantz, Y. and Noiman, S. (2003) Modeling the 3D structure of GPCRs: advances and application to drug discovery. *Current Opinion in Drug Discovery and Development*, **6**, 3353–3361.
12. Shacham, S., Marantz, Y., Bar-Haim, S., Kalid, O., Warshaviak, D., Avisar, N., Inbal, B., Heifetz, A., Fichman, M., Topf, M., Naor, Z., Noiman, S. and Becker, O.M. (2004) PREDICT modeling and in-silico screening for G-protein coupled receptors. *Proteins*, **57**, 51–86.
13. Becker, O.M., Marantz, Y., Shacham, S., Inbal, B., Heifetz, A., Kalid, O., Bar-Haim, S., Warshaviak, D., Fichman, M. and Noiman, S. (2004) 3D G-protein coupled receptors: *in silico* driven drug discovery. *Proceedings of the National Academy of Sciences of the United States of America*, **101**, 11304–11309.
14. Becker, O.M., Dhanoa, D.S., Marantz, Y., Chen, D., Shacham, S., Cheruku, S., Heifetz, A., Mohanty, P., Fichman, M., Sharadendu, A., Nudelman, R., Kauffman, M.G. and Noiman, S. (2006) PRX-00023: *in silico* 3D model-driven discovery of a novel, potent and selective $5HT_{1A}$ agonist for the treatment of anxiety and depression. *Journal of Medicinal Chemistry*, **49**, 3116–3135.
15. Olivier, B., Soudijn, W. and van Wijngaarden, I. (1999) The $5\text{-}HT_{1A}$ receptor and it's ligands: structure and function. *Progress in Drug Research*, **52**, 103–165.

16 Nelson, D.L. (1991) Structure–activity relationships at 5-HT$_{1A}$ receptors: binding profiles and intrinsic activity. *Pharmacology, Biochemistry, and Behavior*, **40**, 1041–1051.

17 Lopez-Rodrigues, M.L., Ayala, D., Benhamú, B., Morcillo, M.J. and Viso, A. (2002) Arylpiperazine derivatives acting at 5-HT$_{1A}$ receptors. *Current Medicinal Chemistry*, **9**, 443–469.

18 Glennon, R.A. (1992) Concepts for the design of 5-HT$_{1A}$ serotonin agonists and antagonists. *Drug Development Research*, **26**, 251–274.

19 Peglion, J.-L., Goument, B., Despaux, N., Charlot, V., Giraud, H., Nisole, C., Newman-Tancredi, A., Dekeyne, A., Bertrand, M., Genissel, P. and Millan, M.J. (2002) Improvement in the selectivity and metabolic stability of the serotonin 5-HT$_{1A}$ ligand, S 15535: a series of *cis*- and *trans*-2-(arylcycloalkylamine)-1-indanols. *Journal of Medicinal Chemistry*, **44**, 165–176.

20 Bojarski, A.J., Kowalski, P., Kowalska, T., Duszynska, B., Charakchieva-Minol, S., Tatarczynska, E., Klodzinska, A. and Chojnacka-Wojcik, E. (2002) Synthesis and pharmacological evaluation of new arylpiperazines. 3.-{4-[4-(3-Chlorophenyl)-1-piperazinyl]butyl}-quinazolidin-4-one: a dual serotonin 5-HT$_{1A}$/5-HT$_{2A}$ receptor ligand with an anxiolytic-like activity. *Bioorganic and Medicinal Chemistry*, **10**, 3817–3827.

21 Heinrich, T., Bottcher, H., Bartoszyk, G.D., Griner, H.E., Seyfried, C.A. and van Amsterdam, C. (2004) Indolebutylamines as selective 5-HT$_{1A}$ agonists. *Journal of Medicinal Chemistry*, **47**, 4677–4683.

22 Heinrich, T., Bottcher, H., Gericke, R., Bartoszyk, G.D., Anzali, S., Seyfried, C.A., Griner, H.E. and van Amsterdam, C. (2004) Synthesis and structure–activity relationship in a class of indolebutylpiperazines as dual 5-HT$_{1A}$ receptor agonists and serotonin reuptake inhibitors. *Journal of Medicinal Chemistry*, **47**, 4684–4692.

23 Lopez-Rodriguez, M.L., Rosado, M.L., Benhamú, B., Morcillo, M.J., Sanz, A.M., Orensanz, L., Beneitez, M.E., Fuentes, J.A. and Manzanares, J. (1996) Synthesis and structure–activity relationships of a new model of arylpiperazines. 1.2-[4-(o-Methoxy-phenyl) piperazin-1-ylmethyl]-1,3-dioxoperhydroimidazo[1,5-a]-pyridine: a selective 5-HT$_{1A}$ receptor agonist. *Journal of Medicinal Chemistry*, **39**, 4439–4450.

24 Lopez-Rodriguez, M.L., Rosado, M.L., Benhamú, B., Morcillo, M.J., Fernandez, E. and Schaper, K.-J. (1997) Synthesis and structure–activity relationships of a new model of arylpiperazines. 2. Three-dimensional quantitative structure–activity relationships of new hydantoin–phenylpiperazine derivatives with affinity for 5-HT1A and α_1 receptors. A comparison of CoMFA models. *Journal of Medicinal Chemistry*, **40**, 1648–1656.

25 Lopez-Rodriguez, M.L., Morcillo, M.J., Fernandez, E., Porras, E., Murcia, M., Sanz, A.M. and Orensanz, L. (1997) Synthesis and structure–activity relationships of a new model of arylpiperazines. 3.2-[o-(4-Arylpiperazin-1-yl)alkyl]perhydropyrrolo-[1,2-c] imidazoles and -perhydroimidazo[1,5-a] pyridines: study of the influence of the terminal amide fragment on 5-HT$_{1A}$ affinity/selectivity. *Journal of Medicinal Chemistry*, **40**, 2653–2656.

26 Lopez-Rodriguez, M.L., Morcillo, M.J., Rovat, T.K., Fernandez, E., Vicente, B., Sanz, A.M., Hernández, M. and Orensanz, L. (1999) Synthesis and structure–activity relationships of a new model of arylpiperazines. 4.1-[o-(4-Arylpiperazin-1-yl)alkyl]-3-di phenylmethylene-2,5-pyrrolidinediones and -3-(9H-fluoren-9-ylidene)-2,5-pyrrolidinediones: study of the steric requirements of the terminal amide fragment on 5-HT$_{1A}$ affinity/selectivity. *Journal of Medicinal Chemistry*, **42**, 36–49.

27 Lopez-Rodriguez, M.L., Morcillo, M.J., Fernandez, E., Porras, E., Orensanz, L., Beneytez, M.E., Manzanares, J. and Fuentes, J.A. (2001) Synthesis and structure–activity relationships of a new model of arylpiperazines. 5. Study of the physicochemical influence of the pharmacophore on 5-HT$_{1A}$/α_1-adrenergic receptor affinity. Synthesis of a new selective derivative with mixed 5-HT$_{1A}$/D$_2$ antagonist properties. *Journal of Medicinal Chemistry*, **44**, 186–197.

28 Lopez-Rodriguez, M.L., Morcillo, M.J., Fernandez, E., Rosado, M.L., Pardo, L. and Schaper, K.-J. (2001) Synthesis and structure–activity relationships of a new model of arylpiperazines. 6. Study of the 5-HT$_{1A}$/α_1-adrenergic receptor affinity by classical Hansch analysis, artificial neural networks, and computational simulation of ligand recognition. *Journal of Medicinal Chemistry*, **44**, 198–207.

29 Lopez-Rodriguez, M.L., Ayala, D., Viso, A., Benhamú, B., Fernandez de la Pradilla, R., Zarza, F. and Ramos, J.A. (2004) Synthesis and structure–activity relationships of a new model of arylpiperazines. 7. Study of the influence of lipophilic factors at the terminal amide fragment on 5-HT$_{1A}$ affinity/selectivity. *Bioorganic and Medicinal Chemistry*, **12**, 1551–1557.

30 Lopez-Rodriguez, M.L., Morcillo, M.J., Fernandez, E., Benhamú, B., Tejada, I., Ayala, D., Viso, A., Campillo, M., Pardo, L., Delgado, M., Manzanares, J. and Fuentes, J.A. (2005) Synthesis and structure–activity relationships of a new model of arylpiperazines. 8. Computational simulation of ligand-receptor interaction of 5-HT(1A)R agonists with selectivity over alpha1-adrenoceptors. *Journal of Medicinal Chemistry*, **48**, 2548–2558.

31 Evers, A. and Klabunde, T. (2005) Structure-based drug discovery using GPCR homology modeling: successful virtual screening for antagonists of the alpha1A adrenergic receptor. *Journal of Medicinal Chemistry*, **48**, 1088–1097.

32 Viskin, S. (1999) Long QT syndromes and torsade de pointes. *The Lancet*, **354**, 1625–1633.

33 Mitcheson, J.S., Chen, J., Lin, M., Culberson, C. and Sanguinetti, M.C. (2000) A structural basis for drug-induced long QT syndrome. *Proceedings of the National Academy of Sciences of the United States of America*, **97**, 12329–12333.

34 Pearlstein, R., Vaz, R. and Rampe, D. (2003) Understanding the structure–activity relationship of the human ether-a-go-go-related gene cardiac K(+) channel. A model for bad behavior. *Journal of Medicinal Chemistry*, **46**, 2017–2022.

35 Doyle, D.A., Morais Cabral, J., Pfuetzner, R.A., Kuo, A., Gulbis, J.M. et al.(1998) The structure of the potassium channel: molecular basis of K$^+$ conduction and selectivity. *Science*, **280**, 69–77.

Index

a

absorption, distribution, metabolism, excretion and toxicity (ADMET) 16, 23, 278
action potential duration (APD) 58
adrenergic receptors 30
α1-adrenergic receptors (α1-ARs) 130, 156
adrenocorticotropic hormone (ACTH) 304
adverse drug reactions (ADRs) 3, 23
affinity fingerprint method 25
alanine-scanning mutagenesis 59
amineptine 4, 5
aminophenazone 5
aminosteroid 9
artificial neural networks analysis 464
aryl hydrocarbon receptor (AhR) 199, 330
astemizole 5
atorvastatin acid 345
ATP-binding cassette (ABC) 341, 367
– efflux transporters 351
azidothymidine monophosphate (AZT) 348

b

Ballosteros–Weinstein nomenclature 135
bilinear regression models 177
bilirubin glucuronides 348
biogenic amine-binding GPCRs 182
BioPrint biological fingerprints 23, 25, 28, 36, 39, 41
blood–brain barrier (BBB) 350, 471
breast cancer resistance protein (BCRP) family 341
bromfenac sodium 6

c

calcium channel blockers 43
calcium-gated potassium channel Mth K 90
cardiovascular toxicity 11
catecholamine neurotransmitters 156
cerivastatin 6, 346

chemical reactivity 249
Cheng–Prusoff equation 213
meta-chlorophenylpiperazine (mCPP) 149
chlormezanone 6
cholesteryl ester transfer protein (CETP) 14
chronic hypoxia 147
CODESSA method 115
comparative molecular field analysis (CoMFA) 464
comparative structural connectivity spectral analysis (CoSCoSA) 330
congestive heart failure (CHF) 7
constitutive androstane receptor (CAR) 325
– endogenous CAR ligands 325
Crigler–Najjar (CN) syndrome 296, 303
cross-chemotype pharmacophore 130, 183
cyclooxygenase (COX)-2 inhibitor 10
CYP (cytochrome P450) enzymes 277
– CYP2C5 255
– catalyzed reactions 268
– crystal structures 253, 258
– CYP2D6-mediated metabolic pathways 201
– induction 318
– iron-containing porphyrin system 247
– irreversible inhibition 217, 268, 269, 279
– ligand interactions 277
– mechanism-based inhibition 270, 271
– mechanism-based inhibitor (MBI) 218, 220, 222, 225, 267, 285
– membrane-attached cytochrome P450 enzymes 277
– metabolite-based inhibition (MBI) 278
– metabolism prediction 289
– reversible inhibition 203, 221

d

dexamethasone (DEX) 304
dexfenfluramine 7
divalent bile salts 348

Antitargets. Edited by R. J. Vaz and T. Klabunde
Copyright © 2008 WILEY-VCH Verlag GmbH & Co. KGaA, Weinheim
ISBN: 978-3-527-31821-6

DNA-binding domain (DBD) 319
dopaminergic receptors 157, 457
drug–drug interactions 267, 269, 300, 317, 318
– quantitative prediction of 269
drug–hormone interactions 303
drug metabolism 197, 247
– drug-metabolizing enzymes 198, 201, 295
– drug-metabolizing transporters 295
drug transporter regulation 298, 389
– computational models for 389
Dubin–Johnson syndrome 298, 303, 348

e
efflux transporters 341
electrophiles 267
electrotopological state (ES) values 382
epinephrine 156
epirubicin 281
erythema nodosum leprosum (ENL) 17
European Agency for the Evaluation of Medicinal Products (EMEA) 7

f
farnesoid X receptor (FXR) 297, 326
fatty acid binding protein (FABP) 307
fatty acid elongase (FAE) 304
fatty acid synthase (FAS) 304
fenfluramine 7
fexofenadine 94, 127
fialuridine 12
fibroblast activation protein (FAP) 404
fingerprints for ligands and proteins (FLAP) method 287
flosequinan 7
Food and Drug Administration (FDA) 3

g
gain-of-function mutation 96
Genetic Algorithm Similarity Program (GASP) 380
Gilbert's disease 303
glafenine 7, 8
glucose-dependent insulinotropic polypeptide (GIP) 401
glucuronic acid conjugation 284
GLUE program 259
glutathione S-transferase (GST) 296, 322
G-protein-coupled receptors (GPCRs) 23, 29, 127, 154, 461
– antitarget affinity 130
– antitarget modeling 127
– heterodimerization 157
– ligands 157

Gram-negative pathogens
– pulmonary infections 10
grepafloxacin 8
GRID flexible molecular interaction fields (GRID-MIFs) 282
GRIND technology 284
GST regulation 297

h
Hansch approach 159, 169
hepatic failure 13
hepatotoxicity 5
heterocyclic aromatic amines (HAAs) 199
HIV-protease inhibitor 351
HMG-CoA reductase inhibitor 6, 345
hormone response elements (HREs) 293
human ether-a-go-go-related gene (hERG) channel 16, 89, 90, 91, 93, 109, 119
– activation gating 95
– binding site 100
– blocking drugs 109, 111, 113
– C-type inactivation gating 93
– drug-binding site of 95
– homology model 119, 120
– ligand-derived models 119
– mutations in 93
– neuroendocrine glands 89
– smooth muscle 93
– target versus antitarget 56
– voltage-sensing domain 92
human liver microsomes (HLMs) 206
human UDP-glucuronosyl transferases, see phase II conjugative metabolism enzymes
hydrophilic metabolites 277
hyperbilirubinemia II, see Dubin–Johnson syndrome
hypersensitivity reaction 269
hypothalamus–pituitary–adrenal (HPA) axis 304

i
immunomodulatory drugs 17
International Conference on Harmonization (ICH) 66
in vitro–in vivo relationships 23

k
KcsA channel X-ray structure 431
Kohonen self-organizing maps 118, 383

l
lactic acidosis 13
LC/MS techniques 205
levacetylmethadol 8

liberation–absorption–distribution–
  metabolism–excretion (LADME)   15
ligand-binding domain (LBD)   319
linear interaction energy (LIE) method   120
Lineweaver–Burk visualizations   213–214
lithocholic acid (LCA)   302
liver X receptors (LXRs)   297, 326, 328
low-density lipoprotein (LDL)   6, 14
luminogenic assays   205

m
machine learning   324
major depressive disorder (MDD)   459
mass spectrometry   205, 227
mdr1 knockout mouse model   349
metabolic biotransformations   277
MetaSite   252, 259, 282, 290
mibefradil   9
Michaelis–Menten kinetic   206, 318
mitogen-activated protein kinase (MAPK)   145
molecular dynamics (MD) simulations   120
molecular interaction fields (MIFs)   383
Monte Carlo conformational analysis   180
multidimensional optimization paradigm   459
multidrug resistance protein (MRP)   298, 341, 346
multidrug resistance protein 1 (Mdr1)   298
– resistant transporter 1 (MDR1)   322–323
multilinear regression models   169
multiple copies simultaneous search (MCSS)   182
muscarinic acetylcholine receptors   30

n
neural network   116, 127, 171
neuropsychotic disorders   5
neurotransmitter receptors   29
new chemical entities (NCEs)   3, 197
nonsteroidal anti-inflammatory drug (NSAID)   6
norfenfluramine   149
N-terminal DNA binding domain (DBD)   293
nuclear hormone receptors (NHRs)   23
nuclear receptors
– activation of   319
nucleotide binding domain (NBD)   346, 368

o
organic anion transporting polypeptide (OATP)   298, 341, 342, 390
organic anion transporting polypeptide 2 (OATP2)   323

organic cation transporter family (OCT)   341, 389
oxidative biotransformations   284

p
Parkinson's disease   129
partial least squares (PLS)   383
– regression-based models   324
– partial least squares discriminant (PLSD)   384
Passe–Partout model   184
peripheral blood mononuclear cells (PBMCs)   352
P-glycoprotein (P-gp), *see* ATP-binding cassette
– crystal structure   371
– role of   349
– *in vitro* models   369
– pharmacophore-based descriptors   40
– receptor-based   158
pharmacophore models   132, 135
phase II conjugative metabolism enzymes   277
phase II conjugation reactions   281
phenobarbital response element (PBRE)   299
polycyclic aromatic hydrocarbons (PAHs)   199
polypharmacology
– BioPrint analysis   26
– of drugs   26
pore-forming α-subunit   89
potassium channels   89
potential ADR liabilities   49
PREDICT technology   433
pregnane-activated receptor (PAR)   293
pregnane X receptor (PXR)   202, 322
– characterization of   293
– in lipid metabolism   304
– in phase II enzyme regulation   296
Protein Data Bank   255, 256
protein–protein interactions   157, 208
pseudo-Cushing's syndrome   304
pseudoxanthoma elasticum (PXE)   349
pulmonary hypertension   143

q
QT prolongation   60, 65
– dose–response relationship for   67
– long QT syndrome (LQTS)   93, 94, 109
– drug-induced long QT syndrome   58, 94
– regulatory aspects of   66
– QT-prolonging drugs   56, 94
quantitative structure–activity relationship (QSAR)   100, 156, 159, 170, 171, 323, 382, 391
– supermolecule-based QSAR   172

quantitative structure–property relationship (QSPR) 161
quantum chemistry 252

r

rapacuronium bromide 9
reactive metabolites, see electrophiles
receiver operating characteristic (ROC) 122
– receptor-based 158
receptorome screening 143, 144
receptor theory 155
recursive partitioning (RP) 116
retinoic acid receptor (RAR) 328
retinoid X receptor (RXR) 202, 293
rhodopsin-based models 182
right ventricular systolic pressure (RVSP) 147
rofecoxib 10

s

Sammon nonlinear maps 118
seasonal rhinitis 127
selective serotonin reuptake inhibitors (SSRIs) 458
serotonin noradrenaline reuptake inhibitors (SNRIs) 458
short QT syndrome (SQTS) 94
sick sinus syndrome 9
single nucleotide polymorphism (SNP) 147
small cell lung cancer (SCLC) 347
stearoyl CoA desaturase-1 (SCD-1) 304
steroid and xenobiotic receptor (SXR) 202, 293
sterol regulatory element binding protein 1c (SREBP-1c) 304
Stevens–Johnson syndrome 7
structure–activity relationships (SAR) 110, 327, 457
structure-based descriptors 23
structure-based molecular design approaches 457, 460
structure–property relationship (SPR) studies 372
succinylcholine 9
support vector machine (SVM) 115, 116
System for Thalidomide Education and Prescribing Safety (STEPS) 17

t

temafloxacin 10
terfenadine 127
thalidomide
– revival of 17
prospective randomized flosequinan longevity evaluation (PROFILE) 7
thymidine kinase promoter 299
time-dependent inhibition (TDI) 221
topology-based methods 118
torcetrapib 14
total molecular surface area (TMSA) 166
traditional Chinese medicines (TCMs) 301
transporter-mediated drug disposition 342
troglitazone 10

u

UDP-glucuronosyltransferases (UGTs) 296, 322
UGT–drug recognition 277
UGT-mediated glucuronidation 281
uridine-5-diphosphate-α-d-glucuronic acid (UDPGA) 284
UV–Vis spectroscopy 274

v

vacuum cleaner model 368
valvular heart disease (VHD) 143
van der Waals energy parameters 121
van der Waals surface area 379
Van de Water's formula 71
vectorial drug transport 352
venous thromboembolism (VTE) 11
ventricular action potential 89
ventricular arrhythmia 109
ventricular cardiac action potential 91
VHD-associated drugs 145, 148, 149
vitamin D receptor (VDR) 326
voltage-gated channels 90, 95
– voltage-gated potassium channels 89

w

warfarin therapy 200

x

xenobiotic biotransformation processes 198
xenobiotic receptor 302
xenobiotic response elements (XREs) 295
xenobiotics 197
ximelagatran 11
X-ray crystallography 120, 282, 371

z

zwitterion approach 425, 434